Food and Natural Resources

Food and Natural Resources

EDITED BY

David Pimentel
College of Agriculture
and Life Sciences
Cornell University
Ithaca, New York

Carl W. Hall
Directorate for Engineering
National Science Foundation
Washington, D.C.

Academic Press, Inc.
Harcourt Brace Jovanovich, Publishers
San Diego   New York   Berkeley   Boston
London   Sydney   Tokyo   Toronto

COPYRIGHT © 1989 BY ACADEMIC PRESS, INC.
ALL RIGHTS RESERVED.
NO PART OF THIS PUBLICATION MAY BE REPRODUCED OR
TRANSMITTED IN ANY FORM OR BY ANY MEANS, ELECTRONIC
OR MECHANICAL, INCLUDING PHOTOCOPY, RECORDING, OR
ANY INFORMATION STORAGE AND RETRIEVAL SYSTEM, WITHOUT
PERMISSION IN WRITING FROM THE PUBLISHER.

ACADEMIC PRESS, INC.
San Diego, California 92101

United Kingdom Edition published by
ACADEMIC PRESS LIMITED
24-28 Oval Road, London NW1 7DX

Library of Congress Cataloging-in-Publication Data

Food and natural resources / edited by David Pimentel, Carl W. Hall.
    p.   cm.
    Includes index.
    ISBN   0-12-556555-0 (alk. paper)
    1. Food supply. 2. Agriculture. 3. Natural resources.
I. Pimentel, David, Date.   II. Hall, Carl W.
TX353.F593   1988
338.1'9—dc19   88-12123
                                                   CIP

PRINTED IN THE UNITED STATES OF AMERICA
89 90 91 92   9 8 7 6 5 4 3 2 1

Contents

CONTRIBUTORS xiii
PREFACE xv

1. **ECOLOGICAL SYSTEMS, NATURAL RESOURCES, AND FOOD SUPPLIES**
   David Pimentel

   I. Introduction 2
   II. The Structure and Functioning of Ecosystems 2
   III. Evolution of Living Systems 6
   IV. Biogeochemical Cycles 7
   V. Aquatic Ecosystems 9
   VI. Terrestrial Ecosystems 10
   VII. Human Food Gathered and Hunted from the Wild 11
   VIII. Agricultural Ecosystems 12
   IX. Resource Constraints in World Food Production 17
   X. Food Needs for Future Generations 21
   XI. Requirements for Solving Food Problems 22
   XII. Conclusion 24
   References 25

2. **INTERDEPENDENCE OF FOOD AND NATURAL RESOURCES**
   David Pimentel, Laura E. Armstrong, Christine A. Flass, Frederic W. Hopf, Ronald B. Landy, and Marcia H. Pimentel

   I. Introduction 32
   II. World Population Growth 32

vi   Contents

    III. Energy Constraints   34
    IV. Arable Land—Quality and Quantity   37
    V. Water Constraints   40
    VI. Biological Diversity   42
    VII. Looking to the Future   43
         References   44

3. **LOSS OF BIOLOGICAL DIVERSITY AND ITS POTENTIAL IMPACT ON AGRICULTURE AND FOOD PRODUCTION**
Norman Myers

    I. Introduction   50
    II. Species' Contributions to Modern Agriculture   50
    III. Extinction Processes   62
    IV. Conclusion   65
         References   65

4. **AVAILABILITY OF AGRICULTURAL LAND FOR CROP AND LIVESTOCK PRODUCTION**
Pieter Buringh

    I. The World Land Area and Its Utilization   70
    II. Food Production and Land   74
    III. Livestock Production and Land   79
    IV. Conclusion   81
         References   82

5. **LAND DEGRADATION AND ITS IMPACT ON FOOD AND OTHER RESOURCES**
R. Lal

    I. Introduction   86
    II. Land Degradation   87
    III. Technological Options for Minimizing Soil Degradation   121
    IV. Land Degradation and World Food Production   128
    V. Need to Restore Productivity of Degraded Lands   131
    VI. Conclusions   132
         References   133

6. **WATER USE IN AGRICULTURE**
   Vashek Cervinka

   I. Introduction   142
   II. Global Perspective   142
   III. Water Sources and Systems   144
   IV. Water in Crop Production   145
   V. Water in Animal Production   148
   VI. Water in Food Processing   149
   VII. Water Quality   152
   VIII. Irrigation Systems   158
   References   162

7. **WATER SCARCITY AND FOOD PRODUCTION IN AFRICA**
   Malin Falkenmark

   I. Introduction   164
   II. Present Resource Crisis   166
   III. Water and Development   173
   IV. Water Availability Limits Increase of Carrying Capacity   180
   V. The Combined Picture   185
   VI. Conclusions   187
   References   189

8. **AGRICULTURAL CHEMICALS: FOOD AND ENVIRONMENT**
   David A. Andow and David P. Davis

   I. Introduction   192
   II. Use of Agricultural Chemicals   192
   III. Nitrogen in Agroecosystems   195
   IV. Pesticides   207
   V. Rational Use of Agricultural Chemicals   224
   References   227

9. **NATURAL GAS AS A RESOURCE AND CATALYST FOR AGROINDUSTRIAL DEVELOPMENT**
   Walter Vergara

   I. Background   236
   II. Monteagudo Agroindustrial Project   240

III. Food Processing Plants   243
IV. Energy Supply   247
V. Alternatives for the Supply of Electricity and Steam   251
VI. Water Supply   253
VII. Gas Pipeline   254
VIII. Transportation System   255
IX. Impact of the Project   255
X. Conclusions   258
References   259

10. **MECHANIZATION AND FOOD AVAILABILITY**
Carl W. Hall

I. Introduction   262
II. Farming and Agriculture   263
III. Mechanization, Tractorization, and Electrification   263
IV. Summary   272
References   272

11. **POPULATION, FOOD, AND THE ECONOMY OF NATIONS**
William J. Hudson

I. Is Population Limited by Food?   276
II. Is Food Driven by Population or by the Economy of Nations?   279
III. Can the Worst Fears of Environmentalists be Substantiated?   284
IV. Vision 2020   293
References   298

12. **ECOLOGICAL RESOURCE MANAGEMENT FOR A PRODUCTIVE, SUSTAINABLE AGRICULTURE**
David Pimentel, Thomas W. Culliney, Imo W. Buttler, Douglas J. Reinemann, and Kenneth B. Beckman

I. Introduction   302
II. Principles for a Productive, Sustainable Agriculture   303
III. Soil Nutrient and Water Resources   304
IV. Pests and Their Control   307
V. Importance of Biological Resources   309
VI. Environmental and Economic Aspects of Ecological Agricultural Management   311

VII. Conclusion   315
References   316

13. **POPULATION GROWTH, AGRARIAN STRUCTURE, FOOD PRODUCTION, AND FOOD DISTRIBUTION IN THE THIRD WORLD**
    Frederick H. Buttel and Laura T. Raynolds

    I. Introduction   326
    II. The Malnutrition Debate   327
    III. Population Growth and Hunger   330
    IV. The Green Revolution and the Alleviation of Hunger: Contribution and Controversy   341
    V. Agrarian Structure, Food Production, and Hunger   350
    VI. Food Policy, Food Consumption, and Nutrition   354
    VII. Discussion   356
    Appendix: Data Sources and Operationalization of Variables for the Empirical Analysis of Food Access in Third World Countries   357
    References   358

14. **ENVIRONMENT AND POPULATION: CRISES AND POLICIES**
    David Pimentel, Linnea M. Fredrickson, David B. Johnson, John H. McShane, and Hsiao-Wei Yuan

    I. Introduction   364
    II. Human Needs Worldwide   365
    III. Standard of Living and Population Growth   366
    IV. Population Growth   367
    V. Per Capita Use of Resources in the United States and China   368
    VI. State of the Environment in the United States and China   370
    VII. Policy Decisions Concerning Environmental and Population Problems   372
    VIII. National Population Policies in the United States and China   380
    IX. Conclusion   384
    References   385

15. FOOD AVAILABILITY AND NATURAL RESOURCES
Carl W. Hall

    I. Introduction   392
   II. Utilization of Resources   392
  III. Production of Food in the United States and the World   396
  IV. Summary   406
      References   407

16. FOOD AS A RESOURCE
Marcia Pimentel

    I. Introduction   410
   II. Patterns of Human Population Growth   410
  III. Food and Dietary Patterns   412
  IV. Major Nutritional Problems   420
   V. Trends in Food Production   424
  VI. Resources Used in Food Production   427
 VII. Planning Future Policy   431
      References   434

17. POPULATION GROWTH AND THE POVERTY CYCLE IN AFRICA: COLLIDING ECOLOGICAL AND ECONOMIC PROCESSES?
A. R. E. Sinclair and Michael P. Wells

    I. The African Paradox   440
   II. Population   441
  III. The Ecological Crisis   452
  IV. The Economic Decline   463
   V. Foreign Aid   473
  VI. The Poverty Cycle and the Way Ahead   479
      References   483

18. FOOD AND FUEL RESOURCES IN A POOR RURAL AREA IN CHINA
Wen Dazhong

    I. Introduction   486
   II. An Overview of Kazhou County   488
  III. The Agroecosystem: Food and Fuel Production and Consumption System   490
  IV. Energy Flows in the Kazhou Agroecosystem   491

V. Assessment of the Kazhou Agroecosystem 496
VI. Strategies for Improving Food and Household Fuel Supplies in Kazhou 499
VII. Conclusions 503
References 504

INDEX 507

Contributors

Numbers in parentheses indicate the pages on which the authors' contributions begin.

David A. Andow (191), Department of Entomology, University of Minnesota, St. Paul, Minnesota 55108
Laura E. Armstrong (31), College of Agriculture and Life Sciences, Cornell University, Ithaca, New York 14853
Kenneth B. Beckman (301), College of Agriculture and Life Sciences, Cornell University, Ithaca, New York 14853
Pieter Buringh (69), Marterlaan 20, 6705 BA Wageningen, The Netherlands
Frederick H. Buttel (325), Department of Rural Sociology, Cornell University, Ithaca, New York 14853
Imo W. Buttler (301), College of Agriculture and Life Sciences, Cornell University, Ithaca, New York 14853
Vashek Cervinka (141), Agricultural Resources Branch, California Department of Food and Agriculture, Sacramento, California 94271-0001
Thomas W. Culliney (301), College of Agriculture and Life Sciences, Cornell University, Ithaca, New York 14853
David P. Davis (191), Department of Entomology, University of Minnesota, St. Paul, Minnesota 55108
Wen Dazhong (485), Institute of Applied Ecology, Chinese Academy of Sciences, Shenyang, China
Malin Falkenmark (163), Swedish Natural Science Research Council, S-11385, Stockholm, Sweden
Christine A. Flass (31), College of Agriculture and Life Sciences, Cornell University, Ithaca, New York 14853
Linnea M. Fredrickson (363), College of Agriculture and Life Sciences, Cornell University, Ithaca, New York 14853

Contributors

Carl W. Hall (261, 391), Directorate for Engineering, National Science Foundation, Washington, D.C. 20550
Frederic W. Hopf (31), College of Agriculture and Life Sciences, Cornell University, Ithaca, New York 14853
William J. Hudson (275), The Andersons Management Corp., Maumee, Ohio 43537
David B. Johnson (363), College of Agriculture and Life Sciences, Cornell University, Ithaca, New York 14853
R. Lal (85), Department of Agronomy, The Ohio State University, Columbus, Ohio 43210
Ronald B. Landy (31), College of Agriculture and Life Sciences, Cornell University, Ithaca, New York 14853
John H. McShane (363), College of Agriculture and Life Sciences, Cornell University, Ithaca, New York 14853
Norman Myers (49), Consultant in Environment and Development, Upper Meadow, Headington, Oxford OX3 8SZ, United Kingdom
David Pimentel (1, 31, 301, 363), College of Agriculture and Life Sciences, Cornell University, Ithaca, New York 14853
Marcia H. Pimentel (31, 411), Division of Nutritional Sciences, Colleges of Human Ecology and Agriculture and Life Sciences, Cornell University, Ithaca, New York 14853
Laura T. Raynolds (325), Department of Rural Sociology, Cornell University, Ithaca, New York 14853
Douglas J. Reinemann (301), College of Agriculture and Life Sciences, Cornell University, Ithaca, New York 14853
A. R. E. Sinclair (439), Department of Zoology, University of British Columbia, Vancouver V6T 2A9, British Columbia, Canada
Walter Vergara (235), Asia Technical Department, The World Bank, Washington, D.C. 20433
Michael P. Wells (439), Department of Zoology, University of British Columbia, Vancouver V6T 2A9, British Columbia, Canada
Hsiao-Wei Yuan (363), College of Agriculture and Life Sciences, Cornell University, Ithaca, New York 14853

Preface

Food is an essential resource for the more than 5 billion humans who now live on earth and the 233,000 people being added to the human population daily. Adequate food supplies depend on the availability and use of numerous natural resources, including land, water, solar energy, fossil energy, forests, plant and animal species, and fisheries. Whereas solar energy is nearly infinite, fossil energy is finite. The other resources are renewable but only within certain use limits; in a sense they are therefore also finite. The interdependencies and interactions among these various resources are clearly complex. The interrelated factors described here are but some of the many that have reciprocal actions.

Along with solar energy, fertile cropland and sufficient water are the most basic resources used in agricultural production. Information detailed in several chapters points to the growing concern about the supply of these resources. For example, although the United States is fortunate in having ample cropland resources, many countries like Japan and Jordan are already experiencing critical cropland shortages. Further, some countries or regions within countries also have serious water shortages. Even in the United States, land areas in the West and Southwest are now feeling the pressure of declining per capita water supplies. Groundwater pumping from major aquifers in these regions is, in essence, mining this precious resource. An aquifer, which stores water, requires that less than 1% be pumped annually to remain a viable water source.

A twofold problem is emerging in the area of fossil fuel use by all segments of human society. Many nations totally lack fossil fuel deposits and make do with minimal imports—creating a serious curtailment of the production of all goods and services, especially food and fiber. Those nations that either possess ample supplies or can afford to purchase them are using fossil energy at alarming rates. According to projections by the U.S. National Academy of Sciences and other experts, per capita use of fossil fuel will peak at the year 2010 and then decline. Depletion of oil and natural gas resources is projected to occur first. As explained in many chapters, the current world system of agricultural production has been sustained on fossil fuel, and rapid population growth is in part due to fossil energy.

When the world population reaches about 7 billion (circa 2010), what energy source will be available as a substitute for fossil fuel? Do we gamble that some technology will rescue us, or should we proceed cautiously? Several options are discussed by the contributors.

Solar energy is vital for the functioning of the entire ecosystem, specifically for crops, livestock, and natural biota (biological diversity). Many problems are associated with any attempt to convert it for domestic, industrial, and agricultural use if we try to achieve a level of energy similar to that supplied by fossil energy. For instance, enormous amounts of land must be devoted to the collection and concentration of solar energy for human use. In turning to forests to supply a city of 100,000 with *only its electrical needs* we find that we need a self-sustaining forested area of 330,000 ha (Pimentel *et al.*, 1984). To supply a city of the same size with hydropower requires 13,000 ha of available land for the reservoir, assuming there is sufficient rainfall. Thus the conflict between land use for energy production and land use for food production becomes increasingly clear.

At first glance, the biological diversity in the United States of an estimated 200,000 species of plants and animals appears reassuring for anticipating future needs. Yet the importance of maintaining this diversity is often overlooked because of the immense variety of functions that these species perform in our ecosystem. Several contributors explain the essential roles that these natural organisms play in agriculture, forestry, and other segments of human society. For example, some species prevent the accumulation of wastes, others clean water and soil of pollutants, and still others recycle vital chemical elements within the ecosystem, including biotic nitrogen fixation. Additional organisms buffer air pollutants, moderate climate, help conserve soil and water, and serve as sources of medicines, pigments, and spices. Some organisms such as fish and wildlife are directly harvested for food. Last, and of prime importance, this great diversity preserves and provides the genetic material needed for the continued improvement of crops and livestock in agriculture.

Forests, another important component of the ecosystem, help maintain biological diversity of the environment. The world is, however, rapidly losing its forest resources. About 12 million ha per year of forests are being removed annually, and in the past 30 years about half of the world's forests have been destroyed (Pimentel *et al.*, 1986). Interestingly, the rapid expansion of agriculture accounts for about 85% of the annual forest destruction. Even now there are shortages of lumber and fuelwood for the poor in the developing countries, and shortages of fuelwood are having disastrous effects on soil quality because people are forced to burn crop residues and manure for fuel. Under this pressure, fertility of soils decreases, and soil erosion increases. This cycle continues with more forest land cleared to replace the degraded cropland in order to maintain food production.

Food resources supplied from the world's fisheries are discussed in several chapters, where it is reported that the fisheries supply less than 3% of human food (97% is from land-based agriculture). Fish supplies have generally been declining worldwide since 1970 because of both overfishing and pollution. The relatively low productivity of the oceans has recently prompted interest in aquaculture. For example, Cyprus, which is located in the Mediterranean Sea, is producing sea bass and bream. The cost is high—these fish sell for $30–40/kg. Catfish and trout sell for one-third to one-half these prices, but they are still relatively expensive compared with chicken.

As humans crowd the earth, environmental pollution and degradation are expected to grow in intensity, especially in the more densely populated land areas. Several authors emphasize their concern about air pollutants and chemical pollutants in soil and water; these toxins are already severely affecting agriculture and forests in many parts of the world by reducing the productivity of crops, livestock, forests, fisheries, and other biological resources. Added to this is the increase in soil erosion and salinization resulting from irrigation.

Biotechnology is one of the few technologies for which many social benefits have been promised. A few developments have been made, and additional technologies will probably be discovered that will improve public health and agriculture (Pimentel, 1987). At the same time, as is true with most technologies, some social, economic, and environmental problems have already surfaced and more are projected (Pimentel *et al.*, 1988). To date, none of the test protocols are 100% effective in identifying potential environmental disasters. Furthermore, scientists lack the ability to distinguish beneficial plants and animals from potential pests with 100% accuracy. Certainly, as the technology expands, accompanied by the release of large numbers of genetically engineered organisms, the probability increases for the introduction of hazardous organisms into the environment. Our goal should be to maximize the benefits from biotechnology while minimizing its risks.

No one can doubt that we are moving rapidly into a future in which we must balance the feeding of a rapidly growing population and a diminishing per capita supply of natural resources. The outcome will have a major impact on the ability of humans to provide themselves with food and, indeed, to survive.

To investigate the interdependency of food and natural resources that affect society, an outstanding group of scientists and engineers representing several disciplines have shared their data and careful assessments. It is hoped that through these discussions a more complete understanding of these timely issues will emerge. This base of knowledge will help individuals and government leaders to develop and implement the types of programs that will result in the effective use and management of land, water, energy, and biological resources for improved food production and a higher standard of living for everyone.

This book represents the cooperative efforts of the authors and many other people in the scientific community who were generous in sharing their expertise. With sincere appreciation we acknowledge the able assistance of Ms. Nancy Sorrells and Ms. Susan Pohl in assembling the numerous chapters and the editorial assistance of the staff of Academic Press.

David Pimentel
Carl W. Hall

REFERENCES

Pimentel, D., Levitan, L., Heinze, J., Loehr, M., Naegeli, W., Bakker, J., Eder, J., Modelski, B., Morrow, M. (1984). Solar energy, land and biota. *Sunworld* **8,** 70–73, 93–95.

Pimentel, D., Dazhong, W., Eigenbrode, S., Lang, H., Emerson, D., and Karasik, M. (1986). Deforestation: interdependency of fuelwood and agriculture. *Oikos* **46,** 404–412.

Pimentel, D. (1987). Down on the farm: genetic engineering meets technology. *Tech. Rev.* **90,** 24–30.

Pimentel, D., Hunter, M., LaGro, J., Efroymson, R., Landers, J., McCarthy, C., Mervis, F., and Boyd, A. (1988). Genetic engineering and environmental policy. Manuscript submitted to *BioScience*.

1

Ecological Systems, Natural Resources, and Food Supplies

David Pimentel

College of Agriculture and Life Sciences
Cornell University
Ithaca, New York

    I. Introduction
   II. The Structure and Functioning of Ecosystems
  III. Evolution of Living Systems
  IV. Biogeochemical Cycles
   V. Aquatic Ecosystems
  VI. Terrestrial Ecosystems
 VII. Human Food Gathered and Hunted from the Wild
VIII. Agricultural Ecosystems
    A. Water
    B. Nutrients
    C. Pest Controls
    D. Agricultural Ecosystem Stability
    E. Species Diversity
    F. Crop Yields
    G. Annual versus Perennial Crops
  IX. Resource Constraints in World Food Production
    A. Land Resources
    B. Water Resources
    C. Energy Resources
    D. Forest Resources
   X. Food Needs for Future Generations
  XI. Requirements for Solving Food Problems
    A. Safeguarding the Environment
    B. Science and Technology
    C. Population
 XII. Conclusion
     References

I. INTRODUCTION

All basic human needs, including food, energy, shelter, and protection from disease, are obtained from the resources found in the ecosystem. Throughout history, humans learned to modify natural ecosystems to better meet their basic needs and desires. Over time, humans have altered ever larger amounts of the environment for human use and used environmental resources to achieve an adequate, even good, standard of living for many.

Human intelligence and technology have developed rapidly, and this development has enabled humans to manipulate the ecosystem successfully more than any other animal species. This so-called advantage has given humans power to control and destroy other species. And now, with nuclear weapons, humans have the power to destroy themselves.

Humans are but one of many functioning species in the ecosystem; they are still an integral part of the earth's ecosystem. They do not function in isolation. Furthermore, their numbers cannot grow exponentially forever because shortages of food, energy, and space will limit the size of the human population eventually, as has occurred for many other species in the past.

In this chapter, the intrinsic dynamics of natural ecosystems—especially the land, water, atmosphere, energy, and biological components—are examined. The components' interaction and their relationship to agricultural productivity are discussed.

II. THE STRUCTURE AND FUNCTIONING OF ECOSYSTEMS

An ecosystem is a network of energy and mineral flows in which the major functional components are populations of plants, animals, and microorganisms. These organisms perform different specialized functions in the system.

All self-sufficient ecosystems consist of producers (plants), consumers (animals and microbes), and reducers or decomposers (microbes and animals). (See Figure 1.1.) Macro- and microscopic plants collect solar energy and convert it into chemical energy via photosynthesis. Plants use this energy for growth, maintenance, and reproduction. In turn, these plants serve as the primary energy source for all the other living organisms in the ecosystem. Animals and microbes consume plants, animals eat other animals, reducers feed on both plants and animals and recycle, thus conserving chemical resources (C, H, O, N, P, K, Ca, etc.) to be used once again by plants. Thus, consumers, reducers, and decomposers all depend, directly or indirectly, on plants as their food source.

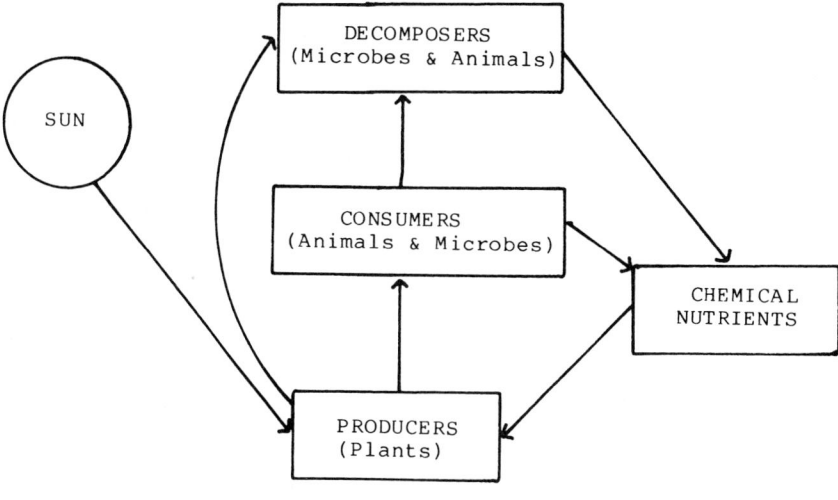

Figure 1.1 Structure of living systems.

The exact number of species needed for a particular self-sufficient ecosystem depends upon many physical and chemical factors, as well as temperature and moisture conditions, and the types of species that make up the ecosystem. At present, our knowledge is insufficient to predict accurately how many and what kinds of species are necessary for the different feeding levels in the ecosystem. For a given ecosystem, species numbers may range from the hundreds to thousands (Andrewartha and Birch, 1954).

In the United States, approximately 200,000 species of plants and animals are vital to the environment and well-being of the plant and animal inhabitants (Pimentel *et al.*, 1980a). No one knows how many of these species can be eliminated before the quality of the ecosystem will be diminished. Therefore, human societies must exercise great care to avoid causing extinction of species. A delicate balance in the natural food system has evolved in each community system and, although there is some redundancy in the system, the linkages in the trophic structure are basic to its functioning.

Elton (1927) pointed out that the "whole structure and activities of the community are dependent upon questions of food supply." Plants are nurtured by the sun and by the essential chemicals (C, H, O, N, P, K, Ca, etc.) they obtain from the atmosphere, soil, and water. The remainder of the species in the ecosystem depend on living or dead plants. About half of all species obtain their resources directly from living hosts (Pimentel, 1968; Price, 1975). The sugarcane plant worldwide, for example, has 1,645 parasitic insect species (Strong et al., 1977) and at least 100 parasitic disease microorganisms (Martin et al., 1961). Oaks in the United States have over 500 known insect species and probably close to 1000 that feed on them (Packard, 1890; de Mesa, 1928; Opler, 1974). One of the major insect herbivore parasites of the oaks in the Northeast is the gypsy moth, which in turn has about 95 parasitic and predaceous species feeding on it (Nichols, 1961; Campbell and Podgwaite, 1971; Podgwaite and Campbell, 1972; Campbell, 1974; Leonard, 1974). Clearly, parasitism and dependence on living food resources constitute a dominant way of life in natural ecosystems.

But a food host population can support only a limited population of parasites before it is so damaged that it no longer can provide the needed amount of food. An individual host utilizes most of its energy resources for its own growth, maintenance, and reproduction. For example, on average plants use 38–71% of their energy resources just for respiration; poikilotherms about 50%; and homeotherms 62–75% (McNeill and Lawton, 1970; Odum, 1971; Humphreys, 1979). In general, only about 10% of the host's resources are passed onto herbivores and other parasitic species (Slobodkin, 1960; Phillipson, 1966; Odum, 1971; Pimentel et al., 1975; Pimm, 1982). In a recent survey of 92 herbivores feeding in nature, only 7% of the plant hosts were consumed (Pimentel, 1988). Because hosts utilize most of their energy resources for themselves and their progeny, even a relatively small amount of herbivore/parasite feeding pressure influences the abundance and distribution of the host. Therefore, from an ecological perspective, host conservation is vital for herbivore/parasite survival.

Many theories exist on how plants survive the attack of herbivore/parasite populations. It is my view that herbivore/parasite populations and plant populations coevolve and function interdependently to achieve a balanced food supply–demand economy. I have proposed that parasites and hosts are dynamic participants in this economy and that control of herbivore/parasite populations generally changes from density-dependent competition and patchiness to the density-dependent genetic feedback and natural enemy (parasite feeding on parasite) controls (Pimentel, 1988). I also postulate that herbivore and other parasite numbers are often controlled by a feedback evolutionary mechanism interdependent with the other density-dependent controls. Feedback evolution limits herbivore/parasite feeding pressure on the host population to some level of "harvestable" energy and conserves the host primarily by individual selection. Essential resources necessary for growth, maintenance, and reproduction account for most of the host's resources, whereas harvestable energy is proposed to be a relatively small portion of host resources. This hypothesis suggests one reason why trees and other plants generally remain green and lush in nature and why herbivores and other parasites are relatively sparse in biomass, especially related to their food hosts.

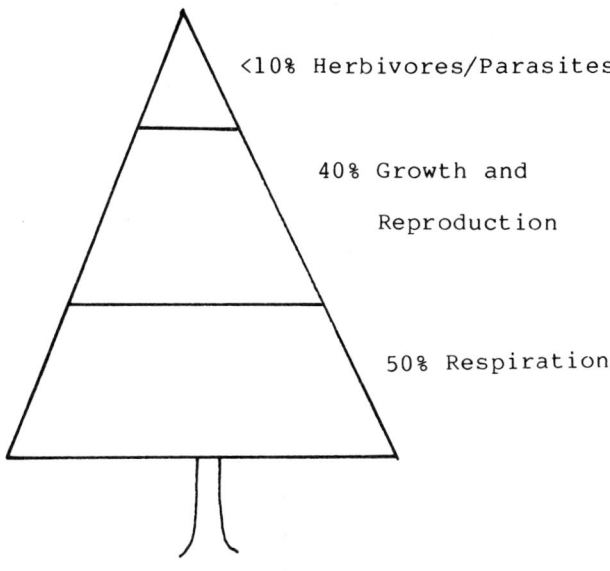

Figure 1.2 Energy and resource budget of plant hosts and the amount of resources consumed by herbivores/parasites.

To achieve a balanced economy in parasite–host systems, individual hosts either evolve defense mechanisms (Figure 1.2), or the herbivore/parasite populations evolve to moderate exploitation of their host population (Pimentel, 1961; Levin and Pimentel, 1981). The amount of resources consumed by herbivores/parasites is often limited to less than 10% (Pimentel, 1988). The defenses that appear in plant hosts include, for example, nutritional, chemical, and physical resistance factors and combinations of these factors (Pimentel, 1968; Whittaker and Feeny, 1970; Levin, 1976; Segal *et al.*, 1980; Berryman, 1982; Coley *et al.*, 1985; Rhoades, 1985). Note, if herbivore numbers are limited by parasites and predators, then the herbivores probably exert little or no selective pressure on the plant host (Hairston *et al.*, 1960; Lawton and McNeill, 1979; Price *et al.*, 1980; Schultz, 1983a,b).

Evolutionary feedback may function as a density-dependent control of herbivore/parasite populations. Thus, when herbivore numbers are abundant and the feeding pressure on the plant host is relatively intense, selection in the plant population will favor allelic frequencies and defenses in the plant population that reduce herbivore rates of increase and eventually herbivore numbers. When slugs and snails, for example, feed heavily on bird's-foot trefoil, the proportion of resistant alleles and level of cyanogenesis increases (Jones, 1966, 1979). This increase tends to reduce feeding pressure on the trefoil.

This relationship can be illustrated further. For simplicity, assume that at one locus in the host there are two alleles, A and A'. The rate of increase of the parasite on a susceptible-type host with AA is >1, whereas on a resistant-type host with A'A'-defenses the rate of increase is <1. Thus, depending upon selection on a proportion of the two alleles in the host population, herbivore or parasite numbers will increase or decrease until eventually some equilibrium ratio of the two plants is approached (Pimentel, 1961).

When the herbivore population is exerting heavy feeding pressure and there is intense selection on the plant host, the frequency of resistant A'-type allele will increase in the plant host population. Natural selection acting on the plant host favors the retention of a sufficient proportion of the A'-defense allele (Levin, 1976; Pimentel *et al.*, 1975). Then, herbivore numbers and feeding pressure will decline. The host population probably can never develop 100% effective defensive mechanisms against all herbivores exerting the selective pressure on it because the production and maintenance of the defensive mechanism must, at some point, become too costly (McKey, 1974; Cates, 1975; Krischik and Denno, 1983; Rhoades, 1985; Rosenthal, 1986). At the point when herbivore numbers have declined to a suitably low level, the host will no longer benefit from spending energy to increase its level of resistance to its parasite.

III. EVOLUTION OF LIVING SYSTEMS

Since the first organism appeared on earth several billion years ago, several basic trends in the evolution of living systems have been apparent. First, the living system has become more *complex,* with an ever-growing number of species. Although the total number of species present on earth at any one time has been growing, over 99% of all species have become extinct and have been replaced in time with new species better adapted to the developing ecosystem (Allee *et al.,* 1949).

Clearly, the growing number of species has increased the complexity of the existing living system and raised the total of living biomass or protoplasm on earth. The growth in living biomass has made it possible to capture energy that can flow through the living system. At the same time, more mineral resources from the environment are being utilized and are flowing through the living system.

Thus, the total size and complexity of the living system has increased its capacity to convert more and more energy and mineral resources into itself. This, in turn, appears to have increased the stability of the living system, making it less susceptible to major fluctuations in the physical/ chemical environment.

Additional stability in the ecosystem has evolved via genetic feedback between the parasites and their food hosts, as mentioned. Because parasites (including herbivores and predators) and hosts are interdependent, stability is essential. Parasites cannot increase their harvest of food from the host species populations indefinitely without eventually destroying the host/food and themselves.

This is not to imply that group selection and self-limitation are dominant activities in natural systems. Hosts under selective pressure may evolve various defense mechanisms to protect themselves from exploitation by parasites (Pimentel, 1988). This evolution takes place primarily by individual selection. Evolution in parasite—host systems together with complexity in general in the ecosystem leads to increased stability. This has survival value for natural living systems.

IV. BIOGEOCHEMICAL CYCLES

Several chemical elements, including carbon, hydrogen, oxygen, nitrogen, phosphorus, potassium, and calcium, are essential to the functioning of living organisms and therefore of ecological systems. Various biogeochemical cycles have evolved to insure that plants, animals, and microbes have suitable amounts of these vital chemical elements. Biogeochemical cycles both conserve the vital elements and keep them in circulation in the ecosystem. Indeed, the mortality of living organisms keeps the vital elements in circulation, enabling the system to evolve and adapt to new and changing environments.

These biogeochemical cycles themselves are a product of evolution in the living system. If the living system had not evolved a way of keeping the vital chemicals in circulation and conserving them for use in the biological system, it would have become extinct long ago.

Every organism, whether a single cell, a tree, or a human, requires nitrogen for its vital structure, function, and reproduction. Although the atmosphere is the major nitrogen reservoir, atmospheric nitrogen cannot be used directly by plants but must be converted into nitrates, which is often accomplished by nitrogen-fixing bacteria and algae (Figure 1.3). Some of these bacteria have a mutualistic relationship with certain plants like legumes. These plants develop nodules and other structures on their roots to protect and feed the bacteria. Some plants, for example, provide the associated bacteria with carbohydrate and other nutrients. In turn, the bacteria fix nitrogen for their own use as well as for the legume plant. In addition, free-living bacteria, like *Azotobacter* and blue-green algae like *Anabaena* fix atmospheric nitrogen for their use. When these bacteria and algae die and are decomposed by other bacteria or algae, their nitrogen is released for use by other plants.

8   David Pimentel

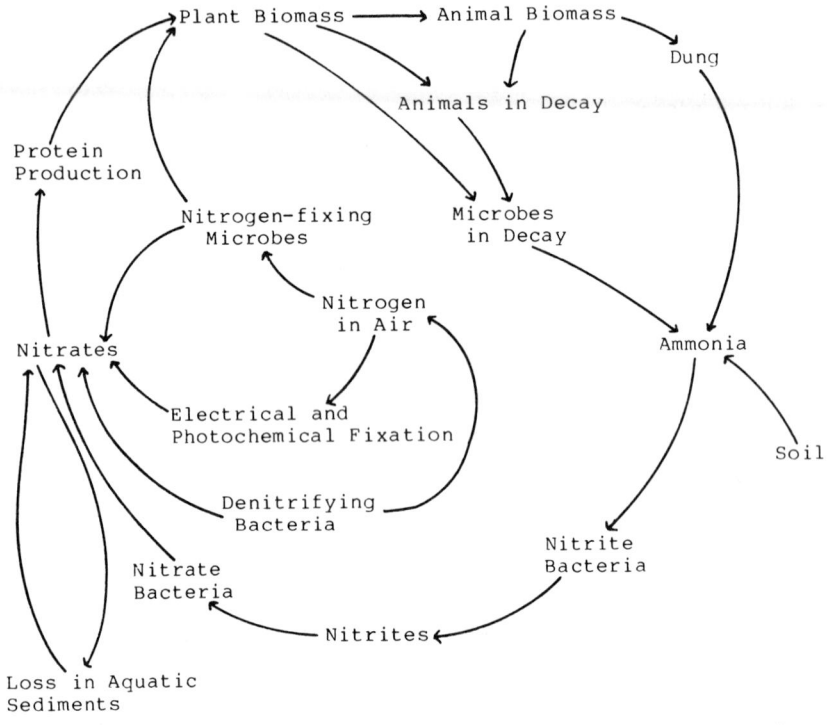

Figure 1.3 The nitrogen biogeochemical cycle.

The decay of plants, animals, and microbes also recycles nitrogen but in the form of ammonia (Figure 1.3). Most of the decomposition of the protoplasm is carried out by microbes. The ammonia released by decomposition of the organic matter is in turn converted by nitrite and nitrate bacteria into nitrates, available for use by plants. Some additional nitrates are produced by electrical storms (Figure 1.3), and some ammonia becomes available to the biological system from volcanic action and igneous rocks.

Phosphorus, another essential chemical element, is recycled by the decomposition of plants, animals, and microbes (Figure 1.4). Additional phosphorus comes from the soil and aquatic systems. At the same time, some phosphorus is continually lost to the aquatic system, especially the marine system, when it is deposited in the sediments.

Like nitrogen and phosphorus, all other essential elements depend on the functioning living system for recycling. Sometimes particular organisms serve special roles in recycling the vital elements. Thus, the living system conserves and recycles the essential elements in the biological system.

1. Ecological Systems, Natural Resources, and Food Supplies

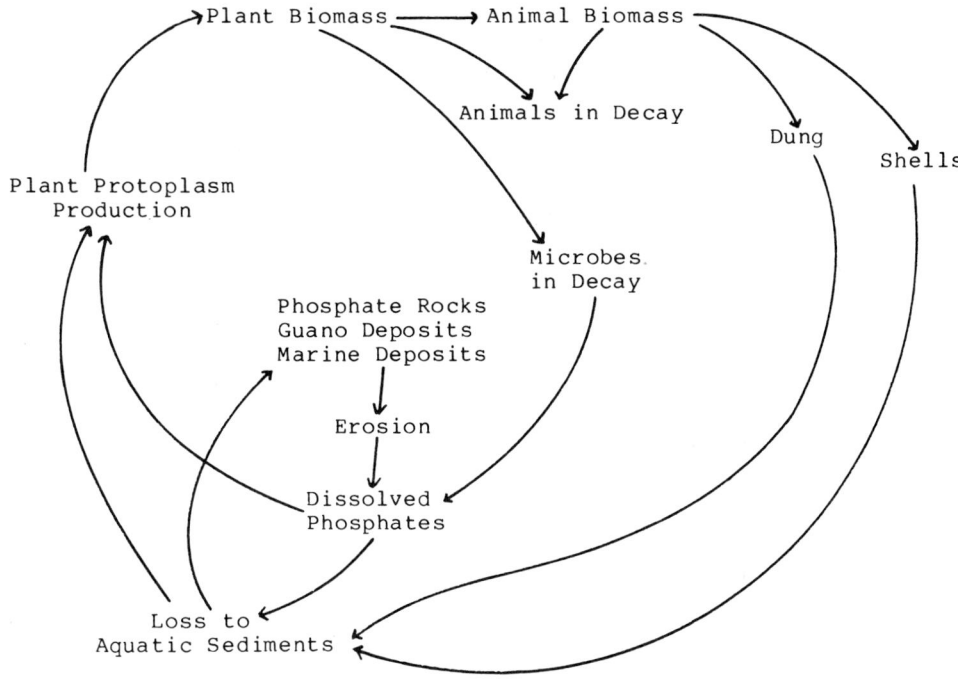

Figure 1.4 The phosphorus biogeochemical cycle.

V. AQUATIC ECOSYSTEMS

Water covers approximately 73% of the earth, but the aquatic system accounts for only 43% of the total biomass produced annually (Odum, 1971). The prime reason for the low productivity is a shortage of nutrients, and the second is lack of sunlight penetration into the aquatic system. However, some shallow aquatic systems with ample nutrients are extremely productive, yielding up to 20 t/ha (metric tons per hectare) of plant biomass under favorable conditions.

Although each aquatic system may be productive in terms of plant biomass, the harvest of fish is quite low. Primary production as phytoplankton must often pass through 3–5 trophic levels before the biomass is harvested as fish (Figure 1.5). Because only about 10% of the resources at each trophic level is generally passed to a higher level, little fish biomass is produced at the top of the food chain. For example, even with a maximum of 20 t/ha of plant biomass, fish harvest is estimated to be only 0.2 kg/ha.

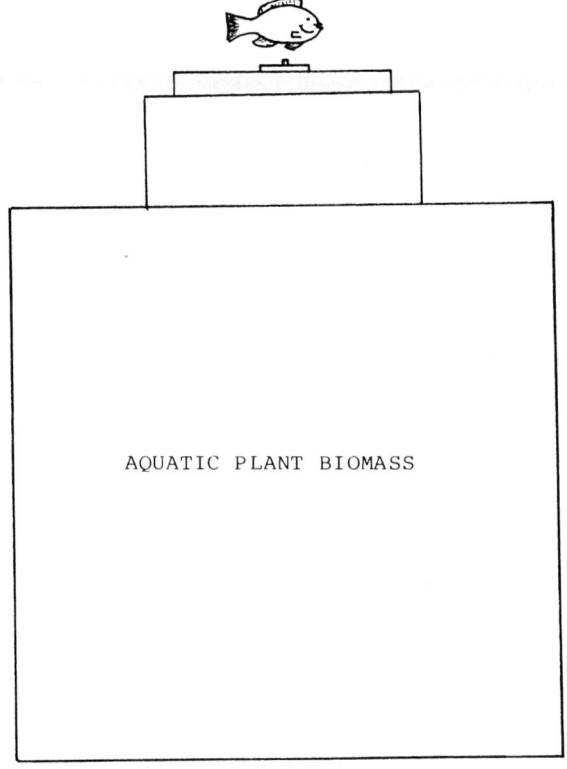

Figure 1.5 Trophic pyramid in an aquatic ecosystem indicating the small quantity of fish that might be harvested from the relatively large quantity of aquatic plant biomass.

Because of its low productivity, humans harvest only 2—3% of their total food from the aquatic system. Thus, it is doubtful that the aquatic system is capable of providing more human food in the future. Further decrease in potential is due to overfishing and pollution (CEQ, 1980).

VI. TERRESTRIAL ECOSYSTEMS

Twenty-seven percent of the earth is covered by land, yet on this small terrestrial system an estimated 57% of the earth's total biomass is produced (Odum, 1971). Forest and agricultural lands account for about 90% of the total biomass production. Considering that more than 97% of human food comes from the terrestrial system, and only about 3% from the aquatic system (Pimentel *et al.*, 1980b) the status of our land is vital to human survival.

Solar energy powers the ecosystem. During one year the solar energy reaching a hectare in temperate North America averages about 14 billion kcal (Reifsnyder and Lull, 1965). During a four-month summer growing season, nearly one-half of this or 7 billion kcal reaches a hectare. Under favorable conditions of moisture and soil nutrients, the annual production of natural plant biomass in North America averages about 2,400 kg (dry) per hectare per year (Pimentel *et al.*, 1978).

The productivity of the terrestrial system depends upon the quality of soil, availability of water, energy, favorable climate, as well as the amount and diversity of biological resources present. Agricultural productivity is affected by the same basic factors that influence the productivity of these natural systems.

VII. HUMAN FOOD GATHERED AND HUNTED FROM THE WILD

For about 99% of the time that humans existed on earth (about 1 million years), they lived as hunter–gatherers. Some of the data on food systems of hunter–gatherers provide a better understanding of ecological systems, natural resources, and food.

Let us assume that a hunter–gatherer requires 2,500 kcal/day for the year to meet food energy needs. By harvesting at least 0.1% of the available animal biomass from 40 ha, it is possible he will provide himself with 88 kcal per day (32,000 kcal per year) in the form of animal food. The remaining 2,412 kcal per day (880,500 kcal per year) of food energy needed would have to come from other sources like seeds, nuts, fruits, roots, and other plant foods. Assuming that 1 kg of digestible plant material yields 3,000 kcal, about 294 kg of plant material per year would have to be harvested from the 40 ha (7 kg/ha) to meet the calorie needs. Although obtaining this amount of edible plant material might be impossible to obtain from a strictly wooded habitat, it might be feasible in an environment containing a mixture of woods, shrubs, and herbs, and a productive stream.

Forty hectares per person may be optimistic, and larger areas probably would be needed by hunter–gatherers even under the most favorable conditions. Clark and Haswell (1970) estimated that at least 150 ha per person of favorable environmental conditions would be needed to provide an adequate food supply. Given a moderately favorable habitat, these scientists estimate that 250 ha per person would be required to supply suitable quantities of food. These estimates are 4–6 times greater than the optimistic model I discussed above.

In marginal environments, such as the cold Canadian Northwest, approximately 14,000 ha are probably necessary to provide each person with about 912,500 kcal of food energy per year (Clark and Haswell, 1970). The land area may range as high as 50,000 ha per person in subarctic lands, and in these cold climates meat and animal products are the predominant food in the diet. Suitable plant foods for humans are relatively rare in these severe environments and do not provide the concentrated energy needed by the people. In fact, animal flesh and fat would comprise up to two-thirds of the food calories consumed.

In all likelihood the solar energy captured by plants and converted into biomass in such marginal habitats would average only 10–200 kg/ha/yr (Whittaker and Likens, 1975), while animal production may average only 4–20 kg/ha/yr. The annual yield of meat protein for humans per hectare may average only 10 g.

In contrast to the 150–50,000 ha of land needed to feed one hunter–gatherer, U.S. agriculture today provides one person a high protein-calorie diet on 1.9 ha (USDA, 1985). Thus, agricultural systems gradually established by human societies were of great benefit as human populations increased above the carrying capacity of the natural ecosystem.

VIII. AGRICULTURAL ECOSYSTEMS

To obtain food, humans manipulate natural ecosystems. In altering the natural system to produce vegetation and/or animal types (livestock) different from that which is typical of the natural systems, a certain amount of human and fossil energy input is necessary. In principle, the greater the change required in the natural system to produce crops and livestock, the greater the energy and labor that must be expended.

This same principle applies in reverse to minimizing the energy inputs into all agricultural systems. That is, the more closely the agricultural system resembles the original natural ecosystem, the fewer the inputs of energy and other factors that will be required in agricultural production. Equally important, the closer the agricultural system is to the natural ecosystem, the more sustainable it is. This is because less environmental degradation takes place in the less intensively managed systems.

The productivity of agricultural plants is limited by the same factors that limit natural plants—sunlight, water, nutrients, favorable temperature, and animal/plant pests. The agriculturalist seeks to optimize the availability of these environmental factors for his crop plants while minimizing the impacts of pests.

A. Water

Water, followed by nutrients, is the principal limiting factor for terrestrial plant productivity, including agriculture. The United States invests about half of its fossil energy input in agricultural production into supplying irrigation water (20%) and fertilizer nutrients (30%) (Pimentel and Wen, 1987). Agricultural practices that help to conserve water and soil nutrients not only contribute to crop productivity but reduce the costly fossil energy inputs in the system (Pimentel *et al.*, 1987).

The availability of water and soil nutrients to crops can best be achieved by controlling soil erosion and water runoff. This will also maximize the amount of soil organic matter present. Maintaining adequate soil organic matter helps maintain nutrients, water, tilth, and the buffering capacity of the soil. All of these characteristics, combined with ample water and soil nutrients, help keep the agroecosystem productive.

Similar to natural ecosystems, the goal in agriculture should be to conserve nutrients and water for optimal production of crops while maintaining the stability of the system. In agriculture, this would mean recycling manure, crop residues, and other wastes.

B. Nutrients

After water, shortages of soil nutrients (nitrogen, phosphorus, potassium, and calcium) are the most important factors limiting crop productivity. Valuable nutrient resources available for recycling include crop residues and livestock manure. Crop residues total about 430 million t per year (ERAB, 1981). This amount of crop residue contains about 4.3 million t of nitrogen, 0.4 million t of phosphorus, 4.0 million t of potassium, and 2.6 million t of calcium. The total amount of livestock manure produced annually in the United States is 1.6 billion t (Anderson, 1972). This manure contains about 80 million t of nitrogen, 20 million t of phosphorus, and 64 million t of potassium (Thompson and Troeh, 1978). These quantities of nutrients in both the residues and manure are significantly greater than the quantities of commercial fertilizer applied annually in the United States, which contain 11 million t of nitrogen, 5 million t of phosphorus, and 6 million t of potassium (USDA, 1983). Except for the extremely small amount of crop residues that are harvested annually, most of the crop residues are recycled on U.S. agricultural lands (ERAB, 1981). However, with manure, Safley *et al.* (1983) calculated that only 2 million t (2.5%) of the total nitrogen in the manure are recoverable and usable with present technology. Some of the difficulty with manure is the uneven distribution of livestock and crop areas. From 30–90% of the nitrogen in manure is often lost through ammonia volatilization when manure and crop residues are left on the surface of croplands and pasturelands (Vanderholm, 1975).

However, less than 5% of the nitrogen is lost as ammonia when the manure is plowed under immediately.

The major loss of soil nutrients in the United States is from soil erosion (Pimentel *et al.*, 1987). Average soil erosion rates are 18 t/ha/yr (Lee, 1984). A ton of rich agricultural soil contains about 4 kg of nitrogen, 1 kg of phosphorus, 20 kg of potassium, and 10 kg of calcium (Alexander, 1977; Greenland and Hayes, 1981). For just nitrogen, 18 t of soil contains 72 kg/ha, which is almost half of the average of 152 kg/ha of nitrogen fertilizer that is applied to U.S. corn (USDA, 1982).

Soil erosion selectively removes different components from the soil. For example, eroded material usually contains 1.3–5 times more organic matter than the remaining soil (Barrows and Kilmer, 1963; Allison, 1973). Soil organic matter is extremely important to the productivity of the land because it helps retain water in the soil, improves soil structure, and the cation exchange capacity of the soil. In addition, organic matter is the major source of nutrients needed by plants (Volk and Loeppert, 1982). About 95% of the nitrogen in the surface soil is stored in the organic matter.

Earlier it was mentioned that 11 million t of nitrogen is applied as commercial fertilizer with a total value of $6 billion annually (USDA, 1983). About 14 million t of nitrogen are biologically fixed by microbes in the United States annually (Delwiche, 1970); this nitrogen has an economic value of $7 billion.

Overall, the harvest of the corn crop itself removes from 25–50% of the total nitrogen applied. Some nitrogen (15–25%) is lost by volatilization and 10–50% by leaching (Schroder, 1985).

C. Pest Controls

Mimicking the natural system also offers many opportunities for pest control. This would include maintaining the genetic resistance of the crop to pests such as insects, plant pathogens, and weeds; encouraging natural enemies; employing crop rotations and other crop diversity patterns; and utilizing natural forage and trees where appropriate (Pimentel, 1986). For example, the spotted alfalfa aphid is under biological control by natural enemies and using alfalfa varieties naturally resistant to the aphid (PSAC, 1965).

Crop rotations can be highly effective in pest control. This is clearly demonstrated with the control of the corn rootworm complex in corn (Pimentel *et al.*, 1977). In addition to aiding in insect control, crop rotations may also help reduce disease and weed problems.

In the United States, most plant pathogen control relies on host plant resistance. It is estimated that nearly 100% of all crops planted in the nation contain some degree of host plant resistance (Pimentel, 1981).

Diseases can also be prevented by planting disease-free propagated material and by other cultural methods that eliminate the source of the inoculum.

Most weed control is accomplished by mechanical tillage, rotations, and employing various polycultural means (Pimentel, 1981). Weed control generally has fewer options than insect and plant pathogen control.

D. Agricultural Ecosystem Stability

A relatively stable natural ecosystem increases the stability of the human food supply. Over time, agricultural stability has been enhanced by selecting crops and livestock that are best adapted to particular environments. In addition, increased amounts of fossil energy inputs are used to enhance or control various aspects of the agricultural environment. For example, nutrient limitations in agriculture have been offset by the addition of fertilizers, water shortages compensated by irrigation, and pest attacks controlled by pesticides and various cultural and biological controls.

E. Species Diversity

Wild plants and animals provide the original sources of genetic material for breeding resistance to pests and improving other crop and livestock features that contribute to increased yields.

Unfortunately, because of the conversion of extensive natural ecosystems into agriculture, thousands of species are being lost each year (Biswas and Biswas, 1985; Eckholm, 1978; Ehrlich and Ehrlich, 1981; Hanks, 1987). The most rapid loss of biological diversity is occurring in tropical forests and savannas, the same regions where most crop and livestock species originated. This loss has alarming implications for future production of human food, important medicines and other products that are obtained from biological resources.

F. Crop Yields

On rich agricultural soils with ample water and fertilizers, average biomass production for several major crops is about 15 t/ha. However, under relatively poor agricultural conditions, biomass yields may range from only 0.5–1 t/ha (Pimentel and Pimentel, 1979).

Forests on good soils, with ample water, will be as productive as agricultural crops, i.e., about 15 t/ha. However, as in agriculture, poor soils with limited moisture may limit forest production to about 0.5 t/ha or less (Pimentel et al., 1978).

Under favorable atmospheric conditions, but with the addition of nitrogen, phosphorus, potassium, and calcium fertilizers, hybrid corn, one of our most productive crops, will yield about 14,000 kg/ha of biomass (dry) or 7,000 kg of grain (Pimentel and Wen, 1988). Wheat production in North America averages about 6,750 kg biomass/ha or 2,700 kg of grain. Both of these yields are much higher than that of natural vegetation. However, many agricultural crops are less productive than either corn or wheat, and overall biomass production in the form of crops probably averages close to that of natural vegetation.

To convert corn biomass to heat energy, the 14,000 kg yield is multiplied by 4500 kcal/kg, yielding 63 million kcal/ha. This represents only 0.5% of the total solar energy reaching a hectare during the year. For wheat the percentage of solar energy harvested as wheat biomass is 0.2%. For natural vegetation production of about 2,400 kg/ha, about 0.1% of solar energy is converted into biomass. This 0.1% is the average production for all natural vegetation in North America, and the 0.1% conversion is also about the average for all U.S. agriculture (Pimentel et al., 1978).

From the total of 14,000 kg/ha of corn biomass, as mentioned, humans are able to harvest approximately half or 7,000 kg of corn grain as food. This is obviously much more than hunter–gatherers were able to harvest per hectare from the natural environment. If a natural ecosystem yields only about 2,400 kg/ha of plant biomass, then only a small portion of this would be converted into animal and microbe biomass.

G. Annual versus Perennial Crops

Most crops cultured in the world are tropical annuals. The fact that most human societies probably originated in the tropics may explain in part why so many crop and livestock species originated in the tropics. Originally, annuals were a practical choice for crops because pest problems, particularly weeds, could be minimized. Planting annuals allowed the land to be cleared of all vegetation by burning and digging, then planted. This gave newly planted crops a head start on weeds and other potential pests (Pimentel, 1977).

At present, 90% of the world's food supply comes from only 15 species of crop plants and eight species of livestock (Pimentel et al., 1986a). This is a very narrow base of crop and animal species, especially considering that there are 5–10 million species of plants and animals in the world today (Ehrlich and Ehrlich, 1981; Myers, 1983).

The human food supply would be enhanced if it could rely on more perennial crops, especially grains (Pimentel et al., 1986a). Since grain crops supply approximately 80% of the total food produced worldwide, the development of perennial grain crops would add stability to food supply and to the agricultural ecosystem as well. Note, I am defining a perennial crop as one that might have to be replanted only once in 5–20 years.

The advantages of perennial grain crops in particular are manifold.

First, the soil would not have to be tilled each year. Annual soil tillage requires an enormous amount of fossil, draft animal, and human energy. The energy required to till 1 ha ranges from 200,000 kcal for hand tillage to nearly 600,000 kcal for tractor tillage (Pimentel and Pimentel, 1979).

Further, less tilling would conserve soil and water resources; both are as important as saving energy. Soil erosion and water runoff are rapidly degrading cropland resources (Pimentel *et al.*, 1987). Erosion and runoff occur primarily when the soil is tilled and exposed to rain and wind energy. Vegetative cover is the principal means of protecting soil and water resources (Pimentel *et al.*, 1987), so a perennial grain crop would be valuable in decreasing erosion in world agriculture.

At present there are no commercial perennial grain crops, and their development will depend in part on genetic engineering, which in turn depends on biological diversity. It is from nature that we hope to obtain the genes that will be used to develop new crop and livestock types. New genetic materials will also be important for use in food processing. Biological diversity also offers the basic resource for new drugs and medicines. Unfortunately, scientists have not had time to investigate the full potential of these natural biological resources.

Clearly, much can be learned from natural systems about maintaining the productivity and sustainability and applied to agricultural systems. Fewer energy inputs would be necessary and fewer problems would occur in agriculture and the system would be more productive and sustainable, if the agricultural production systems could be designed more similarly to natural ecological systems.

IX. RESOURCE CONSTRAINTS IN WORLD FOOD PRODUCTION

Because 230,000 people are added to the world population each day, more food must be produced. In the simplest terms, the production of more food requires more land, water, energy, and biological resources. To gain some insight into the potential capacity of the earth to supply food for a growing population, estimates can be made on the amounts of food that could be produced, based on current technologies and the available natural resources.

A. Land Resources

World arable land resources are approximately 1.5 billion ha (see Chapter 4), and with a world population of 5 billion, the per capita land available is only 0.3 ha. Arable land resources appear to be sufficient to feed the present world population a minimum diet (2,500 kcal/person/day) with current technologies and inputs of fertilizer, irrigation, pesticides, and hybrid seeds. This estimate assumes adequate distribution of food so that the more than 1 billion presently malnourished humans would receive their minimum needs. This is a major assumption given the political and economic status in many countries, especially those in the Third World.

Would arable land be sufficient to feed the current world population of 5 billion a U.S.-type high-protein calorie diet? Presently the United States cultivates about 160 million ha in crops (USDA, 1985). With more than 240 million people, this averages 0.7 ha of cropland per capita.

Therefore, even now, arable land supplies are insufficient to feed the current world population a diet similar to that presently consumed in the United States. This is assuming that sufficient fossil energy supplies also would be available for fertilizers, pesticides, and other inputs everywhere in the world to enhance productivity.

Even if one were to make the optimistic assumption that world arable land resources could be doubled (Buringh, Chapter 4), land would still be insufficient (0.6 ha per capita) to feed the current world population a U.S.-type diet. Given the rapid increase in the human population, the prognosis is grave.

Adding to the problem of land is the alarming rate of soil degradation. A sustainable erosion rate for agricultural soils is about 1 t/ha/yr. Yet soil erosion rates in the United States average 18 t/ha/yr (Lee, 1984), but they are much higher in other nations. For example, in India and China with nearly 40% of the world's population, erosion rates are about 30 and 40 t/ha/yr, respectively (Pimentel *et al.*, 1987). These high erosion rates plus waterlogging of soils, salinization, and soil degradation factors force an estimated 6 million hectares of arable land to be abandoned each year (UNEP, 1980). At the same time, to sustain the productivity of the land that is in use, more fertilizers, pesticides, irrigation, and other inputs are being used.

Certainly not everyone in the world desires to eat a typical U.S. diet, but the above examples clearly illustrate that land is a major constraint in future food production, especially if rapid soil degradation is allowed to continue.

B. Water Resources

Next to sunlight, water is the single most important limiting factor for crop production worldwide. Droughts affect crops annually in some part of the world, and in arid lands water supply is a continual problem. One inescapable aspect of the water problem is the enormous amount of water required by crops. For example, a corn crop yielding 6,500 kg of grain will transpire about 4.2 million liters/ha of water just during the growing season of the crop (Leyton, 1983). To supply this much water to the corn crop requires the application of about 10 million liters of water per hectare. This water must be applied relatively evenly during the growing season if the crop is to have a maximal yield.

Most or about 80% of the water consumed by human societies is directly used for agriculture (Biswas and Biswas, 1985). Industry and public water needs account for the remainder. This high demand by agriculture for water not only will continue but will escalate in the future as more land is put into production.

When it was suggested earlier that land resources for crops might be doubled, the assumption was that a significant amount of the new land could be irrigated. This too is an unrealistic assumption because irrigation has two major limitations. First, water must be available, and second, large amounts of energy are necessary to move the water to croplands. Water shortages exist and groundwater is being mined extensively in the western United States already; also about 20% of the total energy used for direct on-farm use in the United States is for moving irrigation water (USDA, 1974).

The problem of water supply as well as the allocation of supplies for societal needs is by no means limited to the United States. As many as 80 other countries, which account for nearly 40% of the world population, are now seriously beset by droughts (Kovda *et al.*, 1978). Although at present the amount of water pumped per capita/year on a global basis is less than one-third of the amount withdrawn in the United States per capita, the growth in the world population can be expected to double water needs in all areas by the year 2000 (CEQ, 1980). At that time, world agricultural production will require an estimated 64% of all the water withdrawn from aquatic systems and will consume more than 80% of all water.

C. Energy Resources

In the land resources section, fossil energy resources were assumed to be unlimited for a productive agriculture. That is, only arable land was limited, but fertilizers, fuels, and pesticides would be used to enhance the yields. Future projections for food production, however, must be based on a limited and a more expensive fossil energy supply.

The following example illustrates this energy constraint. Seventeen percent of the total per capita energy used each year in the United States is expended for food (Pimentel, 1980). This means that about 1,500 liters of gasoline equivalents will be used for food production, processing, distribution, and preparation per capita per year.

When this example is expanded to include the present world population of 5.0 billion, the equivalent of 7,500 billion liters of gasoline equivalents would be expended to feed them the high-protein-calorie diet of the United States for one year.

Based on this rate of use, how long would it take to deplete the known world petroleum reserves of 113,700 billion liters (Linden, 1980)? Assuming that 76% of the raw petroleum can be converted into fuel (Jiler, 1972), this would provide a usable reserve equal to 86,412 billion liters of gasoline equivalency. Therefore, if petroleum were the only source of energy for food production, and if all petroleum reserves were used just to feed the current world population, the reserve would last less than 12 years.

These estimates indicate that the present world population already has exceeded the capacity of arable land and energy resources to provide a U.S.-type diet produced with U.S. technology. Note that these estimates were based on known arable land and known petroleum resources. If potential arable land and potential petroleum reserves are included, the projection improves, but the eventual end result remains the same. Also, the current world population figures were used in this analysis. Estimates based on various combinations of population size, dietary standards, and production technology are possible and can be expected to give slightly different projections. This example, however, suffices to focus on three of the major factors—land, water, and fossil fuel—that will limit food production in the not too distant future.

D. Forest Resources

Forests are an essential resource for future economic development. They provide lumber for housing, pulp for paper, and biomass for fuel. Although wood is not a major fuel source in the United States, biomass energy is the primary fuel source (about 80%) for the poor people of the world. Approximately half of the biomass is woody material, and the remainder is 50% fuelwood, 33% crop residues, and 17% dung (Pimentel *et al.*, 1986b).

Forests play a vital role in helping to control erosion and water runoff on steep slopes, and thus help conserve soil and water resources. Because rates of erosion and sedimentation are high, the reservoirs and dams are being rapidly filled with sediments. This reduces the generating capacity of hydroelectric plants and irrigation capacity of the dams.

Worldwide deforestation is removing about 11.6 million ha of land annually (FAO, 1982). Most of this deforestation is due to the growing needs of agriculture. About 10 million ha of new land are needed for increased agricultural production each year. Of this, about 6 million ha of the forestland removed is to compensate for soil degradation, while 4 million ha is needed to meet the food needs of the escalating human population.

As the world population continues its rapid growth, more land will be needed for crops and livestock, and most of the "new" land will come from forests. This will continue to put pressure on the forest resources. Loss of forestland diminishes the supply of wood for fuel, and this will increase the removal of crop residues and dung. These activities intensify soil erosion and decrease soil fertility. So the cycle accelerates and more forestland will be sacrificed to offset the loss in soil productivity.

X. FOOD NEEDS FOR FUTURE GENERATIONS

The degradation of agricultural land, forests, and other biological resources greatly affects their productivity. As explained, the productivity of these resources presently is being maintained in large measure by the increased input of fossil energy for fertilizers, pesticides, and irrigation. Thus, it will be a challenge to meet the food needs of the rapidly expanding human population. Food production in all countries, but especially in the developing nations, where both rates are high and generation times short, must increase at a greater rate than ever before.

A study by the National Academy of Sciences (1977) targeted eight food sources for increase: rice, wheat, corn, sugar, cattle, sorghum, millet, and cassava. Currently, these foods provide about 70–90% of all the calories and 66–90% of the protein consumed in the developing countries of the world, with these grains plus others supplying as much as 80% of the total calories.

The NAS report (1977) recommended that developing countries increase food production by 3–4% year until the year 2000. Is this a realistic expectation when the annual increase in food production has been only 2.5% in the last 15 years (NAS, 1977)?

Growing food grain exports in the early 1970s encouraged the United States and other developed countries to expand their production of grains (Webb and Jacobsen, 1982). Because of these encouraging trends, many U.S. farmers purchased more land and invested heavily in new machinery. However, a few years later the situation turned around. OPEC increased oil prices, making it necessary for developing countries to spend their limited funds for imported oil instead of imported food. This change depressed the agricultural markets in most of the developed nations, and the situation continues to date (Hudson, Chapter 12).

Concerning the quantity of food that will have to be produced in the future to meet the food demand of a rapidly growing population, D. Bauman (1982 personal communication) predicted that "an amount of food equal to all the food produced so far in the history of mankind will have to be produced in the next 40 years" to fulfill human food needs. This further confirms the staggering impact of the rapidly growing world population on food and natural resources.

Even if individual dietary patterns are modified to include less animal products and more plant foods like grains, food production must be greatly increased. In conclusion, the message is clear that more food, much more food, will have to be grown to sustain the rapidly growing human population of the future.

XI. REQUIREMENTS FOR SOLVING FOOD PROBLEMS

The approaches to increase food supplies for the current and future human numbers must include protecting the environment, new technologies, and limiting human population growth.

A. Safeguarding the Environment

The environmental resources for food production, including land, water, energy, forests, and other biological resources must be protected if food production is to continue to grow. Over the past four decades, humans have allowed environmental resources to be rapidly degraded and, as mentioned, we have been offsetting this degradation with fertilizers, irrigation, and other massive inputs—all based on fossil energy. Thus, we have been literally substituting a nonrenewable resource for a renewable resource. Clearly, this has been a dangerous, if not a disastrous policy.

B. Science and Technology

Recent decades have witnessed many exciting and productive technological advances that have increased food supplies. For example, the advances in plant genetics that focused on some major crops have raised the "harvest index." In addition, the formulation and use of agricultural chemicals, pesticides, and fertilizers have helped increase yields of food and fiber crops per hectare. Improved processing of foods has enabled the food supply to be safely extended beyond harvest time, and the growing transportation network has moved more food from production sites to far distant markets. In the industrialized nations, this has meant a more abundant, more nutritious, and safer food supply. People living in developing nations, however, have not been as fortunate, even though successful plant breeding products like high-yielding rice have benefited millions in the Far East (Baum, 1986).

The new technology of genetic engineering or biotechnology offers further promise of raising crop and livestock production while improving the use of some major resources. This will be especially true, for example, if we can develop rice, wheat, corn, and other cereal grain crops that will fix nitrogen like legumes. Of the essential nutrients, nitrogen fertilizer requires the largest fossil energy input to produce. Thus, developing cereal grains that fix nitrogen will be a major breakthrough. However, conservative estimates of when this breakthrough will be achieved range from 20–30 years in the future.

Some of the other promises of genetic engineering like growing plants with little or no water, are without sound scientific basis. Even if many of the other promises of biotechnology are forthcoming, it is essential that quality soil, water, and biological resources are maintained. A few biotechnologists have acknowledged that without good soil and water resources biotechnology is a failure.

Biotechnology and other new technologies undoubtedly will help conserve energy resources and facilitate increased food production. Sufficient, reliable energy resources will have to be developed to replace most of the fossil fuels now being rapidly depleted. These new sources will of course be more costly in terms of dollars and the environment. Energy obtained from the sun, from fission, perhaps from fusion, and from the wind will become more viable in the future than they are today. But relying solely on new technological advances is depending and hoping that the "lottery" of science will pay off. These developments may not materialize as rapidly as needed to meet future food and other needs. One has only to observe the plight of millions of people in Calcutta and Mexico City to recognize that science and technology have done little to improve their lives during the last two decades.

C. Population

Thus far, only factors affecting food production have been considered, but production is only one side of the food equation. The other is the demand or rate of consumption. This is determined by the size of the human population. Ultimately, the size of the world population will determine the need for food. When human numbers exceed the capacity of the world to sustain them, then a rapid deterioration of human existence and poverty will follow. As with all forms of life, ultimately nature will control human numbers.

Problems with increasing food production substantially over present levels and those associated with decreasing population growth must be faced now. Both parts of the food equation must be brought into balance if future generations are to have an adequate food supply and live in a world that supports a reasonably acceptable standard of living.

XII. CONCLUSION

Although an estimated one billion humans are malnourished in the world today, sufficient food is produced to feed these malnourished and all other people a minimal adequate diet. This would mean most people subsisting on primarily a vegetarian-type diet and not a high-protein/calories diet typical of that in the United States.

Overpopulation, or a population of five billion humans, is one primary cause of food shortages and general poverty for more than half of the world population. This distressing situation would not be as serious if there were a better distribution of people and resources. For example, some people live where there is less than 0.1 ha of cropland, whereas others live in regions with more than 0.7 ha.

The problem of malnourishment and the human population is more complex than just the distribution of people and resources to produce food. The economic and political structure and priorities of governments also influence the amount of food that is produced and distributed to the people. Most governments of the world do not give high priority to incentives for farmers to produce large quantities of food. One of the typical goals of politicians is "cheap food." Cheap food policies result in minimal food production and especially poor management of land, water, energy, and biological resources that are essential for a productive agriculture.

Instead of most governments giving high priority to food for the people, priority is given to the military. Today, most governments spend from one-third to more than one-half of the nation's economic resources on the military. Spending for weapons and soldiers depletes the resources of a nation and does not serve to supply food or export goods and services for income.

Humans and their governments should start planning for their future generations now by planning and establishing sound policies today. The priorities for food versus military, for a high standard of living with few people or many people with a low standard of living need to be examined.

Earlier we proposed that the ideal population for the world is about 1 billion humans. This is making the assumption that everyone on earth would like to have a high standard of living and that sufficient sustainable resources would be available to maintain a high standard of living now and in the future. Again, the critical resources are arable land, freshwater, energy, and biota. If managed properly in a quality environment, these resources are all renewable.

Of course, no one would expect to move from five billion to one billion in a few years; it would take centuries. This is why policies should be planned and developed now for a quality life for everyone instead of poverty for most of the people on earth.

To cope with the current world problem of malnourishment, several actions would help:

1. Developed and developing nations should improve their economic and political structure to help those people who are malnourished in their nations—including the United States, the Soviet Union, and India.
2. Developed nations should devote about 5% of their resources to help developing nations. Currently, the United States contributes less than 0.25% of its resources for development.
3. It is essential that developing nations control their rapid population growth. For example, Kenya has a 4.2% rate of growth, which means that its population is doubling every 17 years, creating an impossible food and nutrition situation.
4. Soil, water, energy, and biological resources must be conserved for use now and in the future. The loss of 25 mm of topsoil today will require 500 years to replace.
5. Instead of nations wasting 30–50% of their resources for military weapons and soldiers, this should be reduced to approximately 10–15% and the released resources be devoted to meeting the food and other basic needs of the people.
6. There is need for an effective international political and economic system of government that will help achieve peace, control human population growth, and protect the essential environmental resources.

REFERENCES

Alexander, M. (1977). "Introduction to Soil Microbiology," 2nd Ed. Wiley, New York.

Allee, W. C., Emerson, A. E., Park, O., Park, T., and Schmidt, K. P. (1949). "Principles of Animal Ecology." Saunders, Philadelphia, Pennsylvania.

Allison, F. E. (1973). "Soil Organic Matter and Its Role in Crop Production." Elsevier, New York.

Anderson, L. L. (1972). Energy potential from organic wastes: A review of the quantities and sources. *Inf. Circ. U.S. Bur. Mines* No. 8549.

Andrewartha, H. G., and Birch, L. C. (1954). "Distribution and Abundance of Animals." Univ. of Chicago Press, Chicago, Illinois.

Barrows, H. L., and Kilmer, V. J. (1963). Plant nutrient losses from soils by water erosion. *Adv. Agron.* **15**, 303–315.

Baum, W. C. (1986). "Partners Against Hunger." World Bank, Washington, D.C.

Berryman, A. A. (1982). Population dynamics of bark beetles. *In* "Bark Beetles in North American Conifers. A System for the Study of Evolutionary Biology" (J. B. Mitton and K. B. Sturgeon, eds.), pp. 264–314. Univ. of Texas Press, Austin.

Biswas, M. R., and Biswas, A. K. (1985). The global environment. Past, present, and future. *Resour. Policy* **3**, 25–42.

Campbell, R. W. (1974). The gypsy moth and its natural enemies. *Agric. Inf. Bull. (U.S. Dep. Agric.)* No. 381.
Campbell, R. W., and Podgwaite, J. D. (1971). The disease complex of the gypsy moth. I. Major components. *J. Invertebr. Pathol.* **18**, 101–107.
Cates, R. G. (1975). The interface between slugs and wild ginger: some evolutionary aspects. *Ecology* **56**, 391–400.
Clark, C., and Haswell, M. (1970). "The Economics of Subsistence Agriculture." Macmillan, New York.
Coley, P. D., Bryant, J. P., and Chapin, F. S., III. (1985). Resource availability and plant antiherbivore defense. *Science* **230**, 895–899.
Council on Environmental Quality (CEQ). (1980). "The Global 2000 Report to the President," Counc. Environ. Qual. and Dep. State, Vol. 2. U.S. Gov. Print. Off., Washington, D.C.
Delwiche, C. C. (1970). The nitrogen cycle. *Sci. Am.* **223**(3), 137–158.
de Mesa, A. (1928). The insect oak-galls in the vicinity of Ithaca. Thesis, Cornell Univ., Ithaca, New York.
Eckholm, E. P. (1978). "Disappearing Species: The Social Challenge," Worldwatch Paper 22. Worldwatch Inst., Washington, D.C.
Ehrlich, P. R., and Ehrlich, A. H. (1981). "Extinction: The Causes and Consequences of the Disappearance of Species." Random House, New York.
Elton, C. S. (1927). "Animal Ecology." Sidgwick & Jackson, London.
Energy Research Advisory Board (ERAB). (1981). "Biomass Energy." Energy Research Advisory Board, Department of Energy, Washington, D.C.
Food and Agriculture Organization (FAO). (1982). "1981 Production Yearbook." Food Agric. Organ., Rome.
Greenland, D. J., and Hayes, M. H. B. (1981). "The Chemistry of Soil Processes." Wiley, New York.
Hairston, N. G., Smith, F. E., and Slobodkin, L. B. (1960). Community structure, population control and competition. *Am. Nat.* **94**, 421–425.
Hanks, J. (1987). "Human Populations and the World Conservation Strategy." Int. Union Conserv. Nat. Nat. Resour., Gland, Switzerland.
Humphreys, W. F. (1979). Production and respiration in animal populations. *J. Anim. Ecol.* **48**, 427–454.
Jiler, H. (1972). "Community Yearbook." Commodity Res. Bur., New York.
Jones, D. A. (1966). On the polymorphism of cyanogenesis in *Lotus corniculatus*. Section by animals. *Can. J. Genet. Cytol.* **8**, 556–567.
Jones, D. A. (1979). Chemical defense: primary or secondary function? *Am. Nat.* **113**, 445–451.
Kovda, V. A., Rozanov, B. G., and Onishenko, S. K. (1978). On probability of droughts and secondary salinisation of world soils. *In* "Arid Land Irrigation in Developing Countries" (E. G. Worthington, ed.), pp. 237–238. Pergamon, London.
Krischik, V. A., and Denno, R. F. (1983). Individual, population, and geographic patterns of plant defense. *In* "Variable Plants and Herbivores in Natural and Managed Systems" (R. F. Denno and M. S. McClure, eds.), pp. 463–512. Academic Press, New York.
Lawton, J. H., and McNeill, S. (1979). Between the devil and the deep blue sea: on the problem of being a herbivore. *In* "Population Dynamics" (R. M. Anderson, B. D. Turner, and L. R. Taylor, eds.), pp. 223–244. Blackwell, Oxford.
Lee, L. K. (1984). Land use and soil loss: a 1982 update. *J. Soil Water Conserv.* **39**, 226–228.
Leonard, D. E. (1974). Recent developments in ecology and control of the gypsy moth. *Annu. Rev. Entomol.* **19**, 197–229.

Levin, D. A. (1976). The chemical defenses of plants to pathogens and herbivores. *Annu. Rev. Ecol. Syst.* **7**, 121–159.
Levin, S., and Pimentel, D. (1981). Selection of intermediate rates of increase in parasite–host systems. *Am. Nat.* **117**, 308–315.
Leyton, L. (1983). Crop water use: principles and some considerations for agroforestry. *In* "Plant Research and Agroforestry" (P. A. Huxley, ed.), pp. 379–400. Int. Counc. Res. Agrofor., Nairobi, Kenya.
Linden, H. R. (1980). "1980 Assessment of the U.S. and World Energy Situation and Outlook." Gas Res. Inst., Chicago, Illinois.
McKey, D. (1974). Adaptive patterns in alkaloid physiology. *Am. Nat.* **108**, 305–320.
McNeill, S., and Lawton, J. H. (1970). Annual production and respiration in animal populations. *Nature (London)* **225**, 472–474.
Martin, J. P., Abbott, E. V., and Hughes, C. G. (1961). "Sugar-cane Diseases of the World," Vol. 1. Elsevier, Amsterdam.
Myers, N. (1983). "A Wealth of Wild Species." Westview Press, Boulder, Colorado.
National Academy of Sciences (NAS). (1977). "World Food and Nutrition Study." Natl. Acad. Sci., Washington, D.C.
Nichols, J. O. (1961). The gypsy moth in Pennsylvania—Its history and eradication. *Pa. Dept. Agric. Misc. Bull.* No. 4404.
Odum, E. P. (1971). "Fundamentals of Ecology," 3rd Ed. Saunders, Philadelphia, Pennsylvania.
Opler, P. A. (1974). "Biology, Ecology, and Host Specificity of Microlepidoptera Associated with *Quercus agrifolia* (Fagaceae)." Univ. of California Press, Berkeley.
Packard, A. S. (1890). Insects injurious to forest and shade trees. USDA, U.S. Entomol. Comm., *5th Rep., Bull.* No. 7.
Phillipson, J. (1966). "Ecological Energetics." Arnold, London.
Pimentel, D. (1961). Animal population regulation by the genetic feed-back mechanism. *Am. Nat.* **95**, 65–79.
Pimentel, D. (1968). Population regulation and genetic feedback. *Science* **159**, 1432–1437.
Pimentel, D. (1977). Ecological basis of insect pest, pathogen and weed problems. *In* "The Origins of Pest, Parasite, Disease and Weed Problems" (J. M. Cherrett and G. R. Sagar, eds.), pp. 3–31. Blackwell, Oxford.
Pimentel, D., ed. (1980). "Handbook of Energy Utilization in Agriculture." CRC Press, Boca Raton, Florida.
Pimentel, D., ed. (1981). "Handbook of Pest Management in Agriculture," Vols. 1–3. CRC Press, Boca Raton, Florida.
Pimentel, D. (1986). Agroecology and economics. *In* "Ecological Theory and Integrated Pest Management Practice" (M. Kogan, ed.), pp. 299–319. Wiley, New York.
Pimentel, D. (1988). Herbivore population feeding pressure on plant hosts: feedback evolution and host conservation. *Oikos* **53** (in press).
Pimentel, D., Pimentel, M. (1979). "Food, Energy and Society." Arnold, London.
Pimentel, D., and Wen, D. (1988). Technological changes in energy use in U.S. agricultural production. *In* "The Ecology of Agricultural Systems" (C. R. Carroll, J. H. Vandermeer, and P. M. Rosset, eds.), S. R. Gliessman, ed. Macmillan, New York. In press.
Pimentel, D., Levin, S. A., and Soans, A. B. (1975). On the evolution of energy balance in exploiter–victim systems. *Ecology* **56**, 381–390.
Pimentel, D., Shoemaker, C., LaDue, E. L., Rovinsky, R. B., and Russell, N. P. (1977). "Alternatives for Reducing Insecticides on Cotton and Corn: Economic and Environmental Impact." Environ. Res. Lab., Off. Res. Dev., EPA, Athens, Georgia (issued in 1979).
Pimentel, D., Nafus, D., Vergara, W., Papaj, D., Jaconetta, L., Wulfe, M., Olsvig, L.,

Frech, K., Loye, M., and Mendoza, E. (1978). Biological solar energy conversion and U.S. energy policy. *BioScience* **28**, 376–382.

Pimentel, D., Garnick, E., Berkowitz, A., Jacobson, S., Napolitano, S., Black, P., Valdes-Cogliano, S., Vinzant, B., Hudes, E., and Littman, S. (1980a). Environmental quality and natural biota. *BioScience* **30**, 750–755.

Pimentel, D., Oltenacu, P. A., Nesheim, M. C., Krummel, J., Allen, M. S., and Chick, S. (1980b). Grass-fed livestock potential: energy and land constraints. *Science* **207**, 843–848.

Pimentel, D., Jackson, W., Bender, M., and Pickett, W. (1986a). Perennial grains: an ecology of new crops. *Interdiscip. Sci. Rev.* **11**, 42–49.

Pimentel, D., Wen, D., Eigenbrode, S., Lang, H., Emerson, D., and Karasik, M. (1986b). Deforestation: interdependency of fuelwood and agriculture. *Oikos* **46**, 404–412.

Pimentel, D., Allen, J., Beers, A., Guinand, L., Linder, R., McLaughlin, P., Meer, B., Musonda, D., Perdue, D., Poisson, S., Siebert, S., Stoner, K., Salazar, R., and Hawkins, A. (1987). World agriculture and soil erosion. *BioScience* **37**, 277–283.

Pimm, S. L. (1982). "Food Webs." Chapman & Hall, London.

Podgwaite, J. D., and Campbell, R. W. (1972). The disease complex of the gypsy moth. II. Aerobic bacterial pathogens. *J. Invertebr. Pathol.* **20**, 303–308.

President's Science Advisory Committee (PSAC). (1965). "Restoring the Quality of Our Environment." Rep. Environ. Pollut. Panel, President's Sci. Advis. Comm., The White House, Washington, D.C.

Price, P. W. (1975). "Evolutionary Strategies of Parasitic Insects and Mites." Plenum, New York.

Price, P. W., Bouton, C. E., Gross, P., McPheron, B. A., Thompson, J. N., and Weis, A. E. (1980). Interactions among three trophic levels: influence of plants on interactions between insect herbivores and natural enemies. *Annu. Rev. Ecol. Syst.* **11**, 41–65.

Reifsnyder, W. E., and Lull, H. W. (1965). Radiant energy in relation to forests. *U.S. Dept. Agric. Tech. Bull.* No. 1344.

Rhoades, D. F. (1985). Offensive–defensive interactions between herbivores and plants: their relevance in herbivore population dynamics and ecological theory. *Am. Nat.* **125**, 205–238.

Rosenthal, G. A. (1986). The chemical defenses of higher plants. *Sci. Am.* **254**(1), 76–81.

Safley, L. H., Nelson, D. W., and Westermann, P. W. (1983). Conserving manurial nitrogen. *Trans. ASAE* **26**, 1166–1170.

Schroder, H. (1985). Nitrogen losses from Danish agriculture—Trends and consequences. *Agric. Ecosyst. Environ.* **14**, 279–289.

Schultz, J. C. (1983a). Habitat selection and foraging tactics of caterpillars in heterogeneous trees. *In* "Variable Plants and Herbivores in Natural and Managed Systems" (R. F. Denno and M. S. McClure, eds.), pp. 61–90. Academic Press, New York.

Schultz, J. C. (1983b). Impact of variable plant defensive chemistry on susceptibility of insects to natural enemies. *In* "Plant Resistance to Insects" (P. Hedin, ed.), pp. 37–54. Am. Chem. Soc., Washington, D.C.

Segal, A., Manisterski, J., Fischbeck, G., and Wahl, I. (1980). How plant populations defend themselves in natural ecosystems. *In* "Plant Disease" (J. G. Horsfall and E. B. Cowling, eds.), pp. 75–102. Academic Press, New York.

Slobodkin, L. B. (1960). Ecological energy relationships at the population level. *Am. Nat.* **94**, 213–236.

Strong, D. R., McCoy, E. D., and Rey, J. R. (1977). Time and the number of herbivore species: the pests of sugarcane. *Ecology* **58**, 167–175.

Thompson, L. M., and Troeh, F. R. (1978). "Soils and Soil Fertility," 4th Ed. McGraw-Hill, New York.

United Nations Environment Programme (UNEP). (1980). "Annual Review." U.N. Environ. Programme, Nairobi, Kenya.

U.S. Department of Agriculture (USDA). (1974). "Energy and U.S. Agriculture: 1974 Data Base," Vols. 1 and 2. Fed. Energy Adm., Off. Energy Conserv. Environ., State Energy Conserv. Programs, Washington, D.C.

U.S. Department of Agriculture (USDA). (1982). Fertilizer: Outlook and situation. *USDA Econ. Res. Serv.* FS-13.

U.S. Department of Agriculture (USDA). (1983). "Agricultural Statistics 1983." U.S. Gov. Print. Off., Washington, D.C.

U.S. Department of Agriculture (USDA). (1985). "Agricultural Statistics 1985." U.S. Gov. Print. Off., Washington, D.C.

Vanderholm, D. H. (1975). Nutrient losses from livestock waste during storage, treatment and handling. *In* "Managing Livestock Waste," pp. 282–285. Am. Soc. Agric. Eng., St. Joseph, Michigan.

Volk, B. G., and Loeppert, R. H. (1982). Soil organic matter. *In* "Handbook of Soils and Climate in Agriculture" (V. J. Kilmer, ed.), pp. 211–268. CRC Press, Boca Raton, Florida.

Webb, M., and Jacobsen, J. (1982). "U.S. Carrying Capacity, an Introduction." Carrying Capacity, Washington, D.C.

Whittaker, R. H., and Feeny, P. P. (1970). Allelochemicals: chemical interactions between species. *Science* **171,** 757–770.

Whittaker, R. H., and Likens, G. E. (1975). The biosphere and man. *In* "Primary Productivity of the Biosphere" (H. Lieth and R. H. Whittaker, eds.), pp. 305–328. Springer-Verlag, Berlin and New York.

2

Interdependence of Food and Natural Resources

David Pimentel, Laura E. Armstrong, Christine A. Flass, Frederic W. Hopf, Ronald B. Landy, and Marcia H. Pimentel

College of Agriculture and Life Sciences
Cornell University
Ithaca, New York

I. Introduction
II. World Population Growth
III. Energy Constraints
IV. Arable Land—Quality and Quantity
V. Water Constraints
VI. Biological Diversity
VII. Looking to the Future
References

I. INTRODUCTION

Humans, like other animals, appear to have an innate drive to convert a maximum amount of the earth's environmental resources into themselves and their progeny. As a result, the escalating human population is increasing its pressures on natural resources and thereby threatening its ability to supply itself with adequate amounts of food, water, fuel, and other essentials for life (Grant, 1982; Speth et al., 1985). The exact quantities of these natural resources that are still available are not known, but clearly these resources are finite.

The development and use of new technologies is expected to facilitate the more efficient use of limited natural resources that are needed to provide food and other human needs. Some claim that technology will provide for the unlimited economic needs of the world population, whatever its ultimate size (Simon, 1981; Wattenberg and Zinsmeister, 1984). However, many agricultural scientists disagree and caution that new technologies have limited ability to increase productivity (Jensen, 1978; I. N. Oka, 1985 personal communication).

The use of finite natural resources is complicated not only by each society's standard of living, but also by the uneven distribution of essential resources throughout the world (Biswas, 1984). Such constraints emphasize the need to study each major resource and the needs of the human population as parts of the complex natural system, in order to make reasonable projections and devise sound resource management policies for the future (U.S. Congress, 1984; Biswas and Biswas, 1985). This chapter focuses on the availability and interrelationships that exist among population size, arable land, water, energy and other biological resources that function to maintain a quality environment for all humans.

II. WORLD POPULATION GROWTH

The major increase in the population growth rate that occurred about 300 to 400 years ago coincided with the discovery and use of stored fossil energy resources such as coal, oil, and gas. Since then, rapid population growth has closely paralleled the increased use of fossil fuel for agricultural production and improving human health (Figure 2.1).

Never in history have humans, by their sheer numbers, so dominated the earth and its resources. The current world population stands at a high of more than five billion (PRB, 1986). What is more alarming than the numbers is the 1.7% annual growth rate—a rate 1,700 times greater than that of the first two million years of human existence. Such a growth rate adds more than 200,000 people a day to our world population (Salas, 1984). Demographers project that the world population will reach 6.1 billion by

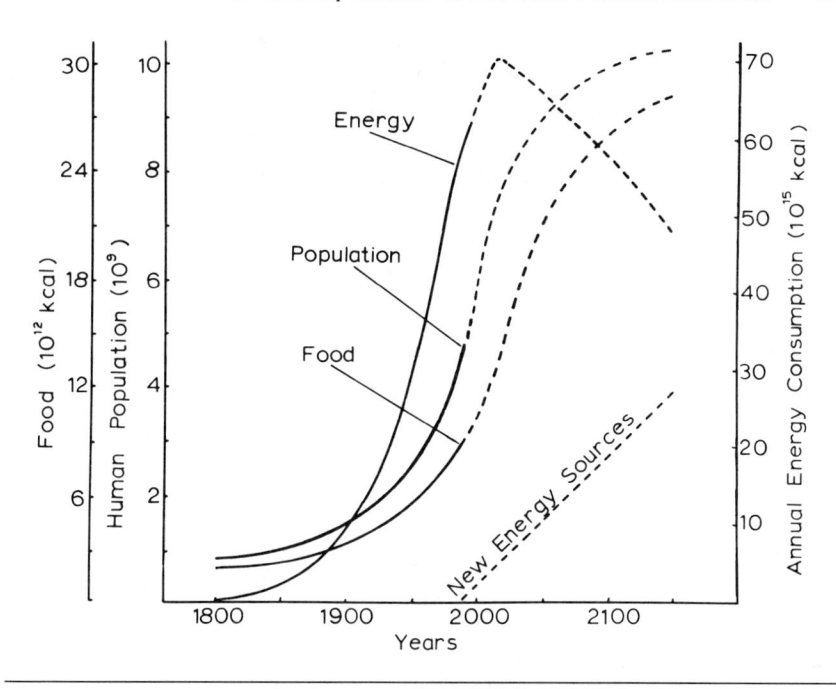

2. Interdependence of Food and Natural Resources   33

Figure 2.1 World population, food energy, and fossil energy use (solid lines) and projected trends (dotted lines) for each (Environmental Fund, 1979; Linden, 1980; USBC, 1982; PRB, 1983).

the year 2000, approach 8.2 billion by 2025 (UN, 1982), and probably reach 12 billion by 2100. Presently there seems to be no generally accepted way to limit this growth (NAS, 1975).

A large proportion of the world population (e.g., 45% of Africa, 40% of Latin America, and 37% of Asia) is now within child-bearing age (PRB, 1983). This young age structure contributes to a rapid population growth rate for the long term. For example, with more than half of the Chinese below the age of twenty years, even limiting births to one child per couple will not stop China's population growth (Coale, 1984). If 70% of the couples have only one child, the current population of slightly more than one billion (Coale, 1984) will still reach 1.2 billion by the year 2000 (Wren, 1982; Zhao, 1982).

Increased economic development is often cited as a possible solution to slowing birthrates. However, except for Japan, Korea, Taiwan, and Singapore (where the economy depends on industry and trade), most overpopulated countries have insufficient natural resources to support economic development similar to that in Europe and North America (Keyfitz, 1976; Hardin, 1985). Most of the 183 nations in the world now require some food imports from other nations (FAO, 1982). Furthermore, the "biological carrying capacity" of the ecosystem in many parts of the world already has been severely stressed and in some regions it has been exceeded (NAS, 1975).

Economic growth and prosperity cannot be relied on to control population growth. Clearly, an urgent need exists for all societies to consider methods to slow the growth of their human population (Reining and Tinker, 1975; Sharma, 1981).

III. ENERGY CONSTRAINTS

The most important factor responsible for rapid population growth and resource destruction has been the availability of cheap fossil energy, which humans use to manipulate natural systems and resources of the earth (Figures 2.1 and 2.2). This energy use has helped provide effective controls of many human diseases like malaria, typhoid, and cholera and also increased food production, all of which have contributed significantly to population growth (NAS, 1971; CEQ, 1980).

Most recent increases in crop yields have been achieved by using enormous amounts of fossil energy to supply fertilizers, pesticides, irrigation, and fuel for machinery (Leach, 1976). This has increased the use of the finite resource, fossil fuel, to compensate for degraded land, perhaps not the wisest choice for the future.

How can food supply and energy expenditures be balanced against the needs of the growing world population? Doubling the food supply during the next 25 years would help offset the serious malnourishment that one billion humans presently endure (Latham, 1984), as well as help feed the additional number of people. However, such an increase, assuming no land degradation and minimal substitution of labor with mechanization, would require about a four-fold increase in the total amount of energy

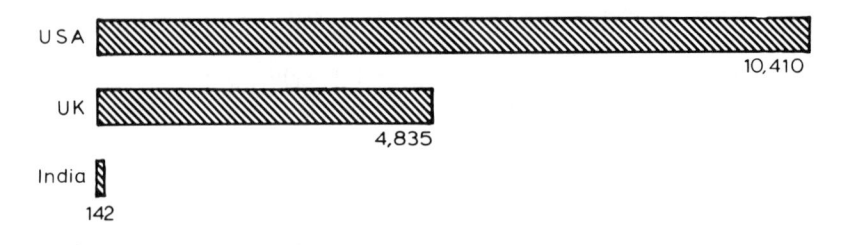

Figure 2.2 Energy consumption rates in kilograms of coal equivalents per capita per year for the United States, United Kingdom, and India (USBC, 1983).

expended for food production. Such a large energy input for food production would be necessary to balance the decreasing return of crop output per fertilizer energy input.

Throughout the world, energy use for food production continues to grow dramatically. Chinese agriculture provides a striking example of the increased reliance on fossil energy in food production. During the past three decades, fossil fuel inputs in Chinese agriculture rose 100-fold and crop yields have tripled (AAC, 1980; Taylor, 1981; Lu *et al.*, 1982).

The United States, like most developed countries, is a heavy energy user. For example, 17% of the total energy supply, or about 1500 liters per person of oil equivalents, is expended on production, processing, distribution, and preparation of food. This contrasts with most developing nations that use less than one-tenth this amount of energy for all their food.

In many developing nations crops are still produced by hand, requiring about 1200 hr of labor (Table 2.1). In contrast, in the highly mechanized U.S. agriculture crops like corn require only about 10 hr of labor (Table 2.2). The energy output to input ratio for the hand-produced corn is about 1:10, whereas in the U.S. mechanized system the ratio is 1:2.2.

Using U.S. agricultural technology to feed the current world population of 5.0 billion a high protein/calorie diet would require 7.1×10^{12} liters/yr of fuel (Pimentel and Hall, 1984). At this rate, assuming petroleum was the only source of energy for food production and all known reserves were used solely for this purpose, world oil reserves would last a mere 12 years. Of course, not all nations desire the diet typical of the United States.

One practical way to increase food supplies with minimal fossil energy inputs and expenditures would be for everyone—especially those living

Table 2.1 Energy Inputs in Corn (Maize) Production in Mexico Using Only Manpower

	Quantity/ha	kcal/ha
Inputs		
Labor	1,144 hr[a]	624,000
Axe and hoe	16,570 kcal[b]	16,570
Seeds	10.4 kg[a]	36,608
		677,178
Outputs		
Corn yield	1,944 kg[a]	6,901,200
kcal output/kcal input		10.19

[a]Lewis (1951).
[b]Estimated.

Table 2.2 Energy Inputs per Hectare for Corn Production in the United States[a]

	Quantity/ha	kcal/ha
Inputs		
Labor	10 hr	5,000
Machinery	55 kg	1,018,000
Gasoline	40 liters	400,000
Diesel fuel	75 liters	855,000
Irrigation	2.25×10^6 kcal	2,250,000
Electricity	35 kwh	100,000
Nitrogen	152 kg	3,192,000
Phosphorus	75 kg	473,000
Potassium	96 kg	240,000
Lime	426 kg	134,000
Seeds	21 kg	520,000
Insecticides	3 kg	300,000
Herbicides	8 kg	800,000
Drying	3,300 kg	660,000
Transportation	300 kg	90,000
		11,037,000
Outputs		
Total yield	7,000 kg	24,500,000
kcal output/kcal input		2.22

[a] After Pimentel and Wen (1988).

in the industrial nations—to consume less animal protein (Pimentel et al., 1980a). This diet modification would reduce energy expenditures and increase food supplies because less edible grain would be fed to livestock to produce costly animal protein.

The average yield from 10 kg of plant protein fed to animals is only 1 kg of animal protein. If the 130 million metric tons of grain that are fed yearly to U.S. livestock were consumed directly as human food, about 400 million people—1.7 times larger than U.S. population—could be sustained for one year (Pimentel et al., 1980a). However, dietary patterns and favorite foods are based not only on availability and economics, but on social, religious and other personal factors as well. For these reasons major diet modifications of any kind are difficult to achieve, especially within a short time span.

In view of the heavy drain on fossil fuel supplies, biomass (grain, sugar bagasse, other crop residues, and fuelwood) has been suggested as a major substitute fuel. To some extent biomass has always been used for fuel, but to use it in the amounts needed to spare fossil fuels needs careful analysis.

One constraint operating against the increased use of biomass for fuel is the amount of land required to grow it. When the amount of cropland needed to feed one person and that required to provide biomass to fuel one average U.S. automobile with ethanol for one year are compared, nine times more cropland is used to fuel the automobile than to feed a person (Pimentel *et al.*, 1984a).

Biomass in the form of wood has long been a major fuel for humans. Presently, over half the world's population depends on firewood as their primary fuel source (Pimentel *et al.*, 1986). But supplies of firewood are diminishing as an estimated 12 million ha of timber are cut and cleared each year primarily to provide more lands for agricultural production (Spears and Ayensu, 1984). As a result, the total amount of wood biomass available in the world per person has declined 10% during the last 12 years (Brown *et al.*, 1985). As more food must be produced to maintain the ever-increasing world population, the supply of fuelwood can be expected to diminish.

If biomass usage is increased, then the effect this would have on available quantities of arable land for food/fiber production must be considered. Further the removal of trees and crop residues is known to decrease soil fertility and facilitate soil erosion. From this, one can conclude that biomass resources are limited in their usefulness as a fossil fuel substitute.

Solar energy technology shows some potential for the future. Photovoltaic and solar thermal energy should supply limited amounts of renewable energy (ERAB, 1982); however, these techniques do require land and also may effect some wildlife (Pimentel *et al.*, 1984b).

IV. ARABLE LAND—QUALITY AND QUANTITY

Next to sunlight, which provides plant growth, land is the most vital natural resource. In the United States, about 160 million ha of cropland are planted to provide food for 240 million people or 0.67 ha/person (USDA, 1983). Considering that more than 20% of U.S. food crops are exported, the arable land planted to feed each person is about 0.5 ha.

Worldwide, with arable land resources estimated to be about 1.5 billion ha (Buringh, 1979) and a world population of 5.0 billion, the available arable land per person amounts to only 0.3 ha. According to FAO predictions (1982), the possible net expansion in total world cropland is 3.9 million ha per year. At this rate world cropland will expand to 2 billion ha by the year 2110.

Undoubtedly, it will be possible to bring some additional land, much of which is considered marginal, into production. Optimistic estimates propose that worldwide cropland can be expanded to 3.4 billion ha (Buringh, 1979). However, others conservatively project that worldwide cropland can be expanded 2 billion ha, but only with the use of large amounts of energy to make the marginal land productive (NAS, 1977). If the world population does increase to 12 billion and expands the land base to 2 billion ha, only 0.2 ha or less of cropland would be available per person (Figure 2.3) and most people would have to consume essentially a vegetarian or plant protein diet.

With excellent soils, favorable temperature and rainfall, heavy fertilization, and effective pest control, it is theoretically possible to provide 20 people with adequate calories and protein on one hectare of land. However, with average land, climate and other factors, it would be optimistic to hope to feed half that number. This is especially true considering the current soil erosion crisis that exists throughout the world (Pimentel *et al.*, 1986). In the United States, for example, several million hectares of marginal land, which is highly susceptible to severe soil erosion, is already being cultivated (OTA, 1982; Naegeli, 1986).

The available arable land, water, and energy resources, as well as the kind of food a particular society desires or can afford, determine the crops produced in a given region. At present, two-thirds of the people in the world consume a primarily vegetarian diet in contrast to industrialized countries where diets are characteristically high calorie/high animal protein. This latter diet requires large amounts of land and energy to produce. Based on all available data concerning future availability of arable land

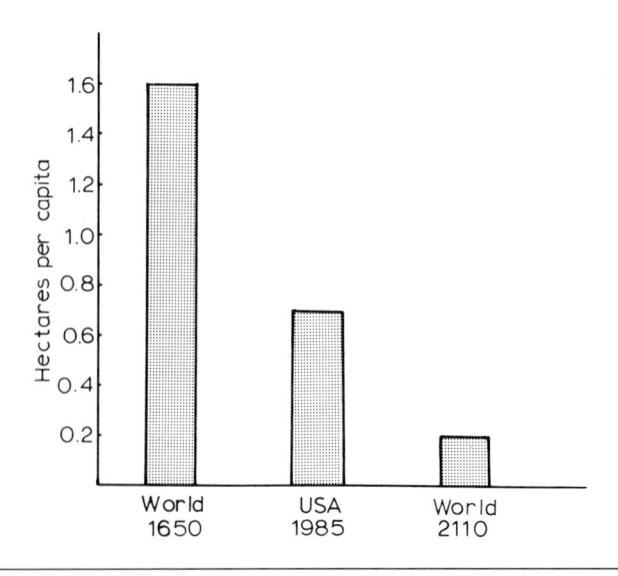

Figure 2.3 Arable land per capita in the world in 1650, in the United States, for 1985 and projected for the year 2110 in the world.

and energy, it will not be possible to provide all people of the world with such a high calorie/high animal protein diet. In fact, if population growth continues, diets in industrialized nations will have to be modified to include larger amounts of plant proteins and less animal protein.

The quality of land is as important as the number of hectares available. As mentioned, to bring land of marginal quality into production, water and energy inputs for fertilizers and/or irrigation are essential. The extent of expansion will depend on the availability of all these resources.

Meanwhile, farmers are using soils that are eroding at an alarming rate (Pimentel et al., 1976; Holdgate et al., 1982; Eckholm, 1983). Soil erosion reduces the natural productivity of soils by removing nutrients and organic matter, and by reducing top soil depth and water availability (Lal, 1984).

Global dimensions of land destruction are a major concern. About 35% of the earth's land-surface is affected (Mabbutt, 1984). The natural productivity of many soils has been reduced 25–100% because of erosion (Langdale et al., 1979; Lal, 1984; Mabbutt, 1984). In the past, several civilizations including those in Mesopotamia and Greece failed in part because of the degradation of their agricultural lands (Jacks and Whyte, 1939; Lowdermilk, 1953; Troeh et al., 1980).

Worldwide, an estimated 6 million ha of agricultural land are irretrievably lost each year because of water and soil erosion, salinization from irrigation, and other factors (UNEP, 1980; Dudal, 1981; Kovda, 1983). In addition, every year crop productivity on about 20 million ha either has a negative net economic return or is reduced to zero because of poor soil quality (UNEP, 1980). Based on current worldwide soil loss, the results of a recent study for the period 1975–2000 project that rain-fed land degradation will depress food production another 15–30% (Shah et al., 1985).

Recent surveys in Iowa, which has some of the richest agricultural land in the world, indicate that about half of the original topsoil has been lost (Risser, 1981). Under normal agricultural conditions, the formation of 2.5 cm (1 inch) of soil requires 200–1,000 years (Hudson, 1981; Larson, 1981; McCormack et al., 1982; Sampson, 1983; Lal, 1984). Fortunately, so far, increased fertilizers and other fossil energy inputs have been available to offset the reduced productivity of some U.S. cropland caused by soil erosion. However, the loss of about 20 cm of topsoil from a topsoil base of 30 cm requires double the usual energy inputs just to maintain crop yields (Pimentel et al., 1981). In future decades these large energy inputs may not be affordable or even available.

In developing countries, the rate of cropland erosion is nearly twice that experienced in the United States (Ingraham, 1975). For example, reports indicate that each year, an average of 38 t/ha of soil is lost from over half of India's total land area because of serious erosion (CSE, 1982).

Erosion is also a problem in China where the sediment load in the Yellow River is equivalent to a loss of about 38 t/ha/yr from the 68-million-ha areas of the river's watershed (Robinson, 1981).

Resources to offset the productivity losses are not always readily available or affordable. In some regions, to obtain more land for crop production, forests are being cut and steep slopes are being used for crops. As vegetation has been removed, erosion has intensified. Often flooding in the lowlands has become a serious environmental problem to the extent that, in India for example, losses from flooding have doubled during the last ten years (USDA, 1965; Sharma, 1981). Certainly, as more marginal land is put into agricultural production, land degradation and flooding will increase (CEQ, 1980).

V. WATER CONSTRAINTS

Along with sunlight and land, water is a vital resource for agricultural production. Sufficient rain falls upon most arable agricultural land, but periodic droughts continue to limit yields in some areas of the world. All crops require and transpire massive amounts of water. For example, a corn crop that produced 6,500 kg/ha of grain will take up and transpire about 4.2 million liters of water per hectare during the growing season (Penman, 1970; Leyton, 1983). To supply this much water each year, about 10 million liters (100 cm) of rain must fall per hectare and furthermore must be evenly distributed during the year and growing season.

A reduction in rainfall of only 5 cm during the growing season reduces corn production by about 15% (Finkel, 1983). Decreasing rainfall to 30 cm per season reduces yields to about one-fifth of those from areas receiving 100 cm rainfall. Sorghum and wheat require less water than corn and could be grown instead of it, but these crops produce about one-third less grain per hectare than corn (USDA, 1984).

Irrigation is essential if rainfall cannot be relied upon to supply the moisture needed for crop production. Irrigated crop production requires the movement of large quantities of water. For example, using irrigation to produce 1 kg of the following food or fiber products requires: 1,400 liters of water for corn, 4,700 for rice, and 17,000 for cotton (Ritschard and Tsao, 1978). In the United States, agricultural irrigation presently consumes 83% of the total 360 billion liters per day that is consumed by all sectors of society (Murray and Reeves, 1977).

Irrigation water is either piped long distances from rainfed reservoirs or pumped from wells supplied by aquifers. These large underground storage areas are slowly filled by rainfall. Intense and prolonged irrigation stresses aquifers to such an extent that, in some areas of the world, water is being "mined" and used more quickly than it can be replaced by rainfall

(CEQ, 1980). Even now in the United States water overdraft exceeds replenishment by at least 25% (USWRC, 1979).

In addition to its large water usage, irrigated crop production is costly in terms of energy consumption. Nearly one-fifth of all energy expended in U.S. agricultural production is used to move irrigation water (USDA, 1974). For example, in Nebraska, rainfed corn production requires about 630 liters/ha of oil equivalents, whereas irrigated corn requires an expenditure of about three times more energy (Pimentel and Burgess, 1980).

Periods of drought, whether for a season or extending over many years, also influence crop production. All countries experience water shortages at times and this influences agricultural production (Ambroggi, 1980). The current severe drought in the Sahel, which extends through much of Africa south of the Sahara, clearly illustrates the complexity of the problem. A substantial decrease in the normal rainfall results in reduced crops (Hare, 1977). The effect, however, has been magnified by the large human population per available arable land as well as the long-time mismanagement of all land resources, including the forest areas (Biswas, 1984).

Water resources are becoming a global problem. From 1940–80 worldwide water use has more than doubled. The anticipated growth in world population and agricultural production can be expected to again double water needs in the coming two decades (Ambroggi, 1980).

By the year 2000 it is estimated that the world agricultural production will consume nearly 80% of total water withdrawn (Biswas and Biswas, 1985). Because more water will be needed to support agricultural production, both the extent and location of water supplies will become major constraints on increased crop production. Considering that about one-third of the major world river basins are shared by three or more countries (CEQ, 1980), water availability is certain to cause competition and conflicts between countries and even within countries (Biswas, 1983).

In contrast to limited water supplies is the problem of too much water, accompanied by rapid water runoff. Both may cause waterlogging of soils and flooding. Fast water runoff decreases crop yield per hectare by reducing the water available to the crop, by removing soil nutrients, and even by washing away the soil and crops themselves (OTA, 1982).

When forests on slopes are replaced with crops to augment food supplies, runoff and soil erosion increase and serious flood damage occurs to valuable crops and pasture (Beasley, 1972). The economic costs of these losses can be significant. In the United States damage from sediments and flooding to surrounding areas causes an estimated $6 billion each year (Clark, 1985). In Bangladesh in 1974, a severe flood diminished the productivity of the soil significantly and reduced the rice harvest, which ultimately led to severe food shortages and famine (Brown, 1976).

In general, most water runoff, which brings with it loose soil, has negative effects on agriculture. But where annual floods can be managed like those characteristic of the Nile River basin, the water and rich soil sediment benefit the crops growing in the flood plains (Biswas, 1984).

VI. BIOLOGICAL DIVERSITY

Almost 90% of human food comes from just 15 species of plants and eight species of livestock, all of which were domesticated from the wild (Mangelsdorf, 1966; Myers, 1979). However, high agricultural productivity as well as human health depend upon the activity of a myriad of natural systems composed of an estimated 10 million species of plants and animals that inhabit the world ecosystem (Pimentel et al., 1980b; Ehrlich and Ehrlich, 1981; Myers, 1983). We know that humans cannot survive with only the presence of their crop and livestock species, but exactly how many and which plant and animal species are essential is unknown.

Natural species are eliminated in many ways. Those having the greatest impact include clearing land of natural vegetation for agriculture, urban areas, and chemical pollution (pesticides, etc.) of the environment.

Of great concern is the growing rate of species elimination and the subsequent loss of genetic diversity that human encroachment is causing (Biswas and Biswas, 1985). Predictions are that an estimated one million species of plants and animals will be exterminated by the end of this century (Eckholm, 1978; Ehrlich and Ehrlich, 1981). Even now on the Indian sub-continent, 10% of the plant species are threatened or endangered (Sharma, 1981). This high rate of extinction is alarming because it is not known how many of these organisms may be necessary to maintain food production and other vital human activities (Myers, 1984, 1985).

Natural biota perform many functions essential for agriculture, forestry, and other sectors of the environment. Some of these include providing genetic diversity basic to successful crop breeding; recycling vital chemical elements such as carbon and nitrogen within the ecosystem; moderating climates; conserving soil and water; serving as sources of certain medicines, pigments, and spices; and supplying fish and other wildlife (Myers, 1979, 1984; Pimentel et al., 1980b; Ehrlich and Ehrlich, 1981). In addition, some natural biota like bacteria and protozoans help prevent human diseases. Some prey on and destroy human pathogens, others degrade wastes, while still others remove toxic pollutants from water and soil and help buffer the impact of air pollutants (Pimentel et al., 1980b).

The relative impact that a species has in the environment can be judged in part by its biomass per unit area. Groups such as insects, earthworms, protozoa, bacteria, fungi, and algae average 6400 kg/ha, more than 350 times the average human biomass of 18 kg/ha in the United States. Reducing or exterminating some species groups could seriously disturb the normal or natural functioning of important environmental systems in nature. For example, soil quality and agricultural productivity jointly depend upon soil biota, organic matter, and the presence of certain inorganic elements (Brady, 1984). Each year various soil organisms break down and degrade about 20 t of organic matter per hectare (Alexander, 1977). This degradation is essential for the release of bound nutrients and their subsequent recycling for further use in the ecosystem (Golley, 1983). Some soil biota help control pests in crops, and others "fix" atmospheric nitrogen for use by plants (Pimentel et al., 1980b). Earthworms, insects, and other biota open holes, improve water percolation, and loosen the soil to enhance new soil formation. Because the complex roles of these biota are not clearly understood, great care should be taken to prevent the extinction of any species.

VII. LOOKING TO THE FUTURE

The earth contains resources of land, water, energy, and natural biota, which are needed and used by humans for their survival. All these resources are interrelated and the availability of one influences the quality, quantity, and usefulness of another. Thus, fossil energy can be used to increase production on land, but supplies of both land and energy can be stressed in the process.

The growing human population, with its needs for food and fuel, is using up important environmental resources, many of which are finite and cannot be resupplied. Population pressures in some parts of the world already are well above the capacity of resources to provide all people living there with a prosperous standard of living (Keyfitz, 1976), and outright shortages of basic resources now exist in some places (Ehrlich et al., 1977; CEQ, 1980).

Soon, if not at present, the people of the world will have to decide whether they want a relatively prosperous life style with ample supplies of food, water, and energy for a population of one to two billion or whether they will be satisfied with limited resources and a meager existence because they allowed the world population to grow to ten times this number.

Future development will require both technology and natural resources—technology alone is insufficient (Speth et al., 1985). Future breakthroughs in science and technology, however, can enable humans to make more efficient use of land, water, energy, and biological resources to meet their growing needs for food, fiber, and shelter. Ultimately the limitations of natural resources will constrain expansion of crop productivity (Jensen, 1978; I. N. Oka, 1985 personal communication). Meanwhile, immediate *conservation* of these resources is vital not only to protect them but also to extend their availability for future use. The present degradation of land and water resources and increased use of chemicals in agriculture are adversely affecting natural biota. Maintaining a diversity of natural biota is essential to successful agricultural production, and clearly, the loss of species diversity will make increased future food production more difficult.

Reducing human birthrates to limit human numbers promises to be the most difficult task ever undertaken by humankind. Controlling birthrates can be carried out successfully only if the direct costs of having children increase and social prestige for small families also increases (Douglas, 1966; NAS, 1975; Harris, 1977; Mayer, 1985). Ultimately, each individual must assume personal responsibility for reducing birthrates. Within each society, these difficult societal changes need to be encouraged in conjunction with scientific and governmental policies that help augment food supplies, improve health, and protect natural resources.

REFERENCES

Agricultural Almanac of China (AAC). (1980). "Zhonggue Nongye Nianjian." Agric. Press, Beijing.
Alexander, M. (1977). "Introduction to Soil Microbiology," 2nd Ed. Wiley, New York.
Ambroggi, R. P. (1980). Water. *Sci. Am.* **243**(3), 101–106, 111–114, 116.
Beasley, R. P. (1972). "Erosion and Sediment Pollution Control." Iowa State Univ. Press. Ames.
Biswas, A. K. (1983). Shared natural resources: future conflicts or peaceful development? *In* "The Settlement of Disputes on the New Natural Resources" (R.-J. Dupuy, ed.), pp. 197–215. Nijhoff, The Hague.
Biswas, A. K. (1984). "Climate and Development." Tycooly Int. Publ., Dublin.
Biswas, M. R., and Biswas, A. K. (1985). The global environment. Past, present and future. *Resour. Policy* **3**, 25–42.
Brady, N. C. (1984). "The Nature and Properties of Soils," 9th Ed. Macmillan, New York.
Brown, L. R. (1976). "World Population Trends: Signs of Hope, Signs of Stress," Worldwatch Pap. No. 8. Worldwatch Inst., Washington, D.C.
Brown, L. R., Chandler, W. U., Flavin, C., Pollock, C., Postel, S., Starke, L., and Wolf, E. C. (1985). "State of the World 1985." Norton, New York.
Buringh, P. (1979). "Introduction to the Study of Soils in Tropical and Subtropical Regions." Cent. Agric. Publ. Doc., Pudoc, Wageningen, Netherlands.
Centre for Science and Environment (CSE). (1982). "The State of India's Environment 1982," A Citizens' Report. Cent. Sci. Environ., New Delhi.
Clark, E. H., II. (1985). The off-site costs of soil erosion. *J. Soil Water Conserv.* **40**, 19–22.
Coale, A. J. (1984). "Rapid Population Change in China." Nat. Acad. Press, Washington, D.C.
Council on Environmental Quality (CEQ). (1980). "The Global 2000 Report to the President," Counc. Environ. Qual. and Dep. State, Vol. 2. U.S. Gov. Print. Off., Washington, D.C.
Douglas, M. (1966). Population control in primitive groups. *Br. J. Sociol.* **17**, 263–273.
Dudal, R. (1981). An evaluation of conservation needs. *In* "Soil Conservation. Problems and Prospects" (R. P. C. Morgan, ed.), pp. 3–12. Wiley, New York.

Eckholm, E. P. (1978). "Disappearing Species: The Social Challenge," Worldwatch Pap. No. 22. Worldwatch Inst., Washington, D.C.

Eckholm, E. P. (1983). "Down to Earth. Environment and Human Needs." Norton, New York.

Ehrlich, P. R., and Ehrlich, A. H. (1981). "Extinction: The Causes and Consequences of the Disappearance of Species." Random House, New York.

Ehrlich, P. R., Ehrlich, A. H., and Holdren, J. P. (1977). "Ecoscience. Population, Resources, Environment." Freeman, San Francisco, California.

Energy Research Advisory Board (ERAB). (1982). "Solar Energy Research and Development: Federal and Private Sector Roles." Rep. Energy Res. Advis. Board to U.S. Dep. Energy, Dep. Energy, Washington, D.C.

Environmental Fund. (1979). "World Population Estimates." Environ. Fund, Washington, D.C.

Food and Agriculture Organization (FAO). (1982). "1981 Production Yearbook." Food Agric. Organ. U. N., Rome.

Finkel, H. J. (1983). Irrigation of cereal crops. In "CRC Handbook of Irrigation Technology" (H. J. Finkel, ed.), Vol. 2, pp. 159–189. CRC Press, Boca Raton, Florida.

Golley, F. B., ed. (1983). "Tropical Rainforest Ecosystems: Structure and Function." Elsevier, Amsterdam.

Grant, L. (1982). "The Cornucopian Fallacies." Environ. Fund, Washington, D.C.

Hardin, G. (1985). Preventing famine. Letter to the editor. *Science* **227**, 1284.

Hare, F. K. (1977). Climate and desertification. In "Desertification: Its Causes and Consequences," pp. 63–129. Pergamon, Oxford.

Harris, M. (1977). Murders in Eden. In "Cannibals and Kings: The Origins of Cultures" (M. Harris, ed.), pp. 89–93. Random House, New York.

Holdgate, M. W., Kassas, M., and White, G. F. (1982). "The World Environment. 1972–1982," Rep. U. N. Environ. Programme. Tycooly Int. Publ., Dublin.

Hudson, N. (1981). "Soil Conservation," 2nd Ed. Cornell Univ. Press, Ithaca, New York.

Ingraham, E. W. (1975). "A Query into the Quarter Century. On the Interrelationships of Food, People, Environment, Land and Climate." Wright-Ingraham Inst., Colorado Springs, Colorado.

Jacks, G. V., and Whyte, R. O. (1939). "Vanishing Lands. A World Survey of Soil Erosion." Doubleday, New York.

Jensen, N. F. (1978). Limits to growth in world food production. *Science* **201**, 317–320.

Keyfitz, N. (1976). World resources and the world middle class. *Sci. Am.* **235**(1), 28–35.

Kovda, V. A. (1983). Loss of productive land due to salinization. *Ambio* **12**, 91–93.

Lal, R. (1984). Productivity assessment of tropical soils and the effects of erosion. In "Quantification of the Effect of Erosion on Soil Productivity in an International Context" (F. R. Rijsberman and M. G. Wolman, eds.), pp. 70–94. Delft Hydraul. Lab., Delft, Netherlands.

Langdale, G. W., Leonard, R. A., Fleming, W. G., and Jackson, W. A. (1979). Nitrogen and chloride movement in small upland Piedmont watersheds: II. Nitrogen and chloride transport in runoff. *J. Environ. Qual.* **8**, 57–63.

Larson, W. E. (1981). Protecting the soil resource base. *J. Soil Water Conserv.* **36**, 13–16.

Latham, M. C. (1984). International nutrition and problems and policies. In "World Food Issues," pp. 55–64. Cent. Anal. World Food Issues, Int. Agric., Cornell Univ., Ithaca, New York.

Leach, G. (1976). "Energy and Food Production," IPC Sci. Technol. Press, Guilford, Surrey, England.

Lewis, O. (1951). "Life in a Mexican Village: Tepoztlan Restudied." Univ. of Illinois Press, Urbana.

Leyton, L. (1983). Crop water use: principles and some considerations for agroforestry. *In* "Plant Research and Agroforestry" (P. A. Huxley, ed.), pp. 379–400. Int. Counc. Res. Agrofor., Nairobi.

Linden, H. R. (1980). 1980 assessment of the U.S. and world energy situation and outlook. Gas Res. Inst., Chicago, Illinois.

Lowdermilk, W. C. (1953). Conquest of the land through seven thousand years. *Agric. Inf. Bull. (U.S. Dep. Agric.)* No. 99.

Lu, M., Ysheng, J., and Chenyueng, S. (1982). Typical analysis of rural energy consumption in China. [Zhonggue Nongchun Nengliang Xiaofei diansing fenshi.] *Nongye Jingji Luenchung* **4**, 216–223.

Mabbutt, J. A. (1984). A new global assessment of the status and trends of desertification. *Environ. Conserv.* **11**, 103–113.

McCormack, D. E., Young, K. K., and Kimberlim, L. W. (1982) "Current Criteria for Determining Soil Loss Tolerance," ASA Spec. Publ. No. 45. Am. Soc. Agron., Madison, Wisconsin.

Mangelsdorf, P. C. (1966). Genetic potentials for increasing yields of food crops and animals. *In* "Prospects of the World Food Supply," Symp. Proc. Natl. Acad. Sci., Washington, D.C.

Mayer, J. (1985). Preventing famine. Letter to the editor. *Science* **227**, 1284.

Murray, R. C., and Reeves, E. B. (1977). Estimated use of water in the United States in 1975. *Geol. Surv. Circ. (U.S.)* No. 765.

Myers, N. (1979). "The Sinking Ark." Pergamon, New York.

Myers, N. (1983). "A Wealth of Wild Species." Westview Press, Boulder, Colorado.

Myers, N. (1984). Genetic resources in jeopardy. *Ambio* **13**, 171–174.

Myers, N. (1985). The end of the lines. *Nat. Hist.* **94**, 2, 6, 12.

Naegeli, W. N. (1986). Interpreting the National Resources Inventory for regional planners and decision makers: A case study for the Tennessee Valley Region. Ph.D. Thesis, Cornell Univ., Ithaca, New York.

National Academy of Sciences (NAS). (1971). "Rapid Population Growth," Vols. 1 and 2. Johns Hopkins Press, Baltimore, Maryland.

National Academy of Sciences (NAS). (1975). "Population and Food: Crucial Issues." Nat. Acad. Sci., Washington, D.C.

National Academy of Sciences (NAS) (1977). "Supporting Papers: World Food and Nutrition Study," Vol. 2. Nat. Acad. Sci., Washington, D.C.

Office of Technology Assessment (OTA). (1982). "Impacts of Technology on Productivity of the Croplands and Rangelands of the United States." Off. Technol. Assess., Washington, D.C.

Penman, H. L. (1970). The water cycle. *Sci. Am* **223**(3), 99–108.

Pimentel, D., and Burgess, M. (1980). Energy inputs in corn production. *In* "Handbook of Energy Utilization in Agriculture" (D. Pimentel, ed.), pp. 67–84. CRC Press, Boca Raton, Florida.

Pimentel, D., and Hall, C. W., eds. (1984). "Food and Energy Resources." Academic Press, New York.

Pimentel, D., and Wen, D. (1988). Technological changes in energy use in U.S. agricultural production. *In* "The Ecology of Agricultural Systems" (C. R. Carroll, J. H. Vandermeer, and P. M. Rosset, eds.). Macmillan, New York. In press.

Pimentel, D., Terhune, E. C., Dyson-Hudson, R., Rochereau, S., Samis, R., Smith, E., Denman, D., Reifschneider, D., and Shepard, M. (1976). Land degradation: effects on food and energy resources. *Science* **194**, 149–155.

Pimentel, D., Oltenacu, P. A., Nesheim, M. C., Krummel, J., Allen, M. S., and Chick, S. (1980a). Grass-fed livestock potential: energy and land constraints. *Science* **207**, 843–848.

Pimentel, D., Garnick, E., Berkowitz, A., Jacobson, S., Napolitano, S., Black, P., Valdes-Cogliano, S., Vinzant, B., Hudes, E., and Littman, S. (1980b). Environmental quality and natural biota. *BioScience* **30**, 750–755.

Pimentel, D., Moran, M. A., Fast, S., Weber, G., Bukantis, R., Balliett, L., Boveng, P., Cleveland, C., Hindman, S., and Young, M. (1981). Biomass energy from crop and forest residues. *Science* **212**, 1110–1115.

Pimentel, D., Fried, C., Olson, L., Schmidt, S., Wagner-Johnson, K., Westman, A., Whelan, A., Foglia, K., Poole, P., Klein, T., Sobin, R., and Bochner, A. (1984a). Environmental and social costs of biomass energy. *BioScience* **34**, 89–94.

Pimentel, D., Levitan, L., Heinze, J., Loehr, M., Naegeli, W., Bakker, J., Eder, J., Modelski, B., and Morrow, M. (1984b). Solar energy, land and biota. *SunWorld* **8**, 70–73, 93–95.

Pimentel, D., Wen, D., Eigenbrode, S., Lang, H., Emerson, D., and Karasik, M. (1986). Deforestation: interdependency of fuelwood and agriculture. *Oikos* **46**, 404–412.

Population Reference Bureau (PRB). (1983). "1983 World Population Data Sheet." Popul. Ref. Bur., Washington, D.C.

Population Reference Bureau (PRB). (1986). "World Population Data Sheet." Popul. Ref. Bur., Washington, D.C.

Reining, P., and Tinker, I., eds. (1975). "Population: Dynamics, Ethics and Policy." Am. Assoc. Adv. Sci., Washington, D.C.

Risser, J. (1981). A renewed threat of soil erosion: it's worse than the dust bowl. *Smithsonian* **11**, 121–131.

Ritschard, R. L., and Tsao, K. (1978). "Energy and Water in Irrigated Agriculture during Drought Conditions," US DOE LBL-7866. Lawrence Berkeley Lab., Univ. of California, Berkeley.

Robinson, A. R. (1981). Erosion and sediment control in China's Yellow River basin. *J. Soil Water Conserv.* **36**, 125–127.

Salas, R. M. (1984). Population, resources and the environment: some crucial issues at the conference on population. *Ambio* **13**, 143–148.

Sampson, R. N. (1983). Soil conservation. *Sierra Club Bull.* **68**(6), 40–44.

Shah, M. M., Fischer, G., Higgins, G. M., Kassam, A. H., and Naiken, L. (1985). "People, Land and Food Production—Potentials in the Developing World." CP-85-11. Int. Inst. Appl. Syst. Anal., Laxenburg, Austria.

Sharma, A. K. (1981). Impact of the development of science and technology on environment. *Proc. Indian Sci. Congr., 68th, Varanasi, India* pp. 1–43.

Simon, J. (1981). "The Ultimate Resource." Princeton Univ. Press, Princeton, New Jersey.

Spears, J., and Ayensu, E. S. (1984). Resources development and the new century: sectoral paper on forestry. *Global Possible Conf.* World Resour. Inst., Washington, D.C.

Speth, J. G., Fernandez, L., and Yost, N. C. (1985). "Protecting Our Environment: Toward a New Agenda." Cent. Natl. Policy, Washington, D.C.

Taylor, R. P. (1981). "Rural Energy Develpment in China." Resources for the Future, Washington, D.C.

Troeh, F. R., Hobbs, J. A., and Donahue, R. L. (1980). "Soil and Water Conservation for Productivity and Environmental Protection." Prentice-Hall, Englewood Cliffs, New Jersey.

United Nations (UN). (1982). "World Population Trends and Policies, 1981 Monitoring Report, Vol. 1: Population Trends." United Nations, New York.

United Nations Environment Programme (UNEP). (1980). "Annual Review." U. N. Environ. Programme, Nairobi, Kenya.

U.S. Bureau of the Census (USBC). (1982). "Statistical Abstract of the United States 1982," 103rd Ed., U.S. Bur. Census. U.S. Gov. Print. Off., Washington, D.C.

U.S. Bureau of the Census (USBC). (1983). "Statistical Abstract of the United States 1983," 104th Ed., U.S. Bureau of the Census. U.S. Gov. Print. Off., Washington, D.C.

U.S. Congress. (1984). "Feeding the World's Population: Developments in the Decade Following the World Food Conference of 1974: Report," 98th Congress, 2nd Session. U.S. Gov. Print. Off., Washington, D.C.

U.S. Department of Agriculture (USDA). (1965). Losses in agriculture. *U.S. Dep. Agric., Agric. Handb.* No. 291.

U.S. Department of Agriculture (USDA). (1974). "Energy and U.S. Agriculture: 1974 Data Base," Vols. 1 and 2. Fed. Energy Adm., Off. Energy Conserv. Environ., State Energy Conserv. Programs, Washington, D.C.

U.S. Department of Agriculture (USDA). (1983). "Agricultural Statistics 1983." U.S. Gov. Print. Off., Washington, D.C.

U.S. Department of Agriculture (USDA). (1984). "Agricultural Statistics 1984." U.S. Gov. Print. Off., Washington, D.C.

U.S. Water Resources Council (USWRC). (1979). "The Nation's Water Resources. 1975–2000," Vols. 1–4, Second Natl. Water Assess., U.S. Water Resour. Counc. U.S. Gov. Print. Off., Washington, D.C.

Wattenberg, B., and Zinsmeister, K. (1984). "Are World Population Trends a Problem?" Am. Enterprise Inst. Public Policy Res., Washington, D.C.

Wren, C.S. (1982). China plans a new drive to limit birth rate. *New York Times* Nov. 7.

Zhao, Z. (1982). Report about the sixth five-year plan. *Renmin Ribao* [People's Daily] Nov. 14.

3

Loss of Biological Diversity and Its Potential Impact on Agriculture and Food Production

Norman Myers

Consultant in Environment and Development
Upper Meadow, Old Road
Headington, Oxford, United Kingdom

I. Introduction
II. Species' Contributions to Modern Agriculture
    A. Improved Forms of Existing Foods
    B. Entirely New Foods
    C. Adaptive Agriculture
III. Extinction Processes
    A. Present Rate of Extinction
    B. Projected Rate of Extinction
IV. Conclusion
    References

I. INTRODUCTION

In this chapter we consider the prospect that we are losing wild species of plants and animals at rates unprecedented in history. Indeed, we are witnessing—and causing—a mass extinction of species. By the middle of the next century, supposing present trends persist (they are likely to accelerate), we shall surely lose one-quarter of Earth's 5–10 million species (minimum estimate), possibly one-third, and conceivably one-half.

As species disappear, so do their genetic materials. Species help us to maintain and even enhance the productivity of modern crops, supply us with altogether new foods, and further assist our agriculture via "environmentally attuned" versions of established crops. This utilitarian value should increase greatly as we make more systematic use of the genetic attributes of species.

This makes the impending demise of millions of species all the more regrettable. As the worth of genetic resources becomes better known, political leaders and policymakers may be persuaded to place a premium on expanded conservation strategies, both as a cost-effective endeavor and as a competitive form of land use.

II. SPECIES' CONTRIBUTIONS TO MODERN AGRICULTURE

For all its bountiful productivity, our agriculture remains essentially a neolithic agriculture insofar as we still depend on virtually the same array of crop plants that were developed by Stone Age farmers. About 30 crops supply virtually all our food—even though they continue to fall somewhat short of meeting all our needs. If we are to keep up with increasing human population and increasing human aspirations, we must double our food output during the last two decades of this century. How shall we best confront this challenge?

A strategy of "the same as before, only more so" is likely to fail us; it will probably prove too expensive. Modern agriculture has been developed during a phase of abundant and readily available supplies of fossil fuels at relatively low cost (Pimentel, 1980). These fossil-fuel materials have served not only to power our tractors and irrigation pumps, they also provide the manufacturing basis for chemical fertilizers, pesticides, and other artificial additives. The result is a "resource based" agriculture which has met the problem of diminishing returns. Adding 25% more fertilizer or pesticides will often generate not nearly so much additional food as did an extra 25% 20 years earlier.

This chapter reflects on the theme of a "crossroads" in modern agriculture. We must formulate some fundamentally different strategies if we are to move from a resource-based agriculture to a science-based, particularly a gene-based, agriculture. Among the various strategies available to meet this goal we shall consider three. The first lies with improved forms of existing foods, the second with entirely new foods, and the third with "adaptive agriculture." This last term means that we should no longer emphasize our efforts to "bend" natural environments to suit the needs of our selected plant varieties. Rather, we should seek to complement these established procedures by developing plant types that suit a much broader range of environmental conditions.

A. Improved Forms of Existing Foods

The productivity of our major crops cannot be maintained, let alone expanded, without a constant infusion of fresh genetic variability. Indeed it is the skills of plant geneticists, even more than the use of artificial additives such as fertilizers and pesticides, that have yielded one record crop after another in North America and elsewhere.

It is difficult to separate the contributions of genetic materials from other factors that enable modern crop variants to flourish. As a rule of thumb—an exceptionally rough one—agriculturalists estimate that we can attribute about 40% of expanded productivity to genetic breeding, and hence to the germ plasm and other genetic materials on which crop breeders depend (National Academy of Sciences, 1972; Myers, 1983; Oldfield, 1984). Due to this regular "topping up" of the genetic or hereditary constitution of the United States' main crops, the Department of Agriculture estimates that germ plasm contributions lead to increases in productivity that average around 1% annually, with a farm-gate value that now tops $1 billion (U.S. Department of Agriculture, 1976). Similar growth-rate gains can be documented for Canada, the United Kingdom, the Soviet Union, India, Pakistan, Philippines, and many other nations. To this extent we enjoy our daily bread by partial grace of the genetic variability that we find in plant relatives, both wild and primitive forms, of modern crop species.

During the past few decades plant geneticists, supported by agronomists, have enabled American farmers to produce over three times more corn per hectare, until production now averages as much as 200 bu/ha (bushels per hectare) and occasionally reaches 600 bu. When we compare this with the current world record for corn production, over 1,000 bu/ha, we can see what a gap there is between present performance and future potential. Furthermore, the theoretical maximum yield for corn is thought to be well above 2,500 bu/ha. Therefore, the present average U.S. yield could be increased five times before it matches the present world record for corn. Similarly large scope applied to other American crops, namely

3 times for soybeans, 4.8 for barley, 5.8 for rice, 6.2 for oats, 6.5 for sorghum, and 7 times for wheat and cassava (Wittwer, 1981).

The heyday of the plant geneticist plainly lies ahead. The Green Revolution, which has brought us outsize harvests during the past two decades, may shortly be supplemented (not supplanted) by a still more remarkable phase in agricultural innovation, the Gene Revolution, as we grow more expert in isolating and manipulating the genes that constitute the hereditary materials of each species. This breakthrough in plant-breeding technology may soon enable us to harvest crops from deserts, to farm tomatoes in seawater, and to grow super-potatoes in many new localities (as well as to enjoy entirely new crops such as a "pomato").

At the same time, we should recall that high productivity of modern types of corn is often accompanied by high vulnerability to diseases. The corn varieties that presently thrive tend to be planted everywhere that corn will grow. By 1970, the parentage of seed corn used in the United States had become dangerously susceptible to disease. No less than 70% of the seed owed its ancestry to only six in-bred lines. Not surprisingly, a severe leaf blight struck, eliminating one-seventh of the entire crop and as much as one-half of the crop in seven states. As a result, corn prices rose by 20% and the cost to farmers passed on to consumers totalled more than $2 billion (Tatum, 1971). Fortunately the situation was corrected thanks to blight-resistant germ plasm whose genetic ancestry originally derived from Mexico. Although we cannot say that the critical genetic material was worth $2 billion, since other factors contributed to the turnaround (such as research facilities and the professional skills of plant breeders), the figure still gives us an idea of the huge sums of money that can be saved or lost.

A more graphic illustration of the value of genetic variability to modern corn growing can be further illustrated by a recent discovery of a wild corn species in a montane forest of South–Central Mexico (Iltis et al., 1979). This plant is the most primitive known relative of modern corn. At the time of its discovery it was surviving in only three tiny patches covering a mere 4 ha, a habitat that was threatened with imminent destruction by squatter cultivators and commercial loggers. The wild species is a perennial, unlike all other forms of corn which are annuals. Crossbreeding with established commercial varieties of corn opens up the prospect that corn growers (and corn consumers) can be spared the seasonal expense of ploughing and sowing, since the plant would spring up again of its own accord like grass or daffodils.

Even more important, the wild corn offers resistance to four of eight major viruses and mycoplasmas that have hitherto baffled corn breeders (Nault and Findley, 1981). These four diseases cause at least a 1% loss to the world's corn harvest each year, worth more than $500 million. Equally to the point, the wild corn was discovered at elevations between

2500 and 3300 m and is adapted to habitats that are cooler and damper than established corn lands. This offers scope for expanding the cultivation range of corn by as much as one-tenth. All in all, the genetic benefits supplied by this wild plant, surviving in the form of no more than a few thousand stalks, could total several billion dollars per year (Fisher, 1982).

The contributions of this wild corn will assist in many aspects of daily welfare. As long as corn and corn products are available in abundant supplies and at cheap prices, we shall benefit in many more ways than breakfast cereal and popcorn. We enjoy the exceptional productivity of modern corn each time we read a magazine, since cornstarch is used in the manufacture of sizing for paper (the reader of this book is enjoying corn by virtue of the "finish" of the page he or she is looking at right now). The same cornstarch contributes to our lifestyles every time we put on a shirt or a blouse. Cornstarch likewise contributes to glue, so we benefit from corn each time we post a letter. And the same applies, through different applications of corn products, whenever we wash our face, apply cosmetics, take an aspirin or penicillin, chew gum, eat ice cream (or jams, jellies, tomato ketchup, pie fillings, salad dressings, marshmallows, or chocolates), and whenever we take a photograph, draw with crayons, or use explosives. Corn products also turn up in the manufacture of tires, molding of plastics, drilling for oil, electroplating of iron, and preservation of human blood plasma.

It is ironic that precisely at a time when we are heavily dependent on the genetic underpinnings of our modern crops, in fact more dependent than ever, we are allowing Earth's stocks of genetic materials to erode at ever-faster rates. The problem lies in part with sheer attrition of natural environments through spread of human activities (Myers, 1987). In part, too, a major threat lies with the trend for subsistence farming to give way to commercial agriculture, whereupon food plants that during centuries have evolved adaptations to their local ecological conditions are supplanted by more productive varieties, often from foreign sources. In this latter sense the Green Revolution, while an admirable achievement in many respects, is proving a disaster for local genetic diversity.

The critically narrow genetic base underpinning U.S. agriculture was described well over a decade ago as follows: "The situation is serious, potentially dangerous to the welfare of the nation, and appears to be getting worse" (National Academy of Sciences, 1972). And more recently: "Genetic resources stand between us and catastrophic starvation on a scale that we cannot imagine. In a very real sense, the future of the human race rides on these materials. Yet most of our so-called world collections (of crop genetic resources) are sadly deficient in wild races. As we obtain more and more useful results from incorporating wild genetic resources, the principle bottleneck lies in the paucity of wild germ plasm in our collections" (Harlan, 1976). And still more recently: "Genetic material is being eroded in many parts of the world to the extent that in the case of many crops we are facing a crisis situation" (Walsh, 1981).

B. Entirely New Foods

During the course of human history we have used around 3,000 plants for food. Yet Earth contains at least another 75,000 edible plants. Of this cornucopia of plant foods only about 150 have ever been cultivated on a large scale, and fewer than 20 now produce 90% of our food.

Nonetheless, there are numerous instances of under-exploited food plants with proven potential (National Academy of Sciences, 1975a; Ritchie, 1979; Vietmeyer, 1986). For instance, aborigines in Australia have used scores and possibly hundreds of plants, especially fruits and bulbs, as food. They favor certain yams that are well adapted to dry conditions, opening up the possibility that crossbreeding with forms in the tropics could allow this important crop to be extended to several further regions, notably the tropics themselves. Another dryland plant, the yeheb nut bush of Somalia *(Cordeauxia edulis),* grows prolific bunches of pods that contain seeds the size of peanuts (though they taste more like cashew nuts) that make a nutritious food that Somalis prefer to staples such as corn and sorghum (Westphal, 1974). In addition, the yeheb's foliage supplies tasty fodder for livestock. Being adapted to arid environments, the yeheb could assist desert dwellers in many parts of the tropics. It is being brought back from the verge of extinction in the wild through domestication efforts in Somalia and Kenya.

Much the same applies to plants in other ecological zones. A marine plant from the west coast of Mexico, *Zostera marina,* sometimes known as eelgrass, produces grain that the Seri Indians grind into flour. This plant opens up the prospect that we can use the seas to grow bread (Felger and Moser, 1985). From the highlands of Ethiopia, leafy, grassy vegetables prove a promising source of plant protein, yielding as much as alfalfa or soybean (Vaughan, 1977). When considering the potential of these Ethiopian vegetables, we might recall that a single wild species of the same genus has provided us, through plant breeding, with cabbage, kale, broccoli, cauliflower, and brussels sprouts.

Many other leafy food plants are important on a local scale while remaining unknown elsewhere (Fleuret, 1979; Herklots, 1972; Oomen and Grubben, 1977). Many leafy plants of the tropics, at least 1,650 of them in tropical forests alone, are reputed to contain roughly as much protein as legumes. They also feature from five to ten times more calcium than legumes and fruits, from two to six times as much iron, and ten to 100 times more carotene (a yellow pigment in the green chlorphyll). In addition, these leafy vegetables often contain as much vitamin C as the best fruits, together with an abundance of vitamin A.

Many other examples of wild foods are available. The wax gourd vegetable *(Benincasa hispida)* grows in the Asian tropics, but could be extended to many parts of Africa and Latin America (Morton, 1971). A creeping vine that looks somewhat like a pumpkin, the wax gourd can be raised more easily than any other curcurbit (pumpkin, squash, melon and so on). It grows rapidly; one shoot grows 2.3 cm every 3 hr during the course of four days. This vigorous growth rate allows three or four crops to be grown each year. A full-grown gourd reaches 35 kg in weight and measures 2 long by 1 in diameter. Fortunately, the fruit's pulp—a thick white flesh that is crisp and juicy—can be eaten at any stage of growth. Possessing a mild flavor, it is used as a cooked vegetable, as a base for soup, as a sweet when mixed with syrup, and as a food extender. A unique feature of the plant that is very pertinent to the humind tropics is that the gourd's waxy coating preserves the food inside from attack by microorganisms, thus allowing the vegetable to be stored for as long as one year without refrigeration.

Another type of gourd vegetable, the buffalo gourd *(Curcurbita foetidissima)*, could supply starch, edible oil, and other foods (Bemis *et al.*, 1975; Johnson and Hinman, 1980). Able to tolerate extreme drought, the buffalo gourd is a potential crop for arid lands, thus offering no competition with conventional agriculture. Experimental plantings of the gourd in the southwestern United States match the performance of traditional sources of protein and oil such as soybeans and peanuts in well-watered lands. A vigorous perennial, the buffalo gourd survives, even flourishes, in its harsh environments by virtue of its large, fleshy tubers that penetrate as deep as 5 to reach groundwater. The main root can grow to 30 kg after just two seasons, whereupon it constitutes 70% moisture. An occasional root can reach 45 kg of which the 25 kg of starchy content equals the amount produced by a score of potato plants growing under favorable conditions. So resilient is the buffalo gourd that some plants reportedly live as long as 40 years. The round yellow fruit, as much as 8 cm in diameter, has earned the buffalo gourd its popular name of "mock orange." Each season the plant produces, through its extensive vine growths, as many as 200 fruits or gourds, each of which contains 200–300 seeds. These seeds consist of one-third protein and one-third oil.

In the United States, domestication of the buffalo gourd is being attempted by the Office of Arid Lands Studies at the University of Arizona (Hinman *et al.*, 1985). This research program is revealing that the buffalo gourd, with seed yields of up to 3 t/ha, can generate at least 1 t/ha of vegetable oil, which, together with half this much protein and much crude starch, leads to a potential crop value of almost $700/ha.

Next, let us review some fruits. Temperate-zone plants have provided only about ten fruit species whereas the tropics have supplied us with almost 200 species. The main tropical source is the rainforest biome, particularly the Southeast Asian sector (Soepadmo, 1979; Williams *et al.*, 1975). Around 125 species of fruit plants are cultivated in Southeast Asia, and many of them originated in the forest. More than 100 other fruit trees grow wild in the forests. several produce edible fruits, and others offer potential for crossbreeding with established crop species. A notable instance is the durian *(Durio zibethinus)*, with delectable taste and execrable smell; the experience of consuming a durian can be described as eating an almond-flavored custard in a public toilet. Also from Southeast Asia comes the rambutan *(Nephelium lappaceum)*, a table fruit that is bright red and covered with whiskers. Perhaps tastiest of all fruits from Southeast Asia is the mangosteen *(Garcinia mangostana)*, though regrettably the plant appears to offer little genetic variability. For those people who favor citrus fruits such as oranges and tangerines, the pummelo *(Citrus grandis)* offers a suitable stimulating taste. It yields a larger harvest than most citrus crops, and it can grow in saline conditions.

Finally, let us take a brief look at a selection of vegetables. Again, Southeast Asia is a principal center with at least 300 species having been used in native cultures, of which about 80 still grow only in the wild in forest habitats (Soepadmo, 1979; Williams *et al.*, 1975). A notable example is the winged bean *(Psophocarpus tetragonolobus)*, also known as the four-angled bean, and the asparagus pea (National Academy of Sciences, 1975b, 1981). Known to forest tribes of New Guinea for centuries, the plant is not to be decried as a "poor man's crop" or to be dismissed as something second-rate for peasant communities. The vine-like plant contains far more protein than potato, cassava, and other crops that serve as principal sources of food for millions of people in the tropics. The winged bean offers a nutritional value equivalent to soybean, with 40% protein and 17% edible oil, plus vitamins and other nutrients. Its capacity to match the soybean might remind us that the United States grew sporadic patches of the soybean for at least a century before the plant was finally upgraded into a widespread crop; today the soybean is the premier protein crop in the world, flourishing in dozens of temperate-zone countries. Could not a similar prospect be in store for the winged bean, scheduled to become the long-sought "soybean of the tropics"? As a result of genetic improvement, the winged bean is now helping to upgrade diets in more than 50 countries of the developing tropics.

Many other little-known crops have exceptional potential, notably the amaranth *(Amaranthus* spp.), a grain crop of the Andes (Downtown, 1973), and spirulina *(Spirulina platensis)* (Protein Advisory Group, 1973).

Moreover, certain animal species present promise as sources of new foods (Evans and Hollaender, 1986; FAO, 1984; Peel and Tribe, 1983). Several dozen wild antelopes and other herbivores of African savannahs are prime examples (Myers, 1972), as are certain species from Amazonia (Wetterberg *et al.*, 1976). The kouprey is a secretive cow-like creature that inhabits the forests of the Thailand/Kampuchea border. The animal is believed to have been one of the wild ancestors of the humped zebu cattle of southern Asia, suggesting that fresh crossbreeding between the two bovids could boost cattle raising throughout the entire region. Regrettably, the kouprey's survival is doubtful due to military activities within its habitats during the past 20 years. Other wild bovids of Southeast Asia's forests, such as the selatang, the tamarau, and the anoa, could help cattle husbandry. Like the kouprey, their numbers have all been severely reduced through human disruption of their life-support systems.

In a similar manner, cattle raising in Africa can be assisted through a highly localized breed of cattle that lives around Lake Tchad. This Kuri breed of cattle is able to swim and feeds off lake-bottom vegetation. The breed is threatened with "genetic swamping" through excessive and haphazard crossbreeding with local zebu cattle. Also in West Africa is a dwarf shorthorn breed of cattle, the N'dama, with tolerance for trypanosomiasis disease that limits cattle raising in some 10 million km^2 of Africa, or one-third of the continent. Yet the N'dama is in danger of disappearing.

Finally let us consider aquaculture, a form of agriculture that may prove to be one of the fastest growing sectors of all agriculture in the foreseeable future. It may also represent the most promising means for us to grow large amounts of that critical form of food, animal protein (Borgese, 1980; Lovell, 1979).

More than 90 percent of the fish we consume is obtained through "hunting" of wild species. Yet fish have been reared in enclosed structures, usually ponds, in Asia for at least 4000 years. Of global fish consumption today, amounting to some 70 million metric tons (60 million from the oceans and 10 million from fresh waters), only a little over 6 million tons are derived from aquaculture, including four million of finfish and more than one million of mollusks such as oysters, mussels, clams, and other high-priced gourmet items (plus more than 1 million tons of seaweeds). This means that aquaculture accounts for about 40% of freshwater output, and 3% of saltwater output. Small as this aquaculture proportion may be, it represents a marked advance over the mid-1960s when the annual harvest amounted to no more than 1 million tons. Today's leaders in aquaculture are China, with 2 million tons, and Indonesia and the Philippines with around one million each. Aquaculture supplies 40% of fish eaten in China, 22% in Indonesia, and 10% in the Philippines. Israel now obtains almost 60% of its total catch of both marine and freshwater fish from culture ponds.

So great is the scope for increasing aquaculture that the Food and Agriculture Organization believes this practice could contribute, by the year 2000, at least three times more animal protein than at present, and possibly six times more. Aquaculturalists in various parts of the world have raised more than 300 species of finfish. However, of the 4 million tons of finfish cultivated in fish ponds today, 85% comes from a few carp species, notably common carp, Chinese carp, and Indian carp. These carp species may soon give way to new contenders from the *Tilapia* genus of finfish (Fryer, 1980; Legner, 1978). Originally raised by the ancient Egyptians, tilapias are now grown widely in eastern Asia. Their production doubled between 1970 and 1975 and is continuing to grow fast. So proliferative are tilapias that a hybrid species raised in garbage-enriched ponds can generate 3 t/ha in 180 days. This performance may be compared with that of common carp, which yields almost 0.4 ton with no supplemental feeding, 1.5 tons with grain as a supplement, and 3.3 tons with a protein-fortified diet.

There are large numbers of wild *Tilapia* species that can be used for selective breeding. Lake Tanganyika contains 126 endemic species, Lake Victoria 164, and Lake Malawi 196 (Lowe-McConnell 1977). These totals may rise since new species are being found all the time. These *Tilapia* species differ from one another in their diets and breeding patterns, which suggests that a systematic approach to aquaculture could utilize combinations of *Tilapia* species in order to expand the protein yield. Different species could divide the food supplies of a pond much as they divide the food supplies in each of the three great African lakes, efficiently exploiting many food types that would not be consumed by a single species.

However, wild *Tilapia* stocks are threatened. Lake Malawi in Central Africa holds more than 500 Cichlidae species, the bulk of them in the *Tilapia* genus—and 99 percent of them endemic. The lake is only one-eighth the size of North America's Great Lakes, which feature only 173 species, fewer than 10 percent of them endemic. Lake Malawi is threatened through pollution from industrial installations and proposed introduction of alien species (Barel *et al.*, 1985). In Lake Victoria, with only 300 endemic Cichlidae species, introduced predators and other problems are reducing the endemics so fast that they may well decline by 80–90% within another decade.

C. Adaptive Agriculture

We have looked at some ways to generate a "science-based" agriculture through development of improved forms and entirely new forms of crop plants and domestic livestock. Let us now look at a third principal way to bring science to bear in support of modern agriculture through an assessment of ways to develop plant variants that better "fit" their environments. To date we have been inclined to cause natural environments to fit the needs of our crop plants. If soils have been deficient in nutrients, we have added fertilizer. If the environment has been too dry, we have added irrigation water. If the environment produces pests to attack crops, we have sprayed pesticides. What prospect that we can reverse this process, and "bend" our crop plants to better suit their environments?

Consider the case of fertilizer. Synthetic fertilizer, especially in the form of chemically-fixed nitrogen, is the most frequent limiting factor in agriculture after water—and it is also the single most costly item in modern agriculture. If Third World farmers are to increase their food output by almost four percent a year, as they must to keep up with demand, they will need to expand their use of fertilizer five times during the last quarter of this century.

A promising approach lies with a technique known as biological nitrogen fixation (Brill, 1979; Felker and Bandurski, 1979; Phillips, 1980). The secret lies with certain bacteria that possess the ability to fix nitrogen from the atmosphere. Generally speaking, these microorganisms set themselves up in nodules, or "nitrogen factories," on the roots of plants. The host plant provides food and energy that the bacteria use, and the bacteria fix nitrogen that the plant uses. So efficient is this process that nitrogen-fixing microorganisms are estimated to fix 175 million tons of nitrogen per year. Synthetic nitrogen fertilizer now totals rather more than 50 million tons a year. When used as fertilizer, this biologically fixed nitrogen proves as effective in boosting crop yields as does commercial fertilizer, and it costs only one-tenth as much.

Biologically fixed nitrogen can be made available in various ways. One is through the use of those leguminous plants that, by virtue of the bacteria-harboring nodules on their roots, can pluck nitrogen from the atmosphere. Legumes, notably species such as alfalfa and clover, fix anywhere between 100–600 kg of nitrogen/ha/yr. All the farmer has to do is to "interplant" his conventional crops with legumes (e.g., between rows of corn stalks, or between seasonal crops of corn), enabling a "free" supply of fixed nitrogen to be introduced into the soil. Corn growers in the United States, using a rotation system of alfalfa with corn, find they can cut their use of commercially manufactured nitrogen fertilizer by as much as 60 kg/ha for a saving of one-fifth of their costs. (At present, U.S. cornlands absorb a full one-half of all the nitrogen fertilizer that American farmers apply to their crops, or one-eighth of all such fertilizer applied worldwide. In similar fashion, peasant farmers in Asia find that by using a leguminous tree, the ipilipil *(Leucaena leucocephala),* as a rotational fertilizer crop they can obtain as much nitrogen from six bags of tree leaves as from one bag of ammonium sulfate.

Legumes are estimated to fix at least 35 million metric tons of nitrogen annually. This service is estimated to be worth several billion dollars. Yet we have only begun to explore the potential benefits of the legume-nodule association. The legume family totals at least 13,000 known species, and we utilize only about 100 of them for commercial purposes. Plainly there are opportunities to mobilize the nitrogen-fixing services of many wild legumes awaiting the attention of plant breeders and geneticists.

We next look at environmental stresses and ways to devise plant varieties that overcome these stresses. Of the Earth's land surface outside Antarctica, almost 90% is less than suitable for modern agriculture due to too little or too much water, excess heat or cold, and soils that are shallow or are infertile because of too few nutrients or too many toxic elements. In various parts of the Earth, notably in the tropics, some 29 million km^2 of land are limited by mineral stresses (not only nutrient deficiencies and toxicities, but also salinity, acidity, and alkalinity, plus air pollution in the form of acid rain). A further 32 million km^2 are limited by shallow soils, together with another 37 million km^2 that suffer from too little water and 16 million km^2 that suffer from too much water. These territories altogether amount to 114 million km^2, to be compared with our present croplands covering only 15 million km^2 (an area equivalent to the United States and half of Canada combined).

Fortunately, we can respond to this challenge through "tolerant" germ plasm (Christiansen and Lewis, 1982; Wittwer, 1982). Already we can take heart at the success of hybrid sorghum and millets that prosper in areas that, until very recently, were considered too hot and dry (Epstein, 1980; Goodin and Northington, 1985; Nabhan, 1985; Wickens et al., 1985). What if we could come up with a type of wheat that would grow in the lowland humid tropics, territories which, with their many diseases, still defy wheat breeders? If "environmental proofing" were to be applied to more crops in more ways, we could open up millions of hectares of land that now remain little utilized.

Moreover, modern crops must be able to cope with changes in climate, such as unexpected fluctuations in rainfall and temperature. This attribute is all the more significant in that crops today, with their narrow genetic bases, are adapted to what may have been the most moist 30-year period during the past 1,000 years (Bryson and Murray, 1977). Yet climatic dislocations may soon emerge on exceptional scale. Within the next half century at most, we may witness greater climatic changes than at any period during the past several thousand years, due to carbon dioxide and other "greenhouse gases."

An increase in greenhouse gases could cause U.S. farmers to experience growing-season conditions akin to the heat and drought of 1980, which reduced certain categories of grain output by one-fifth. If these climatic changes become more pronounced, and if adaptable crop types are not available, Americans farmers in certain areas could find themselves facing "dust bowl" conditions. At the same time, of course, we should recognize that warmer weather will allow the range of many major crops to be expanded into areas that have hitherto been too cold. In the United States, an increase of 1°C. could allow the corn belt to extend 175 km northeastwards. Furthermore, an increase in carbon dioxide may, in certain circumstances, prove beneficial to some crops in some sectors of the world because an "enriched" atmosphere can generate a "fertilizing impact" on photosynthesis.

Salt-tolerant plants are a particular category of environmentally adapted crop varieties. Many natural environments are too saline for conventional crops. Saline soils around the world amount to 9.5 million km^2, in comparison with total world croplands of 15 million km^2. Hence, humankind faces a problem in the form of its saline environments—a problem that is growing greater as salinization spreads. Fortunately we may soon be able to look upon the problem as an opportunity, thanks to recent research that highlights the possibilities of salt-tolerant plants, otherwise known as halophytes (from the Greek "halo" meaning salt and "phyte" meaning plant) (Aronson, 1985; Epstein *et al.*, 1980).

Several halophytic species could qualify as candidates for this futuristic type of agriculture. Some are wild relatives of commercial barley, wheat, sorghum, rice, several types of millet, sugarbeet, tomato, date palm, and pistachio (a kind of nut that is popular in many parts of the United States and Europe). They also include certain kinds of forage plants that serve the needs of livestock; examples are alfalfa, ladino clover, creeping bentgrass, Bermuda grass, and various reeds and rushes. Any of these salt-tolerant plants could be grown in desert areas through irrigation with brackish water. They could be grown in salinized zones such as the Central Valley of California; they could be grown in those areas where fresh water is becoming a scarce commodity, and where seawater is within easy pumping distance, again as in the Central Valley of California; and they could be grown along coastlines irrigated with seawater. For example, a strain of barley has been discovered that produces almost 1.2 t/ha while deriving all its moisture from seawater. A similar prospect appears in store with a number of wheat strains. As for rice, several hundred salt-tolerant varieties have been identified but their grain yield is low and they prove susceptible to insect pests and diseases. Nonetheless, plant geneticists hope to find salt-tolerant varieties among wild rices that will meet the needs of rice growers.

Finally, let us look at the scope for developing forms of crop plants with "built-in" resistance to pests (Greathead and Waage, 1983; Jacobson, 1982; Rice, 1983). Worldwide we lose an average of 14% of our crops to insect pests, while the United States alone loses crops to insects worth more than several billion dollars a year (Pimentel and Edwards, 1982). Fortunately we can call upon the aid of a number of wild plants, notably tropical species that produce chemical compounds that repel insects or inhibit their feeding. These toxic compounds occur in two main categories—the pyrethrins from chrysanthemum-type plants and the rotenoids from roots of rainforest legumes (Secoy and Smith, 1983). Both categories are biodegradable and do not accumulate in organisms; hence, they cause little harm to higher animals such as birds and mammals (including humans). Many plants and insects, notably those of tropical zones, have evolved together, so we can expect that many other similar insect-repelling substances must be available in wild plants. At the same time, however, we should remember that insect pests include variations that can multiply in numbers to overcome plant defenses in only ten years, sometimes as little as three years. So there is constant need to derive further genetic combinations of crop plants to stay ahead of immune varieties of insects. Regrettably, very few wild plants have been screened for this purpose (or for any other purpose), so we can surmise little about their potential for controlling insect pests—except that it is surely very large.

III. EXTINCTION PROCESSES

At least two-thirds, and conceivably as many as nine-tenths, of all species occur in the tropics. This is highly significant for our efforts to safeguard the planetary spectrum of genetic variability, insofar as Third World nations generally do not possess the conservation resources (scientific skills, institutional capacities, and above all funding) to safeguard their wild gene reservoirs, even if they possess the motivation.

Although the tropics cover only a limited portion of the Earth's land surface, we need not be surprised at the concentration of species there. With their year-round warmth and often year-round moisture, the tropics have served as the planet's main powerhouse of evolution. By contrast, the temperate zones and the remainder of the planet are relatively deficient in species concentrations and gene reservoirs. Indeed the developed world can be viewed as genetically depauperate. Yet it is the developed nations that possess the technological capacity to exploit species and their genetic resources for economic advantage. This situation raises several issues salient to North–South relations, and in particular to economic questions addressed in negotiations within the North–South dialogue. The need to preserve germ plasm resources is but one of several global resource and environmental issues that have emerged during the past 20 years, and that are likely to receive increasing attention during the foreseeable future. They highlight the interdependent nature of society at large, and the need for collective action on the part of the community of nations. Moreover, many of these problems are interrelated; progress can be made on one front only by tackling several others simultaneously.

Characteristic of problems that affect most if not all nations, and that thus can be characterized as intrinsically international if not supranational, is the problem of biotic impoverishment. Plainly this is not merely an issue for wildlife enthusiasts. The demise of a single species represents an irreversible loss of a unique natural resource. The planet is currently afflicted with various severe forms of environmental degradation, such as desertification and pollution. But whereas these forms of degradation can generally be reversed, extinction of species cannot. When a species is eliminated, it is gone for good—and, in strictly utilitarian terms, that will frequently turn out to be for bad.

A. Present Rate of Extinction

Tropical forests cover only 6% of Earth's land surface, yet are reputed to harbor at least half, and possibly a far greater share (even as much as 90%), of Earth's total stock of species. How fast are tropical forests in fact being depleted? In terms of complete and permanent removal of forest cover, those instances where all trees have been eliminated and the area has been given over to rice cultivation or cattle ranching or urbanization, the rate postulated for the late 1970s is somewhere between 76,000 and 92,000 km^2 a year (FAO UNEP, 1982; Myers, 1980). In terms of gross disruption of forest ecosystems, with significant degradation of their capacity to support a primary-forest complement of species, the rate postulated is around 100,000 km^2 a year. Overall, then, almost 200,000 km^2 of tropical forest are being degraded or destroyed a year, or just over 2% of a biome that now totals around 9 million km^2.

In light of several key sectors of tropical forests, such as western Ecuador, Atlantic-coast Brazil and Madagascar, we are already losing several species a year in tropical forests (Ehrlich, 1986; Myers, 1986; see also Raven, 1986; Soule, 1986; Western and Pearl, 1988; Wilson, 1988).

B. Projected Rate of Extinction

As for the future, the outlook seems all the more adverse, though its detailed dimensions are still less clear than those of the present. Despite the uncertainty, however, it is worthwhile to delineate the nature and compass of what lies ahead in order to grasp the scope of the extinction spasm that impends. Let us look again at tropical forests. We have already seen what is happening to three critical areas. We can identify a good number of other sectors of the biome that are similarly ultra-rich in species, and that likewise face severe threat of destruction. They include the Mosquitia Forest of Central America; the Chocó forest of Colombia; the Napo center of diversity in Peruvian Amazonia, plus six other centers (out of about 20 centers of diversity in Amazonia) that lie around the fringes of the basin and hence are unusually threatened by settlement programs and various other forms of "development." Other sectors are the Tai Forest of Ivory Coast; the montane forests of East Africa; the relict wet forest of Sri Lanka; the monsoon forests of the Himalayan foothills; Sumatra; northwestern Borneo; certain lowland areas of the Philippines; and several islands of the South Pacific (New Caledonia, for instance, with some 18,500 km^2, or about the size of New Hampshire, contains 3,000 plant species, 80% of them endemic).

These sectors of the tropical forest biome amount to roughly 1 million km^2 (only two and a half times the size of California), or only one-tenth of remaining undisturbed forests. So far as we can best judge from their documented numbers of plant species (Conservation Monitoring Centre, 1986), and by making substantiated assumptions about the numbers of associated animal species, we can reckon that these areas surely harbor one million species (assuming a low planetary total of 5–7 million species). If present land use patterns and exploitation trends persist, there will be little left of these forest tracts, except in the form of degraded remnants, by the end of this century or shortly thereafter. Thus deforestation in these areas alone could well eliminate very large numbers of species, surely hundreds of thousands, within the next 20 years at most (Myers, 1986).

How about the prognosis for the longer-term future to the effect that eventually we could lose at least one-quarter, possibly one-third, and conceivably a still larger share of all extant species? Let us take a quick look at the case of Amazonia (Simberloff, 1986). If deforestation continues at present rates (it is likely to accelerate) until the year 2000, but then were to halt completely, we should anticipate a loss of about 15% of plant species. The calculation behind this loss figure is entirely reasonable and documentable, based as it is on the well-established theory of island biogeography (Soule, 1986) and abundant evidence of pervasive deforestation patterns in Amazonia. Were Amazonia's forest cover to be ultimately reduced to those areas now set aside as parks and reserves, we should anticipate that 66% of plant species would eventually disappear, together with almost 69% of bird species, and similar proportions of all other major categories of species.

Of course, we may learn how to manipulate habitats to enhance survival prospects, or to propagate threatened species in captivity. We may be able to apply other emergent conservation techniques, all of which could help to relieve the adverse repercussions of broadscale deforestation. But in the main, the damage will have been done. For reasons of island biogeography, and of "ecological equilibriation" (delayed fall-out effects), some extinctions in Amazonia will not occur until well into the 22nd century, or even further into the future. A major extinction spasm in Amazonia is entirely possible—indeed plausible, if not probable.

This writer hazards a best-judgement estimate that we shall surely lose a full one-quarter of all species that now share the Earth with us (unless of course we move swiftly to implement conservation measures with much broader scope). A loss of one quarter is an optimistic prognosis;

it is possible that we will lose one third, and conceivable that we will lose one half of all species. Moreover, the surviving species may well lose a great part of their genetic variability. This would be a biological debacle as great, in its compressed timescale, as any during the entire course of evolution.

In contrast to the environmental damage caused by pollution or soil loss, which may be reversed over the long run, the loss of species diversity represents an essentially irreversible process within relevant time scales. Judging by the recoveries following the "species crashes" which ended the Permian and Cretaceous Periods, it will surely take tens of millions of years for evolutionary processes to generate a complement of species comparable to that which exists today.

IV. CONCLUSION

This chapter has presented a few of the strategies we can pursue to develop an innovative agriculture to meet the new challenges of the future. Not only do we need to grow much more food, we also need to initiate a basic rethinking of certain agricultural technologies. While the resource-based strategies of the past have served us exceptionally well, they no longer offer as much promise for the future. We must seek to expand into a gene-based agriculture. The raw materials are available in the form of genetic variability and other resources of the natural world with its wealth of wild species. All we need to do is to mobilize our scientific expertise to realize the potential of the future and, on the conservation front, to ensure species remain in existence to serve our material welfare at several critical points throughout our agricultural sectors.

REFERENCES

Aronson, J. (1985). Economic halophytes—A global review. In "Plants for Arid Lands" (G. E. Wickens, J. R. Goodin, and D. V. Field, eds.), pp. 177–188.
Barel, C. D. N., et al. (1985). Destruction of fisheries in Africa's lakes. Nature (London) **315**, 19–20.
Bemis, W. P., et al. (1975). "The Buffalo Gourd: A Potential Crop for the Production of Protein, Oil and Starch on Aridlands. Off. Agric. Agency Int. Dev., Washington, D.C.
Borgese, E. M. (1980). "Seafarm: The Story of Aquaculture." Abrams, New York.
Brill, W. J. (1979). Nitrogen fixation: Basic to applied. Am. Sci. **67**, 458–465.
Bryson, R. A., and Murray, T. J. (1977). "Climates of Hunger." Univ. of Wisconsin Press, Madison.
Christiansen, M. N., and Lewis, C. F., eds. (1982). "Breeding Plants for Less Favourable Environments." Wiley (Interscience), New York.
Conservation Monitoring Centre. (1986). "Plants in Danger." Conserv. Monit. Cent. (under IUCN), Cambridge, England.

Downton, W. J. S. (1973). *Amaranthus edulis:* A high lysine grain amaranth. *World Crops* **25**(1), 20.
Ehrlich, P. R. (1986). "The Machinery of Nature." Simon & Schuster, New York.
Epstein, E. (1980). Responses of plants to saline environments. *In* "Genetic Engineering of Osomregulation" (D. W. Rains, R. C. Valentine, and A. Hollaender, eds.), pp. 7–21. Plenum, New York.
Epstein, E., *et al.* (1980). Saline culture of crops: A genetic approach. *Science* **210**, 399–404.
Evans, J. W., and Hollaender, A., eds. (1986). "Genetic Engineering of Animals." Plenum, New York.
Felger, R. S., and Moser, M. B. (1985). "People of the Desert and Sea: Ethnobotany of the Seri Indians." Univ. of Arizona Press, Tucson.
Felker, P., and Bandurski, R. S. (1979). Uses and potential uses of leguminous trees for minimal energy input agriculture. *Econ. Bot.* **33**(2), 172–184.
Fisher, A. C. (1982). "Economic Analysis and the Extinction of Species." Dep. Energy Resour., Univ. of California, Berkeley.
Fleuret, A. (1979). The role of wild foliage in the diet: A case study from Lushoto, Tanzania. *Ecol. Food Nutr.* **8**, 87–93.
Food and Agriculture Organization (FAO). (1984). "Animal Genetic Resources." Data Banks and Training, Food Agric. Organ., Rome.
Food and Agriculture Organization and United Nations Environment Programme (FAO UNEP). (1982). "Tropical Forest Resources." Food Agric. Organ., Rome and U.N. Environ. Programme, Nairobi, Kenya.
Fryer, G. (1980). "Conserving and Exploiting the Biota of Africa's Great Lakes." Freshwater Biol. Assoc., Ambleside, Cumbria, England.
Goodin, J. R., and Northington D. K., eds. (1985). "Plant Resources of Arid and Semiarid Lands." Academic Press, Orlando, Florida.
Greathead, D. J., and Waage, J. K. (1983). "Opportunities for Biological Control of Agricultural Pests in Developing Countries," Tech. Pap. No. 11. World Bank, Washington, D.C.
Harlan, J. R. (1976). Genetic resources in wild relatives of crops. *Crop Sci.* **16**, 329–333.
Herklots, G. A. C. (1972). "Vegetables in Southeast Asia." Allen & Unwin, London.
Hinman, C. W., Cooke, A., and Smith, R. I. (1985). Five potential new crops for arid lands. *Environ. Conserv.* **12**, 309–315.
Iltis, H. H., Doebley, J. F., Guzman, R. M., and Pazy, B. (1979). *Zeadiploperennis* (Gramineae), a new teosinte from Mexico. *Science* **203**, 186–188.
Jacobson, M. (1982). Plants, insects, and man—Their interrelationships. *Econ. Bot.* **36**, 346–354.
Johnson, J. D., and Hinman, T. W. (1980). Oils and rubber from arid land plants. *Science* **208**, 460–464.
Legner, E. F. (1978). Mass culture of *Tilapia zillii* (Cichlidae) in pond ecosystems. *Entomophaga* **23**(1), 51–55.
Lovell, R. T. (1979). Fish culture in the United States. *Science* **206**, 1368–1372.
Lowe-McConnell, R. H. (1977). "Ecology of Fishes in Tropical Waters," Studies in Biology, No. 76. Arnold, London.
Morton, J. F. (1971). The wax gourd—A year-round Florida vegetable with unusual keeping quality. *Proc. Fla. State Hortic. Soc.* **84**, 104–109.
Myers, N. (1972). "The Long African Day." Macmillan, New York.
Myers, N. (1980). "Conversion of Tropical Moist Forests," Report to National Academy of Sciences. Natl. Res. Counc., Washington, D. C.
Myers, N. (1983). "A Wealth of Wild Species." Westview Press, Boulder, Colorado.

Myers, N. (1986). Tropical forests: An overview assessment, with impacts on extinctions. In "Conservation Biology: Science of Scarcity and Diversity" (M. E. Soule, ed.), pp. 394–409. Sinauer, Sunderland, Massachusetts.
Myers, N. (1987). Mass extinction of species: A great creative challenge. Albright Lect., Univ. of California, Berkeley.
Nabhan, G. P. (1985). "Gathering the Desert." Univ. of Arizona Press, Tucson.
National Academy of Sciences. (1972). "Genetic Vulnerability of Crops," pp. 126–172. Natl. Acad. Sci./ Am. Assoc. Adv. Sci. Washington, D.C.
National Academy of Sciences. (1975a). "Underexploited Tropical Plants with Promising Economic Value." Natl. Acad. Sci., Washington, D.C.
National Academy of Sciences. (1975b). "The Winged Bean: A High Protein Crop for the Tropics." Natl. Acad. Sci., Washington, D.C.
National Academy of Sciences. (1981). "The Winged Bean: A High Protein Crop for the Tropics," 2nd Ed. Natl. Acad. Sci., Washington, D.C.
Nault, L. R., and Findley, W. R. (1981). Primitive relative offers new traits for corn improvement. *Ohio Rep.* **66**(6), 90–92.
Oldfield, M. L. (1984). "The Value of Conserving Genetic Resources." Natl. Parks Serv. U.S. Dep. Inter., Washington, D.C.
Oomen, H. A. P. C., and Grubben, G. J. H. (1977). Tropical leaf vegetables in human nutrition. *Communication* **69**, 24–41, 51–55. Dep. Agric. Res., K. Inst. Trop., Amsterdam.
Peel, L., and Tribe, D. E. (1983). "Domestication, Conservation and Use of Animal Resources." Elsevier, New York.
Phillips, D. A. (1980). Efficiency of symbiotic nitrogen fixation in legumes. *Annu. Rev. Plant Physiol.* **31**, 29–49.
Pimentel, D., ed. (1980). "Handbook of Energy Utilization in Agriculture." CRC Press, Boca Raton, Florida.
Pimentel, D., and Edwards, C. A. (1982). Pesticides and ecosystems. *BioScience* **32**(7), 595–599.
Protein Advisory Group of the United Nations. (1973). Proteins from microalgae and microfungi. *Trop. Sci.* **15**, 77–81.
Raven, P. H. (1986). Biological resources and global stability. Speech delivered at presentation of International Prize for Biology, Kyoto. Missouri Bot. Gard., St. Louis.
Rice, E. L. (1983). "Pest Control with Nature's Chemicals." Univ. of Oklahoma Press, Norman.
Ritchie, G. A., ed. (1979). "New Agricultural Crops." Westview Press, Boulder, Colorado.
Secoy, D. M., and Smith, A. E. (1983). Use of plants in control of agricultural and domestic pests. *Econ. Bot.* **37**, 28–57.
Simberloff, D. (1986). Are we on the verge of a mass extinction in tropical rain forests? In "Dynamics of Extinction" (D. K. Elliott, ed.), pp. 165–180. Wiley, New York.
Soepadmo, E. (1979). The role of tropical botanic gardens in the conservation of threatened valuable plant genetic resources in Southeast Asia. In "Survival or Extinction" (H. Synge and H. Townsend, eds.), pp. 63–74. Royal Bot. Gard., Richmond, Surrey, England.
Soule, M. E., ed. (1986). "Conservation Biology: Science of Scarcity and Diversity." Sinauer, Sunderland, Massachusetts.
Tatum, L. A. (1971). The southern corn leaf blight epidemic. *Science* **171**, 1113–1116.
U.S. Department of Agriculture. (1976). "Introduction, Classification, Maintenance, Evaluation and Documentation of Plant Germplasm," Natural Resources Prog. Rep. No. 20160. Agric. Res. Serv., U.S. Dep. Agric., Washington, D. C.
Vaughan, J. G. (1977). A multidisciplinary study of the taxonomy and origin of *Brassica* crops. *BioScience* **27**(1), 35–38.

Vietmeyer, N. D. (1986). Lesser-known plants of potential use in agriculture and forestry. *Science* **232**, 1379–1384.

Walsh, J. (1981). Germplasm resources are losing ground. *Science* **214**, 421–423.

Western, D., and Pearl, M., eds. (1988). "Conservation 2100," Proc. Conf., Wildl. Conserv. Int. and N.Y. Zool. Soc., 1986. Oxford, New York.

Westphal, E. (1974). "Pulses in Ethiopia, Their Taxonomy and Agricultural Significance." Cent. Agric. Publ. Doc., Univ. of Wageningen, Wageningen, Netherlands.

Wetterberg, G. B., *et al.* (1976). "An Analysis of Nature Conservation Priorities in the Amazon," Tech. Ser. No. 8. Brazilian Inst. For. Dev., Brasilia.

Wickens, G. E., Godin, J. R., and Field, D. V., eds. (1985). "Plants for Arid Lands." Allen & Unwin, London.

Williams, J. T., Lamourex, C. H. and Wulijarni-Soetjipto, N., eds. (1975). "Southeast Asian Plant Genetic Resources." BIOTROP, Bogor, Indonesia.

Wilson, E. O., ed. (1988). *Biodiversity*. Natl. Acad. Sci., Washington, D.C.

Wittwer, S. H. (1981). "The Further Frontiers: Research and Technology for Global Food Production in the 21st Century." Michigan Agric. Exp. Stn., East Lansing.

Wittwer, S. H. (1982). "Worldwide Influences on U.S. Farm Production." Michigan Agric. Exp. Stn., East Lansing.

4

Availability of Agricultural Land for Crop and Livestock Production

Pieter Buringh

Marterlaan 20
6705 BA Wageningen
The Netherlands

I. The World Land Area and Its Utilization
II. Food Production and Land
III. Livestock Production and Land
IV. Conclusion
   References

I. THE WORLD LAND AREA AND ITS UTILIZATION

This chapter discusses land resource availability on a global scale, with a focus on cropland and grassland. Land and water are the principal natural resources required for food production. The quality and productive capacity of land vary, and many important regional and local differences exist in land resources around the world. Consequently, the potential for food production differs within and among all countries. Thus, the results of global investigations cannot be projected from data based on a specific region or country. Ideally, this global study should be based on specific studies of various countries, but such studies are scarce at present.

A problem with all global assessments, calculations, and projections on land and its related aspects is that they currently cannot be verified. All data on land use, land productivity, degradation, and losses of land, for example, contain inaccuracies because in many countries no reliable statistics are available on land and soils. Data presented by international organizations like the Food and Agriculture Organization (FAO) are often based on estimations, but they are used by researchers because they are the best available. The problem of reliability of the information used becomes evident when computers analyze data and identify the various discrepancies. For example, a discussion on the land area of the world that is suitable for agricultural production depends on the "Soil Map of the World" (FAO/UNESCO, 1974–1981). This provides the best information available; however, only one-quarter of the land area is covered by real soil surveys, and the rest is based on scarce information picked up during various tours (Dudal, 1982). Moreover, no uniform definitions are given for many land and soil resources in world statistics. For example, when is an area covered by grass and trees considered pasture? And when is it forestland? This not only depends on the definition of both types of land use, but also on how such definitions are interpreted. Statistical data and information on maps are often used to support political situations. Thus, some countries present data specifically adapted for political reasons.

Another difficulty with global assessments is the enormous numbers that must be used in world analyses. In an effort to simplify the presentation where possible, percentages instead of millions of hectares will be used to illustrate trends.

The question of interest in this chapter is how much land is available for crop and livestock production? Agricultural production depends primarily on climate and soil conditions, plus the farming practices employed. Thus, about 10% of the total world land area (149 billion ha) is covered by ice, 15% is too cold to grow crops, and 17% is too dry to grow crops.

In addition, some 18% of the total land area is too steep for farming, 9% is too rocky and stony or soils are too shallow, 4% is too wet, and 5% is too poor for other reasons. This means that approximately 78% of the world land area lacks potential for growing crops. However, part of it can be used as poor-quality grazing land. The area suitable for crop production is approximately 22% of all land, or 24% of the land area not covered by ice, some 3.3 million ha. Most recent assessments of land resources agree with these numbers (±1%). The productive capacity of this land varies. More than half of the 22% of suitable cropland (13%) has a low capacity, 6% a medium, and only 3% a high capacity for crop production.

Not all land suitable for crop production is currently being used to grow crops. Table 4.1 compares land usage for crops, forage, and forest production 100 years ago with land usage today. A total area of 13.1 billion ha is considered available for use, or 13.4 billion ha if lakes and rivers are included. These numbers are smaller than the total land area presented in the beginning, because all land covered by ice is excluded. It is interesting to observe that only about 11% of the land area is used for growing crops. This is approximately half of the area that is suitable for crop production. The other half is currently used for pasture or forest production.

Clearly, a large area of the world's land resources can be used for growing crops in the future. However, significant land area is lost from production each year because of erosion, salinization, degradation, and non-agricultural uses (housing, industries, roads, highways, parks, sport grounds, etc.). I estimate the annual total loss is four million ha. Various specialists (see, e.g., Eckholm, 1976) estimate that the annual loss of land is 7 million ha, of which 3 million is the result of soil erosion, 2 million is due to salinization and 2 million is caused by desertification. In a recent article, UNESCO reported an annual loss of 21 million ha of fertile land and a transformation of 6 million ha into desert (UNESCO, 1985)! When

Table 4.1 World Land Use, 1882 and 1984[a]

Category[b]	1882 (mha)	(%)	1984 (mha)	(%)
Arable land	860	7	1,477	11
Pasture	1,500	11	3,151	24
Forestland	5,200	40	4,091	31
Other land	5,581	42	4,362	33
	13,081	100	13,081	99

[a]Source: 1882, various data and estimations; 1984, FAO (1986).

[b]"Arable land" includes 100 million ha with permanent crops. "Other land" is mainly land in polar regions, desert land, and stony and rocky land in mountains.

studying all the information on misues and loss of land, it appears that much of the data are inaccurate and exaggerated. One loss that is not exaggerated is the ever increasing area taken out of agricultural production to be used for non-agricultural purposes. The human population adds 80 million people per year, and an average of 0.1 ha is needed per capita, thus 8 million ha of additional land are needed for non-agricultural purposes worldwide. Frequently, this productive land is located near large cities. The problem of land degradation, erosion, salinization, and desertification is discussed in other chapters in this book.

How is arable cropland used? The best information is from FAO (1986) crop and yield data. Crops are grown on only 75% of the arable land. One-quarter of the total land is fallow, often because of dry conditions. On such land crops are grown only once every two or three years. On the other hand, a relatively small area of land (mainly irrigated land) can produce two or three crops each year. Approximately 220 million ha, or 15%, of the area of arable land is irrigated. More than half of this land is situated in Asia, and of this two-thirds is used for growing rice. The FAO Production Yearbooks provide average yields per hectare of various crops for almost all countries. When this information is combined with knowledge on soil and climate conditions, it is evident that the yields of the food crops in a great many countries are extremely low. These yields could be two to four or more times higher if high-yield varieties were employed along with fertilizers and pest control (fossil energy resources). Modern energy intensive farming is practiced on only 25% of all cropland and on 5% of all grassland. These percentages are calculated using information on soils and application of chemical fertilizers and crop yields.

Two important conclusions concerning land resources and food production are:

1. There is an enormous reserve of land suitable for crop production in some parts of the world.
2. A large amount of productive land is lost every year because of soil degradation and non-agricultural use.

These two subjects are dealt with extensively in the literature (PSAC, 1967; Buringh et al., 1975; FAO, 1978–1981; FAO/UNFPA/IIASA, 1982; Dudal, 1987; Pimental et al., 1987). Although the annual losses of land are considerable, much land is still available, and crop production can be increased with greater use of fossil energy resources as demonstrated in many countries during recent decades (Pimental and Pimentel, 1979). The total production of staple crops such as grains in some regions like the United States and Europe has resulted in surpluses and low prices. This may explain why some governments are not concerned with soil losses due to erosion, salinization, degradation and non-agricultural land use. Soil conservation is no better in countries with serious food shortages

(Pimentel *et al.*, 1987). However, what will happen in the future when the world population increases and these problems become more serious?

A global study is being conducted in cooperation with specialists from FAO, UNESCO, and UNEP (Buringh, 1982, 1987) to investigate these problems. This study is based on the following considerations:

1. It includes the total world land area.
2. The variability of the productivity of land for growing crops should be taken into account because it is much more serious to lose highly productive land than to lose land of low productivity.
3. Most degraded land is not totally lost for food production, since, for example, eroded cropland may be used as grassland for livestock. Thus, attention should be given to land transformation.
4. Assessments should be made at least over 25 years, in this case 1975–2000.

The main conclusions of this analysis are:

1. The area of potential productive agricultural land that will be lost is about 4%, which is less than often reported; however, the loss of highly productive land will be 22%.
2. The reserve of cropland (now used as grassland or forest land) will be reduced by 24%; however, the reserve of highly productive cropland will be reduced by 33%.
3. The total area of forest land will be reduced by 15% (0.6% per annum), which is less than that reported by most foresters (Eckholm, 1979).
4. The forest areas on productive agricultural land will be reduced by 70%.
5. The land area needed for non-agricultural purposes will increase by 50%.

This global analysis has been checked by comparing the results with those of other studies, which give higher figures for land losses. Our study had all the disadvantages of other global studies in that it does not present data on regional differences. For example, it does not show that there is no reserve of productive land in Egypt and a large reserve in Brazil. The results, however, seem to be more realistic than previous studies. The introduction of three classes of potentially productive land is a new approach.

Most studies of potential agricultural land present similar results, although regions vary slightly. The largest reserves of arable land are located in Africa and South America, where only 21% and 15%, respectively, of the potential agricultural land is currently used (Dudal, 1982). An interesting detailed assessment of the population-supporting capacity of land has been carried out by FAO/UNFPA/IIASA (1982), covering 117 developing countries. The report concluded that 19 countries cannot produce enough food for their projected population for the year 2000, even at a high level of farm inputs.

Thus, the loss of highly productive land is significant, and within one century there will be serious food production problems. Considering that agriculture has been practiced for about 10,000 years, one century is a short period.

We can calculate how much land on average is currently available per person. Approximately 0.3 ha of arable land (of which 75% is harvested), 0.6 ha of grassland, 0.8 ha of forest and woodland, and 0.1 ha of non-agricultural land is utilized per person in the world today.

II. FOOD PRODUCTION AND LAND

Of the total biomass produced annually by all plants in the world, only one percent is actually consumed by humans and livestock. Approximately 300 types of plants are used in world agriculture, yet only 24 of these are important for food production. More than 85% of our food comes from eight types of plants (mainly cereals, pulses, and tuber crops). The main crops for food production are wheat, corn and rice, which supply more than 50% of human food. For all cereals this is some 79% and for tuber crops 7% (Harlan, 1976). Therefore, cereal crops are given most attention when calculations and estimations are made in connection with studies on the world food problem.

Table 4.2 lists the total area and the average yields of the principal food crops worldwide. This table comprises all crops, which are grown on an area of more than one percent of the world's arable land. The average yields of most crops have increased during the last 15 years, and wheat, rice and maize are the principal food crops. The average yield of all cereals is about 2500 kg/ha. Taking in to account an underestimation of the average yield of 10% and a similar percentage for losses on the farm, this figure can be used for some simple calculations. More detailed information (per country) is given in the FAO production yearbooks. These figures are not accurate, some believe that the average yield figures are 10% too low (Poleman, 1977). When we want to know how much is available for human food, we must also know how much is needed for seed for the next season, and how much is eaten by animals or used as raw material by industries. Moreover, not all products are consumed, because part of the yield is lost during transportation and storage.

In most countries with low yields, the increase in the total agricultural production is often accomplished by cultivating more land. According to

4. Availability of Agricultural Land for Crops and Livestock

Table 4.2 Area (1985) and Average Yield (1969–1971 and 1985) of the Main Food Crops of the World.[a]

Crops	Area, 1985 (mha)	Yield (t/ha) 1969–1971	Yield (t/ha) 1985
Total cereals	730	2.1	2.5
Wheat	230	1.6	2.2
Rice (paddy)	145	2.4	3.2
Maize	133	2.5	3.7
Barley	79	1.9	2.2
Sorghum	50	1.2	1.5
Millet	45	0.7	0.7
Oats	26	1.8	1.8
Rye	17	1.5	1.5
Total roots and tubers	47	10.5	12.4
Potatoes	20	13.7	14.8
Sweet potatoes	8	8.1	13.9
Cassava	14	8.9	9.6
Total pulses	68	0.7	0.7
Soybeans	52	1.4	1.9
Beans (dry)	25	0.5	0.6

[a]Source: FAO (1986).

Richards et al. (1983), the land area expanded by some 432 million ha in the period 1860–1920, and by some 419 million ha in the period 1920–1978. Another cause of the increase in production is the result of the introduction of new crop varieties, application of chemical fertilizers and pesticides, mostly called the "Green Revolution" technologies. India, for example, was formerly a food-importing country but is now a cereal-exporting country. This, however, does not mean that the hunger problem in India is solved. The hunger problem is not a food supply problem but a poverty problem, a fact that is documented in recent literature and newspapers, although it has been known for more than a decade since American, Dutch, and FAO specialists have published the results of their studies (PSAC, 1967; Buringh et al., 1975; FAO, 1978–1981).

It is interesting to note what is happening with crop production in some western European countries, Japan, and the United States, where crop yields have increased about three-fold during the last decades as a result of new technologies. In the period 1981–1985 the average yield for wheat in The Netherlands was 7.1 t/ha (world average, 2.1), for rice in Japan 5.9 t/ha (world average, 3.0), and for maize in the United States 6.7 t/ha (world average, 3.4). These figures demonstrate clearly that yields of these crops can be greatly increased and that the so-called "Green Revolution" in developing countries is only a beginning, compared with what is going on in some industrialized countries. All crop production specialists agree that crop yields in many countries can be several times higher than they are now; therefore, many of them recommend not to clear more forestland for growing crops, but to concentrate on increasing crop production per hectare (Pimentel *et al.*, 1986).

In addition to the two important points mentioned earlier in this chapter (the large reserve of land suitable for crop production and the high annual losses of productive land), there is now a third point, i.e., the possibility of increasing the yield per hectare on most land that is currently used as cropland. From a technical and agricultural point of view, this is easily said, but can it be accomplished? Farmers are not stupid, and if they had the opportunity to grow more food on their land most of them would so so—if it would bring them increased profit! This leads to a new series of problems: the socioeconomic and political conditions of countries and in particular of farm families. Three conclusions are evident:

1. development of agriculture is a slow process;
2. most farmers cannot afford to invest heavily and take risks;
3. higher yields do not always mean higher profits for farmers because fertilizers and other fossil energy inputs are expensive, and prices tend to decrease.

Agriculture started about 10,000 years ago in the Middle East. Before this, and for nearly two million years, the people of the world obtained their food as hunter-gatherers. Thus, for 99% of their existence, humans have survived without agriculture! I estimate that in general about 80 ha of land were needed to feed one person in a hunter-gatherer society.

I believe that it is worthwhile to examine food production in the past and potential for the future. The growth of agriculture in many western European countries appears to have gone through seven stages. Similar stages can be recognized in other countries.

The seven stages in the development of agriculture in nations are characterized by the following:

1. land rotation or shifting cultivation or bush-fallow (a few years of crop cultivation followed by 10, 20, or more years of regrowth of forest);
2. low traditional (mainly one crop-year and one fallow-year);
3. moderate traditional (mainly two crop-years and one fallow-year);
4. improved traditional (continuous cropping of cereals in rotation with legumes, root crops or grass);
5. moderate technological (continuous cropping with application of some chemical fertilizers and simple mechanization);
6. highly technological (similar to 5, but using more fertilizers and increased mechanization);

7. specialized technological (similar to 6, but with heavy applications of fertilizers, full mechanization, heavy applications of pesticides).

Each of these stages includes more factors; for example, in traditional production, more than one crop is grown at the same time on the same plot of land (so-called mixed cropping). In general, traditional production modes have little or no input from outside the farm, whereas the technological modes depend mostly on inputs from outside the farm, primarily fossil energy. This raises the interesting question of what will happen to world food production as fossil fuel resources decline (Pimentel and Pimentel, 1979; Chapman and Barker, 1987).

The land rotation or shifting cultivation mode of production is at present of minor importance from the point of view of total world food production, although some 200 million people in the tropics still depend on it. Farming with the traditional mode of production results in low yields, because it is limited by the natural supply of nitrogen. In this low mode, little manure is available. In the moderate traditional mode more manure is used, because of a somewhat improved animal husbandry system. The improved traditional mode of production is a transitional stage to the technological modes, when much more manure is available as fallow-years are replaced by growing feed and forage for farm animals. Giving up the fallow periods means that much more land becomes available for growing food crops. As soon as technological modes of production are used, external inputs become essential to the system. As tractors replace draught animals, more land becomes available for food production, as approximately one hectare of land is needed to produce feed and forage for one horse. Again, however, more fossil fuel resources are needed.

Each consecutive production mode requires more labor input and/or capital and fossil fuel, which means higher production costs. The important point is that more people can be fed from the same land area. In most countries several of these modes of agricultural production are used simultaneously. Table 4.3 lists the world areas in which each mode of agricultural production is currently practiced. The information is based on studies of the modes of production and some estimates. Note that approximately 75% of all cropland is traditional farming, whereas some technology is used on only 25%. In Table 4.3 the area of arable land (medium productivity) needed per capita for each mode of production was calculated (Buringh, 1984). Currently, approximately 0.3 ha of arable land (of which 75% is harvested) is used per capita, which also demonstrates that the traditional modes of production are mostly practiced at the present time. Moreover, this is a clear indication that there are possibilities to improve food production through better farm management.

The transition from one mode of production to the next is always a slow process, except may be for the last ones. Many factors influencing this process are mentioned in the literature on the history of agricultural

Table 4.3 Yields in Grain Equivalents and Percentages of Cropland for Various Modes of Agricultural Production in the World

Mode of production	Yield[a] (kg/ha)	Cropland (%)	Average area of arable land needed (ha/capita)
Land rotation	?	2	2.65
Low traditional	800	28	1.20
Moderate traditional	1,200	35	0.60
Improved traditional	2,000	10	0.17
Moderate technological	3,000	10	0.11
High technological	5,000	10	0.08
Specialized technological	7,000	5	0.05

[a]Yields are expressed as grain equivalents calculated over the harvested area plus the fallow land.

development. These include, for example, the rise of cities, transportation, and trade. International trade and transportation associated with food began about 850 years ago when the Baltic states, Poland and the Ukraine exported grain to western Europe by ship. Since 1950, there has been a rapid rise in international trade and transportation of food products for those who can pay!

Can enough food be produced for the growing world population? The question is answered in various studies carried out by American and western European specialists and by some international organizations such as FAO and the World Bank. Most of these studies report that enough food can be produced, even for a world population of as many 12 billion people. However, to accomplish this, socioeconomic and political conditions have to be changed in a great many countries, before new agricultural production technologies can be employed throughout the world. Since there is at present a surplus of food products in the Western world (particularly grains) and a shortage in the USSR and in many developing countries, this task will not be easy. Attempts to increase food production in developing countries through foreign aid have not always been successful. Since calculations on food availability are mainly based on cereals, mistakes can be made for countries in which root and tuber crops are most important food crops, as shown in Fresco's (1986) study on Zaire.

Increasing food production involves the application of chemical fertilizers, including lime, and breeding of new crop varieties. Are adequate fertilizers available for future agriculture? This question has been studied by many fertilizer specialists, who conclude that there will be no shortage for several decades (Aller, 1977; Roth, 1978; Stangel, 1976). Breeding new crop varieties is essential for new production technologies. Unfortunately, little attention has been given to breeding important food crops like millet, sorghum, and cassava for developing countries.

Maximum crop production can be estimated under the most favorable conditions. This includes sufficient water and minerals, no diseases, no insect pests, and no weeds. The maximum production depends only on solar irradiance intercepted by the crop, on the temperature of the environment, and on the physiological properties of the crop. The maximum total biomass production and grain production can be calculated. Clearly, maximum crop production does not exist in nature, therefore, we must take into account the actual water and nutrient availability in the soil at specific sites, which makes the model calculation extremely difficult. Daily climatic conditions (solar radiation, temperature, precipitation, etc.) differ in various parts of the world. The calculation must also take into account nutrient availability, weeds, insects and plant pathogens. The production of a specific crop cultivated at a specific site with known weather and soil conditions and farming technique can, therefore, be simulated using models. These new techniques employ computers as tools for modeling.

In this way maximum crop production for a specific crop and for a specific site can be calculated. The result can be verified by comparing it with crop yields obtained on experimental fields under the most favorable conditions. (For more details see Keulen and Wolf (1986), which describes all the factors dealt with in this hierarchical, dynamic simulation model with several sub-models.) The first part of this model is used in a simplified way to calculate maximum average yields of wheat in several countries (Buringh, 1987). The conclusion is again that yields in many countries can be increased by three or more times. This is not new. What is new is that we now have a method to obtain more reliable, quantitative figures, which are important for quantified land evaluation (Driessen, 1986) and quantified calculation of potential world food production.

Although crop yields can be increased, this does not mean that farmers can make the necessary profit and desire to increase yields. For example, in the period 1900–1980 the gross income of wheat farmers in the Netherlands hardly increased, despite the higher average yields and higher wheat prices, because of the inflation rate (Buringh, 1985).

III. LIVESTOCK PRODUCTION AND LAND

Livestock products account for about 7% of the world food supply (Harlan, 1976). However, many animals also supply draft power and valuable nonfood products such as wool, bones, etc. Table 4.4 reports the total number of livestock worldwide. These animals are mainly fed by forage from grasslands and forest lands. In addition, about 13% of the total cereal production is fed to livestock (FAO, 1981). Chickens, for example, eat mainly products of arable land, while cows and the like eat grass along

Table 4.4 Numbers (Millions) of Livestock in the World in 1985[a]

Livestock	No.	Livestock	No.
Cattle	1,269	Asses	41
Sheep	1,122	Camels	17
Pigs	791	Mules	15
Goats	460	Chickens	8,287
Buffaloes	129	Turkeys	216
Horses	65	Ducks	169

[a]Source: FAO (1986).

roadways. Wild animals, not counted in statistics on livestock production, also eat grass. This demonstrates that it is impossible to calculate which part of the livestock comes from which type of land use and exactly how much of the world's grassland contributes to our food supply.

The approximately 3,200 million ha of grassland in the world account for about 24% of the total land area and more than twice the present arable land. The productivity of grassland is highly variable. Some grassland can support one cow or one horse for one year on one hectare. On the other hand, 20 or more hectares of grasslands are needed to feed one cow in other situations.

As mentioned, some of the land currently in grass is suitable for cultivating crops. Most grassland is natural–only 5% receives fertilizers (Buringh, 1982). Often grassland is community property, owned by inhabitants of a village or by a tribe. This seldom encourages care and sound management.

The forage on grassland is eaten by both domestic and wild animals. Although the fertility is increased by the manure from these animals, it is generally low because the natural supply of nitrogen is short.

Many types of forage grasses exist with varying nutritional value. In the tropics some grasses can produce more than 60 t/ha/yr of dry matter. In the temperate regions a high level is 20 t/ha/yr. Unfortunately, many grasses are grown on shallow, poor soils or in regions with long dry periods and, thus, the yields are only a few tons per hectare of dry material per year. Some grasslands are regularly flooded, and grasslands in some river valleys may be too wet and others have a ground water table that is too high. The evidence suggests that most grassland in the world could produce several times greater yields than at present, if pasture management were improved. This is clearly demonstrated on experimental fields in countries all over the world. As suggested by field experiments, grass production could be increased serveral times even in dry regions like the Sahel in Africa by applying fertilizers, particularly phosphorus (Penning de Vries and Djitèye, 1982).

Damage and loss of grassland due to soil erosion is less serious than on arable lands. Overgrazing results in soil erosion, which is a serious problem in many countries. Moreover, much grassland in semi-arid regions is damaged when attempts are made to transform grassland into arable land. When this occurs it is extremely difficult to restore the natural grass vegetation on the land. Without vegetation this land is exposed to wind erosion.

IV. CONCLUSION

At present, arable land is clearly sufficient to produce adequate supplies of food for humans and part of forage for livestock. In the future, employing currently available technologies, we should be able to produce sufficient food to feed a population of 12 billion a minimal diet.

The primary problem for increasing the supply of food in most countries is not land and technology but socioeconomic and political will. Most governments favor low-priced food for consumers, which keeps farmers poor and forces them to use simple management practices. Farmers could be provided with the necessary incentives to produce more and better food if the governments adopted appropriate policies.

Often insufficient attention is given to the development of agriculture and the resources on which agriculture depends. This is especially true in protection soil and water resources for use in crop production. Few nations in the world have sound, effective policies for conserving soil and water resources. Reserves of good, productive cropland are rapidly decreasing, and within one century a world shortage of cropland is projected. In several countries cropland shortages already exist.

Agriculturalists have documented that it is more economically and ecologically sound to intensively manage highly productive cropland and make it more productive than to clear forestland and convert it into agricultural land. Marginal land requires larger amounts of energy inputs to achieve minimal yields than highly productive land.

When suggesting that more people can be fed on existing cropland by increasing the input of fertilizers, irrigation water, pesticides, and other inputs, this clearly implies heavy dependence on fossil fuels, particularly oil and gas supplies. Shortly after the year 2000 more than one-half of the petroleum reserves will be consumed, and the remaining supplies will be rapidly depleted. Currently, no low-priced substitute liquid fuels are available. Thus, increasing food production with a greater dependence on fossil will face serious difficulties in the future (Chapman and Barker, 1987).

Will the energy resource limitation be overcome to allow for the development of agriculture? The situation is not encouraging, and development is always a difficult, slow process even without serious resource limitations. Certainly before agriculture can be improved, the local environmental conditions must be known in detail. Then with proper incentives farmers who are provided with high-yield varieties and fertilizers and other supporting inputs can make their land highly productive. The various stages of agricultural development (Table 4.3) cannot be neglected.

In summary, although arable land appears adequate to produce the needed food for the next 25 years, farmers must have sufficient fertilizers, irrigation, pesticides, and other energy inputs. Potential problems are projected with fossil fuel supplies. Equally important, the future of agriculture depends on socioeconomic and political support and incentives for farmers. Let us hope that land, water, energy and biological resources are managed effectively for socioeconomic development and the future of humankind.

REFERENCES

In addition to the literature cited, some references to recent books and articles dealing with the problem concerned are presented for readers who are interested in more information.

Aller, G. R. (1977). The world fertilizer situation *World Dev.* **5**(5/7), 525–536.

Blaxter, K., and Fowden, L. (1985). "Technology in the 1990s: Agriculture and Food." Royal Soc., London.

Brown, L. R., and Wolf, E. C. (1984). "Soil Erosion: Quiet Crisis in the World Economy," Worldwatch Pap. No. 60. Worldwatch Inst., Washington, D.C.

Brown, L. R., Chandler, W., Flavin, C., Postel, S., Starke, L., and Wolf, E. (1984). "State of the World 1984." Norton, New York.

Brown, L. R., Chandler, W. U., Flavin, C., Pollock, C., Postel, S., Starke, L., and Wolf, E. (1985). "State of the World 1985." Norton, New York.

Buringh, P. (1982). Potentials of world soils for agricultural production. *Trans. Int. Congr. Soil Sci., 12th, New Delhi* Vol. 1, Plenary Sess. Pap., pp. 33–41.

Buringh, P. (1984). The capacity of the world land area to produce agricultural products. *Options Méditerranéennes, Cihean IAMZ, Zaragossa* **84/1**, 15–33.

Buringh, P. (1985). The land resource for agriculture. *In* "Technology in the 1990s: Agriculture and Food" (K. Blaxter and L. Lowden, eds.), pp. 5–14. Royal Soc., London. (Also, *Philos. Trans. R. Soc. London, Ser. B* **310**, 151–159.)

Buringh, P. (1987). Bioproductivity and land potential. *In* "Biomass: Regenerable Energy" (D. O. Hall and R. P. Overend, eds.), pp. 27–46. Wiley, New York.

Buringh, P., and Dudal, R. (1987). Agricultural land use in space and time. *In* "Land Transformation in Agriculture" (M. G. Wolman and F. Fournier, eds.), pp. 9–44. SCOPE, Paris and Wiley, New York.

Buringh, P., van Heemst, H. J. D., and Staring, G. J. (1975). "Computation of the Maximum Food Production of the World." Agric. Univ., Wageningen, Netherlands. (Also, *in* H. Linnemann *et al.*, "MOIRA, Model of Agricultural Relations in Agriculture," Chap. 2. North-Holland Publ., Amsterdam, 1979.)

Chapman, D., and Barker, R. (1987). Resource depletion, agricultural research, and development. *U.S. Natl. Acad. Sci.–Czech. Acad. Sci. Workshop Agric. Dev. Environ. Res., Ceske Budejovice, Czech.*

4. Availability of Agricultural Land for Crops and Livestock 83

Driessen, P. M. (1986). Quantified land evaluation (QLE) procedures, a new tool for land use planning. *Neth J. Agric. Sci.* **34**, 295–300.

Dudal, R. (1981). Land resources and production potential for a growing world population. *Proc Int. Potash Inst., 12th, Bern,* pp. 277–288.

Dudal, R. (1982). Land degradation in world perspective. *J. Soil Water Conserv.* **37**(5), 245–249.

Dudal, R. (1987). Land resources for plant production. *In* "Resources and World Development" (D. J. McLaren and B. J. Skinner, eds.), Dahlem Konferenzen, pp. 659–670. Wiley, New York.

Dudal, R., *et al.* (1982). Land resources for the world's food production. *Proc. Int. Congr. Soil Sci., 12th, New Delhi.*

Eckholm, E. P. (1976). "Losing Ground." Norton, New York.

Eckholm, E. P. (1979). "Planting for the Future: Forestry for Human Needs," Worldwatch pap. No. 26. Worldwatch Inst., Washington, D.C.

FAO. (1978–1981). "Reports on the Agro-Ecological Zones Projects," World Soil Resources Rep. No. 48, Vols. 1–4. FAO, Rome.

FAO. (1981). "Agriculture: Toward 2000." FAO, Rome.

FAO. (1986). "FAO Production Yearbook," Vol. 39. FAO, Rome.

FAO/UNESCO. (1974–1981). "Soil Map of the World, 1:5,000,000," Vols. I–X. UNESCO, Paris.

FAO/UNFPA/IIASA. (1982). "Potential Population Supporting Capacities of Land in the Developing World." FAO, Rome.

Fresco, L. O. (1986). "Cassava in Shifting Cultivation." Royal Trop. Inst., Amsterdam.

Harlan, J. R. (1976). The plants and animals that nourish man. *Sci. Am.* **235**(3), 88–97.

Keulen, H. van, and Wolf, J. (1986). "Modelling of Agricultural Production: Weather, Soils and Crops." Pudoc, Wageningen, Netherlands.

McLaren, D. J., and Skinner, B. J., eds. (1987). "Resources and World Development: Energy and Minerals, Water and Land," Dahlem Konferenzen. Wiley, New York.

McMains, H. J., ed. (1978). "Alternatives to Growth: The Engineering and Economics of Natural Resources Development." Ballinger, Cambridge, Massachusetts.

Penning de Vries, F. W. T., and Djitèye, M. J. (1982). "La Productivité des Pâturages Sahéliens. Une étude de sols, des végétations et de l'exploitation de cette ressource naturelle." Pudoc, Wageningen, Netherlands.

Pimentel, D., and Pimentel, M. (1979). "Food, Energy, and Society." Arnold, London.

Pimentel, D., Dazhong, W., Eigenbrode, S., Lang, H., Emerson, D., and Karasik, M. (1986). Deforestation: interdependency of fuelwood and agriculture. *Oikos* **46**, 404–412.

Pimentel, D., Allen, J., Beers, A., Guinand, L., Linder, R., McLaughlin, P., Meer, B., Musonda, D., Perdue, D., Poisson, S., Siebert, S., Stoner, K., Salazar, R., and Hawkins, A. (1987). Word agriculture and soil erosion *BioScience* **37**, 277–283.

Poleman, T. T. (1977). World food: myth and reality. *World Dev.* **5**(57), 383–394.

PSAC (President's Science Advisory Committee). (1967). "The World Food Problem." Washington, D.C.

Richards, J. F., Olson, J. S., and Rotty, R. M. (1983). "Development of a Database for Carbon Dioxide Releases Resulting from Conversion of Land to Agricultural Uses." Oakridge Assoc. Univ., Oak Ridge, Tennessee.

Roth, W. (1978). "Nutzbare Rohstoffvorräte für die Düngung," No. 22, pp. 89–107. Sonderreihe, Umwelttagung. Univ. Hohenheim.

Scientific American. (1976). Food and agriculture. Vol. 235, No. 3.

Stangel, P. J. (1976). World fertilizer reserves in relation to future demand. *In* "Plant Adaptation to Mineral Stress in Problem Soils." Workshop Beltsville, Ithaca, New York.

UNESCO. (1985). *Unescokoerier,* Maandblad No. 144, p. 11. Keesing, Deurne-Antwerpen, Belgium.

5

Land Degradation and Its Impact on Food and Other Resources

R. Lal

Department of Agronomy
The Ohio State University
Columbus, Ohio

I. Introduction
II. Land Degradation
    A. Vegetation Degradation
    B. Soil Degradation: Definition and Processes
    C. Environmental Degradation
    D. Climatic Change
III. Technological Options for Minimizing Soil Degradation
    A. Soil Erosion Management
    B. Preventing Laterization
    C. Salinity Control and Water Management
    D. Fertility Maintenance
IV. Land Degradation and World Food Production
V. Need to Restore Productivity of Degraded Lands
VI. Conclusions
    References

I. INTRODUCTION

Despite the ever-increasing demand for prime agricultural land, the earth's land resources are not only finite but also nonrenewable relative to present civilization. The total land area of the earth is 134.01 × 10^8 ha of which 11% (14.39 × 10^8 ha) is cropped, 21.5% is under meadows and pastures, 29.7% is under forests, and 37.8% is potentially available for future production. The potentially productive land is marginal for agricultural use because most of it is either inaccessible, too steep, too shallow, or is in regions with too little or too much water, and other essential ingredients for crop production are not available. Presently most cropped areas lie in regions with favorable rainfall regimes. For example, 40% of the cropped area is in regions with mean annual rainfall of 1000–1500 mm, another 40% in regions of 500–1000 mm rainfall, 15% in 250–500 mm annual rainfall, and only 5% in regions with mean annual rainfall of less than 250 mm (Fukuda, 1976).

Considering the total world population of five billion in 1986, the per capita arable land amounts to a meager one-third of a hectare. It is estimated that the minimum per capita arable land needed for an adequate diet is 0.5 ha; assuming, of course, that the land is of good quality and that its productive capacity will be conserved or preserved.

Conserving soil's productive capacity, however, is easier said than done. Conserving or preserving of soil's productivity must be defined in terms of resource use, i.e., managing the soil to yield maximum sustainable benefits to mankind. It is the sustainable utilization of the soil's productivity that is at stake. In 1982, for example, the average annual output of major food crops reached an estimated 1830 million metric tons (Paulino, 1986). The mean per capita crop production was 400 kg, although the per capita production for developed agricultural economies was three times (780 kg) that of the developing economies (260 kg). With ever-decreasing land resources, can per capita food production be sustained at 400 kg? What are the possibilities for improvements?

Maintaining the per capita arable land area at the minimum level of 0.5 ha is being challenged by the rapid increase in world population from five billion in 1986 to a predicted 6.2 billion in 2000, 9.3 billion in 2055 and eventually to a stable 10.5 billion in 2110 (Salas, 1981; Dudal, 1982). If the arable land area is maintained at 14.39 × 10^8 ha, the per captia arable land will progressively decline from 0.33 now to 0.23 ha, 0.15 ha, and 0.14 ha in 2000, 2055, and 2100, respectively. Assuming, of course that no new land is brought under cultivation and existing land is prevented from being degraded. The production of the minimum dietary requirements from 0.14 ha of per capita arable land can be met by technological innovations that may bring about a quantum jump in food production. If not, widespread malnutrition is inevitable. Grigg (1985) estimated that in 1960 about 10–15% of the world population was undernourished. In the mid-1970s (1975–1977), some 62–300 million people were undernourished, and at least 455 million people had food intakes below the 1.2 times the basal metabolic rate (Grigg, 1985).

Like the availability of land, the population increase is uneven. Population growth rates for 1980–85 were 3.01, 2.30, 2.20, and 1.20%, respectively, for Africa, Latin America, South Asia, and East Asia (McNamara, 1985). The population of sub-Saharan Africa, a region with a perpetual food crisis, is expected to increase to 678 million in 2000, 1,202 million in 2025; 1,658 million in 2050, and 2,041 million in 2100. Such an imbalance between land and people makes the degradation processes self-reinforcing.

The land–population scenario is faced with three challenges. First, the available land resources are unevenly distributed. Second, regions with high demographic pressures often have the least available land reserves, and technological inputs are also limited. Consequently, the existing land is subject to severe degradation. Third, available technological options often do not consider the socioeconomic constraints. The rate of land degradation is often accelerated when technological options are considered without involving those for whom the technology is supposed to serve.

II. LAND DEGRADATION

Degradation is a vaguely used term and implies decline in the quality of an ecosystem through its misuse by humans. The term "land degradation," therefore, implies deterioration in quality and capacity of the life-supporting processes of land. Land, in the context of this chapter, means an ecosystem comprising micro- and mesoclimate, vegetation, soil, and water resources. Land degradation, therefore, means deterioration or decline in the productive capacity of an ecosystem through adverse changes in the life-supporting processes of its *climate,* vegetation, soil or water resources. It is often argued that for about 14.49×10^8 ha of currently cropped land, an additional 20×10^8 ha of once biologically productive lands have been rendered unproductive through irreversible degradation of their life-support systems (UNEP, 1986). FAO/UNEP (1983) estimated that at present 5–7 million ha of cultivated area (0.3–0.5%) are being lost every year through soil degradation. The projected loss by 2000 is feared to be 10 million ha annually (0.7% of the area presently cultivated). If these estimates are correct, planners must develop systems to restore the disturbed ecosystems and ensure that productivity of existing lands is conserved, and enhanced.

Without any anthropogenic perturbations, different components of an ecosystem are in steady-state equilibrium. Alterations in one component, vegetation, for example, can cause drastic changes in others, e.g., soil, microclimate, and water resources. If the alterations are slight and temporary, the system usually can recover its original state when the perturbation is removed. The system may, however, undergo irreversible changes, if the degree or duration of perturbations are drastic and intense.

Not all alterations are necessarily bad. Simplification of a natural and a complex ecosystem to an agroecosystem results in higher net productivity. Sustaining the higher productivity level, however, presupposes adoption of improved managerial skills and technological innovations, and continued efforts to replace the essential ingredients harvested through products for human use. The key to successful sustained land use lies in balancing the ecological equilibrium for all components (i.e., climate, vegetation, water, and soil) and for all processes (energy, nutrients, flora and fauna, organic matter). It is failure to preserve equilibrium, through neglect and greed, that causes degradation.

A. Vegetation Degradation

Removing of the native vegetation cover was an inevitable consequence of adopting settled agriculture, which may have begun as early as 20,000 years ago (Goudie, 1981). Mass-scale deforestation and altering of vegetation to simplified agricultural ecosystems caused by rapid and dramatic increases in human population, however, have drastically altered the landscape. Landscapes of Asia, the Middle East, and Europe have long been transformed, as have more recently been the cases with North America, Australia, and Africa. Removal of vegetative cover alters the water and energy balance, disrupts in cycles of major plant nutrients, biological activity, and diversity of a system (Figure 5.1). Some of these alterations, no doubt, have adverse ecological effects. Yet, agriculture cannot be practiced without replacing relatively unproductive vegetation with a simplified, but more productive, system.

Deforestation in the tropics is one of the major environmental issues of modern times. It is feared by some that the tropical rainforest is being removed at an annual rate of about 11 million ha (Scott, 1978; Lal, 1986b). Some of the forested lands converted for cropping or pastures have been used judiciously and have sustained economic production without causing severe ecological problems. Other lands, through misuse and greed and lack of knowledge of soil, have been replaced soon after clearance with barren and unproductive lands where lush, green forest once prevailed. The forest cover is slow and difficult to reestablish if the ecosystem has been drastically disturbed.

5. Impact of Land Degradation

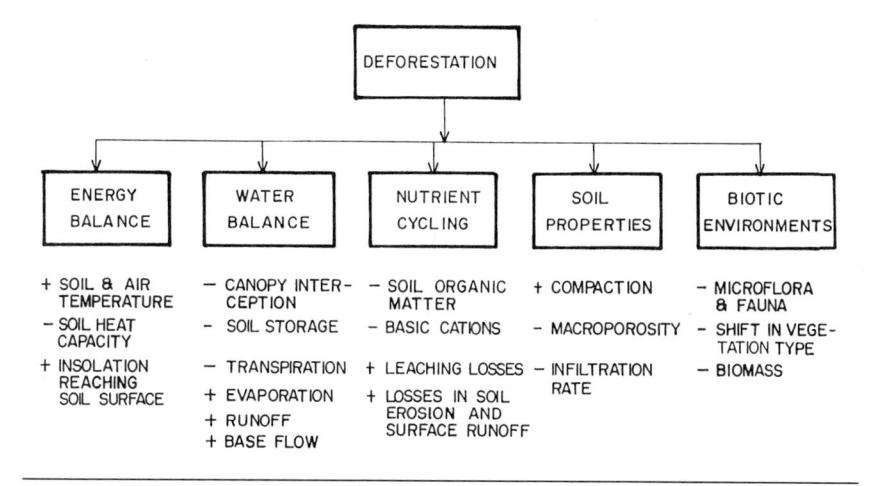

```
                          DEFORESTATION
    ┌─────────────┬──────────────┬──────────────┬──────────────┐
 ENERGY         WATER        NUTRIENT         SOIL          BIOTIC
 BALANCE       BALANCE       CYCLING       PROPERTIES    ENVIRONMENTS
```

Energy Balance	Water Balance	Nutrient Cycling	Soil Properties	Biotic Environments
+ SOIL & AIR TEMPERATURE	− CANOPY INTERCEPTION	− SOIL ORGANIC MATTER	+ COMPACTION	− MICROFLORA & FAUNA
− SOIL HEAT CAPACITY	− SOIL STORAGE	− BASIC CATIONS	− MACROPOROSITY	− SHIFT IN VEGETATION TYPE
+ INSOLATION REACHING SOIL SURFACE	− TRANSPIRATION	+ LEACHING LOSSES	− INFILTRATION RATE	− BIOMASS
	+ EVAPORATION	+ LOSSES IN SOIL EROSION AND SURFACE RUNOFF		
	+ RUNOFF			
	+ BASE FLOW			

Figure 5.1 Ecological effects of removal of deforestation on soil, climate, and vegetation. Minus sign indicates decrease; plus sign, increase.

1. Consequences of Deforestation and Altered Vegetation

The magnitude and trends of alterations in different parts of an ecosystem depend on deforestation methods, land use, and management systems. Methods of deforestation, manual versus mechanized, significantly affect soil properties, runoff, and erosion, and forest regeneration (Lal and Cummings, 1979, Ghuman and Lal, 1987). The data in Table 5.1 and Figures 5.2 show that deforestation increased the total water yield from a watershed and also increased surface runoff and soil erosion. The increased water yield was more from increase in surface than subsurface flow. Streams flowed for a shorter time during the first year after clearing than

Table 5.1 Effects of Methods of Deforestation on Runoff and Soil Erosion (Lal, 1981)

Treatment	Runoff (mm/yr)	Soil erosion (t/ha/yr)
Traditional farming	3	0.01
Manual clearing	35	2.5
Shear blade	86	3.8
Tree pusher, root-rake	202	17.5

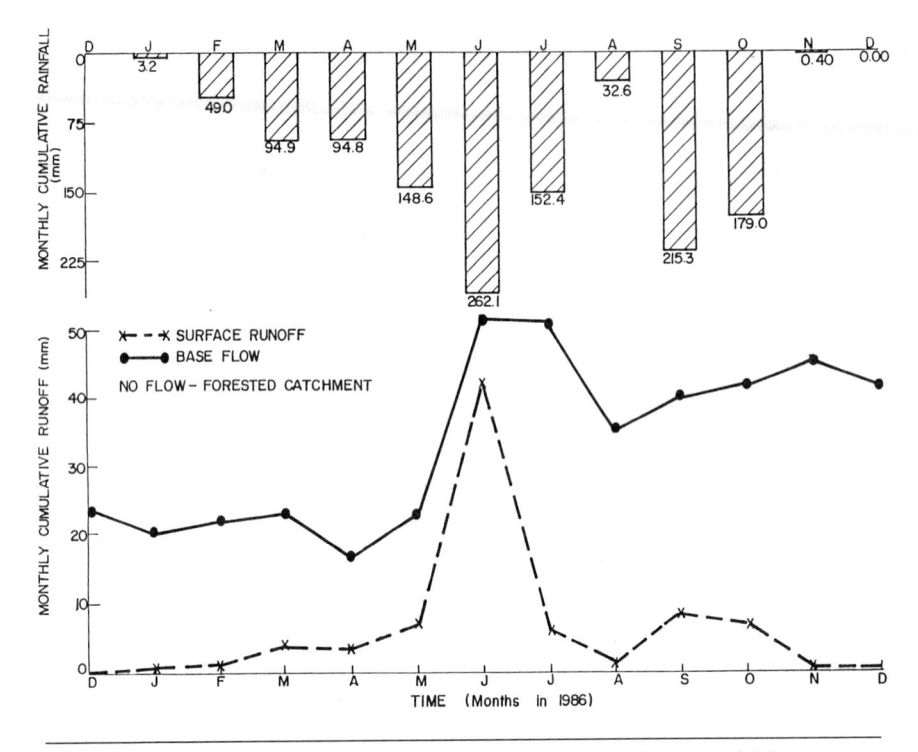

Figure 5.2 Effects of deforestation in the tropics on total water yield in 1986 from a watershed growing seasonal crops at the International Institute of Tropical Agriculture, Ibadan, Nigeria.

in later years. Furthermore, runoff and soil erosion were greater with mechanized rather than manual clearing. There were also differences in runoff and soil erosion among two methods of mechanized land clearing. The method that least disturbs the soil surface causes less runoff and less erosion. Clearing of the deep rooted perennial vegetation and replacement with shallow rooted annual crops and pastures causes major disturbance of water, salt, and energy balance.

The conversion of forest vegetation to savanna, called the man-made savanna, has been the subject of much research and debate. Fire, both natural and human-induced, has been the single-most important cause of degeneration of forest into savanna. Fire has been the most ancient agricultural tool, and it is still used widely to clear the forest, manage the residue, and improve pasture quality. Although controlled burning can be a useful tool, untold damages have been done by uncontrolled voluntary fires.

Effects of fire on soil and vegetation depend on the fire's intensity and duration. Depending on the biomass available, the temperature may be as high as 500–1000°C (Lal, 1987a; Gouldie, 1981). Soil properties, especially soil moisture retention and water-transmission properties, are greatly altered by fire. Excessive burning causes loss of nitrogen and other nutrients by volatilization and makes nutrients available for leaching during the rainy season. Le Houerou (1977b) estimated that bush fires in the African grasslands burn more than 80 million tons of forage per year. Not only could that amount feed about 20 million cattle every year, but considerable amounts of nutrients released in the ash are either washed away or leached from the root zone. Development of savannas in mid-latitudes and shrublands are attributed to repeated cycles of natural or man-caused fires. Fire alters vegetation through evolution of some fire-tolerant species. Fire helps break seed dormancy and stimulates vegetative regrowth of many species. Repeated cycles of fire are responsible for replacement of fast- growing, light-loving trees and shrubs with fire-tolerant grasses and scrub vegetation—the pyrophytes. Deep-rooted trees that help in nutrient recycling are eliminated from a fire-prone ecology. Some people maintain that in some parts of the African Sahel the transition from wooded savanna with cattle to sand dunes with camels has occurred within living memory (Cloudsley-Thompson, 1977).

One of the environmental consequences of fire is ejection of smoke and other greenhouse gases into the atmosphere. Bryson (1947a,b) and Kovda (1980) estimated that about 100–250 million tons of smoke are annually ejected into the atmosphere. The smoke comprises a lot of pollutants, e.g., CO, N_2O, and methanol (Crutzen et al., 1985). Long-term global effects of these aerosols are not known, but they may contribute to the greenhouse effect.

Excessive grazing is another factor responsible for vegetation degradation. Uncontrolled grazing depletes vegetation, changes floral composition, causes soil compaction, and accelerates runoff and erosion. Nowhere else are the adverse effects of uncontrolled and excessive grazing more obvious than in the West African Sahel. The cattle population in the Sahel has increased drastically since 1940. Gallais (1979) and NRC (1984) estimated that the western Sahel witnessed a five-fold increase in cattle during the 25 years preceding the 1968 drought (Table 5.2). Grazing can cause drastic changes in vegetation, soil, and hydrological properties. In East Africa, Pereira et al. (1961) reported the effects of grazing of planted pastures on soil structure, runoff, and erosion. They observed that under violent rainfall, a characteristic of tropical climates, sealing of the exposed and trampled soil surface resulted in heavy runoff and severe sheet and gully erosion. In fact, rainfall infiltration rates in heavily grazed pastures were much lower than in soil under seasonal crops (Table 5.3). Recent studies on grazing at the International Institute of Tropical Agriculture also have shown more soil compaction, low infiltration rate, and

Table 5.2 Sahelian Cattle Populations[a]

Country	\multicolumn{4}{c}{Number of cattle (thousands)}
	1940	1968–1970	1974	1978
Chad	—[b]	4,630	3,250	3,600
Mali	1,174	5,300	3,640	3,800
Mauritania	850	2,100	1,175	1,200
Niger	754	4,200	2,200	2,850
Senegal	440	2,615	2,318	2,500
Upper Volta	491	2,900	2,300	2,600

[a]Source: Gallais (1979); National Research Council (1984).
[b]Not available.

higher runoff and erosion than under grain crops (Figure 5.3). Grazing and overstocking are severe causes of vegetation and soil degradation in Australia. Perrens (1986) reported that in that country out of a total grazing area of 3.4 million km² in arid region, 55% of the area needs restorative measures for land degradation. He found that there are about 1.337 million km² of grazing land in the non-arid region, of which 36% require restorative measures against severe degradation. About 51% of the total land area of 5.2 million km² in Australia requires treatments for land degradation (Perrens, 1986).

Another important consequence of deforestation in the tropics is the increase of the atmospheric concentration of CO_2, which has been increasing steadily (Bach, 1986; see Figure 5.4). Deforestation may add CO_2 by releasing the vast amount of carbon immobilized in the biomass. Fur-

Table 5.3 Effects of Grazing and Food Crop Production on Soil Structure and Rainfall Acceptance[a]

Parameter	In grass after 4 years of heavy grazing, 1954			In arable soil after three crop seasons, 1956		
	Rhodes	Cenchrus	Veld	Rhodes	Cenchrus	Veld
Acceptance of a storm 2.5 cm in 10 min (%)	19.3	24.9	28.5	39.3	32.7	34.5
Rainfall acceptance rate (cm/h)	2.95	3.78	4.34	5.99	4.98	5.26
Percolation (cm/h) under 1.25 cm static head	6.86	9.91	13.21	36.75	24.89	27.94

[a]Source: Pereira et al. (1961).

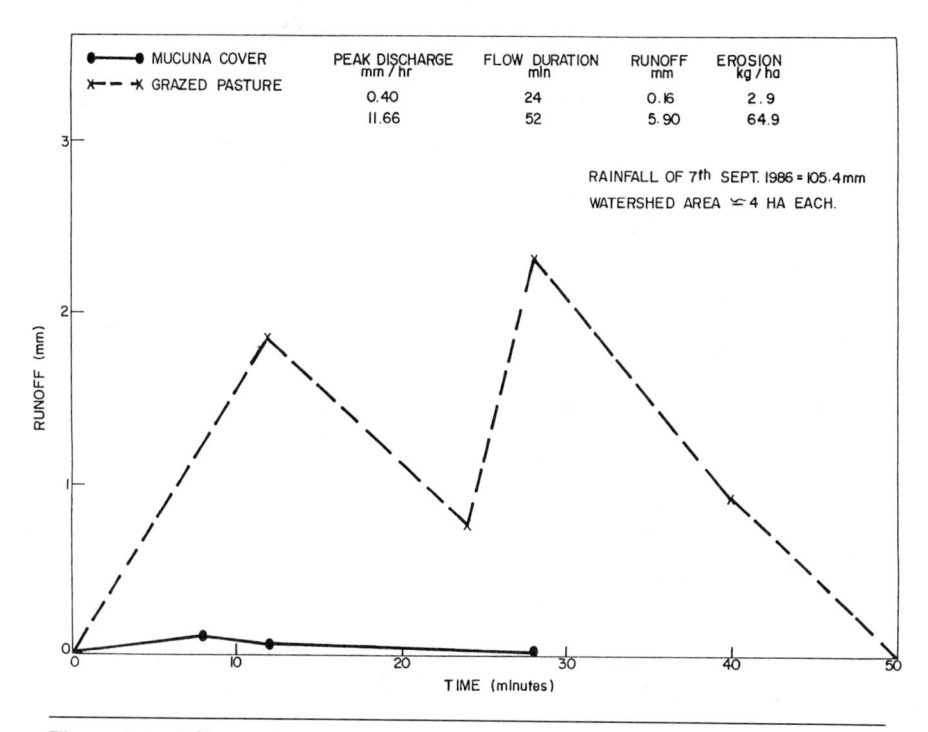

Figure 5.3 Effects of grazing on runoff and soil erosion from an Alfisol at the International Institute of Tropical Agriculture, Ibadan, Nigeria (Lal, 1986b).

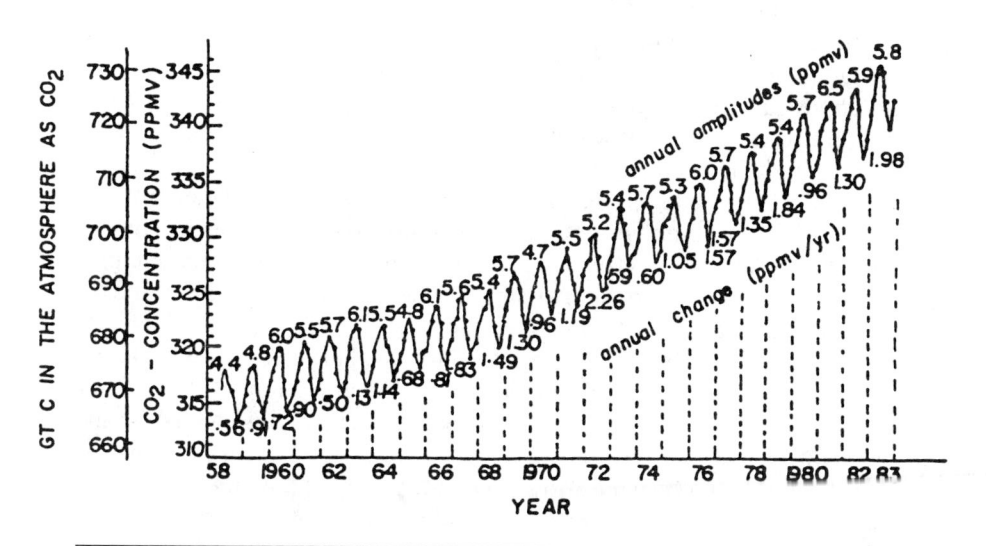

Figure 5.4 Increase in atmospheric CO_2 concentration on Mauna Loa, Hawaii, 1958–1983 (Bach, 1986).

thermore, the carbon normally fixed by the biomass as a photosynthate is released into the atmosphere. Woodwell *et al.* (1978) believed that deforestation in the tropics could be a major source of CO_2 being added into the atmosphere. Recent studies, however, indicate that CO_2 is released not only by the fire but by oxidation of biomass following deforestation.

2. Management Strategies

Although tropical forest resources should be preserved, some countries have no choice but to expand their land bases by converting tropical rainforest to arable land. Planners in these countries should be provided the needed data on appropriate land use and ecologically compatible soil and crop management practices. Appropriate research information is available but requires validation and on-site adaptation (Lal *et al.*, 1987).

There are at least two other strategies to reduce demand for deforestation in the tropics. One is to substantially increase production from existing lands by adopting high-yielding but resource-efficient technologies. Technological options are available that can significantly increase production (Sanchez *et al.*, 1982; Lal, 1987b). The second strategy would be to restore the productivity of degraded lands and bring them into production.

B. Soil Degradation: Definition and Processes

Soil degradation, caused partly by deforestation and degradation of vegetation but mostly by soil misuse and over-exploitation, is a severe global problem. Soil degradation is defined as "the decline in soil quality caused through its use by humans. Soil degradation includes physical, biological and chemical deterioration such as decline in soil fertility, decline in structural condition, erosion, adverse changes in salinity, acidity or alkalinity, and the effect of toxic chemicals, pollutants, or excessive inundation" (UNEP, 1982). Specifically, soil degradation is the diminution of the soil's current and/or potential capability to produce quantitative or qualitative goods or services as a result of one or more degradative processes. Soil depletion, in comparison, is a less drastic process that means the eluviation of nutrients by water moving through the soil, depletion due to removal of crops, or through the produce harvested. Soil depletion however, is, an initial stage of advanced soil degradation.

The term soil degradation is used vaguely. To avoid ambiguity and confusion, it is important that soil degradation by different processes is defined quantitatively. To do so is to identify critical limits, or upper and lower limits, of the soil properties beyond which crops will not grow. The critical limits are affected by interactions, i.e., organic matter and structure, that are hard to quantify. Nevertheless, these critical limits should be defined to delineate different levels of degradation. The limits vary for different soils, antecedent soil moisture regimes, climatic conditions, land use, crops, and agroecological regions. Lack of knowledge regarding these critical limits makes one uncertain about the validity of statistical data on the extent of soil degradations. For example, if the critical limits of organic matter content, water and nutrient status, porosity, and compaction are not known for major soils and crops, it is difficult to judge whether a soil is degraded, and if so to what degree. What are the critical organic matter levels for major soils of the world necessary to maintain or improve soil structure? While significant progress has been made in defining the critical limits for salt contents of alkaline and saline soils in relation to crop growth, and of toxic levels of Al and Mn for acid soils, such information is not yet available for certain soil physical processes. For example, there is a paucity of basic information on crop growth in relation to degree of soil degradation, effective rooting depth, plant-available water reserves, etc. The quantity and quality of organic matter necessary to maintain an adequate structural condition differ from different soils and environments and are not known.

Soil degradation is the result of alterations in soil properties caused by intensive land use, accelerated erosion by water or wind, compaction through traffic, hard setting and laterization through exposure and ultra-desiccation, decline in biotic activity and biomass carbon. The global extent of these processes and their effects are briefly outlined next.

1. Soil Erosion by Water

Soil erosion by water is a severe global problem and a major environmental concern. In spite of the voluminous literature, however, quantitative and reliable data on the magnitude of the problem are few. Furthermore, there exist few, if any, checks to verify the validity of available statistics on the magnitude of the soil-erosion problem. Most of the available data on the global extent of the erosion problem are based on reconnaissance surveys that lack a strong data base. There is a lack of precise data on both global and national levels. Estimates of denudation rates often differ by several orders of magnitude. Such an information may be good for creating public awareness but is of little value for developing and implementing erosion-prevention or erosion-control strategies. Such non-verifiable statistics may also create credibility problems among professionals.

While recognizing the limitations of such data, it is important to appraise the known information. Brown (1984) estimated that the world is now losing some 23 billion t (metric tons) of soil per year from uplands in excess of new soil formation. UNEP (1982) has estimated that by the year 2000, more than one-third of the world's arable land will be lost or destroyed. Similar warnings were sounded by Brown (1981a), who estimated that about one-fifth to one-third of the world's cropland is now being steadily degraded. Buringh (1981) estimated the annual global loss of agricultural land to be 3 million ha due to soil erosion and 2 million ha due to desertification.

In the United States, Pimentel *et al.* (1976) observed that as much as one-third of the topsoil from arable land has been lost over the past 200 years. In another survey, Pimentel *et al.* (1983) further observed that the average soil loss from U.S. cropland, the land assigned to seasonal crops, is about 18 t/ha/yr. This implies that the gross soil loss from USA arable land is about 5 billion t annually. Brown (1981b) estimated that in the USA, 34% of all cropland is losing topsoil at a rate that is undermining long term productivity. The mean rate of soil erosion by water from cropland is estimated to be 14.1 t/acre/yr in Tennessee, 10.9 in Mississippi, and 9.9 in Iowa. The cost of replacing the major nutrients (N, P, K) lost through runoff and erosion alone has been estimated at US $6.8 billion annually.

Some of these statistics of gloom and doom have, fortunately, not come true. For example, in the United States nearly 40 million ha had supposedly been ruined for agricultural purposes by erosion up to 1935 (FAO/UNEP, 1983), and 50–60% of the topsoil had been lost on an additional 40 million ha. Yet U.S agricultural production has gone up by several orders of magnitude over the last 50 years because of the addition of fertilizers, irrigation, and pesticides plus the use of hybrid varieties.

Severe erosion is also reported from tropical Africa. Brown (1981b) estimated that as much as 1 billion t of topsoil is lost from the Ethiopian highlands each year. The average annual soil loss from Madagascar is estimated to be 25–40 t/ha/yr over the whole country (Finn, 1983). Severe erosion also occurs in most parts of sub-Saharan Africa. FAO/UNEP (1983) estimated that a total of 87% of the Near East and Africa north of the equator are in the grip of accelerated erosion. In Zimbabwe, the FAO/UNEP reports that in 1974, only 50 years after most of it was first opened up for cultivation, 41% of the land was already affected by erosion and of this 12% was moderately to severely damaged.

In south Asia, the Himalyan–Tibetan mountain ecosystem is one of the most severely eroded. In India, it is estimated that 150 million ha of land were subject to accelerated soil erosion by the year 1975 (FAO/UNEP, 1983), while an additional 270,000 km^2 (out of a total land area of 3.3 million km^2) are being degraded by floods, salinity, and alkalinity (Bali and Kanwar, 1977). Dent (1984) estimated that siltation of land in reservoirs in northern India is about 200% more than anticipated in their design. In Nepal, it is estimated that 63% of the Shivalik zone, 86% of the Middle Mountain zone, 48% of the transition zone, and 22% of the high Himalayas have been reduced to poor and fair watershed conditions. In Pakistan, a survey of the Upper Indus Basin revealed that 84% of the area had moderate to severe erosion problems even in the 1960s (Dent, 1984).

The literature reveals that accelerated erosion is equally ruinous in China. It is estimated that as much as 46 million ha of the loess plateau that drain into the Yellow River are subject to erosion-caused degradation. The bed of the Yellow River is raised an average 10 cm annually. Severe degradation problems also exist in the watersheds of Yangtze, Huaihe, Pearl, Liaolie, and Songhua rivers (Dent, 1984).

In South America, soil erosion has caused severe losses in the Andean region and in the Caribbeans, especially in Haiti and the Dominican Republic. Sanchez *et al.* (1982) estimated that 39 million ha or 8% of the entire Amazon Basin is characterized by soils of high erodibility. In Argentina, 18.3 million ha, or 13% of its total cultivated area is affected by water erosion (FAO/UNEP, 1983).

Water erosion is equally severe in the USSR. Brown (1981b) estimated that about 2% of the south-central Soviet Union suffers from severe gully erosion. Erosion is very severe in southern Europe, e.g., Spain, Portugal, and France.

Although statistics regarding the global extent of erosion and erosion-caused degradation are hard to verify, some field data on erosion measurements are available for different ecological regions. Measurements of erosion rates from field runoff have been made for francofone West Africa by Roose (1977a), for western Nigeria by Lal (1976a,b) and Wilkinson (1975), for southern Africa by Elwell and Stocking (1982) and Hudson (1971), and for eastern Africa by Rapp *et al.* (1972). These data, summarized in Table 5.4, show that erosion rates of 100 t/ha/yr from arable lands are not uncommon. Similarly, high rates have been observed for Asia and South America.

2. Soil Erosion by Wind

Wind erosion is a severe problem in extremely arid, arid, and semiarid regions where the following conditions prevail: (1) loose, dry, finely divided soil; (2) smooth soil surface devoid of vegetative cover, (3) large fields, and (4) strong winds (FAO, 1960). Arid lands comprise about 36% of the world's total land area covering about 50 million km^2 of the earth's surface, of which 30 million km^2 are susceptible to severe wind erosion. Arid climates are classfied according to Thornthwaite's (1978) aridity index:

$$J_m = \frac{100\,s - 60d}{n}$$

where s is moisture surplus, d is moisture deficit, and n is moisture need. By this index arid climates are divided into three groups: (1) semiarid

Table 5.4 Magnitude of Soil Erosion from Croplands in Various Countries

Country	Rate of erosion (t/ha/yr)
Argentina, Paraguay, and Brazil	18.8
Belgium	10–25
Benin	17–28
Burkina Faso	10–20
China	11–251
Ecuador	210–564
Ethiopia	34
Guatemala	5–35
Guinea	17.9–24.5
India	75
Ivory Coast	60–570
Jamaica	90
Kenya	5–47.1
Lesotho	40
Madagascar	25–250
Nepal	40
Niger	35–70
Nigeria	14.4
Papua New Guinea[a]	6–320
Peru	15
Senegal	14.9–55
United States	9.6
Tanzania	10.1–92.8
Zimbabwe	50

[a]Source: Humphreys (1984); Barber (1983); Fournier (1967); Lal (1976a,b); Roose (1977b); World Resources Institute (1986); Ngatunga et al. (1984).

regions with J_m between -20 and -40, (2) arid regions with J_m between -40 and -57, and (3) extremely arid regions with $J_m < -57$. The geographical distribution of land area in these regions is shown in Table 5.5.

Climatic factors responsible for severe wind erosion are (a) the quantity, distribution, and nature of precipitation, (b) the temperature regime, and (c) the wind velocity. The desiccation of soil surface and structual degradation facilitate easy displacement of soil by the wind. Wind transports soil particles either by suspension, saltation, or mass drift. Wind erosion is caused by turbulent wind whose speed normally exceeds a threshold of about 20 to 50 km/hr, depending on the history of the field (Chepil and Woodruff, 1963). Generally, a wind speed of 20 km/hr is considered nonerosive.

The extent of damage caused by wind erosion is even harder to estimate than that by water. The desertification map of the world prepared by UNEP showed that areas affected by wind erosion include 43% of the

Table 5.5 Estimates of Areas (million km^2) Subject to Desertification[a]

Class of hazard	Arid Area	Arid % of zone	Semiarid Area	Semiarid % of zone	Subhumid Area	Subhumid % of zone	Total area
Very high	1.1	6.4	2.2	12.1	0.2	1.2	3.5
High	13.4	77.3	2.4	13.6	0.6	4.3	16.4
Moderate	2.1	12.1	12.5	69.4	3.2	23.3	17.8
Total	16.6	95.8	17.1	95.1	4.0	28.8	37.7

[a] Source: Mabbutt (1978).

nondesert area of Africa, 32% of Asia, and 19% of South America. UNEP has estimated that globally 80% of 3700 million ha of rangeland, 60% of the 570 million ha of cropland, and 30% of 131 million ha of irrigated lands are affected by wind erosion.

Wind erosion is a well known phenomenon in the Sahara. Wind erosion rates of 10 mm of topsoil removed per year have been measured in southern Tunisia (Floret and Le Floch, 1973; Le Houerou, 1977). Wind-blown dust from the Sahara is known to cross the Atlantic Ocean, and "sand rains" are reported from northern Europe (Le Houerou, 1977a). Rapp (1974) reported that the concentration of dust of Sahelian origin traced in Barbados was 6 g/m^3 in 1966, 8 g/m^3 in 1967–68, 25 g/m^3 in 1972, and 24 g/m^3 in 1973. The maximum dust concentration occurred at a height of 1500–3700 m. It is estimated that between 25 and 37 million tons of African soil are carried out annually over the Atlantic Ocean (Rapp, 1974; Prospero and Carlson, 1972).

In Ukraine, USSR, the normal rate of wind erosion is estimated to be 2–3 t/ha/yr. In exceedingly bad years, however, erosion rates of 300–400 t/ha/yr are commonly observed (Dolgilevich, 1972). Kovda (1980) reported wind erosion rates of 370 t/hr/yr of fine earth removed during 1974–1975 in the Azov and Dnieper steppe region of the USSR. This severe erosion was caused by massive plowing of vast territories. Grigoriev et al. (1976) observed that the critical wind speed to cause erosion in Uzbekistan region is 9.3 m/s for sandy loam, 24.0 m/s for light loam, and 36.9 m/s for loam. The destruction of aggregates by tractor cultivation, however, reduces the indices by 1.5 to 2.0 times.

There is a history of severe wind erosion in the central plains of the United States. Beasley (1973) estimated that 28 million ha suffer from pronounced wind erosion, and an average of 2.7 million ha are damaged by wind erosion annually. Brown (1981b) estimated the annual wind erosion rate to be 14.9 ton/acre/yr in Texas and 8.9 t/acre/yr in Colorado.

3. Consequences of Accelerated Erosion

Effects of wind and water erosion can be broadly grouped into off-site and on-site effects.

a. Pollution. Off-site effects on environments are those related to sedimentation and pollution of natural water. Both wind and water erosion are the major pollutants of environments. The dust ejected into the atmosphere is a major health hazard, and a risk to civil aviation. The quantity of dust added to the atmosphere each year is estimated by Bryson (1974a,b) and Kovda (1980) as follows: Volcanic dust, 4×10^6 tons; anthropogenic dust, 296×10^6 tons; smoke, $40–60 \times 10^6$ tons; and dust storms from arable lands, $100–250 \times 10^6$ tons. Many fear that high dust concentration of 300–600 µg/m^3 may have a significant cooling effect (Figure 5.5), that changes rainfall amounts and distribution.

Similar to wind-blown dust, sedimentation is a major pollutant of natural waters. Agricultural chemicals, either dissolved in water runoff or absorbed on solid particles, lead to eutrophication of natural waters. Siltation of reservoirs and lakes reduces their capacity and causes severe floods. In India, floods affect about 4.9 million ha of land annually (Dent, 1984). The most severe floods in recorded history were experienced in Bangladesh and northern India during the monsoon of 1988.

b. Desertification. A problem caused by severe wind erosion in arid and semiarid climates is "desertification," an important environmental issue of recent years. Some environmentalists prefer to call the spread of desert-like conditions in and around semiarid areas "desertization" (Rapp, 1974). It is in the fringes of arid lands and deserts where desertization is feared spreading.

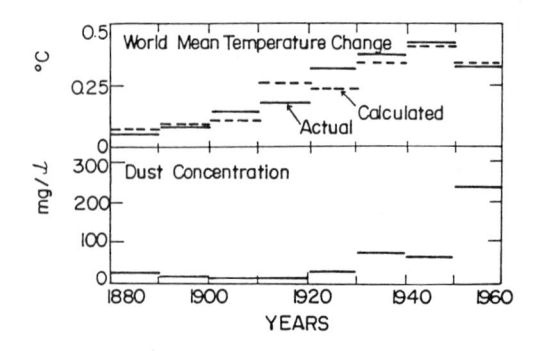

Figure 5.5 Atmospheric dust concentration changes and their effects on world mean temperature, 1880–1960 (Kovda, 1980).

5. Impact of Land Degradation 101

Desertification is defined as the "impoverishment of arid, semiarid, and subhumid ecosystem by the impact of man's activities. This process leads to reduced productivity of desirable plants, alterations in the biomass and in the diversity of life forms, accelerated soil degradation, and increased hazards for human occupancy" (UNEP, 1977). It often means "change in the character of land to more desert condition" (Mabbutt, 1978).

Estimates of areas subject to desertifiation vary widely among authors, change from year to year and are hard to validate. The data in Table 5.5 show that desertification is a moderate to very high hazard for 16.6 million km^2 in arid regions, 17.1 million km^2 in semiarid regions, and 4.0 million km^2 in subhumid regions. The global area subject to desertification is estimated to be 37.7 million km^2. The global loss to desertification, the land irretrievably lost or degraded to desert-like conditions, is estimated at 6 million ha/yr (UNEP, 1984). In addition, the land reduced to zero or negative net economic productivity is 20–21 million ha annually. Areas affected by at least moderate desertifiation are: 3,100 million ha of rangeland, 335 million ha of rainfed croplands, and 40 million ha of irrigated land. UNEP (1984) estimated that the rural population severely affected by desertification is 135 million.

In the United States, 2 billion ha of land have been identified where the risk of desertification is extremely high (Council on Environmental Quality, 1980). In India, Singh (1977) estimated that Rajsthan desert has been spreading outward about half a mile per year encroaching about 50 square miles of fertile land every year.

The process of desertification is at work in many parts of the west African Sahelian and Sudanian zones. Sahel is a zone approximately 200–400 km wide, centered on latidue 15°N in sub-Saharan Africa. Included in the west African Sahel are parts of Cape Verde, Senegal, the Gambia, Mauritania, Mali, Burkina Faso, Niger, and Chad. In these countries Sahel proper constitutes approximately 2 million km^2, constituting 27% of Senegal, 39% of Mauritania, 40% of Mali, 7% of Burkina Faso, 50% of Niger, and 32% of Chad (National Research Council, 1984). The process is aggravated in the recent past by low, erratic and variable rainfall, and by soils of low inherent fertility. The process of desertification is the most severe in the Sahelo–Sudanian zone with a mean annual rainfall of 350–600 mm (World Bank, 1985).

Floret *et al.* (1977) observed in Tunisia that the process of desertification is most marked in those areas where the average annual rainfall is between 100–200 mm. These authors evaluated the extent of desertification by using the productivity criterion—a land was considered desertified if a large proportion of its productivity could not be restored within 25 years of restorative management. Using this criterion, it was observed that about

25% of the land area had been desertified from the point of view of grazing and about 12% from the point of view of crop production. Declining agricultural production is a severe consequence of desertification.

c. Reduced crop productivity. Available statistics on global soil erosion, rough and unreliable as they may be, have created awareness among policy makers and the general public. On the basis of the quality of available data, however, the fears of environmental and soil degradation are likely to be exaggerated. The scientific community should initiate a coordinated program of monitoring sediment transport through principal watersheds and relating it to the field rates of erosion measured at some representative sites.

One of the severe on-site effects of erosion is the decline in crop yield. The magnitude of the adverse effects of erosion on yield depend on many factors: the depth of topsoil, physical and nutritional properties of the exposed subsoil, crop grown, soil and crop management, the prevailing micro- and meso-climate, and the availability of improved technologies. It is, therefore, difficult to generalize the effects of erosion on crop yield.

Depending on the conditions listed, erosion has been reported to enhance yields, to have no effect on yields, to slightly reduce yields, and to cause complete crop failure. Three contrasting examples of erosion-caused enhancement of yield, slight yield reduction, and severe yield reductions are shown in Tables 5.6, 5.7, and 5.8, respectively. Yield increase, if any, by erosion can occur but in exceptional cases such as a buried soil. For fertile and deep soils, however modest levels of yield reductions (e.g., 5–10% per 2.5 cm of soil lost) are commonly observed. Most soils being severely eroded, however, are similar to those for which the yield response is shown in Table 5.8. For such marginal soils, severe erosion can cause drastic yield reductions. For example, for shallow Alfisols and Ultisols in southern Nigeria, Mbagwu *et al.* (1984) observed that the loss

Table 5.6 Relative Yield of Cereal Crops on Eroded and Uneroded Parabraunerde Loess in Germany[a]

Crop	Difference in yield from severely eroded vs. uneroded soil (%)	
	Grain yield	Stover yield
Rye	+0.4	−14.5
Barley, site A	−1.7	+10.5
Barley, site B	−3.3	−3.6
Winter wheat, site A	+4.7	+10.9
Winter wheat, site B	+5.7	+3.2

[a]Source: Grosse (1967).

Table 5.7 Effect of Topsoil Loss on Wheat Yields at Various U. S. Locations[a]

Location	Wheat yield reduction per inch of topsoil lost (%)
Akron, Colorado	2.0
Geary County, Kansas	6.2
Manhattan, Kansas	4.3
Columbus, Ohio	5.3
Oregon, site A	2.2
Oregon, site B	5.8
Palouse area, Washington	6.9

[a]Source: Lyle (1975).

of the top 20 cm of soil caused a complete crop failure in spite of adding chemical fertilizers. In the Azov and Dnieper steppes region of the USSR, Kovda (1980) observed that severe dust storms drastically reduced winter-wheat yields over an area of 1 million ha. The mean yield of the affected crop was 1.9 t/ha compared with yields of 3.1–3.8 t/ha for the crop raised with stubble mulch to prevent wind erosion.

Erosion also affects productivity by increasing the cost of production. Pimentel and Levitan (1986) estimated that U.S. agriculture uses about 49 million t of commercial fertilizer and 350,000 t of pesticides annually. Undoubtedly, some proportion of those inputs is to compensate for nutrients lost in runoff and erosion. The effects of erosion on crop yield are masked by improved technologies, are cumulative, and may be observed only after a long time.

3. Laterization

Laterite is a hard sheet of iron- and or aluminium-rich duricrust. Iron- or aluminium-rich layers are the result of natural soil-evolutionary processes

Table 5.8 Effect of Soil Removal Depth on Cassava Yield from a Tropical Alfisol[a]

Depth of soil removed (cm)	Cassava tuber yield (t/ha) With fertilizer	Without fertilizer
Control	36.0	39.5
10	21.4	12.7
20	17.1	7.8

[a]Source: Unpublished data, Lal (1981).

that remove silica and accumulates sesquioxides. While within the soil body and protected by vegetation cover, such layers are often soft. When exposed and desiccated by deforestation and erosion, they harden into rock-like extensive sheets unfavourable for crop production. Laterized horizons are hard, compact, and cannot be cultivated. Once the laterized horizon is exposed to the surface, the soil is irretrievably lost for the purpose of crop production.

Like the problem of soil erosion, the extent and regional distribution of soils already laterized or those that are in imminent danger of being laterized is not known. Some have warned that most tropical areas, when cleared of vegetation, will become worthless brick pavement in a few years (McNeil, 1964). Others have reiterated that laterites have a limited aerial extent in the tropics (Hardy, 1933). In support of the latter argument, Sanchez and Buol (1975) reported that there may be only 21 million ha (6% of the total tropical land area) that may become laterized if the subsoil is exposed.

Sanchez and Buol (1975) and others (Prescott and Pendleton, 1952; Segalen, 1970) argued that fears of laterization are exaggerated. They estimated that soils already laterized or those prone to laterization cover 2% of tropical America, 5% of central Brazil, 7% of the tropical parts of Indian subcontinent, 11% of tropical Africa, and 15% of sub-Saharan Africa. Although reliable statistics are not available, hardened plinthite is widely distributed in the subhumid, semiarid, and arid regions of West Africa. Obeng (1978) estimated about 250 million ha of iron-pan soils in the West African savanna. I have travelled extensively in northern Burkina Faso and central Niger and estimated that about 50% of the landscape may be covered by massive laterized sheet in the region of 200–500 mm annual rainfall.

The climatic factors responsible for laterization are those that cause (1) intense weathering that leads to removal of silica and accumulation of sesquioxides, (2) wet/dry climate characterized by an intense rainy season followed by a prolonged dry season, and (3) accelerated soil erosion. The tropical rainforest is supposedly the original vegetation of the Plinthite soils. The presently observed savanna cover is the result of forest degradation. In fact, the transformation of forest into savanna vegetation contributes to formation of extremely hard crusts.

4. Compaction and Hard Setting

Intensive cultivation and decline in soil organic matter content cause deterioration of soil structure leading to crusting, compaction, hard setting, poor aeration, waterlogging, and physical degradation. Soil structure is extremely vulnerable to mechanical forces involved in normal farm operations. Increasing use of mechanical power, a major ingredient of modern mechanized agriculture, is a necessary evil at best and a disaster at worst. Kuipers (1982) reviewed the effects of mechanical forces involved in soil degradation.

Structurally inert soils, containing low organic matter and predominantly low-activity clays are prone to physical degradation. The surface layer of soils in tropical savanna is often compacted with resulting high bulk density and low infiltration. Experiments at IITA have shown a drastic decline in infiltration rate of soil with mechanized intensive cultivation. The data in Figure 5.6 show an infiltration rate decline by a factor of ten over 5 years. Compaction is particularly severe in the headlands where machinery turns the soil (Fig. 5.6). In southern Brazil, Klamt et al. (1986) also observed a rapid decline in water infiltration rate. Cumulative water infiltration in 2 hr was more than 250 cm, 55 cm, 10 cm, and less than 5 cm, respectively, for soil under native forest, cultivated with animal traction for 7 yr, 1 yr of conventional tillage after clearing with bulldozer, and conventional tillage for more than 20 yr. This rapid decline in water acceptance is attributed to structural collapse caused by machinery and decline in organic matter content. The problems caused by wheel traffic are equally severe in the temperate-zone agriculture (Davies, 1982; Voohees et al., 1978; Soane, 1981).

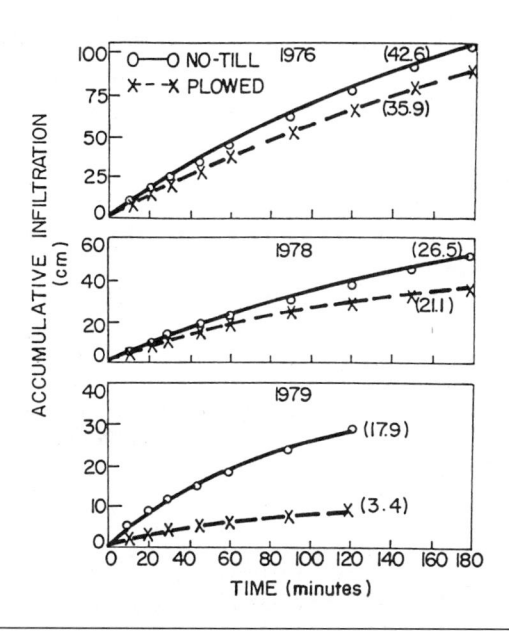

Figure 5.6 Changes in water infiltration rate with duration of cropping in no-tillage and conventionally plowed watershed. Figures in parentheses show the standard error of the mean (Lal, 1985).

Table 5.9 Effects of Compaction and Erosion-Caused Soil Degradation on Decline in Maize Grain Yield over 5 Years[a]

Tillage treatment	Maize grain yield (Mg/ha)					
	1975	1976	1977	1978	1979	1980
No-till, unterraced	2.8	4.5	4.8	5.0	3.8	3.0
Plowed, terraced	2.7	4.0	3.9	4.0	2.9	1.0

[a]Source: Lal (1984a).

Soils prone to physical degradation are widespread in seasonally moist regions with intense summers. Some soils containing prodominatly low activity clays and those low in organic matter content set hard on drying and desiccation, and limit crop productivity. Hard-setting soils are widespread in dryland regions of Australia, in west African savanna (Charreau and Nicou, 1971; Jones and Wild, 1975), in Botswana (Sinclair, 1985), and in Zambia (Veldkamp, 1986).

Physical degradation involving compaction and hard setting leads to high runoff losses, and accelerated soil erosion. Lal (1984a) observed poor crop stand, and low yields on eroded and compacted watersheds. The data in Table 5.9 show yield decline in maize from the plowed watershed from 4.0 t/ha in 1976 to 1.0 t/ha in 1980. In Australia, Wood (1984) observed that sugarcane yield in the Herbert Valley declined at a rate of 0.12 tons of sugar per ha/yr due to soil physical degradation (e.g., compaction, decline in organic matter content). The data in Table 5.10 from northeast Thailand show severe yield reduction in maize and mungbean over a short period of three years. Significant yield reductions are also observed in cassava. Drastic reductions in cassava yield due to physical degradation are, however, observed over a long period of 10–15 years (Figure 5.7).

Table 5.10 Decline in Crop Yield Due to Continuous Cultivation without Added Fertilizer, Sa Nan Research Station, Thailand, 1967–1969[a]

Year after clearing	Crop yield (kg/rai)				
	Cotton	Mung beans	Rice	Maize	Peanuts
First (1967)	175	153	406	428	100
Second (1968)	127	178	302	101	185
Third (1969)	196	69	222	55	108

[a]Source: Chapman (1978).

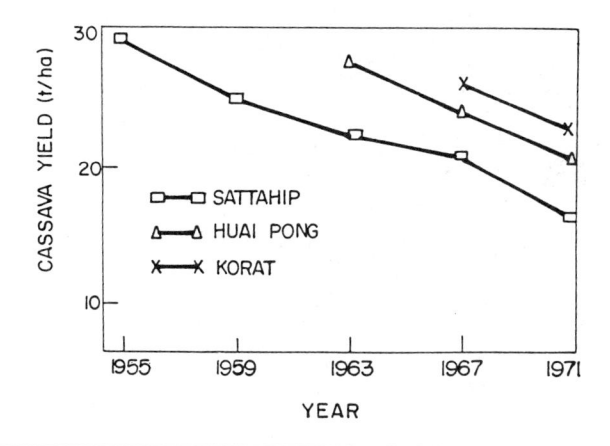

Figure 5.7 Effects of continuous cultivation on cassava yield at three locations in Thailand (Kubota *et al.*, 1982).

Amelioration of physically degraded soils may involve frequent mechanical loosening. Subsoiling or chiseling is an energy-intensive process that most small-scale farmers cannot afford. An alternative is to fallow the land but land shortage makes that increasingly difficult. Fallowing with deep-rooted perennials (e.g., *Cajanus cajan*) is an alternative (Hulugalle *et al.*, 1986). Another strategy is to adopt mulch-farming systems with low machine traffic, and low cropping intensity. Mulch farming reduces physical degradation by improving soil organic matter content and enhancing activity of soil fauna, e.g., earthworms. Regular addition of organic matter content improves soil structure by serving as food for soil fauna, which create the desired porosity and promote formation of polysaccharides and other compounds that stabilize the microaggregates (Oades, 1984).

5. Biological Degradation

The declining soil organic matter content, reduced biomass carbon, decrease in the biological activity, and reduced diversity of soil fauna are the basic factors responsible for soil biological degradation. Because of prevailing high soil and air temperatures throughout the year, biological soil degradation is more severe in the tropics than in the temperate zone. The rate at which soil organic matter content decomposes is doubled for every 10°C increase in temperature. Consequently, the organic matter content of soil is usually lower in the tropics than in the temperate regions. There are also differences in soil temperature and moisture regimes. Decrease in food availability and diversity are important factors that reduce both the activity an diversity of soil fauna. Indiscriminate use of chemicals,

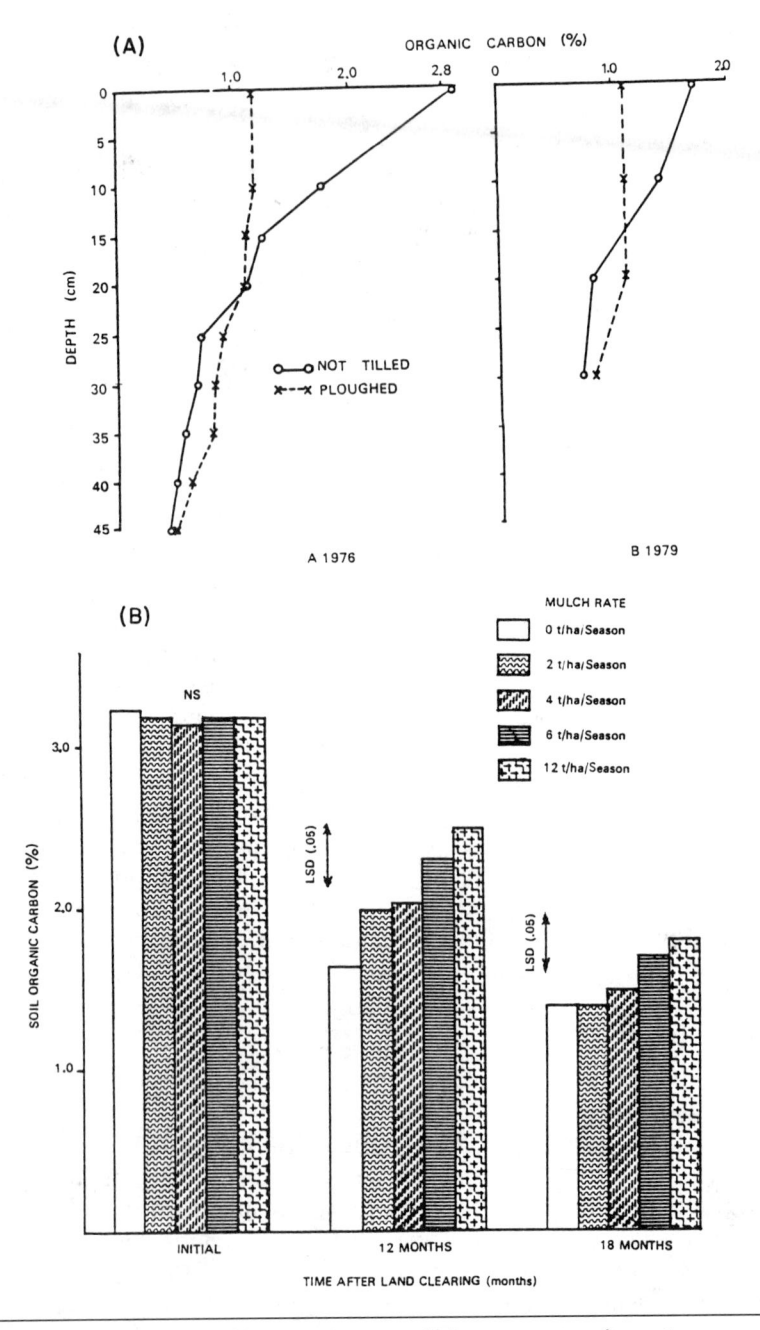

Figure 5.8 Effects of soil management techniques on organic matter content: (A), tillage effects; (B), mulch rate (Lal, 1976; 1982).

such as ammonium sulphate, furadan, and other pesticides, also reduces soil faunal activity. A soil devoid of macro- and micro-fauna is easily crusted and compacted.

Soil organic matter content depends on soil- and crop-management systems. Agronomic techniques based on liberal use of crop residue mulch, planted fallows, and no-till farming help build up organic matter in soil (Figure 5.8). The organic matter content, both quality and quantity, significantly affect soil structure (Hamblin, 1985; Hayes, 1986). Decline in organic matter content increases the soil's susceptibility to erosion and enhances the processes of soil degradation (Newbould, 1982; Lal, 1984b). Soil erodibility increases exponentially with decreasing soil organic matter content (Figure 5.9).

6. Salinization, Alkalization, and Waterlogging

Accumulation of excessive soluble salts in the root zone to levels that are toxic to plant growth occurs in arid and semi-arid regions where the mean annual evaporation grossly exceeds the precipitation. Climatic aridity, an important factor for salt enrichment, is attributed to low precipitation, high evapotranspiration, high temperatures, and low humidity. Saline soils are easily formed in regions where the ratio P/PET is less than 0.75 (P is precipitation and PET is potential evapotranspiration). The accumulated salts usually comprise chlorides, sulphates, and cabonates of sodium, magnesium, and calcium. If the predominant cations accumulated in the soil is sodium, the process is also called salinization or alkalization (Buol

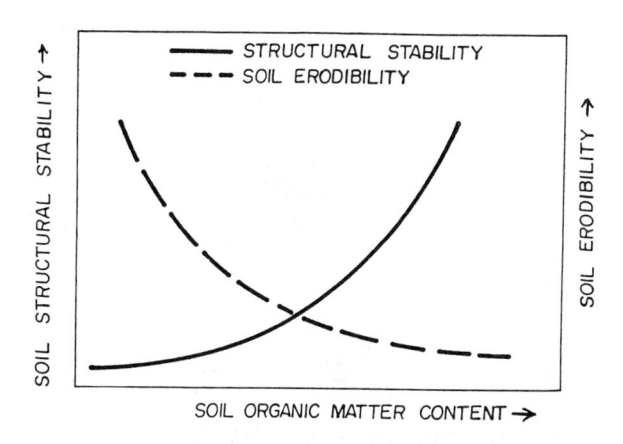

Figure 5.9 Effects of decline in organic matter content on soil structural stability and erodibility.

et al., 1973). High concentrations of Na⁺ on the exchange complexes disperses the clay that may eventually eluviate to subsoil and form a horizon of massive structure with low permeability to water. The source of soluble salts may be indigenous from either parent material or ground water. Salts may also be brought in with irrigation water, as fertilizers, and in other amendments. Watson (1984) observed for some irrigated projects in Australia that the salinity levels of some river waters used for irrigation is high. Particularly high salinity levels are observed for the lower Murray River in south Australia and in the Wellington Reservoir in the Western Australia. In coastal areas, salt comes from sea encroachment or is blown inland by wind.

Whatever the source, the enrichment of soluble salts in the soil alters the soil's physical and chemical properties. Soil is characterized as "saline" if the conductivity of saturated soil paste exceeds 4 s/cm, and "alkaline" if the conductivity exceeds 8 s/cm at 25°C. Alkali soils occur extensively in regions with mean annual rainfalls between 550–1000 mm. In addition to altering soil structure, high osmotic pressure severely curtails plant-water availability.

The climatic factors responsible for salt imbalance include those that reduce or eliminate leaching. Salt accumulation does not occur in humid and subhumid conditions because soluble salts are leached from the root zone during the rainy season.

Salt-affected soils are widely distributed in different continents. The data in Table 5.11 show that 322.9 million ha are either presently affected by excessive salts or will soon be affected. Asia and Australia have more than their share of such soils. The loss in productivity of salf-affected soils can be slight to complete. By adapting reclamative technologies that comprise flushing the salts through drainage and following appropriate cropping

Table 5.11 Global Distribution of Salt-Affected Soils[a]

Region	Area (million ha)
Africa	69.5
Near and Middle East	53.1
Asia and Far East	19.5
Latin America	59.4
Australia	84.7
North America	16.0
Europe	20.7
World total	322.9

[a]Source: Beck *et al.* (1980).

systems with balanced fertilizers, one can restore the productivity of some salt-affected soils. Providing drainage to leach salts from the drainage basin, however, is difficult. Salts, though temporarily leached from the root zone, accumulate in the groundwater and eventually into the river. Salts are easily recycled to the root zone through irrigation. Some of the irrigated areas of Australia (south Australia, Victoria, and southern New South Wales), India, Pakistan, China, and the Middle East, are highly prone to developing salinity problems.

The irrigated land area is likely to increase in the future. At present only 15–20% of the world's arable land is irrigated, providing 30–40% of the total agricultural production. Yields of irrigated crops are 2–2.5 times as high as those of rainfed crops. Expansion of irrigated agriculture in semiarid and arid regions is likely to aggravate the problem of salinization. The total irrigated land area in the world has increased dramatically in the 20th century, so has the area of salt-affected soils. The data in Table 5.12 show that the irrigated land area was merely 48 million ha in 1900 but it is expected to be 300 million ha at the turn of the century. Irrigated land area in 1900 was 15.5 million ha in the Indian subcontinent, 3.80 million ha in Russia, 3.0 million ha in the United States, and 2.0 million ha in Egypt.

Fukuda (1976) estimated that 75% of the cultivable land area available in arid, semiarid, and subhumid regions requires irrigation for successful crop growth (Table 5.13). The productivity of land in the humid regions can also be increased with supplemental irrigation (Greenland and Murray-Rust, 1986). The lack of good-quality irrigation water, however, poses a serious problem. Fresh water resources of the earth are estimated at 37,000 km^3 of annual stream flow, of which only 3,200 km^3 are used by man (Kovda, 1980). Streamflows fluctuate as much as 450% annually, making the available water reserves highly erratic.

Table 5.12 Irrigated Land Area in the World[a]

Year	Irrigated land area (10^6 ha)
1800	8
1900	48
1949	92
1949	149
1980	235
2000	300

[a]Source: Fukuda (1976); Kovda (1980); Szabolcs (1986).

Table 5.13 Arable Land Area in Different Rainfall Regimes[a]

Climate/ecology	Mean annual rainfall (mm)	Area (%)
Arid	<250	25
Semiarid	250–500	30
Subhumid	500–1,000	20
Humid	1,000–1,500	11
Perhumid	1,500–2,000	9
Superhumid	>2000	5

[a]Source: Fukuda (1976).

Food production in semiarid Africa can be increased if small-scale irrigation projects are developed (Lal, 1987b). Greenland and Murray-Rust (1986) estimated that about 4 million ha of land can be developed for irrigated agriculture in sub- Saharan Africa.

Depending on the water quality, structure of irrigated soils is easily destroyed by intensification of landuse through supplementary irrigation. Structural degradation is the first step in development of waterlogging and accumulation of salts in the root zone. Experiments conducted in Morocco by Mathieu (1982) indicated that irrigation decreases the clod and aggregate porosity. The data in Table 5.14 show that in non-irrigated soils that aggregates are more porous at 0–5 cm depth (33.8% v/v) than at 45–65 cm depth (31.3% v/v). In irrigated soils, however, the aggregate porosity decreased up to 65 cm depth. The total decrease in porosity of elementary aggregates caused by irrigation was about 4.6% (v/v) in the 0–25 cm layer, and about 3.0% (v/v) at 25–45 cm depth. Significant differences in clod porosity were also observed between 15 and 35 cm depth. The total porosity of the surface 0–15 cm layer was, however, more for irrigated than non-irrigated soils. The magnitude of the effect on aggregate and clod porosity depends on soil type, quality of irrigation water and the management. In northern Victoria, Australia, Mason et al. (1984) observed that intensive fodder crop production with irrigation increased slaking percentage from 8 to 10% from November 1981 to April 1983. In New South Wales, McKenzie et al. (1984) reported that irrigated cotton production has caused deterioration in soil physical fertility of a Vertisol, particularly between 15 and 30 cm in depth.

Proper management of the soil surface, by maintaining favorable salt and water balance, is the key to preserving and increasing productivity of irrigated soils. Regular addition of farmyard manure, compost, and other organic materials is an important method to decrease slaking and

Table 5.14 Effects of Irrigation on Porosity (%, v/v) of a Soil in Northeast Morocco[a]

Depth (cm)	Non-irrigated soils	Irrigated soils Mean	Statistical analysis[b]
Total porosity			
5–5	50.0	56.0	**
5–15	49.6	53.0	*
25–35	49.3	47.2	n.s.
35–45	46.8	48.2	n.s.
45–65	45.4	47.0	n.s.
Clod porosity			
0–5	40.4	39.6	n.s.
5–15	39.7	38.7	n.s.
15–25	39.5	35.5	**
25–35	38.5	36.3	*
35–45	37.2	37.8	n.s.
45–65	37.4	38.4	n.s.
Aggregate porosity			
0–5	33.8	29.2	***
5–5	33.2	28.9	***
15–25	33.7	28.8	***
25–35	32.8	29.6	***
35–45	33.2	30.5	***
45–65	31.3	31.2	n.s.

[a] Adapted from Mathieu (1982).
[b] Statistical analysis is between non-irrigated and irrigated mean. $*p < .10$; $**p < .05$; $***p < .01$; n.s., not significant.

improve soil fertility (Mason et al., 1984). In addition to the effects on land quality, there are also health hazards, such as malaria, river blindness, and schistosomiasis, caused by swamps and waterlogged soils. e.g.

7. Leaching and Acidification

Leaching is the reverse of salt accumulation. It occurs in very humid conditions, and in soils with predominately low activity clays and free drainage. Leaching is a natural process in soil evolution. Similar to soil erosion, however, excessive leaching leads to soil degradation by depleting bases and change and degradation of clay minerals. Some clay is leached from the profile but in most cases it is translocated from one horizon to another. The accelerated leaching process involves the loss of such bases as calcium, sodium, and magnesium rendering the soil acidic in reaction and often replacing bases with exchangeable aluminum.

The soluble cations may leach from the root zone either in solution or as absorbed/adsorbed cations on the exchange complex of eluviating coloids, mainly clay. The leaching may be vertical or oblique, depending on the relative proportion of percolating water moving vertically down the soil profile or horizontally along a gradient. Lateral or oblique drainage often occurs in layered profiles with differences in permeability among horizons. The eluviated or leached horizons lose bases and a part or most of their clay. The loss of bases and clay alters both physical and chemical properties. The leached soil, therefore, not only has low chemical fertility but also low plant-available water reserves. The exchange complex is dominated by aluminum.

Some examples of leached soils occur in diverse climatic regions (WMO, 193): (1) temperate Atlantic climates in soils with differentiation of textural B horizon, (2) the process of podzolization in humid temperate climates caused by acid litter, (3) subtropical climates with more rainfall received in winter, (4) in tropical climates with seasonally humid moisture regime that leads to the formation of Alfisols, and (5) in equatorial climates that form highly leached, tropical Ultisols and Oxisols. Substantial areas of acid tropical soils occur in the Amazon Basin, in the Cerrados and llanos of Brazil and Colombia, the Congo Basin, and the humid regions of Southeast Asia.

Leaching is one of the principal mechanisms of inorganic N loss in tropical soils. The loss of NO_3-N with the mass flow of water causes significant leaching in the tropics. Reliable data on leaching losses monitored over a long period using monolithic lysimeters are few from tropical regions. Losses of 70–107 kg/ha/yr of N have been reported from bare fallow, unfertilized plots in India, compared with 329–511 kg/ha from bare fallow and 3–156 kg/ha from cropped plots in Peradeniya, Sri Lanka (Martin and Skyrning, 1962). Suarez de Castro and Rodriguez (1958) reported from lysimetric investigations in the high rainfall regions of Colombia that an average of 360 kg/ha/yr of inorganic N was lost in the leachate from bare soil, whereas, only 62 kg/ha/yr was lost when the legume *Indiofera endecaphylla* was grown. High leaching losses of N have also been reported from a sandy soil in Malaya (Bolton, 1968). Martin and Cox (1956) reported leaching losses of 27 kg N/ha from a black earth in a subhumid environment of Queensland, Australia.

The effect of soil type on the amounts of fertilizers leached from the profile in the subhumid tropical climate of Ibadan, Nigeria, is shown in Table 5.15. Fertilizer losses from sandy Apomu soil were significantly more than those from heavy textured Egbeda. Similar experiments were conducted on sandy soils in northeast Thailand by Yoshioka *et al.* (1987). These authors observed quick leaching of N under field conditions. Using plastic tubes (23 cm^2, 15 cm long) filled with different soils and kept under natural rainfall conditions, Yoshioka *et al.* (1987) observed that out of 27.5 mg of N applied on the surface, little nitrogen remained after 100 mm of rain except in two clayey soils (Figure 5.10). The stage of crop growth and the time fertilizer is applied can also affect the quality of percolating

Table 5.15 Leaching Losses of Plant Nutrients from Different Soils Growing Maize[a,b]

Soil and treatment	NO$_3$–N	NH$_4$–N	P	K	Ca	Mg
Egbeda						
F	7.4	0.66	0.42	36.1	32.4	1.6
UF	0.7	0.04	0.07	2.7	8.2	0.2
Alagba						
F	1.3	0.16	1.3	4.5	2.0	0.9
UF	0.3	T	0.5	1.3	1.5	0.4
Onne						
F	0.2	T	0.5	0.9	0.7	0.3
UF	T	T	0.04	0.4	0.2	0.1
Apomu						
UF	50.6	0.79	3.5	12.7	41.0	13.5
F	58.3	0.78	2.9	12.9	78.6	11.0

[a]Source, Lal (1977).

[b]F: fertilizer applied at the rate equivalent to 100 kg N, 13 kg P, and 30 kg K per ha. UF: unfertilized. T: trace.

water. Alberts *et al.* (1978) observed that most of the average annual total N and P losses occurred during the initial establishment period of crop growth.

C. Environmental Degradation

Intensive land use necessitates inputs of fertilizers and other chemicals including herbicides and pesticides. Indiscriminate use of these chemicals has been responsible for pollution and eutrophication of natural waters (Stewart and Rohlich, 1977). The recovery of nitrogenous fertilizers by crops is less than 50%, and a maximum of 10% is recovered by a succeeding

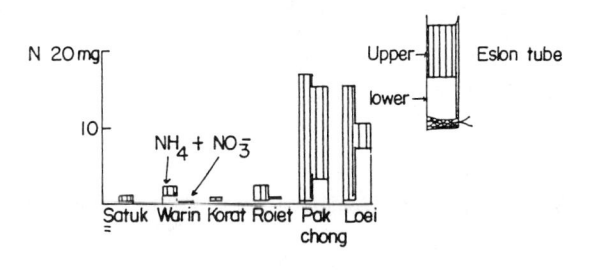

Figure 5.10 Leaching losses of N from sandy soils of northeast Thailand (Yoshioka *et al.*, 1987).

crop. The unrecovered fertilizer is easily lost in runoff and seepage flow. In addition to polluting the environment, the economic loss of expensive inputs can be substantial. It is estimated that agricultural lands in the United States lose more than 50 million t of plant nutrients annually with an estimated cost of about $6.8 to $7.75 billion each year (Biswas and Biswas, 1978).

Nutrient concentrations in water runoff depend on land use, cropping systems, and soil and crop management. In a forested land, decomposing leaf litter on the surface may increase the concentration of basic cations and organic nitrogen in runoff. Timmons and Holt (1977) and Timmons et al. (1977) observed that organic N and P comprised 68 and 82% of the respective annual losses in water runoff from a forested prairie and an aspen birch forest. The quantities of cations in surface runoff were in the following order: Ca > K > Mg > Na.

The nutrient concentration in water runoff and eroded soil from an underutilized forested plot on an Alfisol in Ibadan, Nigeria, is compared in Table 5.16 with a cleared plot growing maize with commercial fertilizer. The nutrient concentration in water runoff and eroded soil from the fertilized maize plot was more than that of the unfertilized forested plot.

The quantity of pollutants added to natural waters from croplands is related to soil properties, and soil and crop management systems. Tillage systems and other soil conservation measures affect the quality of water runoff and eroded sediments. Barnett et al. (1972) reported from studies on some Puerto Rican soils that the average concentration of N in runoff ranged from 0.01 to 0.02 ppm, and that of K from 0.01 to 2.29 ppm. In northern Nigeria, Kowal (1972) reported average annual losses of Ca, Mg, and Na in runoff water and eroded soil to be 14–30 kg/ha, depending on soil and crop management practices. Maximum nutrient loss always occurs from bare, unprotected soil. The data in Table 5.17 show nutrient losses of about 60 kg/ha/yr in water runoff and about an additional 270 kg/ha/yr

Table 5.16 Nutrient Concentration in Water and Soil Particles in the Runoff from Maize Plots Located on Natural Slopes[a]

Slope %	Water runoff (ppm)								Soil runoff (ppm)				
	Maize				Forest				Maize			Forest	
	P	Na	Ca	K	P	Na	Ca	K	P	Ca	K	P	Ca
1	0.1	1.5	21.8	5.2	0.1	1.3	2.7	1.7	18.1	985	58	3.9	475
5	0.2	2.3	14.5	8.4	0.1	1.5	2.6	2.1	18.1	788	98	5.5	725
10	0.7	1.7	5.6	7.9	0.4	2.2	1.6	2.8	18.1	995	102	8.2	790
15	0.5	1.1	1.8	3.0	0.6	1.6	1.4	2.3	65.3	1515	154	14.7	1135

[a] Source: Lal (1976a).

Table 5.17 Relative Nutrient Loss in Water Runoff and Eroded Soil from a Bare Fallow Alfisol on 15% Slope[a]

Nutrient	Nutrient loss (kg/ha/yr)	
	Water runoff	Eroded soil
N	9.6	3.4
P	2.9	13.1
K	13.2	29.4
Ca	29.0	203.1
Mg	7.3	18.1

[a]Source: Lal (1976a).

in eroded soil. Nutrient losses are generally associated with eroded sediments. Nutrients are transported as absorbed elements on the exchange complex. Nitrogen loss in eroded sediments may be as much as 92% of the total loss (Schuman et al., 1973).

Similar investigations in temperate regions indicate a significant effect of soil conservation measures and tillage systems on the quality of water runoff (Burwell et al., 1974; Klausner et al., 1974). Nutrient losses can be held low by suitable conservation practices. Schuman et al. (1973) reported that terracing reduced runoff and sediment yields from an agricultural watershed in Missouri Valley loess. With contour cultivation, 92% of the N loss was associated with eroded sediments. Romkens et al. (1973) observed significant effects of tillage systems on N and P concentrations in surface runoff. The coulter and chisel system controlled soil loss, but runoff water contained high levels of soluble N and P from surface applied fertilizers. Conventional tillage systems had high losses of soil and water but lower concentrations of nutrient losses.

Data in Figure 5.11 show the significant effect of surface mulch on nutrient concentration in water runoff. The concentrations of Ca^{2+} and Mg^{2+} were generally more under bare fallow and that of K was higher in mulched than in other treatments. Relative nutrient concentrations in water runoff and eroded soil are associated with crop, fertility level, and the type of fertilizer used (White and Williamson, 1973; Dunigan et al., 1976; Shelton and Lessman, 1978; Klepper, 1978). Edwards et al. (1972) reported NO_3–N concentration of <2 mg/liter in barn-lot runoff water. Most of the soluble N was in a reduced form with maximum monthly concentration of <70 mg/liter. Long et al. (1975) observed that NO_3 levels in runoff water were not affected by manure application at 45.10^3 kg/ha/yr and were <2 mg/liter. There are also differences in water quality of surface runoff compared with subsurface flow (Jackson et al., 1973; Lal, 1976a). Burwell

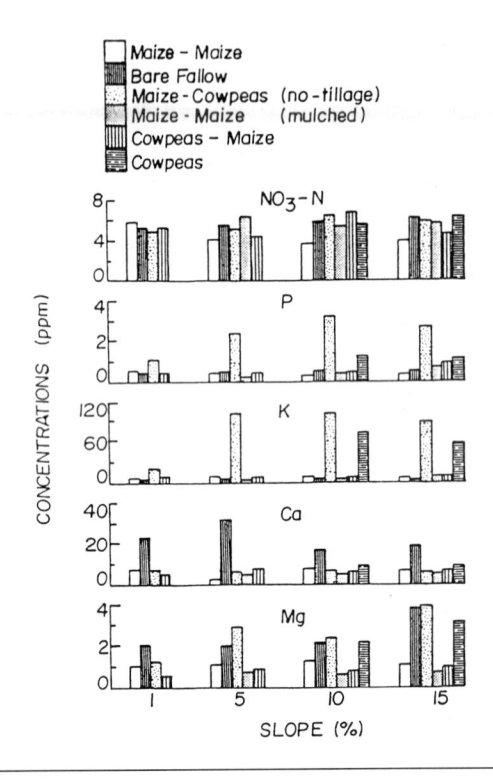

Figure 5.11 Effects of cropping systems and residue management on nutrient concentration in water runoff (Lal, 1976a).

et al. (1976) reported that NO$_3$ in subsurface discharge accounted for 84–95% of the total annual soluble N discharge in stream flow. The data in Table 5.15 show that leaching losses of nutrients from soils in Nigeria range from 10 kg/ha/season for a crop receiving no chemical fertilizers to 120 kg/ha/season from one receiving normal rates of chemical fertilizers. The nutrients, leached from the root zone, eventually contaminate the ground water used widely for human consumption.

Judicious use of chemicals is desirable to minimize the pollution hazard. Conservation farming techniques (e.g. no-till, mulch farming, or use of cover crops) and minimizing the use of agrochemicals are appropriate systems to reduce environmental pollution.

Environmental degradation is equally serious when industrial wastes are dumped in rivers or ejected into the atmosphere. One such example of pollution of natural waters of the Amman Zerka region in Jordan is described by Bandel and Salameh (1981). There are numerous such examples from both industriliazed and developing economies. This report cannot address this important issue.

D. Climatic Change

Drastic changes in micro-, meso-, and macroclimate can occur from anthropogenic factors. The global warming trend has been speculated to stem from the possible greenhouse effect caused by the increasing CO_2 content of the atmosphere (Revelle, 1982). Data in Figure 5.4 show the annual amplitude of CO_2 concentration in the atmosphere. The gradual increase at 0.4% per year in the past 10 years, is attributed primarily to burning fossil fuel, but somewhat to deforestation in the tropics. The decline in soil organic matter constent and its degradation also contribute to atmospheric CO_2. The most important anthropogenic factors include changes in the gaseous composition of the atmosphere. In addition to CO_2, there are a range of other gases that also contribute to the warming trend. These gases include carbon monoxide, methane, nitrous oxide, ozone, and chlorofluroromethane (Bach, 1976, 1986). Burning and forest fires contribute considerably to the greenhouse gases. Crutzen *et al.* (1985) reported significant increase in CO content of the atmosphere in the dry season when fire was being used to clear new land in the Brazilian Amazon. Nitrous oxide and methane are also contributed by forest burning, application of chemical fertilizers and other related anthropogenic activities (Woodwell *et al.*, 1978).

Another effect of mass-scale deforestation in the tropic is the possible reduction in water vapor that is recycled into the atmosphere as rainfall. Deforestation, done on a large scale covering hundreds of thousands of km^2, is expected to increase the total water yield from a watershed. Most of the rainwater received may escape to the ocean without being recycled as rain. Although there is no concrete evidence of reduced rainfall due to deforestation, Salati *et al.* (1983) showed that water vapors for at least 60% of the rainfall received over the Amazon is contributed directly by evapotranspiration over the basin.

The anthropogenic effects on microclimate are much more easily demonstrated than those of meso- or macroclimate. For example, deforestation in the tropics increases maximum air and soil temperatures (Figures 5.12 and 5.13) and decreases maximum relative humidity (Lal and Cummings, 1979). The evaporative demand of the atmosphere, or the climatic aridity, is higher on deforested than on forested land. Consequently, shallow-rooted seasonal crops may suffer from frequent drought stress even during the rainy season. The severity of drought stress is further aggravated by high growth rates and the soil's generally low water reserves available to plants.

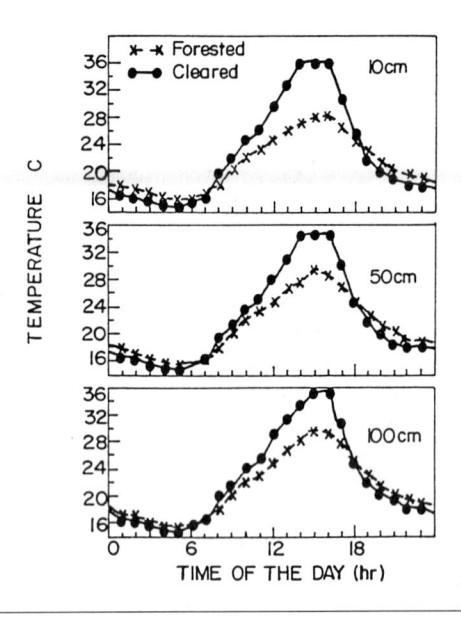

Figure 5.12 Effects of removal of a semideciduous rain forest at Ibadan, Nigeria, on air temperature (Lal and Cummings, 1979).

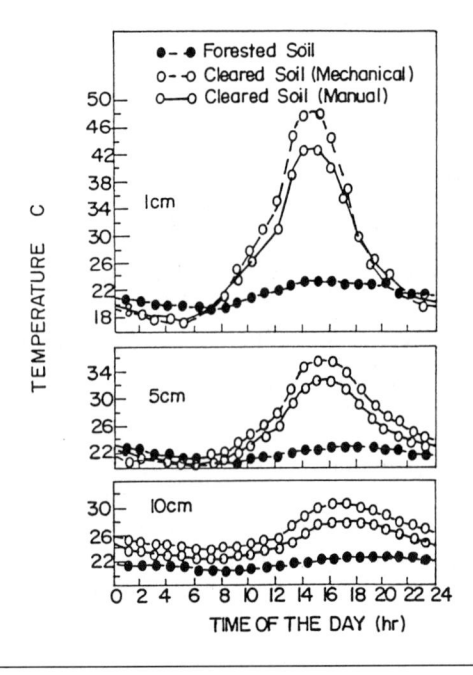

Figure 5.13 Effects of removal of a semideciduous rain forest at Ibadan on soil temperature (Lal and Cummings, 1979).

III. TECHNOLOGICAL OPTIONS FOR MINIMIZING SOIL DEGRADATION

On the basis of scientific knowledge available, the world is capable of bringing about substantial increases in global food production. This can be done without causing additional soil degradation. Many soil scientists agree that the single most important factor limiting crop yield on a worldwide scale is soil infertility caused by inadequate physical, nutritional, and biological properties (Brady, 1982). Soil degradation is mostly caused by land mismanagement. Choosing an appropriate land use, adopting ecologically compatible soil and crop management systems, and maintaining modest ambitions for crop yields should preserve the favorable balance between land and people and provide sustainable returns. Some basic principles of soil and crop management with wider applicability in different ecological regions are briefly outlined below.

A. Soil Erosion Management

Preventing erosion implies reducing soil splash or raindrop impact, and decreasing both runoff rate and velocity. Conservation-effective farming systems to curtail erosion involve soil and crop management practices cause minimal disturbance of the soil surface, maintain a layer of crop residue mulch, and provide continuous vegetative cover throughout the rainy season. Cultural practices that have been proven successful include no-till farming, liberal use of crop residue mulch, mixed and/or relay cropping, and agroforestry systems (e.g., alley cropping). Ridge-furrow and tied-ridges also have been effective in semiarid and arid regions of Africa. Terracing and other engineering techniques for safe disposal of excess water runoff are generally the last resort.

No-till and reduced tillage systems are very effective in controlling erosion and are rapidly spreading in the United States, western Europe, and Australia (Sprague and Triplett, 1986). Mulch farming techniques have also proven successful and effective in soil and water conservation in the tropics (Figure 5.14; Lal, 1985). In northern Nigeria, Lawes (1962) observed that unmulched soil, even with cultivation at fortnightly intervals to break the crust, lost 50–70% of the rainfall as runoff. Infiltration of crusted soil was less than 2.5 cm/hr. Infiltration rate with residue mulch and without any cultivation exceeded 12.5 cm/hr, and there was no runoff. One of the obstacles to widespread adoption of no-till farming is dependence on herbicides, which millions of small farmers of Africa and elsewhere in the tropics cannot afford. Another is the nonavailability of crop residue mulch in adequate amounts. Agronomists have, however, developed cropping systems and rotations based on appropriate cover crops

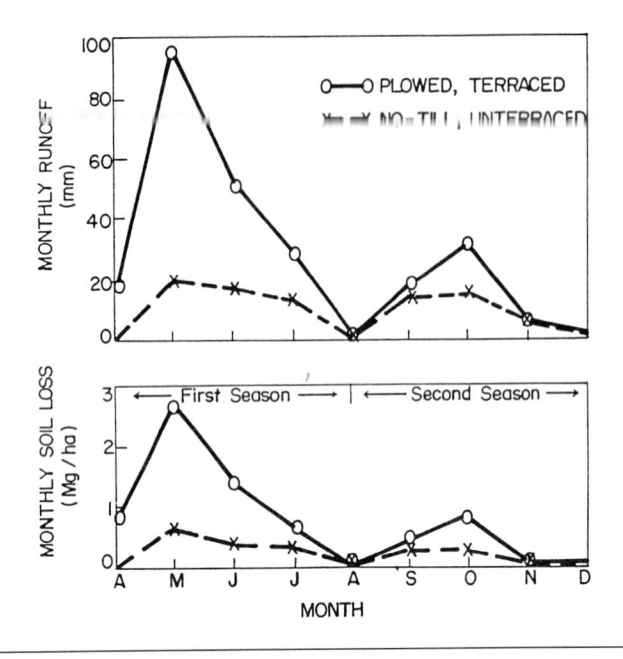

Figure 5.14 Effects of no-tillage and plowed seedbeds on water runoff and soil erosion from watersheds growing maize (Lal, 1985).

that minimize dependence on herbicides (Wilson and Lal, 1986). Another alternative is using perennial shrubs, grown as contour hedges, to supply mulch and control erosion. *Leucaena* hedges grown every 2 or 4 m apart can effectively control runoff and soil erosion (Table 5.18). With appropriate soil management, crop rotations, and vegetative cover, Lowedermilk (1953) estimated that soil life can be increased from a few years to hundreds of thousands of year. The data in Table 5.19 show that useful life of soil can be greatly increased when erosion is controlled by mulch farming, crop rotations, and afforestation.

Controlling wind erosion involves reducing the wind velocity close to the soil surface. No-till and stubble-mulch farming, rough and cloddy seedbed, and erecting shelter belts as wind breaks are some of the systems that reduce wind erosion. Improving soil structure, by regularly adding organic matter and by using soil conditioners and other amendments, is also an important strategy. Soil structure can be improved by activity of roots and soil animals. Long-lasting improvements in soil structure are brought about by regulation of humic reserves through the production of assimilable carbon, and production of exudates in the rhizosphere and mucopolysaccharides by earthworms in the drilosphere (Lavelle, 1983).

Table 5.18 Effect of Alley Cropping on Runoff and Soil Erosion under Maize–Cowpea Rotation, 1984[a]

Treatment	Runoff (mm)	Soil erosion (t/ha/yr)	Maize grain yield (t/ha)	Cowpea grain yield (t/ha)
Plowed	232	14.9	4.2	0.5
No-till	6	0.03	4.3	1.1
Leucaena planted 4 m apart	10	0.2	3.9	0.6
Leucaena planted 2 m apart	13	0.1	4.0	0.4
Gliricidia planted 4 m apart	20	1.7	4.0	0.7
Gliricidia planted 2 m apart	38	3.3	3.8	0.6

[a] Each value is an average of 4 years of data. Alleys were plowed before crop plants to incorporate the mulch into the soil. Effects were measured at the International Institute of Tropical Agriculture, Ibadan.

Organic substances and polysaccharides play an important role in producing stable aggregates (Hamblin, 1985; Hayes, 1986).

Controlling wind erosion and spread of desertification are important issues for many regions lying on the fringes of the existing deserts. Technologies for dune stabilization are well documented (Hagedorn *et al.*, 1977). Van der Poel and Timberlake (1980) observed that regulating grazing is a major factor in stabilizing sand dunes in Botswana. They observed that dunes could be stabilized by natural regeneration if the area is protected

Table 5.19 Effect of Land Management on Runoff and Erosion on an 8% Sloping Soil at Stateville, North Carolina[a]

Land management	Average runoff (%)	Average soil loss (t/ha)	Time to deplete 17.8 cm of topsoil (years)
Fallow, clean tillage without cropping	29	143	18
Continuous cotton, clean tillage	10	49	44
Crops in rotation	9	—	109
Grass cover	<1	—	96,000
Forest, burned annually	3.5	0.1	1,800
Forest, not burned	<0.3	—	>500,000

[a] Source: Lowdermilk (1953).

from grazing by livestock. The restoration process can be hastened by afforestation with native and introduced tree species.

Establishing wind breaks and fences is an important step in controlling wind erosion from croplands. Types of material used and methods of construction of fences are described in detail by Hagedorn *et al.* (1977). Once the dunes are stabilized, cultivation for food grain production should be done through no-tillage and mulch farming techniques. The crop residue should be returned to the soil surface as mulch (Woodruff, 1972).

Perennials and trees that can be successfully established in the Sahel include *Acacia ehrenbergiana, A. Laeta. A. nilotica, A. senegal, A. tortilis, Balanites aegyptiaca, Maerua crassifolia, Salvadora persica, Zizyphys mauritiana* (National Research Council, 1984). Once the trees are established and the area is fenced out against stray animals, it is easy to establish annuals—both grasses and legumes. Annuals such as *Aristida adscensionis, A. funiculata, Panicum laetum,* and *Schoenefeldia gracilis* are found on silty soils; *Aristida mutabilis, Cenchrus biflorus,* and *Tribulus terrestris* are found on sandy soils (National Research Council, 1984).

B. Preventing Laterization

Hardening of plinthite is caused by ultradesiccation through exposure of the subsoil. Preventing extremes of soil moisture and soil temperature regimes, minimizing soil erosion, and maintaining a continuous vegetative cover are the general principles that should limit the hardening process. Soils with a potential risk of being laterized should be maintained under a forest cover. If clearing is inevitable, these soils should not be intensively used. Soil moisture and temperature regimes can be regulated through mulch farming techniques involving conservation tillage, no-till or reduced tillage, frequent use of cover crops, agroforestry, and so on. Most of these systems are not necessarily high yielding; however, they facilitate economic crop production without causing further soil degradation.

C. Salinity Control and Water Management

Water management in arid regions is the key to salinity control. It is overirrigation and use of poor quality water that cause waterlogging and salt accumulation in the root zone. The irrigation schedule should meet the minimum water requirement and no more. Wherever economically feasible, drip irrigation should be used. Reclamation of saline soils, however, requires flood irrigation and a good drainage system to flush salts through the soil.

Crop yields in sub-Saharan Africa and elsewhere in the semiarid regions can be substantially improved by providing supplementary irrigation.

It is desirable to give small-scale irrigation projects high priority for resource-poor farmers. This may involve building farm-ponds and other small reservoirs for runoff storage and possible gravity irrigation. Exploiting the groundwater reserves through bore holes for irrigating 1–2 ha is another alternative. Multimillion-dollar grandiose schemes, if considered at all, should be given lowest priority. Irrigation must be expanded in the form of highly efficient small-scale wells or farm ponds.

Before designing and installing any irrigation system, due consideration should be given to providing adequate surface and subsurface drainage. The lack of adequate drainage has caused severe salinization in northwest India and in Pakistan, southern and southeastern Australia, and in the western United States. In Pakistan, 1.9 million ha have already been lost to salinity; another 4.5 million ha have developed saline patches since the inauguration of canal colonies in the 1930s (Snelgrove, 1967). Alkali soils have occurred extensively in Indo-Gangetic plains and now cover approximately 2.5 million ha (Narayana and Abrol, 1981). Worthington (1977) observed that salt-affected soils amount to 50% of the irrigated area in Iraq, 23% in Pakistan, 50% in the Euphrates valley of Syria, 30% in Egypt, and more than 15% in Iran. These problems could have been minimized or entirely prevented, if provisions for drainage were made at the planning stage, seepage losses were reduced from the distributory canals, and irrigation scheduled and controlled to supply the minimum water required.

Existing swamps can be made highly productive by providing proper drainage and primary health facilities to the people. Rice production in west Africa can be substantially increased by drainining large swamp areas. Once again providing drainage and water control systems should be done through simple techniques involving ditches, dikes, and water outlets. Soil surveys are, however, necessary to delinate "acid sulfate" soils. There are approximately 13 million ha of acid sulphate soils (Beek *et al.*, 1980), most of which lie in the tropics of Asia and Africa. The older, deeply developed soils can be used with balanced fertilizer application and proper water management. The newly developed, acid-sulfate soils are better left alone.

Although rice culture or paddy cultivation is a stable land use system and has sustained high population densities in Asia over centuries, continuous rice cultivation on the same land can cause yield reductions. Decline in rice grain yield with continuous cultivation has been reported from the Philippines (Figure 5.15) and the USSR (Figure 5.16). Yield decline may be attributed to build up of pests and diseases. Buildup of harmful organic compounds in the soil due to anaerobic conditions and nutrient imbalance in soil and plant are important contributors. Declining rice yields with continuous cultivation at IRRI, Philippines, are thought to be due to boron toxicity, zinc deficiency, increased disease and insect pressure and

126   R. Lal

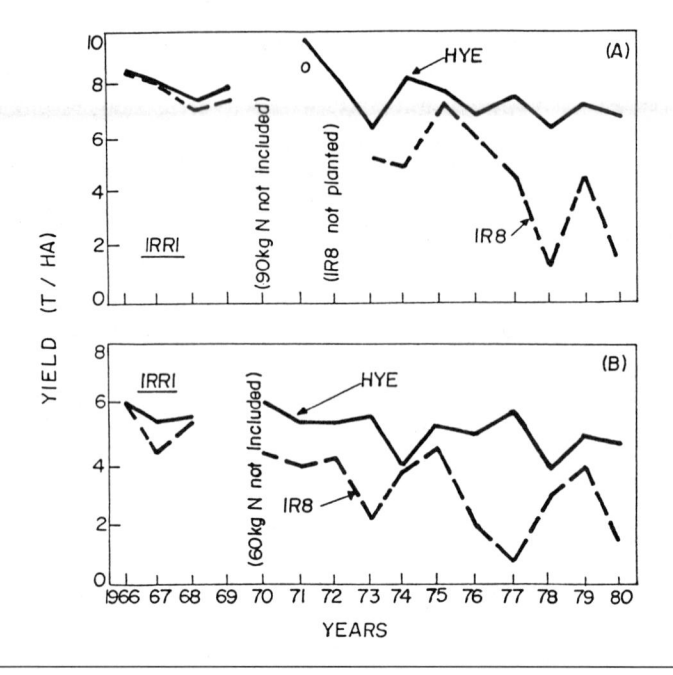

Figure 5.15 Decline in rice grain yield at Los Banos, Philippines, due to continuous cultivation (Flinn and De Datta, 1984). (A) and (B) are two separate but adjacent fields.

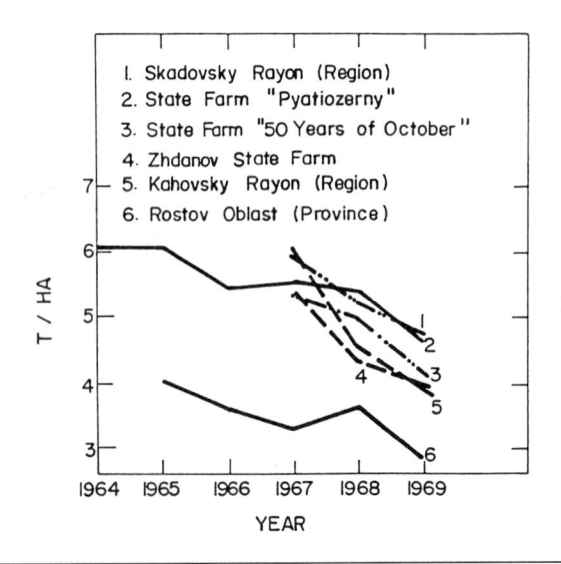

Figure 5.16 Declining trends in rice grain yields on some irrigated lands in the southern USSR, 1964–1969 (Kovda, 1980).

loding (Flinn et al., 1982; Flinn and De Datta, 1984). Soil and crop management techniques based on appropriate rotations should be used to minimize the adverse effects of anaerobic conditions.

Water management is crucial to the use of Vertisols, the clayey soils often called "black cotton soils." There are about 260 million ha of these sticky and cracking soils distributed in semiarid parts of India, Australia, Sudan, and the United States (Beek et al., 1980). When dry, Vertisols have a very hard consistency. When wet, they are very plastic and very sticky. Their optimum range of moisture content for preparation of seedbeds is very narrow. Because of low permeability, these soils are susceptible to erosion, waterlogging, and salinization. Surface-soil management techniques developed at ICRISAT can enable double-cropping of these lands which hitherto were underutilized or unutilized (Kampen et al., 1981). The Gezira scheme in Sudan is a successful gravity irrigation project on some 700,000 ha of soils. The success of this large-scale irrigated project is partly from the good quality of the irrigation water from the Blue Nile.

Technological innovations are also being developed and successfully adapted for management of peat soils. There are some 240 million ha of these organic soils around the world. In the tropics, there are examples of successful cultivation of these soils to grow pineapple and vegetables. If improperly managed, however, they can be easily oxidized and eroded or otherwise lost.

D. Fertility Maintenance

There is a heavy dependence on fertilizers, especially in North American and west European agriculture. Energy and environmental costs of such a farming strategy are high. The standards of western agriculture (involving plowing, and application of fertilizers, pesticides, and irrigation) are often wasteful. Rather than depending solely on chemical fertilizers, the farming systems that are adopted should be those that do not deplete soil fertility. Frequent use of leguminous crops in rotation with cereals, and liberal use of compost, farmyard manure, and city wastes, should reduce dependence on chemical fertilizers. The efficiency of fertilizer use should also be improved by decreasing losses in water runoff, eroded soil, in seepage water, and by volatilization. The intensive agricultural systems, followed for hundreds to thousands of years in many regions of south and southeastern Asia, have useful lessons to offer western agriculture.

Fertilizer use in African agriculture is the lowest in the world; so, regrettably, are yields. Africa uses fertilizers at about 18 kg/ha of arable land and land in permanent crops, compared with an average of 67 kg/ha used in Asia and 218 kg/ha in western Europe (FAO, 1981; Sheldon and

McCune, 1984). The whole African continent uses about the same total nutrients as West Germany, but has 119 times the land area and 78 times the population. While there is an urgent need to increase African yields per unit area per unit time, it should be accomplished by judicious use of chemical fertilizers. Increasing crop yields definitely means increasing the nutrient inputs, because low soil fertility is an important factor responsible for low yields in African agriculture (Breman and de Wit, 1983). The inputs may be in the form of biological nitrogen fixation by growing legumes in rotation with or in association with nitrophilic crops (Lal *et al.*, 1978, 1980; Wilson and Lal, 1986), mixed and relay cropping (Okigbo, 1978); agroforestry systems including alley cropping (Kang *et al.*, 1985), or ley farming techniques (Sheldon and McCune, 1984), and supplemented by chemical fertilizers. The Australian ley farming system has wide application for soils of Africa, Spain, and Portugal (Avery, 1985).

The objective is to provide balanced nutrition for crops to be grown. Productivity of about 300 million ha of acid soils of Cerrado in Brazil can be substantially increased by adding phosphatic fertilizers, and adopting improved cultivars and appropriate farming systems. The Oxisols and Ultisols of the Amazon basin can also be put to continuous and intensive land use by using balanced fertilizers, mulch-farming techniques, and the right crop rotations (Sanchez and Salinas, 1981). Sanchez *et al.* (1982) reported that three grain crops can be grown annually with appropriate fertilizer inputs.

IV. LAND DEGRADATION AND WORLD FOOD PRODUCTION

On the basis of the limited quantitative data available, effects of land degradation on soil productivity are hard to generalize. The magnitude of effect depends on inherent soil properties, the stage or degree of degradation, and management. The generalized soil degradation pattern and its effect on yield are shown in Figure 5.17 for two soils. All soils follow three phases of degradation. The first phase (represented by segments AB and ab) represents none or slight degradation. The second phase (represented by segments BC and bc) represents the rapid rate of soil degradation whereby crop yields experience a drastic decline. The segment CD and cd represents the third phase with a slow rate of degradation. The soil is so degraded that it does not degrade any further. The duration of each stage depends on soil properties and management. Crop yields within each phase also vary with crop and management system. For example, in northeast Thailand Attaviroj (1986) observed differences in yield reduction

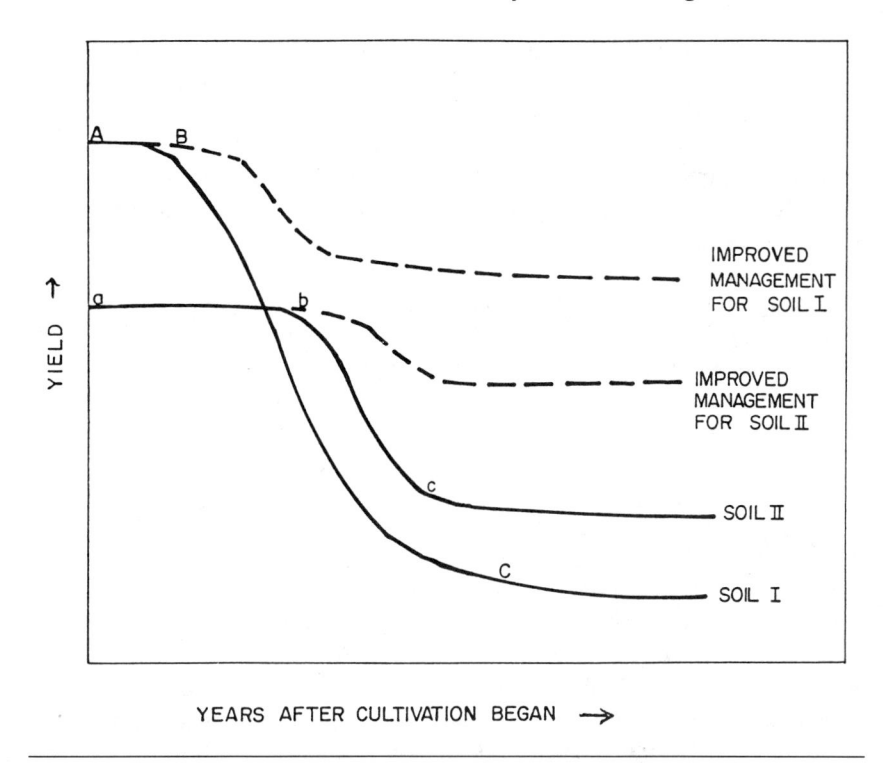

Figure 5.17 Generalized diagram of a soil degradation pattern and its effects on yield.

among maize and rice. In severe cases, when the soil is at the third phase of degradation, the yield may be as low as 10% of that obtained from an undegraded soil.

Land degradation has plagued mankind for thousands of years. Some thriving civilization in the past have undoubtedly succumbed to this self-imposed menace. Archaeological evidences have shown that land degradation was responsible for the extinction of the Riparian and Harappan Kalibangan civilization in western India, the Mayan culture in central America, the ancient kingdoms of Lydia and Mesopotamia, and many others (Olson, 1981). In spite of all these factors, global production has experienced quantum jumps during the last three to five decades. No doubt there are pockets of insufficient and erratic food supply, but the world agricultural production at present is higher than ever before. Also gaining momentum are the rates of accelerated soil erosion, compaction, and waterlogging observed in newly developed and existing farmlands. A disturbing but relevant and timely question is, Are we headed toward the same fate as some of the extinct civilizations? Is the world running out of good arable lands?

Seeking an objective answer to these questions is a challenge, because it is often difficult to separate emotions from facts. If statistics on land degradation are correct, there is definite cause for concern, and the challenge to mankind and to agricultural scientistics is the greatest. If the statistics on land degradation are nearly correct and their consequences as drastic as often portrayed, it is difficult to comprehend why urgency is lacking on the part of policy makers to do something about it? Perhaps it is because the scientific community not only over-dramatized the issue, thereby creating a credibility problem but also scientists did not prioritize the ecological regions where risks of gross ecological imbalance are the greatest.

It is true, however, that accelerated soil erosion can cause drastic reductions in crop yields. Equally drastic reductions of agrucultural production are brought about by accumulation of salts in the root zone, soil compaction, laterization, and waterlogging. Furthermore, arable lands belonging to productivity classes I and II with no or a few limitations toward intensive landuse are finite and relatively small. For example, Dregne (1982) reported that 65% of land in Africa, 80% in Asia, 76% in Australia and New Zealand, 94% in Europe, 84% in North and Central America, 37.5% in South America, and 73% in the world as a whole belong to classes IV, V, and VI, with respective productivity ratings of medium to low and very low. If these statistics are correct, it implies that most of the available land that can still be brought under cultivation is rather marginal, belonging to classes V and VI and with severe limitations to intensive land use.

There is another, more positive, view of this most basic of all problems facing agricultural scientists and planners. Rapid advances have been made in the past three to four decades in agricultural technologies, some of which have only briefly been described in the previous sections. Whenever applied, these innovations have brought about quantum jumps in food production, e.g., the Green Revolution in Asia. The technological advances, however, have not reached all the places they are needed. Improved crop production technologies are awaiting successful transfer in west and central Africa and in some countries of South and Central America. Agricultural scientists are also at the threshold of making innovative discoveries. Scientific advances are imminent in (1) methods of recycling agricultural wastes; (2) reducing climatic vulnerability by developing weather prediction technologies and climatic data collation and evaluation systems; (3) minimizing environmental contamination by developing integrated pest management systems, including increasing use of biological pest control, and adopting a systems approach to agricultural management;

(4) reducing energy costs by diversifying sources of energy for agriculture, e.g., wind and solar power, biological fertilizers, and hydroponics; (5) expanding agriculture to arid regions by developing new sources of fresh irrigable water; and (6) controlling soil erosion and degradation by expanding the use of no-till and conservation tillage techniques. Although the role of biotechnology is not in the scope of this review, the pressure on land may be drastically reduced by applying biotechnology to increasing agricultural production.

But adopting a positive approach to agricultural research and development and to restoring our ever-dwindling land productivity does not justify complacency. The challenge for conserving our soils, reclaiming salt-affected lands, preventing laterization, avoiding soil compaction and waterlogging, and draining swamps and waterlogged areas will continue increasing well into the 22nd century, when the world population is expected to stabilize at 10.5 billion. We can no longer afford to lose even an acre of our precious land. To be negligent is to call for our demise, just as happened to Mayan and other ancient cultures. The perpetual food deficit and mass starvation and undernourishment in sub-Saharan Africa results at least partly from misuse of land resources leading to severe land degradation. Population pressure *per se* is certainly not the only cause of human suffering. In 1981, the population pressure in Africa was only 16 persons/km^2 compared with 98 in Europe, 316 in Japan, 1151 in Bermuda and 4932 in Hong Kong (UN, 1982). The problem lies in realizing that maintaining high productivity of land resources is the first priority of any successful culture or economic and political system.

V. NEED TO RESTORE PRODUCTIVITY OF DEGRADED LANDS

The world has the capacity to feed itself and to maintain the minimum required standard of living for each of its expected 10.5 billion inhabitants. To achieve this, however, implies adopting a strict code of conduct in managing the world's presently cultivable land resources, and in restoring the productivity of lands that have been rendered unproductive by past mismanagement. The production base can be vastly expanded even if only 50% of the supposedly 2 billion ha of now degraded lands can be brought under cultivation. Although coercive measures have rarely been successful in the past, regional and global legislation will be required for proper use of existing lands, to restore productivity of lands disturbed by nonagricultural activities (mining, urbanization, etc.), and to rehabilitate lands that have been abandoned for declines in agricultural productivity. Although there is a need to develop new conservation-effective measures, some of the available technologies should be immediately implemented.

Land resources are not to be preserved; they must be used, restored, and improved. The policy is to conserve, sustain, and enhance productivity for human welfare and use. For example, it is estimated that there are some 800 million ha of potentially arable land reserves in the humid tropical forest region of South America, central and western Africa, and in southeastern Asia and Oceania. The reduction in fallow period for the traditional shifting cultivation followed in these regions is an inevitable fact of life that we have to learn to live with. Appropriate land use and soil and crop management systems have to be adapted to adjust to these changing realities.

Assessing the degree of land degradation, establishing the cause-effect relationships, and developing methods of restoring productivity of degraded lands are important research priorities. Although the world has the capacity to feed itself, it is still up to agricultural scientists and planners to make this dream come true.

VI. CONCLUSIONS

The intensive use of agricultural land, a necessity for increasing agricultural production, increases the risks of environmental and land degradation. Land degradation may cause alterations in the climax vegetation from forest to savanna and to scrub grasses, increase risks of soil erosion by wind and water through increasing susceptibility of a soil to crusting and compaction, decline in soils organic matter content and in plant-available nutrient reserves, increase in the probability of nutrient-imbalance (e.g., Al^{+3} or Mn^{+2} toxicity), and waterlogging and salt accumulation in the root zone. Technological innovations are available, however, that may decrease the risks of land degradation while sustaining an economic level of production. An ecological approach to resource management should be adopted, one that not only includes appropriate technical methods of soil and crop management but also considers the socioeconomic aspects of implementation.

Technological options that provide effective conservation may not be high yielding, but they do provide yields that are sustainable over long time horizons. These measures include no-till and mulch farming, planted fallows and cover crops, agroforestry, balanced use of fertilizer, and appropriate crop rotations and crop sequences. It is also important that appropriate steps be taken to restore the productivity of degraded lands. Land resources are finite and non-renewable. They should be used, improved, and restored.

ACKNOWLEDGMENT

Salaries and research support were provided by state and federal funds appropriated to the OARDC, The Ohio State University. Manuscript #217-87.

REFERENCES

Adelhelm, R., and Kotschi, J. (1986). Environmental protection and sustainable landuse: implications for technical cooperation in the rural tropics. *Q. J. Int. Agric.* **25**, 100–111.
Alberts, E. E., Schuman, G. E., and Burwell, R. E. (1978). Seasonal runoff losses of nitrogen and phosphorus from Missouri Valley loess watershed. *J. Environ. Qual.* **7**, 208–220.
Attaviroj, P. (1986). Soil erosion and degradation in northern Thai uplands. *Proc. Int. Conf. Econ. Dryland Degradation Rehabil., Canberra, Aust.* Annex.
Avery, D (1985). U.S. farm dilemma: The global bad news is wrong. *Science* **230**, 408–412.
Bach, W. (1976). Global air pollution and climatic change. *Rev. Geophys. Space Phys.* **14**, 429–474.
Bach, W. (1986). Trace gases and their influence on climate. *Nat. Res. Dev.* **24**, 9–124.
Bali, Y. P., and Kanwar, J. S. (1977). Soil degradation in India. In FAO, Assessing Soil Degradation, FAOUNEP consultation, Rome. *FAO Soils Bull.* No. 34.
Bandel, K., and Salameh, E. (1981). "Hydrochemical and Hydrobiological Research of the Pollution of the Waters of the Amman Zerka Area, Jordan." Ger. Agency Tech. Coop., Eschborn, Germany.
Barber, R. G. (1983). The magnitude and sources of soil erosion in some humid and semi-arid parts of Kenya, and the significance of soil loss tolerance values in soil conservation in Kenya. *In* "Soil and Water Conservation in Kenya" (D. B. Thomas and W. M. Senga, eds.), Occasional Paper No. 42, pp. 20–46. Univ. of Nairobi, Nairobi, Kenya.
Barnett, A. P., Carreker, J. R., and Abruna, F. (1972). Soil and nutrient losses in runoff with selected cropping treatments in tropical soils. *Agron. J.* **64**, 391–395.
Beasley, R. P. (1973). "Erosion and Sediment Pollution Control." Iowa State Univ. Press, Ames.
Beek, K. J., Blokhuis, W. A., Driessen, P. M., Van Breemen, N., Brinkman, R., and Pons, L. J. (1980). "Problem Soils: Their Reclamation and Management," ILRI Publ. No. 27, pp. 47–72. ILRI, Wageningen, Netherlands.
Biswas, M. R., and Biswas, A. K. (1978). Loss of productive soil. *Int. J. Environ. Stud.* **12**, 189–197.
Bolton, J. (1968). Leaching of fertilizers applied to a latosol in lysimeters. *J. Rubber Res. Inst. Malaya* **20**, 274–284.
Brady, N. C. (1982). Chemistry and world food supply. *Science* **218**, 847–853.
Breman, H., and de Wit, C. T. (1983). Rangeland productivity and exploitation in the Sahel. *Science* **221**, 1341.
Brown, L. R. (1981a). "Building a Sustainable Society." Worldwatch Inst., Washington, D.C.
Brown, L. R. (1981b). World population growth, soil erosion and food security. *Science* **214**, 995–1002.
Brown, L. R. (1984). "State of the World, 1984." Norton, New York.
Bryson, R. A. (1974a). Climate change and agricultural responses. *In* "A Statement on Research And Technological Priorities between Now and the Year 2000." Inst. Environ. Stud., Univ. of Wisconsin, Madison.

Bryson, R. A. (1974b). A perspective on climatic change. *Science* **184**, 753–760.
Buol, S. W., Hole, F. D., and McCracken, R. J. (1973). "Soil Genesis and Classification." Iowa State Univ. Press, Ames.
Buringh, P. (1981). "An Assessment of Losses and Degradation of Productive Agricultural Land in the World," FAO Working Group Soils Policy. FAO, Rome.
Burwell, R. E., Schuman, G. E., Piest, R. E., Spomer, R. G., and McCalla, T. M. (1974). Quality of water discharged from two agricultural watershed in south western Iowa. *Water Resour. Res.* **10**, 359–365.
Chapman, E. C. (1978). Shifting cultivation and economic development in the lowlands of northern Thailand. *In* "Farmers in the Forest" (P. Kunsfadter *et al.*, eds.), pp. 222–235. East–West Cent., Honolulu, Hawaii.
Charreau, C., and Nicou, R. (1971). L'amelioration du profil cultural dans les sols sableux et sablo argileux de la zone tropicale seche ouest Africaine et ses incidences agronomiques. *Agron. Trop.* **26**, 209–255, 565–631, 903–978, 1184–1247.
Chepil, W. S., and Woodruff, N. P. (1963). The physics of wind erosion and its control. *Adv. Agron.* **15**, 211–302.
Cloudsley-Thompson, J. (1977). "The Desert." Putnam, New York.
Council on Environmental Quality and Department of State. (1980). "The Global 2000 Report to the President, Vol. 1, Entering the Twenty-First Century. U.S. Gov. Print. Off., Washington, D.C.
Crutzen, P. J., Delany, A. C., Greenberg, J., Haagenson, P., Heidt, L., Leub, R., Pollock, W., Seiler, W., Wartburg, R., and Zimmerman, P. (1985). Tropospheric chemical composition measurements in Brazil during the dry season. *J. Atmos. Chem.* **2**, 233–256.
Davies, D. B. (1982). Soil degradation and soil management in Britain. *In* "Soil Degradation" (D. Boels, D. B. Davies, and A. E. Honston, eds.), pp. 19–26. Balkema, Rotterdam.
Dent, F. J. (1984). Land degradation: present status, training and education needs in Asia and the Pacific. *In* "UNEP Investigations on Environmental Education and Training in Asia and the Pacific." FAO Reg. Off., Bangkok.
Dolgilevich, M. I. (1972). Theoretical and experimental investigations aimed at protecting soils in Ukraine from wind erosion. Ph.D. Thesis, Dokutchaev Soil Inst.) Moscow.
Dregne, H. E. (1982). "Impact of Land Degradation on Future World Food Production," ERS-677. USDA-ERS, Washington, D.C.
Dudal, R. (1982). Land degradation in a world perspective. *J. Soil Water Conserv.* **37**, 245–247.
Dunigan, E. P., Phelon, R. A., and Mondart, C. L. (1976). Surface runoff losses of fertilizer elements. *J. Environ. Qual.* **5**, 339–342
Edwards, W. M., Simpson, E. C., and Frere, M. H. (1972). Nutrient content of barnlot runoff water. *J. Environ. Qual.* **1**, 401–405.
Elwell, H. A., and Stocking, M. A. (1982). Developing a simple yet practical method of soil loss estimation. *Trop. Agric.* **59**, 43–48.
FAO. (1960). Wind Erosion." FAO, Rome.
FAO. (1981). "FAO Fertilizer Yearbook." Rome.
FAO/UNEP. (1983). "Guidelines for the Control of Soil Degradation." FAO, Rome.
Finn, D. (1983). Land use and abuse in the East African region. *Ambio* **12**, 296–301.
Flinn, J. C. and De Datta, S. K. (1984). Trends in irrigated rice yields under intensive cropping at Philippine research stations. *Field Crops Res.* **9**, 1–15.
Flinn, J. C., De Datta, S. K., and Labadan, E. (1982). An analysis of longterm rice yields in a wetland soil. *Field Crops Res.* **5**, 201–216.
Floret, C., and Le Floch, E. (1973). "Production, sensibilite et evolution de la vegetation et du milieu en Tunisie presaharienne," CEPE Doc. No. 71. Montpellier, France.
Floret, C., Le Floch, E., Pontainier, R., and Romane, F. (1977). Case study on desertification Oglat Merteba Region, Tunisia. *Proc. UN Conf. Desertif.*, Nairobi, Kenya Mimeo.

Fournier, F. (1967). Research on soil erosion and soil conservation in Africa. *Afr. Soils* **12**, 53–96.
Fukuda, H. (1976). "Irrigation in the World." Univ. of Tokyo Press, Tokyo.
Gallais, J. (1979). La situation de l'elevage bovin et le problem des eleveurs en Afrique occidentale et centrale. *Cah. Outre- Mer* **32**, 113–138.
Ghuman, B. S., and Lal, R. (1987). Effects of partial clearing on microclimate in a humid tropical forest. *Agric. For. Meteorol.* **40**, 17–29.
Glantz, M. H. (1977). "Desertification: Environmental Degradation In and Around Arid Lands." Westview Press, Boulder, Colorado.
Goudie, A. (1981). "The Human Impact: Man's Role in Environmental Change." MIT Press, Cambridge, Massachusetts.
Greenland, D. J. (1986). Soil organic matter in relation to crop nutrition and management. *Int. Conf. Manage. Fertil. Upland Soils, Nanjing, China*.
Greenland, D. J., and Murray-Rust, D. H. (1986). Irrigation demand in humid areas. *Philos. Trans. R. Soc. London, Ser. A* **316**, 275–295.
Grigg, D. (1985). "The World Food Problem, 1950–1980." Blackwell, Oxford.
Grigoriev, V. Y., Kuznetsov, M. S., and Glazunov, G. P. (1976). Experience in the valuation and antideflational stability of soils. *In* "Laws of the Manifestation of Erosional and Fluvial Processes in Different Natural Conditions." Moscow State Univ. Press, Moscow.
Grosse, B. (1967). The productivity of severely eroded Parabraunerde formed from loess under moderate climatic conditions. *Trans. Int. Congr. Soil Sci., 8th* **2**, 729.
Hagedorn, H., Giessner, K., Weise, O., Busche, D., and Grunert, G. (1977). "Dune Stabilization." Ger. Agency Tech. Coop., Eschborn, Germany.
Hamblin, P. A. (1985). The influence of soil structure on water movement, crop root growth and water uptake. *Adv. Agron.* **38**, 95–158.
Hardy, F. (1933). Cultivation properties of tropical red soils. *Emp. J. Exp. Agric.* **1**, 103–112.
Hayes, M. H. B. (1986). Soil organic matter extraction, fractionation, structure and effects on soil structure. *In* "The Role of Organic Matter in Modern Agriculture" (Y. Chen and Y. Avnimelech, eds.), pp. 182–108. Nijhoff, Dordrecht, Netherlands.
Hudson, N. W. (1971). "Soil Conservation." Batsford, London.
Hulugalle, N. R., Lal, R., and ter Kuile, C. H. H. (1986). Amelioration of soil physical properties by Mucuna after mechanized land clearing of a tropical rainforest. *Soil Sci.* **141**, 219–224.
Humphreys, G. S. (1984). "The Environment and Soils of Chimbu Province, Papua New Guinea with Particular Reference to Soil Erosion." Res. Bull. No. 35. Dep. Primary Ind., Port Moresby, Papua New Guinea.
Jackson, W. A., Asmussen, L. E., Hausen, E. W., and White, A. W. (1973). NO_3 in surface and sub-surface flow from a small agricultural watershed. *J. Environ. Qual.* **2**, 480–482.
Jones, M. J., and Wild, A. (1975). "Soils of the West African Savanna." Commonw. Agric. Bur., Harpenden, England.
Kampen, J., Hair Krishna, J., and Pathak, P. (1981). Rainy season cropping on deep Vertisols in the semi-arid tropics—effects on hydrology and soil erosion. *In* "Tropical Agricultural Hydrology" (R. Lal and E. W. Russell, eds.), pp. 257–272. Wiley, New York.
Kang, B. T., Wilson, G. F., and Lawson, T. L. (1985). "Alley Cropping." Int. Inst. Trop. Agric., Ibadan, Nigeria.
Klamt, E., Mielniczuk, J., and Schneider, P. (1986). Degradation of properties of Red Brazilian subtropical soils by management. *Proc. Symp. Red Soils, Nanjing, China*.
Klausner, S. D., Zerman, P. J., and Ellis, D. F. (1974). Surface runoff losses of soluble N and P under two systems of soil management. *J. Environ. Qual.* **3**, 42–46.

Klepper, R. (1978). Nitrogen fertilizer and nitrate concentrations in tributaries of the upper Sangamon river in Illinois. *J. Environ. Qual.* **7,** 13–22.

Kovda, V. A. (1980). "Land Aridization and Drought Control," p. 277. Westview Press, Boulder, Colorado.

Kowal, J. (1972). The hydrology of a small catchment basin at Samaru, Nigeria. IV. Assessment of soil erosion under varied land management and vegetation cover. *Niger. Agric. J.* **7,** 143–147.

Kubota, T., Verakpatananirund, Piyapongse, P., and Phetchawee, S. (1982). Improvement of the moisture regime of upland soils by soil management. In Proc. Symp. on Distribution, Characteristics and Utilization of Problem Soils, Japanese Society of Soil Science and Plant Nutrition. *Trop. Agric. Res. Ser.* No. 15, 351–372.

Kuipers, H. (1982). Processes in physical soil degradation in mechanised agriculture. *In* "Soil Degradation" (D. Boels, D. B. Davies, and A. E. Johnston, eds.), pp. 7–18. Balkema, Rotterdam.

Lal, R. (1976a). "Soil Erosion Problems on an Alfisol in Western Nigeria and their Control," Monogr. No. 1. IITA, Ibadan, Nigeria.

Lal, R. (1976b). Soil erosion on Alfisols in western Nigeria. *Geoderma* **16,** 363–431.

Lal, R. (1977). Losses of plant nutrients in runoff and eroded soil. *In* "Nitrogen Cycling in West African Ecosystems" (T. Russwell, ed.), pp. 31–38. SIDA, Uppsala, Sweden.

Lal, R. (1979). Physical characteristics of the soils of the tropics: determination and management. *In* "Soil Physical Properties and Crop Production in the Tropics" (R. Lal and D. J. Greenland, eds.), pp. 7–40. Wiley, New York.

Lal, R. (1981). Deforestation and hydrological problems. *In* "Tropical Agricultural Hydrology" (R. Lal and E. W. Russell, eds.), pp. 131–140. Wiley, New York.

Lal, R. (1983). "No-Till Farming," IITA Monogr. No. 2. Ibadan, Nigeria.

Lal, R. (1984a). Mechanized tillage systems effects on physical properties of an Alfisol in watershed cropped to maize. *Soil Tillage Res.* **4,** 349–360.

Lal, R. (1984b). Soil erosion from tropical arable lands and its control. *Adv. Agron.* **37,** 183–248.

Lal, R. (1985). Mechanized tillage systems effects on physical properties of an Alfisol in watersheds cropped to maize. *Soil Tillage Res.* **6,** 149–161.

Lal, R. (1986a). No-tillage and surface tillage systems to alleviate soil-related constraints in the tropics. *In* "No-Tillage And Surface Tillage Agriculture" (M. A. Sprague and G. B. Triplett, eds.), pp. 261–318. Wiley, New York.

Lal, R. (1986b). Conversion of tropical rainforest: agronomic potential and ecological consequences. *Adv. Agron.* **39,** 173–264.

Lal, R. (1986c). Soilsurface management in the tropics for intensive landuse and high and sustained production. *Adv. Soil Sci.* **5,** 1–109.

Lal, R. (1987a). "Tropical Ecology and Physical Edaphology." Wiley, New York.

Lal, R. (1987b). Managing soils of sub-Saharan Africa. *Science* **236,** 1069–1076.

Lal, R., and Cummings, D. J. (1979). Clearing a tropical forest. I. Effects on soil and microclimate. *Field Crops Res.* **2,** 91–107.

Lal, R., Wilson, G. F., and Okigbo, B. N. (1978). No-tillage farming after various grasses and leguminous cover crops in tropical Alfisols. I. Crop performance. *Field Crops Res.* **1,** 71–84.

Lal, R., De Vleeschauwer, D., and Nganje, R. M. (1980). Changes in properties of a newly cleared Alfisol as affected by mulching. *Soil Sci. Soc. Am. J.* **44,** 827–833.

Lal, R., Nelson, M., Scharpenseel, H. W., and Sudjadi, M. (1987). "Land Clearing for Sustainable Agriculture." IBSRAM, Bangkok.

Lavelle, P. (1983). The soil system in the humid tropics. *Proc. Symp. Savanna Woodland Ecosyst. Trop. Am. Afr.: Comparison, ICUS/UNDP, Brasilia* pp. 2–17.

Lawes, D. A. (1962). The influence of rainfall conservation on the fertility of the loess plain soils of northern Nigeria. *J. Geogr. Assoc. Nigeria* **5**(1), 33–38.

Lee, L. (1980). The impact of land ownership factors on soil conservation. *Am. J. Agric. Econ.* **62**, 1070–1076.
Le Houerou, H. N. (1977a). The scapegoat. *Ceres* **10**, 14–18.
Le Houerou, H. N. (1977b). The nature and causes of desertization. *In* "Desertification" (M. H. Glantz, ed.), pp. 17–38. Westview Press, Boulder, Colorado.
Long, E. L., Lund, Z. F., and Hermanson, R. E. (1975). Effect of soil incorporated dairy cattle manure on runoff water quality and soil properties. *J. Environ. Qual.* **4**, 163–166.
Lowdermilk, W. C. (1953). Conquest of the land through 7000 years. *Agric. Inf. Bull. (U.S. Dep. Agric.)* No. 99.
Lyle, L. (1975). Possible effects of wind erosion on soil productivity. *J. Soil Water Conserv.* **30**, 279–283.
Mabbutt, J. A. (1978). The impact of desertification as revealed by mapping. *Environ. Conserv.* **5**, 45–56.
Mabbutt, J. A. (1984). A new global assessment of the status and trends of desertification. *Environ. Conserv.* **11**, 103–113.
McCown, R. L. (1984). "An Agro-Ecological Approach to Management of Alfisols in the Semi-Arid Tropics," Mimeo. Div. Trop. Crops Pastures, CSIRO, Aitkenvale, Queensland, Australia.
McCown, R. L., Haaland, G., and de Hann, C. (1979). The interaction between cultivation and livestock production in semi- arid Africa. *In* "Ecological Studies 34, Agriculture in Semi-Arid Environments" (A. R. Hall, G. H. Cannell, and H. W. Lawton, eds.), pp. 297–332. Springer-Verlag, Berlin and New York.
McCown, R. L., Jones, R. K., and Peake, D. C. I. (1985). Evaluation of a no-till tropical legume ley farming strategy. *In* "Agro-Research for Australia's Semiarid Tropics" (R. C. Muchow, ed.), pp. 450–472. Univ. of Queensland, Brisbane, Australia.
McKenzie, D. C., Abbott, T. S., and Higginson, F. R. (1984). The effect of 15 years of irrigated cotton production on the properties of a sodic Vertisol. *Proc. Natl. Soils Conf., Brisbane, Aust.*
McNamara, R. S. (1985). The challenges for sub-Saharan Africa. Sir John Crawford Mem. Lect., Washington, D.C.
McNeil, M. (1964). Lateritic soils. *Sci. Am.* **211**(5), 96.
Martin, A. E., and Cox, J. E. (1956). N studies on black soils from Darling Downs, Queensland, Australia. *J. Agric. Res.* **7**, 169–193.
Martin, A. E., and Skyring, G. W. (1982). Losses of nitrogen from soil-plant system. A review of nitrogen from the tropics with particular reference to pastures. *Commonw. Agric. Bur. Bull.* **46**, 19–34.
Mason, W., Small, D., and Pritchard, K. (1984). Soil and irrigation management for intensive fodder crop production in Northern Victoria. *Proc. Natl. Soils Conf., Brisbane, Aust.*
Mathieu, C. (1982). Effects of irrigation on the structure of heavy clay soils in North-East Morocco. *Soil Tillage Res.* **2**, 311–329.
Mbagwu, J. S. C., Lal, R., and Scott, T. W. (1984). Effects of artificial desurfacing on Alfisols and Ultisols in Southern Nigeria. Parts I & II. *Soil Sci. Soc. Am. J.* **48**, 823–838.
Narayana, V. V. D., and Abrol, I. P. (1981). Effect of reclamation of alkali soils on water balance. *In* "Tropical Agricultural Hydrology" (R. Lal and E. W. Russell, eds.), pp. 283–298. Wiley, New York.
National Research Council (1984). "Environmental Change in the West African Sahel," (NRC). Board on Science & Technology for International Development. Natl. Res. Counc., Natl. Acad. Press, Washington, D.C.
Newbould, P. (1982). Losses and accumulation of organic matter in soils. *In* "Soil Degradation" (D. Boels, D. B. Davies, and A. E. Johnston, eds.), pp. 107–131. Balkema, Rotterdam.

Ngatunga, E. L. N., Lal, R., and Uriyo, A. P. (1984). Effects of surface management on runoff and soil erosion from some plots at Mlingano, Tanzania. *Geoderma* **33**, 1–12.

Nicou, R. (1974). The problem of caking with drying out of sandy and sandy–clay soils in the arid tropical zone. *Agron. Trop.* **30**, 325–343.

Oades, J. M. (1984). Soil organic matter and structural stability: mechanisms and implications for management. *Plant Soil* **76**, 319–337.

Obeng, H. B. (1978). Soil, water, management and mechanization. *Afr. J. Agric. Sci.* **5**, 71–83.

Okigbo, B. N. (1978). "Cropping Systems and Related Research in Africa," 10th Annu. Spec. Publ. AAAS, Addis Ababa, Ethiopia.

Olson, G. W. (1981). Archaeology: lessons on future soil use. *J. Soil Water Conserv.* **36**, 261–264.

Paulino, L. A. (1986). "Food in the Third World, Past Trends and Project ion to 2000," IFPRI Res. Rep. No. 52. Int. Food Policy Res. Inst., Washington, D.C.

Pereira, H. C., Hosegood, P. H., and Thomas, D. B. (1961). The productivity of tropical semiarid thorn-scrub country under intensive management. *Emp. J. Exp. Agric.* **29**, 269–286.

Pereira, H. C., Chenery, E. M., and Mills, W. R. (1954). The transient effects of grasses on the structure of tropical soils. *Emp. J. Exp. Agric.* **22**, 148–160.

Perrens, S. J. (1986). Conversion of forest land to annual crops: Australian experience. *In* "Landuse, Watersheds, and Planning in the Asia-Pacific Region," RAPA Rep. 1986/3, pp. 112–137. FAO Reg. Off. Asia Pac., Bangkok.

Pimentel, D., and Levitan, L. (1986). Pesticides: amounts applied and amounts reaching pests. *Bioscience* **36**, 86–91.

Pimentel, D., Terhaune, E. C., Dyson-Hudson, R., Rochereau, S., Samis, R., Smith, E. A., Denman, D., Reifschmeider, D., and Shepard, M. (1976). Land degradation: effects on food and energy resources. *Science* **194**, 149–155.

Pimentel, D., Fied, C., Olson, L., Schmidt, S., Wagner-Johnson, K., Westman, A., Whelan, A. M., Foglia, K., Poole, P., Klein, T., Sobin, R., and Bochner, A. (1983). "Biomass Energy: Environmental and Social Costs," Rep. No. 83-2. Dep. Entomol. Sect. Ecol. Syst., Cornell Univ., Ithaca, New York.

Prescott, J. A., and Pendleton, R. L. (1952). Commonwealth Bureau. *Soil Sci. Tech. Commun.* **47**.

Prospero, J., and Carlson, T. (1972). Vertical and areal distribution of Saharan dust over the western equatorial North Atlantic Ocean. *J. Geophys. Res.* **77**, 5255–5265.

Rapp, A. (1974). "A Review of Desertization in Africa (Water, Vegetation, Man)." Secretariat Int. Ecol., Stockholm.

Rapp, A. L., Berry, L., and Temple, P. (1972). Erosion and sedimentation in Tanzania. *Geogr. Annu.* **54A**, Nos. 3 and 4.

Revelle, R. (1982). Carbon dioxide and world climate. *Sci. Am.* **247**, 33–41.

Revelle, R., and Suess, H. E. (1957). Carbon dioxide exchange between atmosphere and ocean and the question of an increase of atmospheric CO_2 during past decades. *Tellus* **9**, 18–27.

Romkens, M. J. M., Nelson, D. W., and Mannering, J. V. (1973). Nitrogen and phosphorus composition of surface runoff as affected by tillage methods. *J. Environ. Qual.* **2**, 292–295.

Roose, E. J. (1977a). Adaptation of soil conservation techniques to the ecological and socioeconomic conditions of West Africa. *Agron. Trop.* **32**, 132–140.

Roose, E. J. (1977b). Application of the USLE of Wischmeier and Smith in West Africa. *In* "Soil Conservation and Management in the Humid Tropics" (D. J. Greenland and R. Lal, eds.), pp. 177–188. Wiley, New York.

Salas, R. M. (1981). "The State of the World Population, 1980." U. N. Fund Popul. Act., New York.

Salati, E., Dall'Olio, A., Matsue, E., and Gat, J. R. (1979). Recycling of water in the Amazon Basin: an isotope study. *Water Resour. Res.* **15**, 1250–1258.

Salati, E., Lovejoy, T. E., and Vose, P. B. (1983). Precipitation and water recycling in tropical rainforests. Commission on Ecology, Occasional Paper No. 2. *Environmentalist* **3**, 67–74.

Sanchez, P. A., and Buol, S. W. (1975). Soils of the tropics and the world food crisis. *Science* **188**, 598–603.

Sanchez, P. A., and Salinas, J. G. (1981). Low input technology for managing Oxisols and Ultisols in tropical America. *Adv. Agron.* **34**, 279–406.

Sanchez, P. A., Bandy, D. A., Hugo Villachila, J., and Nicholaides, J. J. (1982). Amazon Basin soils: management for continuous crop prodution. *Science* **216**, 821–827.

Schuman, G. E., Burwell, R. E., Piest, R. E., and Spomer, R. G. (1973). N losses in surface runoff from agricultural watersheds in Missouri valley loess. *J. Environ. Qual.* **2**, 299–302.

Scott, P. A. (1978). Tropical rainforest in recent ecological thought: the reassessment of a non-renewable resource. *Prog. Phys. Geogr.* **2**, 80–98.

Segalen, P. (1970). "Pedologie et Developpement: Techniques Rurales en Afrique," No. 10. Sec. État Aff. Étrangers, Paris.

Seiler, W., and Fishman, J. (1981). The distribution of carbon monoxide and ozone in the free troposphere. *J. Geophys. Res.* **86**, 7255–7265.

Sheldon, V. L., and McCune, D. L. (1984). Fertilizers in Africa's agriculture future. *In* "Advancing Agricultural Production in Africa" (J. Hawkworth, ed.) Soils Bur.,

Shelton, C. H., and Lessman, G. M. (1978). Quality characteristics of agricultural and waste disposal runoff. *J. Soil Water Conserv.* **33**, 134–139.

Sinclair, J. (1985). Crusting, soils strength and seedling emergence in Botswana. Ph.D. Thesis, Univ. of Aberdeen, Aberdeen.

Singh, G. (1977). Climate changes in the Indian desert. *In* "Desertification And Its Control," pp. 25–30. ICAR, New Delhi.

Snelgrove, A. K. (1967). "Geohydrology of the Indus River, West Pakistan." Sind Univ. Press, Hyderabad. Pakistan.

Soane, B. D., Blackwell, P. S., Dickson, J. W., and Painter, D. J. (1981). Compaction by agricultural vehicles: a review. I. Soil and wheel characteristics. *Soil Tillage Res.* **1**, 207–238.

Spears, J. (1982). Rehabilitating watersheds. *Finn. Dev.* **19**(1): 30–33.

Sprague, M. A., and Triplett, G. B. (1986). "No-Tillage and Surface-Tillage Agriculture: The Tillage Revolution." Wiley, New York.

Stewart, K. M., and Rohlich, G. A. (1977). "Entrophication: A Review," Rep. State Water Qual. Control Board. Sacramento, California.

Suarez de Castro, G. K., and Rodriquez, G. A. (1958). Movimiento del agua en el suelo (estudio en Pisimetros nonoliticos). *Bol. Tec. Fed. Nac. Cafeteros Colomb.* No. 2.

Szabolcs, I. (1986). Agronomic and ecological impact of irrigation on soil and water quality. *Adv. Soil Sci.* **4**, 189–218.

Thornthwaite, C. W. (1948). An approach towards a rational classification of climate. *Geogr. Rev.* **38**: 55–94.

Timmons, D. R., and Holt, R. E. (1977). Nutrient losses in surface runoff from a native prairie. *J. Environ. Qual.* **6**, 369–373.

Timmons, D. R., Verry, E. S., Burwell, R. E., and Holt, R. F. (1977). Nutrient transport in surface runoff from an asperbirch forest. *J. Environ. Qual.* **6**, 188–192.

UNEP. (1977). Desertification: an overview. *U. N. Conf. Desertif., Nairobi, Kenya.*

UNEP. (1982). "Worlds Soil Policy." U. N. Environ. Program, Nairobi, Kenya.

UNEP. (1984). "General Assessment of Progress in the Implementation of the Plan of Action to Combat Desertication, 1978–84," Mimeo. Governing Council, 12th Session, Nairobi, Kenya.

UNEP. (1986). "Farming Systems Principles for Improved Food Production and the Control of Soil Degradation in the Arid, Semi- Arid and Humid Tropics," Expert meet. sponsored by UNEP, 1973. ICRISAT, Hyderabad, India.

Van der Poel, P., and Timberlake, J. (1980). "Sand Dune Stabilization Project, Bokspits, Botswana." Minist. Agric., Gabrone, Botswana.

Veldkamp, W. J. (1986). Soil map of Zambia. Soil survey of Mt. Makulu. Cited in C. E. Mullins (1986). "Hard-Setting Soils," Mimeo. Univ. of Aberdeen, Aberdeen.

Voorhees, W. B., Senst, C. G., and Nelson, W. W. (1978). Compaction and soil structure modification by wheel traffic in the northern Corn Belt. *Soil Sci. Soc. Am. J.* **42,** 344–349.

Walsh, J. (1984). Sahel will suffer even if rains come. *Science* **224,** 467–471.

Watson, C. L. (1984). The effect of saline irrigation water in clay soils: a review. *Proc. Natl. Soils Conf., Brisbane, Aust.*

White, E. M., and Williamson, E. J. (1973). Plant nutrient concentrations in runoff from fertilized cultivated erosion plots and prairie in eastern South Dakota. *J. Environ. Qual.* **2,** 453–454.

Wilkinson, G. R. (1975). Effect of grass fallow rotations in Northern Nigeria. *Trop. Agric.* **52,** 97–103.

Wilson, G. F., and Lal, R. (1986). New concepts for post-clearing land management in the tropics. *In* "Land Clearing And Development In The Tropics" (R. Lal, P. A. Sanchez and R. W. Cummings, eds.), pp. 371–381. Balkema, Rotterdam.

WMO. (1983). "Meteorological Aspects of Certain Processes Affecting Soil Degradation—Especially Erosion," Tech. Note No. 178. WMO, Geneva.

Wood, A. W. (1984). Some aspects of soil degradation under intensive sugercane cultivation in the Herbert Valley. *Proc. Natl. Soils Conf., Brisbane, Aust.*

Woodruff, N. P. (1972). Wind erosion as affected by reduced tillage system. *Proc. No-Tillage Syst. Symp., Ohio State Univ., Columbus* pp. 5–20.

Woodwell. G., *et al.* (1978). The biota and the world carbon budget. *Science* **199,** 141–146.

World Bank. (1985). "Desertification in the Sahelian and Sudanian Zones of West Africa." World Bank, Washington, D.C.

World Resources Institute. (1986). "Annual Report," pp. 270–271. Washington, D.C.

Worthington, E. B., ed. (1977). "Arid Land Irrigation in Developing Countries: Environmetal Problems and Effects." Pergamon, Oxford.

Yoshioka, S., Vi boonsukh, N., Wongwiwatchai, C., Paisarnchareon, K., Somnus, P., and Sittiwong, P. (1987). Dynamic behavior of plant nutrients in upland soils in Northeast Thailand. *Proc. Soil Manage. Under Humid Conditions Asia (ASIALAN) IBSRAM, Bangkok.* pp. 251–269.

6

Water Use in Agriculture

Vashek Cervinka

Agriculture Resources Branch
California Department of Food and Agriculture
Sacramento, California

    I. Introduction
   II. Global Perspective
  III. Water Sources and Systems
  IV. Water in Crop Production
   V. Water in Animal Production
  VI. Water in Food Processing
 VII. Water Quality
VIII. Irrigation Systems
       References

I. INTRODUCTION

People, land, and water are interrelated. Throughout history, the prosperity as well as the decline of civilizations have reflected the management of land and water.

Water, when used for food production, is a scarce resource. Many world regions cannot rely upon the availability and reliability of rain; applied ground or surface water is converting desert or semidesert areas into fertile farmlands. However, this usable water represents less than 0.7% of water resources on the earth. The greatest quantity of water (97.13%) exists in the oceans, and the second major quantity (2.24%) exists as ice and snow in the polar regions. Water in the oceans and polar regions plays an essential role in the hydrological cycle by renewing water resources in rivers and lakes, recharging groundwater supplies, and in the evapotranspiration process in soils and crops. Groundwater represents 0.612%, lakes 0.009%, and streams 0.0001% of the total water resources of the world.

In the following chapters, water use in agriculture will be discussed from the perspective of food production. Technical data will be presented on water usage in crop and animal production, and in the food processing industry. Water quality will be discussed in its effects on soil conditions and production levels of crops. Waste water treatment and usage also will be analyzed. Finally, the most common irrigation methods and engineering systems will be described.

II. GLOBAL PERSPECTIVE

The total amount of farmland in the world is 1,471 million ha, which represents all arable land and land with permanent crops (U. N., 1986). The area of irrigated farmland in the world is 213 million ha, representing 14.5% of the total land used for food production. Considering the world population at the level of 5,000 million people, the ratio of farmland to population is 0.29 ha/person (3.45 persons/ha) for total agricultural land, and 0.04 ha/person (25 persons/ha) for irrigated land.

The largest area of agricultural land is in Asia (31.0% of the total), followed by North and Central America (18.5%), and the Soviet Union (15.8%) (U. N., 1986). Asia has the largest proportion the world's irrigated farmland, 62.5% of the total followed by North and Central America with 12.9% (Table 6.1).

In Asia 29.3% of the land is under irrigation, while the world average is only 14.5% (U. N., 1986). Europe and North and Central America have about 10% of their agricultural land irrigated. Table 6.2 provides a list of countries with the largest area of irrigated farmland, headed by China

Table 6.1 Irrigated Land in the World[a]

Region	Total arable and permanent cropland (1,000 ha)	(%)	Irrigated area (1,000 ha)	(%)	Ratio of irrigated to total land (%)
World	1,471,731	100.0	213,376	100.0	14.5
Africa	183,059	12.4	8,227	3.9	4.5
North and Central America	272,989	18.5	27,485	12.9	10.1
South America	138,694	9.4	7,835	3.7	5.6
Asia	456,037	31.0	133,395	62.5	29.3
Europe	140,366	9.5	15,307	7.2	10.9
Oceania	48,196	3.3	1,981	0.9	4.1
USSR	232,390	15.8	19,146	9.0	8.2

[a]Source: U. N. (1986).

(45,144 million ha) and India (39,500 million ha). The United States and the Soviet Union have a similar area under irrigation, about 19 million ha. The highest intensity of irrigation is in Egypt (100% of farmland irrigated), followed by Pakistan (71.8%), and Japan (67.4%).

As indicated above, the average ratio is 25 persons/ha of irrigated

Table 6.2 Irrigated Land by Country[a]

Country	Total arable and permanent cropland (1,000 ha)	Irrigated area (1,000 ha)	Ratio of irrigated to total land
China	100,894	45,144	44.7
India	168,350	39,500	23.5
USA	189,915	19,831	10.4
USSR	232,390	19,146	8.2
Pakistan	20,490	14,720	71.8
Indonesia	20,310	5,418	26.7
Mexico	23,600	5,250	22.2
Iran	13,700	4,000	29.2
Thailand	19,370	3,472	17.9
Japan	4,806	3,240	67.4
Spain	20,508	3,133	15.3
Italy	12,222	2,950	24.1
Afganistan	8,054	2,660	33.0
Egypt	2,471	2,471	100.0
Turkey	26,390	2,120	8.0
Brazil	74,700	2,100	2.8

[a]Source: U. N. (1986).

farmland in the world. This compares with 37 persons/ha of irrigated farmland in Japan, 22.1 persons/ha in China, 18.9 persons/ha in India, 17.1 persons/ha in Egypt, 14.4 persons/ha in the USSR, 11.9 persons/ha in the United States, and 6.7 persons/ha in Pakistan.

III. WATER SOURCES AND SYSTEMS

A basic concept in the understanding of water sources and systems is the hydrologic cycle. This cycle interrelates the flow of all forms of water through the vapor, liquid, and solid phases, and it is essential for studying water systems in agriculture. The hydrologic cycle consists of precipitation, evaporation, transpiration, infiltration, percolation, runoff, and storage.

Precipitation adds water to the surface of the earth from the atmosphere. It may be either liquid (rain and dew) or solid (snow, frost, and hail). Through the process of evaporation, water is changed into its gaseous form. Water evaporates from the seas, oceans, lakes, or soil. Evaporation from the surface of oceans and seas is the main source of water for land areas. Transpiration is the process by which water is absorbed by the root systems of plants and passes through their living structure and then into the atmosphere. Water soaks into, or is absorbed by, the surface soil layers through the process of infiltration. Percolation is the downward flow of water through soil and permeable rock formations to the groundwater storage area. Water runoff represents a difference between water input by precipitation and/or irrigation and output through evapotranspiration and percolation. It is the gravity movement of residual water in surface channels. At different phases of the hydrologic cycle, water is stored in the seas, oceans, rivers, lakes, dams or in the ground.

The availability of water for agriculture varies by geographical location, ecological conditions, and time. Water supply depends on the interaction and characteristics of individual components of the hydrologic cycle. Changes produced by the management or depletion of natural resources, and by technological decisions on one component of the hydrologic cycle, inevitably may affect the whole system. Well known examples are the effects of deforestation or dam construction on agricultural ecology and the functioning of hydrologic cycles.

Various users compete with farmers for water supplies. Water management needs to satisfy the demands for industrial or domestic use, as well as for power generation, fish and wildlife, navigation, and recreational use.

Balancing water supplies and demands requires the knowledge and understanding of natural hydrologic cycles, how these cycles are affected by actions of people, and how water resources are utilized. The system may be a very complicated network that includes water inputs from precipitation, surface supplies, pumping of groundwater, and sources of recycled water, and water uses by agriculture, industry, power generation, residential, wildlife, and recreational use. Inevitable losses occur in the system through the process of water transporation and application; some losses are recoverable, but water is also lost over the boundaries of a given hydrological system.

IV. WATER IN CROP PRODUCTION

Plants are a major component of the hydrologic cycle because of the large volume of water which is absorbed, transmitted, and evaporated by living plants. Water maintains cell rigidity and moves nutrients within plants. Water continuously flows through living plants in amounts that depend upon the plant species growing within given ecological regions—tropical rain forest and desert being two environmental extremes.

Photosynthesis is a basic biological process; when crops open pores to take in carbon dioxide, they transpire water. The efficiency of water use varies among plant species, and this variability may have serious environmental and economic implications in areas of arid farmland.

The complexity of the crop–water–soil relationship exists because crops receive water from three sources: (1) precipitation, (2) groundwater, and (3) irrigation. As discussed in Section II, irrigated areas represent only about 15% of agricultural land in the world; most food production depends upon precipitation. This dependence upon natural precipitation, which is frequently unreliable, contributes to food shortages in many world regions. The crop–water–soil relationship is very dynamic, and its function is affected by numerous factors such as soil type, water source, plant characteristics, farming methods, and climatologic conditions.

Water inflow in a system is by precipitation, water application, and water uptake by plant roots from groundwater tables. Water is used by crops for the production of biomass, and water is also transpired by crops. Water evaporates from the soil surface; it also replenishes groundwater storage by deep percolation. The remaining water runs off from the surface of soil to streams or ponds; runoff water not only represents a loss, but it can also be a cause of soil erosion. In case of soil salinity and/or high water tables, a portion of applied water may be removed by subsurface drainage.

From the point of view of the total hydrologic cycle there are no water losses—only different modes of water transformation. Water may be stored in oceans, lakes, rivers, groundwater aquifers, the atmosphere, distribution pipes, or plant cells. When plants die, water returns to the hydrologic cycle by evaporation. When crops are harvested, water again returns to the cycle either through dehydration/evaporation, processing, or consumption.

From the point of food production, we are concerned with water losses and the efficiency of water use. Water losses in agriculture have various characteristics, such as evaporation from the soil surface, excessive water applied to crops, water use by weeds, ecological suitability of given crop types, deep percolation of water, high outflow of drainage water or surface runoff.

In farming, water management is driven by economic conditions such as the cost of water and its delivery, commodity market prices, and the costs of cultivation methods. If we separate crop production from its economy, a high efficiency of water use can be achieved by the following measures:

1. Optimum water application, considering the evapotranspiration characteristics of a given crop
2. Foliage cover of soil surface, mulching, or cultivation methods to reduce soil evaporation
3. Selection of commercial crops for given ecological conditions
4. Reduced competition of non-commercial plants (weeds)
5. Reduced water surface runoff through application of optimum water amounts, irrigation practices, and field levelling
6. Reduced drain water outflow through optimum water and salt management
7. Reduced deep percolation through application of optimum water amounts

Evapotranspiration of crops is a function of biological and climatological factors. Evapotranspiration can be estimated by direct measurements or by empirical methods. Equipment for direct measurement of evapotranspiration includes lysimeters and neutron probes. Empirical methods employ mathematical models based either upon evaporimeters or monitored climatological data. The first method is using standard evaporation pans; the other method is using climatological parameters of air temperature, humidity, wind, and solar radiation. Based upon experimental work, evapotranspiration coefficients have been developed for different crops and geographic regions, and these can be applied to water management on irrigated farm land.

The efficiency of water use is frequently expressed as a ratio of plant material produced per total water lost by transpiration, or as a ratio of the economic part of plant material produced per water lost by transpiration and evaporation. Both methods are applicable to plant research done under controlled conditions, and they clearly indicate a plant's potential in an open environment.

Farm management practices need to consider the economy of irrigated farming, influenced by costs of water per ton of commodity produced or per dollar of its market value. Water costs are a function of water availability, delivery, and application, and these are different in various countries and regions. Instead of water costs, water volume can be calculated per ton of a farm product and per dollar of its market value. An example of this calculation is given in Table 6.3. The data are a function of crop yields, market price for a given commodity, water requirements depending upon soil and/or climatic conditions, and management practices. Naturally, each farm may achieve its own specific results. The table offers a relative

Table 6.3 Economic Efficiency of Irrigated Crops[a]

Commodity	Water applied, M^3/ha	Crop yield, t/ha	Crop value, $/t	Water use M^3/t	Water use M^3/$
Alfalfa hay	18,000	12.2	93.1	1,473.8	15.8
Corn	9,000	7.7	126.0	1,164.1	9.2
Cotton (lint)	11,100	1.1	1,289.6	10,108.5	7.8
Rice	18,300	5.3	174.2	3,445.6	19.8
Sugar beets	13,500	56.5	38.7	239.1	6.2
Irrigated wheat	6,000	5.5	126.5	1,097.3	8.7
Almonds	12,000	0.6	1,453.3	19,124.1	13.2
Apples	7,500	34.7	208.2	215.9	1.0
Apricots	12,000	13.6	229.5	883.6	3.9
Grapefruit	7,500	15.0	227.4	499.5	2.2
Table grapes	12,000	13.8	253.5	869.3	3.4
Oranges (valencia)	7,500	20.5	258.3	365.4	1.4
Peaches	9,900	27.8	234.0	355.7	1.5
Plums	6,900	13.1	566.1	528.1	0.9
Walnuts	9,000	1.8	812.4	4,958.1	6.1
Asparagus	13,500	3.1	1,669.4	4,397.2	2.6
Broccoli	7,500	9.2	495.6	814.1	1.6
Carrots	8,400	28.5	232.8	294.7	1.3
Cauliflower	6,000	10.4	581.9	574.5	1.0
Lettuce	7,500	30.7	242.5	244.1	1.0
Melons	9,000	17.6	249.1	512.3	2.1
Onions	12,000	41.4	200.6	289.8	1.4
Potatoes	9,900	37.1	179.4	266.8	1.5
Strawberries	13,500	41.9	951.3	322.1	0.3
Tomatoes	10,500	48.8	92.0	215.3	2.3

[a] Source: Crop Budgets, University of California, Statistical Report, California Department of Food and Agriculture.

comparison of individual crops, and indicates that vegetables or fruits are more efficient users of applied water than grains, cotton or hay.

V. WATER IN ANIMAL PRODUCTION

When people consider the problems associated with water and food production, they mostly refer to the production of crops. We may be designing and operating irrigation systems for farmlands, or be concerned with drought conditions and their effect on crop failures.

Meat, eggs, milk, and dairy products are essential components of human nutrition for a great majority of people. Animals are a source of power on farms and for the transportation of farm products in many world regions. Animal hides are a resource for many essential products.

Water requirements for farm animals depend on the environment and the intensity of animal husbandry. Range cattle would have lower water requirements than milk cows on dairy farms, where water is not only consumed by animals but it is also used for cleaning the milking equipment and housing. Typical water requirments for animals are listed in Table 6.4, and these values are affected by the type of feeding and the overall farm operation.

Table 6.5 presents values of estimated water requirements for animals and their relative comparison with the area of farmland which would be irrigated by corresponding volumes of water. By using U. N.–Food and Agriculture Organization data on the number of animals in the world, typical daily water consumption, and an irrigation rate of 0.3 m/ha, the estimated total amount of water used by animals would be sufficient to irrigate only about 6 million ha. When compared to the 200 million ha of irrigated land in the world, animal use of water is very small in relation to the water demand of crops.

Table 6.4 Water Requirements for Animals (Liters/Day/Animal)[a]

Food source	Horse	Cow	Cattle	Sheep	Goat	Pig
Pasture	10–20	20–30	10–30	1–3	2–3	
Barn						
Mostly hay	50–80	60–90	60–80	2–8	4–8	20–50
Silage/forage		30–40				15–20

[a]Source: University of California (personal communication).

Table 6.5 Estimated Total Water Requirements for Animals[a,b]

	Average water use (liters/yr)	Animals in world (1,000)	Total water use (million m³/yr)
Horses	5,475	63,871	350
Mules	5,475	15,279	84
Asses	5,475	39,866	218
Cows	9,125	221,546	2,022
Cattle	9,125	1,272,541	11,612
Buffalo	9,125	126,102	1,151
Camels	9,125	17,207	157
Pigs	1,460	786,668	1,149
Sheep	730	1,139,520	832
Goats	915	459,575	421
			17,994

[a]Source: U. N. (1986).

VI. WATER IN FOOD PROCESSING

Processing preserves farm commodities for off-season use and delivery to distant markets. Farm products are brought to stable, non-perishable conditions by drying, canning, smoking, freezing or radiation and, thus, a large portion of the production is saved for food or industrial markets. The type of processing used is determined by many factors such as the type of commodity, market characteristics, seasonal factors, type of distribution means, consumer preference, tradition, energy availability, and the overall economics of processing and marketing.

In societies where a majority of the population lives in rural areas, food processing would be required mostly for off-season consumption. As the population concentrates in urban areas, farm commodities need to be shipped to distant markets, and food processing is essential for preserving the usable life of farm commodities as well as their quality. Food processing is very important in an urban and industrial society. Table 6.6. indicates the type of marketing of selected food commodities in California; as can be seen, a major portion of some products is processed.

Water is used in food processing operations, and its use depends upon the type of final product. A typical water use pattern in canning operations is presented in Table 6.7. Washing, cooling, and plant cleanup are the major uses.

Water use in the processing of selected fruits and vegetables is pre-

Table 6.6 Percentage of 1978 Crop Tonnage Subjected to Various Types of Primary Processing[a]

Commodity	Fresh market	Canned	Frozen	Dried	Crushed	Chips	Cubed	Cleaning and handling	Dressing	Oil extraction	Ginning	General processing
Alfalfa				0.57			6.68					
Alfalfa seed								100.00				
Almonds	28.00	19.20		10.00								
Apples	5.28	63.41	6.91	12.80	40.00							
Apricots	67.47	14.67	17.86	24.39								
Asparagus	100.00											
Avocados								100.00				
Barley								100.00				
Beans (dry)	19.72	3.82	76.46									
Beans (green)	49.07		50.93									
Broccoli									100.00			
Broilers/fryers												
Carrots	73.45	13.28	13.28									
Cauliflower	47.00	3.00	50.00									
Celery	100.00											
Corn (grain)				100.00								
Cotton (seed cotton)										50.00	32.00	
Dairy												100.00
Eggs	100.00											
Feedlot									100.00			
Grapefruit	47.00		53.00									
Hogs									100.00			

Grapes	10.93	1.46	20.14	67.47
Lemons	63.00		37.00	
Lettuce	100.00			
Melons	100.00			
Nectarines	97.77	2.33		
Oats					100.00
Onions	40.00		28.00	60.00
Oranges	72.00	77.82		2.33
Peaches	15.20	82.03	4.65	
Pears	16.13	1.95		
Plums	98.05	5.00		
Potatoes[a]	80.00			100.00
Prunes				100.00
Rice					15.00	100.00
Safflower						100.00
Sheep/lamb			100.00	
Sorghum						100.00
Strawberries	67.00		33.00	
Sugarbeets						100.00
Tomatoes	5.60	94.40			100.00
Turkeys					100.00
Walnuts						100.00
Wheat				100.00

[a]Source: Cervinka et al. (1981).

[b]All potatoes considered as fresh market; chipping and canning were considered secondary processing and thus no energy figures are included for these processes.

sented in Table 6.8. Water requirements are specific to given commodities and operations. Water is used but not wasted during processing operations; usually water intake is equivalent to water discharge. However, the quality of water changes during processing and its discharge requires specific treatments. Untreated water may cause environmental damage. While this is still tolerated in some regions, in many countries the discharge of water pollutants is controlled. The method of treatment may depend upon the location of individual plants. Available methods of treatment include municipal water treatment plants, ponding, land irrigation, bio-digestion, biological filters, and others.

VII. WATER QUALITY

The hydrologic cycle is both a water recycling and purifying system. Water is purified through evaporation from bodies of water and transpiration from plants, and is returned in the form of rain or snow.

Water quality is related to the environment through water sources in streams, lakes, oceans, and groundwater storages. Water quality may be negatively affected by technologies used in industries and on farms, or, fundamentally, by the mode of living of a society. The concentration of population in urban areas creates a large volume of waste water which needs to be treated and discharged.

The increasing population is facing the fact that water resources are not expandable—only a given volume of water is available to a growing

Table 6.7 Water Use in Canning Operations[a]

Operation	Water use (liter/t)	(%)
Peeling	202	2
Spray washing	1,617	17
Sorting/slicing	504	5
Exhausting of Cans	202	2
Thermal processing	101	1
Container cooling	3,969	37
Plant cleanup	3,528	33
Box washing	294	3
	10,416	100

[a]Source: Townsend and Somers (1949); Katsuyama (1979).

Table 6.8 Water Used in Processing Fruits and Vegetables[a]

Commodity	Processed product	Water used (liter/t) Intake	Water used (liter/t) Discharge
Apple	Sauce	8.8	8.8
	Juice	10.1	9.7
Apricot	Unpeeled	25.6	26.5
	Peeled	31.9	37.4
Asparagus	All	21.4	26.0
Bean	Freeze	15.5	15.1
	Can	18.1	16.8
Carrot	Freeze	34.0	33.6
Cherry	Whole	17.6	17.6
Citrus	Frozen concentrate	10.5	15.1
	Canned juice	12.6	10.9
Corn	Freeze	16.4	15.5
	Can	20.6	18.9
Pea	Freeze	16.4	16.0
	Can	17.6	16.8
Peach	Canned slices	16.8	17.2
Pear	Canned slices	16.0	17.2
Potato	Freeze	12.2	13.0
	Can	18.5	22.3
Tomato	Peeled	14.3	53.3
	Juice	6.3	5.5
	Concentrate	6.3	7.6

[a]Source: Townsend and Somers (1949); USEPA (1977); Katsuyama (1979).

population and its residential, industrial, farming, or recreational needs. Two interrelated hydrological cycles need to be considered—a natural and a technological cycle. As water demands are increasing, water recycling within the system of residential areas, industry, and agriculture is becoming very important.

Water quality is not only an environmental and health problem, it is a social and political problem as well. Throughout the history of mankind, civilizations have declined because of the degradation of water quality and mismanagement of the technological water cycle. The accumulation of salts on irrigated land destroyed the food production base of many civilizations and caused their decline.

Sources of water pollution include municipal waste water, discharges from chemical and other industries, leakage from landfills, agricultural chemicals, intensive accumulation of manure, mismanagement of irrigation practices, waste water from mining operations, leakage of underground petroleum storages, seawater intrusion and others. The degradation of water quality is mostly associated with people's daily living practices and work.

Agriculture can be affected by water quality, and farming operations themselves may also cause water degradation. Water quality interacts with agriculture in the areas of salts, nitrates, residues of agricultural chemicals, soil erosion, and waste discharge from food processing plants. These factors, with their causes and possible measures for reducing and eliminating water pollution, are presented in Table 6.9. It can be seen that technology

Table 6.9 Agriculture and Water Quality

Degradation of water quality	Cause	Corrective measures
Salts	Natural occurrence	Leaching
	Irrigation	Leaching
		Drainage
		Water Management
		Irrigation practices
		Agroforestry
		Selection of crops
Nitrates from manure	Production intensity	Bio-digestion of manure
		Burning of manure (heat/power/generation)
		Transport to other locations
		Composting
		Reduced production intensity
Nitrates from industrial fertilizers	Farming methods and production intensity	Change in farming and application methods
		Use of compost
		Use of legumes
		Use of "green" manure
Agricultural chemicals	Farming methods	Change in farming and application methods
		Improved farm management
		Biological pest control
		Weed control by cultural practices
		Design of a cropping system
		Crop polyculture
Soil erosion	Water or wind	Change in farming methods
		Improved farm management
		Design of a cropping system
		Tillage methods
		Permanent crops
		Windbreaks
		Agroforestry
Industrial pollution	Water water from food processing plants	Water treatment facilities
		Bio-digestion of waste
		Use of waste products
		Waste reduction
		Treatment by biofilter crops
		Irrigation of crops

and management information are available to maintain water at a desirable level of quality.

Salinity is a major water quality problem on irrigated fram land. Salinity generally refers to total dissolved solids (TDS) which includes combinations of sodium, chlorides, calcium, magnesium and other salts. Salts are a part of soil composition and additional salts are brought by irrigation water. If irrigated land is not correctly managed, the accumulated salts cause crop damage, reduced growth, and loss of productivity. In extreme cases, no crops can be grown economically in affected soils. Agriculture has been destroyed through the extreme accumulation of salts in many world regions.

A well designed irrigation and drainage system supplies adequate fresh

Table 6.10 Methods of Converting Saline Water to Fresh Water[a]

Distillation processes	
Examples	Multistage flash distillation
	Vertical tube distillation
	Multieffect multistage distillation
	Solar humidification
Attributes	Most widely used
	Energy intensive and costly
	Results in "ultrapure" water
	Favored by seawater
Membrane processes	
Examples	Reverse osmosis
	Electrodialysis
	Transport depletion
	Piezodialysis
Attributes	Favored for brackish water
	Require pretreatment to remove pollutants
	Potentially energy efficient
	Increasingly popular
Crystalization processes	
Examples	Vacuum freezing-vapor compression
	Secondary refrigerant freezing
	Eutetic freezing
	Hydrate formation
Attributes	Experimental stage
	Minimize corrosion
	Potentially energy efficient
	High recovery without major pretreatment
Chemical processes	
Example	Ion exchange
Attributes	Less costly "ultrapure" water
	Useful for low-salinity water

[a] Source: OTA (1983).

water and has an outflow that removes the high concentration of salts from the land. However, drainage water must be either discharged from the farming system or treated within the system. The oceans or larger inland lakes may be points of drainage water discharge, but these methods may not be always feasible for economic or political reasons. Available methods for the control and treatment of drainage water within an agricultural system include the management of irrigation practices, desalinization plants (Table 6.10), evaporation ponds, the agroforestry system, or, preferably, the combination of all these measures.

Crops have different sensitivites to salt accumulation; their levels of tolerance are listed in Table 6.11. Some plants have a very high tolerance to salts—these include atriplex, mesquite, types of salt grasses, and others. Research is being done to evaluate the suitability of these crops as marketable commodities.

Water quality presents a tremendous challenge to agriculture, in its effect on food production and the environment. Scientific knowledge exists to meet this challenge.

Table 6.11 Salt Tolerance of Agricultural Crops[a]

Crop	Maxium soil salinity without yield loss (threshold)	% Decrease in yield at soil salinities above the threshold
Sensitive crops	dS/m	% per dS/m
Bean	1.0	19
Carrot	1.0	14
Strawberry	1.0	33
Onion	1.2	16
Almond	1.5	19
Blackberry	1.5	22
Boysenberry	1.5	22
Plum; prune	1.5	18
Apricot	1.6	24
Orange	1.7	16
Peach	1.7	21
Grapefruit	1.8	16
Moderately sensitive crops		
Turnip	0.9	9.0
Radish	1.2	13
Lettuce	1.3	13
Clover, berseem	1.5	5.7
Clover, strawberry	1.5	12
Clover, red	1.5	12
Clover, alsike	1.5	12
Clover, ladino	1.5	12
Foxtail, meadow	1.5	9.5

Table 6.11 *Continued*

Crop	Maxium soil salinity without yield loss (threshold)	% Decrease in yield at soil salinities above the threshold
	dS/m	% per dS/m
Grape	1.5	9.6
Orchardgrass	1.5	6.2
Pepper	1.5	14
Sweet potato	1.5	11
Broadbean	1.6	9.6
Corn	1.7	12
Flax	1.7	12
Potato	1.7	12
Sugarcane	1.7	5.9
Cabbage	1.8	9.7
Celery	1.8	6.2
Corn (forage)	1.8	7.4
Alfalfa	2.0	7.3
Spinach	2.0	7.6
Trefoil, big	2.3	19
Cowpea (forage)	2.5	11
Cucumber	2.5	13
Tomato	2.5	9.9
Broccoli	2.8	9.2
Vetch, common	3.0	11
Rice, paddy	3.0	12
Squash, scallop	3.2	16
Moderately tolerant crops		
Wild rye, beardless	2.7	6.0
Sudangrass	2.8	4.3
Wheatgrass, std. crested	3.5	4.0
Fescue, tail	3.9	5.3
Beet, red	4.0	9.0
Harding grass	4.6	7.6
Squash, zucchini	4.7	9.4
Cowpea	4.9	12
Soybean	5.0	20
Trefoil, birdsfoot	5.0	10
Ryegrass, perennial	5.6	7.6
Wheat, durum	5.7	5.4
Barley (forage)	6.0	7.1
Wheat	6.0	7.1
Sorghum	6.8	16
Tolerant crops		
Date palm	4.0	3.6
Bermuda grass	6.9	6.4
Sugarbeet	7.0	5.9
Wheatgrass, fairway crested	7.5	6.9
Wheatgrass, tall	7.5	4.2
Cotton	7.7	5.2
Barley	8.0	5.0

[a]Source: Maas (1984).

VIII. IRRIGATION SYSTEMS

The application of water to crops differs according to the delivery of water and its distribution in a field. Fields can be irrigated by flooding or natural down-slope flow of water. Water can be delivered by manual pumping or by using wind, engines, and grid or solar electricity to drive water pumps. Water is distributed in the fields by controlled flooding or flow, by spraying water over the soil surface, or by delivering water close to the root system of plants.

The choice of irrigation methods is determined by crop types, soil structure, land configuration, power availability, system costs, and general economics of farming. The basic irrigation techniques include basin, border-strip, furrow, sprinklers, and drip irrigation. Basin irrigation is done in a level area of any shape or size. The borders or ridges retain applied

Figure 6.1 Furrow irrigation for a tomato field.

water until it infiltrates. Any water loss is either caused by deep percolation or surface evaporation. Border-strip irrigation can be used in a sloping area; water flow is guided by borders. The area coverage is controlled by water inflow. Water loss is caused by deep percolation, evaporation or water runoff. Furrows distribute water by the slight sloping of land. The uniformity of irrigation is affected by the length of furrows, soil conditions, and rates at which water is applied. Water losses are caused by deep percolation and runoff.

Sprinklers should uniformly apply irrigation water over the soil surface. The efficiency of irrigation is affected by their design and distri-

Figure 6.2 Water delivery for a furrow irrigation system.

160   Vashek Cervinka

bution on the field. Water losses mainly are caused by wind drift and evaporation. Drip irrigation systems discharge water close to the plants through small holes in tubing placed on the soil surface or near to it. Only a portion of soil, surrounding each individual plant, is kept moist by frequent or continuous application. Water losses can be kept at minimum.

Various irrigation systems are presented in Figures 6.1–6.4, including furrow irrigation of a tomato field (Figure 6.1), water discharge to a furrow system (Figure 6.2), movable sprinklers (Figure 6.3), and a center-pivot irrigation system (Figure 6.4). A comparison of the characteristics of alternative irrigation systems is presented in Table 6.12.

Figure 6.3 Hand-movable sprinklers irrigating roses.

Figure 6.4 Center-pivot irrigation of alfalfa.

Table 6.12 Characteristics of Irrigation Systems[a]

Irrigation method	Intensity of Energy	Intensity of Labor	Intensity of Capital	Requirement of Uniform soils	Requirement of Land slope	Requirement of Water supply	Practical efficiency %
Basin	Low	Low	Low	Yes	Level	Large	60–85
Border-strip	Low	Low	Low	Yes	Mild	Large	70–90
Furrow	Low	Medium	Low	Yes	Mild	Medium	70–90
Sprinklers	High	High	Medium	No	Any	Small	65–80
Drip	Medium	Low	High	No	Any	Small	70–90

[a] Source: Modified from Merriam (1977).

REFERENCES

Cervinka, V., et al. (1981). "Energy Requirements for Agriculture in California." California Dep. Food Agric. and Univ. of California, Davis.
Department of Water Resources. (1976). "Water Conservation," Bull. No. 198. Sacramento, California.
Department of Water Resources. (1983). "The California Water Plan," Bull. No. 160–83. Sacramento, California.
Katsuyama, A. M. (1979). "A Guide for Waste Management in the Food Processing Industry." Food Process. Inst., Washington, D.C.
Maas, E. V. (1984). Crop tolerance. *Calif. Agric.* **38**(10), 20–21.
Merriam, J. L. (1977). "Efficient Irrigation." California State Univ., San Luis Obispo.
Merriam, J. L., and Keller, J. (1978). "Farm Irrigation System Evaluation." Utah State Univ., Logan.
Office of Technology Assessment (OTA). (1983). "Water Related Technologies," OTA-F-212. Washington, D.C.
Townsend, C. T., and Somers, I. I. (1949). How to save water in canneries. *Food Ind.* **21**, W11–W12.
U. N.–Food and Agricultural Organization. (1986). "Production Yearbook." Rome.
University of California. (1977). "Water." California Agric., Berkeley.
U.S. Bureau of Reclamation. (1979). "Agricultural Drainage and Salt Management." California.
U.S. Department of Interior (1984). Estimated use of water in the United States. *Geol. Surv. Circ. (U.S.)* No. 10001.
U.S. Environmental Protection Agency (USEPA). (1972). "Managing Irrigated Agriculture." Washington, D.C.
U.S. Environmental Protection Agency (USEPA). (1977). "Pollution Abatement in the Fruit and Vegetable Industry," EPA-PB 299 613. Cincinnati, Ohio.

7

Water Scarcity and Food Production in Africa

Malin Falkenmark

Swedish Natural Science Research Council
Stockholm, Sweden

I. Introduction
II. Present Resource Crisis
   A. An African Outlook
   B. Drought Is Characteristic of the Climate
   C. Drought-Triggered Hunger Disasters
   D. Risk for Collapse of Life-Support Systems
   E. Crucial Role of Water Deficiency
III. Water and Development
   A. Balanced Development of the Environment
   B. A Water Perspective on Rural Development
   C. Terrestrial Water Cycle Divided in Two Branches
   D. Land Productivity a Function of Availability of Soil Water
   E. Prospects for Rain-fed Agriculture
   F. Soil and Water Conservation Success Stories
IV. Water Availability Limits Increase of Carrying Capacity
   A. Population Increase Generates Water Scarcity Problems
   B. Rapid Approach of Large-Scale Water Crisis
V. The Combined Picture
   A. The Next Decade
   B. First Decades of Next Century
VI. Conclusions
   References

I. INTRODUCTION

Water is a substance with great diversity, and water is deeply involved in most sectors of society. The amount of renewable water provided to a country is an important determinant and natural constraint to societal development, unless large-scale technology is available to transfer water from neighboring regions that are better endowed. The amount of water available depends on the country's location in the earth's system of aerial water vapor flux, which distributes evaporated seawater over the continents. Water is provided to a territory as local precipitation (endogenous supply) and by entering aquifers and rivers (exogenous supply) (Figure 7.1). Water availability limits the size of the population because food and fodder production, together with water for human and livestock consumption, requires 1,250 m^3 of water per person per year (Falkenmark, 1984).

Society relies on this amount of water both for its plant production, where water is consumed in the biomass production process, and for technical supplies of habitats, industry, or irrigated agriculture. A number of additional uses rely on the water flowing down the rivers, such as hydropower production, navigation, dilution of waste water, or fishing. The quality of water is influenced by polluted waste waters leaching agricultural lands and by other polluting land uses such as waste deposits and urban activities.

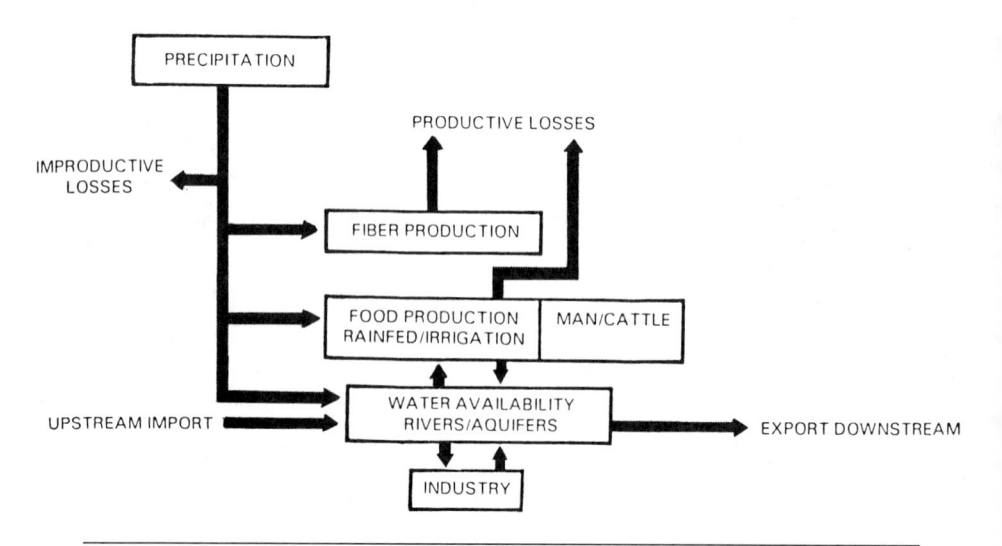

Figure 7.1 Water availability, consumption, and use: the national perspective. [From Falkenmark (1984).]

7. Scarcity and Food Production in Africa 165

By its diversified characteristics, the water supply also causes many problems. Temporary surpluses during the wet season cause inundations and massive damage, especially in densely populated flatland areas. High groundwater tables cause waterlogging. Dry periods cause crop failure and complicate the water supply. Water is also both a physically and chemically active agent, continuously moving in the landscape above and/or below the ground surface and involved in most environmental feedback processes, whereby land use activities are transferred into secondary feedbacks on the resource base. Moreover, man's and biota's dependence on water gives water an important role in the transmission of diseases.

There are great interregional differences in the importance of water-related constraints to development. Awareness of the deeper implications of such climatically caused differences is often low, creating potential risk for biases in problem definition (for example, by international organizations familiar primarily with temperate situations). There is a surprisingly poor understanding of water-related limits of growth, as water is a renewable resource which may be depleted if the rate of use surpasses the limits of regeneration. What is not generally understood is that the naturally circulated water defines a carrying capacity just as forest regeneration and crop production potential defines one. In the case of water the limit emerges from restrictions on our present ability to distribute water to many parallel and competing uses.

An example of climatic bias is the approach generally taken to droughts, phenomena generated by uncontrollable meteorological semi-periodical events. They are often seen as abnormal, although they are regular parts of the semi-arid climate. From a water-related perspective, the societal effects of droughts merely reflect an absence of adaptation to the natural constraints of the semi-arid tropics, caused by large inter-annual fluctuations in the amount of rainwater provided by the planet's system of aerial water vapor flux.

The aim of this book is to assess the ecological and economic status of natural resources use and management, related to food production, as seen in a global perspective. This chapter will focus in particular on Africa's problems with reference to water. The ubiquitous nature of water, on one hand, and the dominance of experts educated in the temperate zone, on the other, have contributed to a broad neglect of problems of water "inadequacy" and water-related constraints to development. It is interesting to note that ecosystem theories pay minimal attention to water flows as a basis for phytomass production and biogeochemical cycling. Likewise, soil moisture problems often vanish under concepts such as droughts and agrometeorology.

This chapter will give a broad macro-scale hydrological perspective to sustainable development of the African continent, given the present population predictions for the coming decades. The basic questions raised will be the following: in what ways will water be involved in the efforts towards sustainable development of Africa? And to what degree will water scarcity develop into a fundamental constraint for agricultural production and for technical supplies in the short- and medium-term perspectives?

II. PRESENT RESOURCE CRISIS

A. An African Outlook

A good overview of the African resource crisis and the way African governments look at ways out of the crisis is provided by a series of nine articles in Kenya's largest newspaper by Dr. Yussuf Ahmad (1986), Director of Special Assignments in UNEP. The overall viewpoint of those articles is presented in this section.

The present crisis is visible through its effects, and has led to a marked reduction in the carrying capacity of the continent. There has been a major reduction of usable pastoral lands. At least 150 km^2 of productive land on the southern Saharan margin has become completely unproductive. There has been a shift in the ecological zones with desertification of the Sahelian zone, Sahelization of the savanna, and a savannization of the woodlands.

Ahmad sees the causes of the crisis as systemic in nature, reflecting a breakdown in a coherent and systematic natural resources management and a lack of defined and practical policies for such management. The rapid growth of the rural population in rangelands, particularly vulnerable to desertification, and a large increase in livestock numbers in the worst-affected areas have contributed to the crisis. Large-scale deforestation is one of the causes, mainly due to clearing for agricultural production and shifting cultivation practices. The land degradation crisis has been triggered by droughts, especially a long period of persistent droughts, which has been concurrent with ongoing population growth. African economic policies have mirrored a "wholesale neglect of rainfed agriculture." There have been no economic or other incentives for rural farmers to produce food for sale.

According to Ahmad, there exists a potential for recovery once the economy is brought under control. The general policy is to strive for self-sufficiency in food and energy. The remaining land potential is considerable—nearly half of the arable land of the continent is still in reserve. A regional food plan has been approved by the governments and by FAO.

It is stressed that better use should be made of the continent's water resources, which are said to present a great potential. Ahmad refers to the fact that 4,200 km^3/yr of river flow is going to the sea. With only 10% of this amount, African food production could be increased by 10% (enough to irrigate 13 million ha of arable land). As many as 200 million domestic animals could also be put to work. And only a small percentage of the enormous hydropower potential is already developed.

In developing the water resources, Ahmad stresses that irrigation projects have to begin on scales that facilitate the transition from previous management systems. In regions with water scarcity, the most efficient use of water would be on a basin-wide scale. It is also possible and advisable to integrate forest and watershed management because the same upland areas are involved in both cases.

Ahmad finally stresses that much of the water resources of the African continent are in transnational rivers. UNEP has strongly emphasized the development of concepts and principles for sharing of natural resources such as water, and at present a case study is under development in the Zambesi basin with the participation of six governments. The basic concept in this program is environmentally sound management of inland waters.

B. Drought Is Characteristic of the Climate

Two-thirds of Africa receives more than half its annual rainfall in only three months (U. N., 1987). The rainfall, therefore, limits the length of the growing season, defined as that part of the year when there is enough water in the soil to satisfy at least half the water demands from a growing crop. For as much as 44% of the African continent drought is a major limitation to rain-fed agriculture. For another 18% of the total land area, the soil composition is a major limitation to agriculture (Bentley et al., 1979). In all, only 16% of the total area is suitable for agriculture and only about half of it is actually cultivated. There should exist a considerable potential for horizontal expansion of agriculture. Unfortunately, endemic diseases such as river blindness and sleeping sickness close off some of the better-watered, potentially arable areas (Levine et al., 1979).

There is great interannual variability of rainfall, which implies that in areas with a short average growing season there is also considerable risk of years with crop failures. The zone with 75–150 days of growing season is considered drought-prone, meaning that there is a risk for recurrent drought hazards above a certain risk level. The zone is largely too dry for secure rain-fed agriculture, and most of the rainfall is consumed in evapotranspiration, leaving only minor amounts for recharge of aquifers and rivers. The map in Figure 7.2(a) shows the crescent-shaped area of sub-Saharan Africa defined as drought-prone. As can be seen from Figure 7.2(b), the countries abnormally dependent on food aid during the drought of 1984–85 were all located in this zone. In addition, Figure 7.2(c) demonstrates that water availability in groundwater and water courses is also vulnerable, because only minor amounts of the rainfall provide a water surplus available for recharge of these waters even during average years.

The three maps clearly show the predicament of this broad zone of

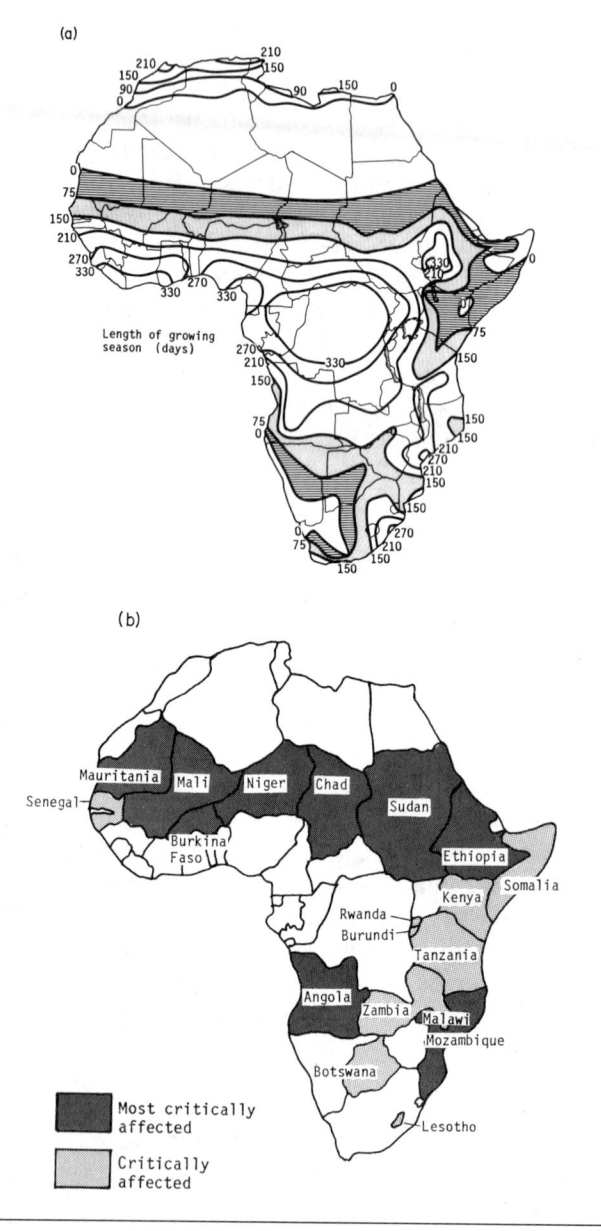

Figure 7.2 Water inadequacy in the African hunger crescent: (a) growing season of 0–75 days (striped) allows pasture only, while 75–150 days (shaded) allows rain-fed agriculture with risk of crop failure during drought years [from U.N. (1987)]; (b) countries critically affected by drought during 1984–1985; (c) amount of rainwater (mm/yr) remaining after evapotranspiration (long-term average). [From Falkenmark (1986).]

(c)

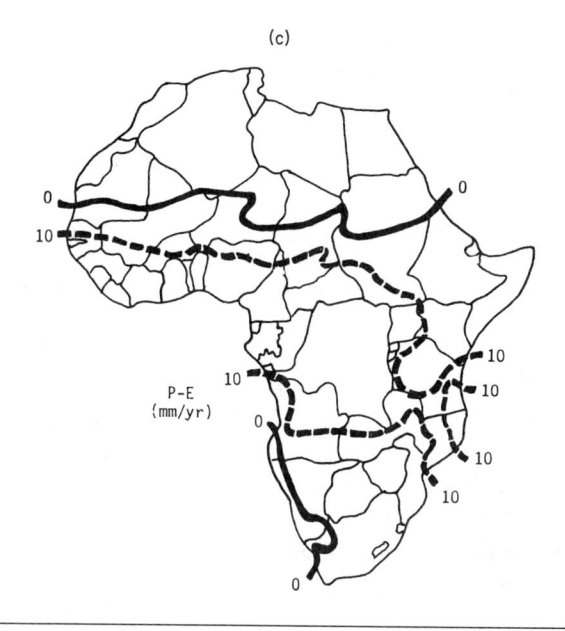

Figure 7.2 (*Continued.*)

Africa. There is a large risk of intermittent drought and crop failures, there is very little recharge of aquifers and rivers even under average conditions, and this recharge is subject to the same interannual fluctuations as the soil moisture, because both phenomena are fed by rainfall. Altogether, 160 million inhabitants lived in these countries in 1982, and the population is predicted to increase to 290 million by 2000, and to 565 million by 2025. In other words, about one-third of the population of the continent live and will be living in these vulnerable regions.

The rainfall variability increases as mean annual rainfall decreases. Irregularity of rainfall is therefore part of the climate in semi-arid and arid areas and life-styles have traditionally developed in accordance. The original adaptation to centuries of rainfall variability was expressed in the nomadic, pastoralist life-style. However, the high rainfall in the 1950s allowed an expansion of rainfed agriculture in the Sahelian region (U. N., 1987). In western Sahel, for instance, there was a five-fold increase in cattle during 25 years. The combined effects of growing human and animal pressure on the vulnerable land lead to environmental degradation. This has triggered desertification and related processes such as reduced infiltration, increasing flash floods, lowered groundwater tables, and loss of topsoil and soil nutrients.

Basically, drought is a term used for rainfall deficiency below a certain threshold, which depends on the type of management. A basic problem is that the degrading management and the high human and animal pressures have caused the threshold to rise (U. N., 1987), i.e., increased the vulnerability to droughts.

Nicholson and Chervin (1983) have shown that there are two different modes of African drought years. The Type A drought where tropic regions differ from equatorial regions (examples are 1931, 1947 and 1968), and the Type B drought where both regions suffer simultanously from abnormally low rainfall (examples 1971, 1972 and 1973). The causes are complicated and the deficiencies are part of large-scale changes in the tropical atmosphere, which act on a continental scale and affect both hemispheres. The large scale of the anomalies seems to indicate the possibility of a feedback between rainfall and the general circulation.

Great scientific interest is being devoted to the problem whether the African rainfall is decreasing or not. One factor of possible relevance is the biophysical feedback expected from an albedo increase due to overgrazing, increasing subsidence, and reducing rainfall. This would make the region less suitable for vegetation (Barnett, 1985).

An even more thought-provoking possibility is tied to the possible link to the global warming in general. Figure 7.3 compares the annual average precipitation anomaly for Africa as a whole, according to Barnett (1985), and the variations of the global mean annual surface temperature according to Bolin *et al.* (1986). Barnett concludes that "the strong trend towards decreasing rainfall over the continent, *if real,* represents a natural low-frequency signal that requires explanation: Is it the result of natural variation or could it be due to increasing CO_2 concentration?"

C. Drought-Triggered Hunger Disasters

At present we have to accept that the great interannual variability of the rainfall in semi-arid Africa, together with the fact that drought years tend to develop persistence patterns, makes it difficult to arrive at a consistent view whether the recent droughts reflect an overall pattern of reduced rainfall or not. Irrespective of whether the drought pattern and the drought severity is changing or not, the way in which droughts hit human societies is largely dependent on the contingency planning and the general life-styles.

The societal drought problems caused by below-average rainfall tend to develop as chain processes from two primary effects: crop failures and below-average recharge of groundwater and wadis, which complicate local people's ability to find water for their basic needs (Swedish Red Cross, 1984). Much work has gone into investigations of the Sahelian drought in the early 1970s. Garcia (1981), in an in-depth study organized by IFIAS, arrived at the conclusion that there had been no profound climatic changes, and that there was no clear evidence that the amplitudes of climatic variables had increased during recent decades. The title of the book gives the

7. Scarcity and Food Production in Africa   171

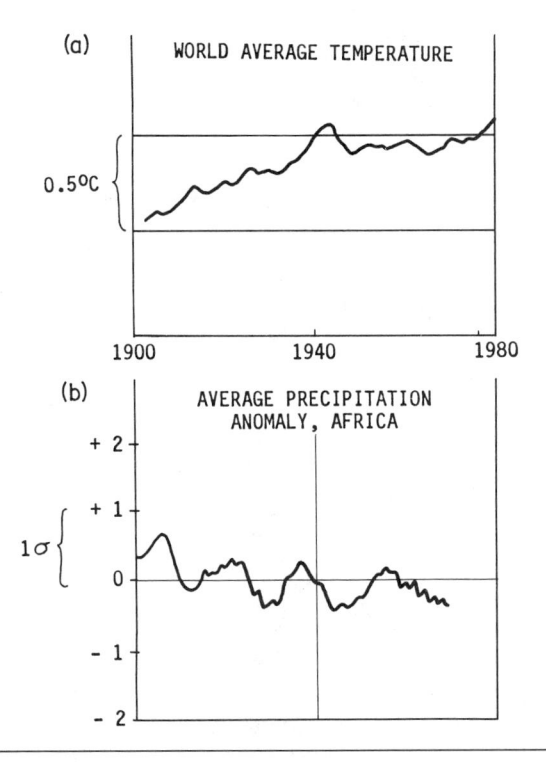

Figure 7.3 Long-term trends in world average temperatures and in average precipitation for Africa: (a) global mean annual surface temperature [data from World Resources Institute (1986)]; (b) temporal variation of annual average precipitation for the African continent, expressed as deviation from the long-term average. [Data from Barnett (1985).]

resulting explanation to the drought disaster: *Nature Pleads Not Guilty*. The explanation offered is that two main determinants are involved: profound changes in the international food market in the early 1970s and events associated with the droughts. Garcia's conclusion is that droughts are not the *cause* of internal disequilibria in society—they merely *reveal* a preexisting disequilibrium. The events developing after the drought has stricken are determined by the structure of the whole socio-ecosystem. The Sahelian countries had passed an evolution in recent decades towards more vulnerable conditions.

The Garcia study concludes that climate and its variations must play an important role in rational planning of a society that is striving for a preservation of adequate minimum standards of living for all of its population. An interesting fact is that many of the hard-hit countries asking for food aid were formerly net exporters of food. Unavailability of food for some segments of the population was primarily a consequence of national policies, and the drought pressures exposed their plight to the rest of the world.

D. Risk for Collapse of Life-Support Systems

Brown and Jacobson (1986) have discussed the ongoing large-scale land degradation and desertification in Africa from a carrying capacity viewpoint, noting that national governments and the international development community have been slow to take carrying capacity into account when formulating economical and population policies. Based on a World Bank study, they analyzed the rural food crop capacity in a number of countries in west Africa and noted that it will soon be surpassed by population increase. The excessive population pressure tends to spread from one support system to the next, producing less protective vegetation cover, which initiates not only erosion but also reduction of soil organic matter, decline of soil structure, and loss of its capacity for water retention. The end result of this process will be a skeletal shell of sand, i.e., a soil lacking fine particles as well as organic matter to make it productive. A byproduct is the reduction of local water supplies by falling water tables.

Brown and Jacobson (1986) conclude that a continued, uncontrolled growth of population in subsistence economies is degrading the resource base throughout Africa, where breaches in the thresholds of carrying capacity of the present foodcrop and fuelwood ecosystems are commonplace. They do not expect any significant increases in carrying capacity without technological breakthroughs. When population demands are growing beyond sustainable yield thresholds, future economic opportunities are in fact sacrificed. A three-stage ecological transition is envisaged in countries with high population growth:

1. Demands remain below the sustainable yield.
2. Demands have increased above the sustainable yield and continue expanding as the biological resource system is being consumed.
3. Demands are forcibly reduced as the biological life- support system collapses.

They conclude that at present many countries in Africa are already in the second stage.

E. Crucial Role of Water Deficiency

As the reader may have noticed, there is indeed an interesting inconsistency in views regarding the availability of water on the African continent. On one hand, the absence of soil water during the dry season is the main factor limiting the growing season as well as the yields that are possible to achieve on African soils. On the other, Ahmad (1986) referred to the rich water resources available in Africa that are flowing from the continent to the sea. These resources are said to provide a considerable reserve for future development. This leads us to an additional question: Is water scarcity just a temporary or even a lasting problem in Africa?

After consulting Korzun (1978), we may conclude that the large amounts of river flow, seen as continental water reserves, are mainly produced in the equatorial zone (draining into the Atlantic, with some additional rivers draining into the Indian Ocean) and the Nile River (draining into the Mediterranean). Of the total flow, 40% is located in the Congo (1,414 km^3/yr), Niger (268 km^3/yr), Nile (73 km^3/yr), and Zambesi (106 km^3/yr) rivers.

The rest of this chapter is devoted to a close look at regional differences in order to find out to what degree water availability will constrain African development during planned efforts toward restoration of degraded ecosystems and the development of food security for the rapidly growing populations.

III. WATER AND DEVELOPMENT

A. Balanced Development of the Environment

The natural environment provides us with water, food, energy (including fuelwood), wood, and minerals. In order to benefit from all the resources provided by the environment to support increasing populations, societies are forced to manipulate the environment (Figure 7.4). Different types of natural constraints that cause problems for societal development (waterlogging, dry soils, obstructing vegetation, etc.) have to be overcome or compensated for. To run the biomass production within agriculture and forestry, the land has to be cleared, drained, fertilized, and prepared in other ways. To get water, we must construct systems for water withdrawal, storage, and supply to households, industry, and irrigated agriculture. Hydroelectric plants are built for energy production.

As all such measures related to soil, vegetation, and hydrology involve interventions with complex natural systems, feedback develops in response. For some reason, we have so far preferred to bring all these secondary effects together under the diffuse concept of "environmental effects." As these secondary effects are impossible to avoid—except possibly for outputs of polluting substances—they have to be managed. The ultimate question in relation to environment and development is, How does one define the ecologically acceptable balance between the benefits for societal development, made possible through such interventions, and the costs in terms of the unintended changes of that environment, caused by those interventions due to the way the natural systems operates?

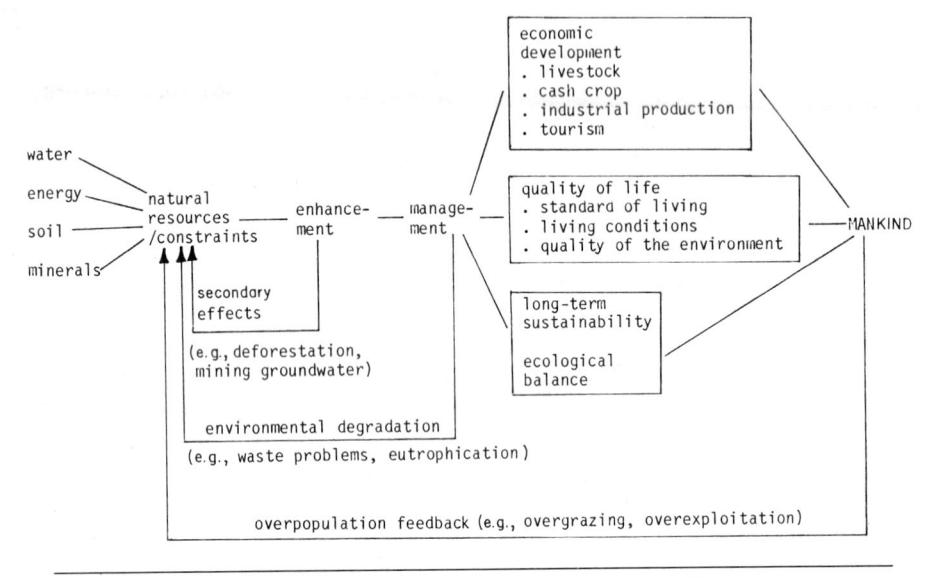

Figure 7.4 Relationships between humankind and water, soil, energy, wood, and mineral resources.

B. A Water Perspective on Rural Development

Rural life, especially under conditions when at least part of the year is arid, is closely related to soil moisture conditions in several parallel ways. There are three basic needs, which are all related to water either in the soil or in aquifers and rivers: daily food, water, and fuelwood supplies. Food is produced by farming fields or home gardens, water is carried to the house by family members, and so is fuelwood. Water sources could be groundwater wells or springs, local washes, or nearby rivers.

Conditions for satisfying these basic needs depend on the characteristics of the natural environment. The main determinants of potential satisfaction of basic needs are fundamental environmental characteristics such as precipitation and soil permeability, together with the presence or absence of nearby rivers and of forests/woodlands. Soil permeability is a crucial component of the system, because it determines the rootzone wetting and therefore not only crop and tree growth but also the recharge of groundwater with surplus percolation water.

Overpopulation creates increasing hazards to soil permeability. Agriculture on marginal lands or overgrazing with too large herds threatens the soil as does overexploitation of local woodlands. The role of the soil surface is thus crucial. When soil is permeable rain infiltrates, feeds the root zone, and recharges groundwater aquifers with surplus water. When soil is impermeable, flash floods are generated which further deteriorate the soil surface in a vicious circle. The presence of insects, earthworms, and similar small organisms play a major role in making holes in the soil for water percolation. The existence of vegetation, including trees, contributes not only to the availability of fuelwood but also helps keep the soil surface permeable.

There is, therefore, a close link between integrated soil and water conservation and the maintenance and protection of a vegetation cover, which keeps the soil permeable to local rain. Local conservation measures include small-scale irrigation with groundwater or water from wadis. Household sullage can be used to water the home garden.

C. Terrestrial Water Cycle Divided in Two Branches

When the precipitation reaches the land surface it is divided in two main flows: under favorable conditions the water infiltrates, recharging the water reservoir of the soil. It is this water that drives the plant production process, returning a considerable amount of water to the atmosphere. This return flow, together with the evaporation from moist surfaces, constitutes

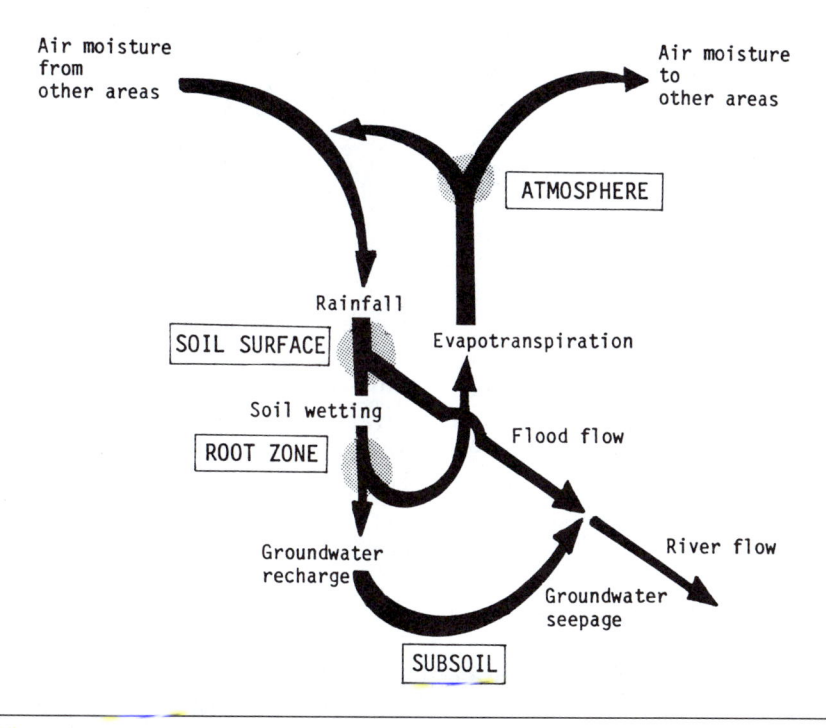

Figure 7.5 Continuity relationships in the terrestrial part of the water cycle. [From Falkenmark (1986a).]

the *short branch* of the water cycle, linking the terrestrial ecosystem with the atmosphere. The water surplus after the diversion back to the atmosphere forms the water recharging aquifers and rivers, a partial flow that is called the *long branch* of the water cycle. This branch contains the water that is available for societal use as it flows through a country—for water supply of houses and industry, for irrigation, for hydropower production, and so on (see Fig. 7.1).

The fact that humans live on the land surface implies that we occupy a very crucial position in the water cycle (Figure 7.5). All human activities that influence either the partial flow returning to the atmosphere, or the infiltration capacity of the soil surface, can be expected to have secondary effects on the water yield in the long branch of the water cycle and on its seasonality. If more rainwater is diverted along the soil surface, the flood flow would be expected to increase at the expense of the dry season flow.

Evidently, conditions must be particularly sensitive in a climate where part of the year is dry, because natural vegetation may be scarce and the soil surface susceptible to erosion. At the same time, overgrazing and chemical processes may produce physical or chemical crustas, making the soil impermeable. The process triggers itself by leaving the former root zone dry, further impeding vegetation. The effects may be devastating, including rapid erosion along the slopes and inundations and sedimentation problems in the downstream areas of the basin.

D. Land Productivity a Function of Availability of Soil Water

Basically, the plant production process takes place in the "factory" provided by the soil, and run by the water (Hyams, 1976). Carbohydrate is produced by this process, and the rate of production is controlled by the supply of raw materials and energy. The sun, besides providing the radiation necessary to keep the photosynthesis process going, also provides the energy needed to force a water-carried nutrient flow from the soil up through the plant. At the same rate that water evaporates from the leaf surfaces, new nutrient solution is sucked in by the roots from the moisture in the root zone. Nutrients and minerals are provided to the plant in, and only in, water solution. The plants are well adapted to local climate. During the dry season, grass needs only a very limited amount of water, whereas woody plants always need some water to survive the prolonged dry season (Balek, 1977).

The amount of water consumed in the phytomass production process varies considerably between vegetation systems, depending on general growing conditions and on the amount of unproductive evaporation that takes place, in addition to the productive water flow through the plant.

As illustrated in a paper with data from L'vovich (1979), there is a difference in the amount of water consumed between ecological zones; 200 m^3/ton biomass or even less in the equatorial forests, as compared to over 1,000 in the semiarid tropics (Falkenmark, 1986).

E. Prospects for Rain-fed Agriculture

Water availability in the root zone puts a limit to the length of the growing season in semiarid regions. By this link water limits the carrying capacity of an area, i.e., the size of the population that can be supported on rainfed agriculture.

To a certain degree, the level of technological input may increase the carrying capacity, for instance by improving soil permeability, but the ultimate water availability constraint remains. The FAO, in cooperation with UNEP and IIASA, has calculated the potential population supporting capacity in 117 countries in Asia, Africa, and Latin America. The study estimated the outcomes from rainfed production on potentially cultivable rainfed lands. An ideal crop composition was assumed, ascertaining maximum calorie production. Three different levels of technological inputs were taken into account: low level inputs (existing methods, no soil conservation), intermediate level inputs (somewhat improved crops mixture, some soil conservation, improved tools, some fertilizers), and high level inputs (complete mechanization, optimum genetic material, necessary farm chemicals and soil conservation measures). The length of the growing season was calculated in terms of temperature and moisture. During that period, the soil moisture storage, calculated from a simple soil moisture model fed by long-term average rainfall, should be able to support at least half the potential evapotranspiration.

The study showed that a great number of African countries are too dry to attain food self-sufficiency by 2000 *even when present irrigation plans are taken into account*. Thus, "the FAO study presents a compelling and sobering portrait of the agricultural capabilities of an increasingly populated developing world, many parts of which need to think in terms of self-sufficiency because they lack the financial resources to purchase food abroad" (World Resources Institute, 1986).

The FAO study gives sharp relief to the importance, from a long-term perspective, of the degrading land-base in the third world. Even if the present rate of soil loss is reduced to half (assumption behind the intermediate level input), many countries will still be suffering. Stopping the ongoing land degradation is therefore a fundamental task; finding ways to make better use of the local rain water another.

The need to push carrying capacity forward is indicated in the map in Figure 7.6 which is based on rain-fed agriculture solely; that is, irrigation is excluded (FAO, 1986). The map shows that rain-fed agriculture is not

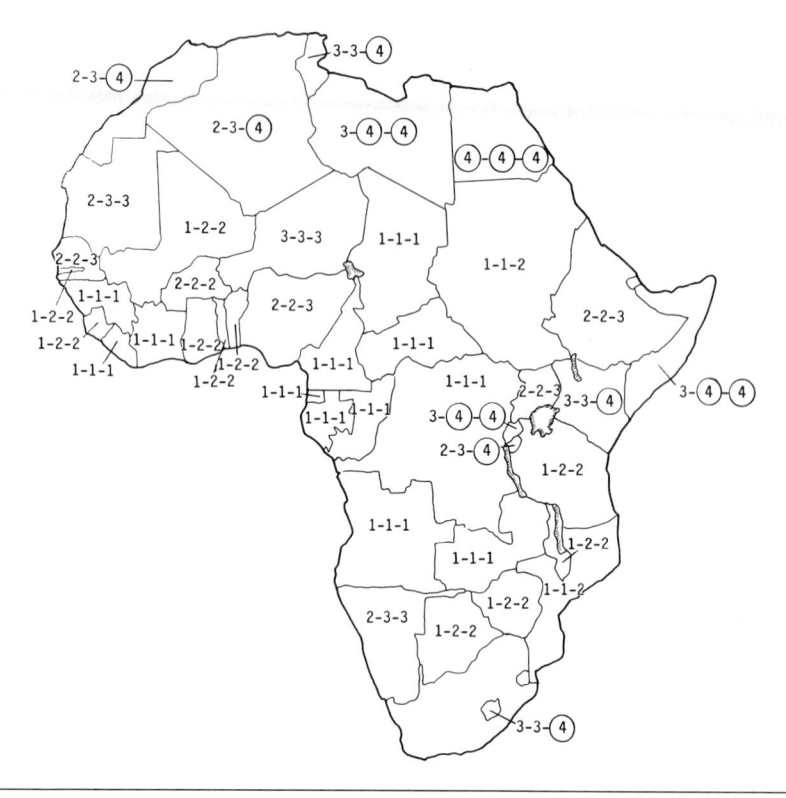

Figure 7.6 Population capacity that can be supported by rain-fed agriculture in African countries. The numbers in each three-digit series refer to 1982, 2000, and 2025, respectively, and indicate the level of yield-increasing technology needed to achieve self-sufficiency: (1) low; (2) intermediate; (3) high; (4) high-level will not suffice. [Based on data from FAO (1986).]

enough to support the present populations even on a subsistence level in north Africa, most countries in east and central Africa, and some of the west African countries, particularly in the Sahel. Two countries in south Africa belong to the same category.

If we look to the conditions estimated for the turn of the century, 13 countries will have to push their carrying capacity at least up to the yield level of modern agriculture in the western world to be able to support their populations, even on a subsistence level. For many countries even these yield levels will not be enough due to the short growing season in the dry climate.

We may therefore conclude that the choice will be either small-scale irrigation to lengthen and secure the growing season, large-scale irrigation to get an additional crop season (provided that there is enough water as well as energy available), or to import food. By 2025, conditions will have intensified due to population growth that will practically double the population of the continent.

Most of the countries in the hunger crescent will be forced to increase their rain-fed carrying capacity to support their populations on a subsistence level, at least to the yield levels of "intermediate level input," by 2000 (exceptions are Chad, Sudan, Mozambique, Zambia, and Angola). Seen in this short perspective, as pointed out by Brown and Jacobson (1986), the costs of increasing the level of technology, especially in view of the high costs of inputs together with the external indebtedness characterizing many of the suffering countries at present, would imply that many of these countries are not likely to afford much beyond the "low level inputs." The conclusion would be that best possible yields from local rain, secured by small-scale irrigation, might turn out to be an extremely interesting way. In other words, to make best possible use of the excess rain water accumulating in local washes in the undulating landscapes.

F. Soil and Water Conservation Success Stories

As indicated earlier, soil conservation is necessary to secure adequate availability of water in the root zone. Consequently, soil conservation and water conservation have to go hand-in-hand to secure the best possible use of local rain. One of the most impressive examples is the system for reallocation of local rainwater, developed in South Maharashtra state in India and spoken of as the Naigaon project (Salunke, 1983). The main components are rainwater collection from a watershed, an infiltration dam for subsurface storage of the rain water collected in a local wash, groundwater retrieval in a system of wells, allocation of the water in relation not to land ownership but to the size of the family up to a maximum of five family members. Only crops with reasonable water consumption are allowed. The system includes loans to farmers for the needed improvements, and a school for extension instructors where students from the village school are trained (to keep down the risk of brain drain).

This elaborated socio-technical system was able to reverse the prior migration to neighboring cities, as the strictly controlled use of local rain implied not only increased yields and food security, but also increased income and increased employment. It rapidly expanded from the initial test village of Naigaon to a large number of villages in the region.

World Resources Institute (1987) has also reported a number of success stories from Africa in the efforts to increase productivity of degraded lands. By placing lines of rocks across slopes, runoff may be slowed enough to infiltrate, allowing barren crusted wastelands to regenerate, as shown by

Oxfam on the Yatenga Plateu in Burkina Faso. The technique spread rapidly through networks of self-help village cooperatives. In Niger, rehabilitation of a degraded natural forest was made possible by various forms of small-scale, simple, rehabilitating measures such as microbasins to catch rainwater, rock dams to divert flash floods in local gullies in over land, and other efforts. In the improved soil, local tree species regenerated spontaneously.

In Kenya, traditional methods of soil conservation have been identified and perfected in a number of SIDA projects with the idea of executing the methods with simple means. Basic techniques are digging narrow trenches, throwing the soil uphill, and piling up ridges or trash lines against which the soil may accumulate when moving downhill, thus forming naturally produced terraces.

Agroforestry, the growing of foodcrops between regularly pruned hedgerows of fast-growing, nitrogen-fixing trees, has many advantages. In addition, trees serve many other purposes such as windbreaks to reduce evapotranspiration, production of fertilizers, and the use of leaves and twigs as mulch.

IV. WATER AVAILABILITY LIMITS INCREASE OF CARRYING CAPACITY

The question of improving the carrying capacity of African agriculture in a short- to medium-term perspective is basically a question of managing unreliable rainfall during the rainy season, so that the average yields may be increased. As just mentioned, small-scale irrigation based on local water storage of the rapid runoff in local washes is a natural way to advance, as few countries will be able to provide or to finance large-scale technological inputs. The techniques to be chosen would primarily be surface water storage in local tanks or groundwater storage behind subsurface dams. There is the additional possibility of micro-scale rainwater harvesting and collection.

Whatever the source, better use of local rainfall implies that excess water, now flowing out of an area, will be put to active use locally. In this section we will discuss the African dilemma from the aspect of availability of water in the landscape in relation to the number of people depending on that water for the development of their quality of life. The scale used in this broader outlook will be the macro-scale defined by the dimensions of the individual countries.

Given the arid conditions in large parts of Africa, the present food supply problems, and the rapidly rising populations, it seems important to analyze the development perspectives seen in relation also to water—especially as this is a quite new perspective, left unattended even by the

World Commission on Environment and Development (1987) in their discussion of sustainable development. To what degree will water limitations act as important constraints to development?

A. Population Increase Generates Water Scarcity Problems

As already indicated, the average amount of water available to a country is principally fixed by the water cycle, unless influenced by natural or human changes. Water arrives in two ways: Endogenous water arrives via local rainfall and is consumed in plant production, with the surplus recharging groundwater aquifers and rivers; and exogenous water arrives via rivers and regional aquifers (Fig. 7.1).

The long-term water availability being fixed, population increase in water-scarce countries generates higher and higher stress on limited resources. Population increase may, in other words, limit the possibility for a country to satisfy their increasing water demands. In fact, hydrological conditions and limitations in the water management capability tend to put a limit on the number of individuals which can be fed on a self-sufficiency basis by the simple fact that water is consumed in crop production.

Figure 7.7 illustrates the water competition level with the present and predicted populations by the year 2000 in a number of African countries, based on water availability data estimated from macro-scale hydrological analyses by L'vovich (1979). National data from the majority of the African countries are unfortunately very incomplete and many countries lack reliable data on their water resources.

It should be observed that the European and North American countries, whose water management problems form the core of an extensive water management literature, represent temperate, generally humid regions with often 300–500 people competing for each flow unit of 1 million m^3/yr (Falkenmark, 1986). The lower Colorado basin, known for water scarcity problems, represents just below 700 people per flow unit (p/unit).

Societal development generally implies rising per capita levels of water demands for improved hygienic conditions, improved quality of life, industrial production, and other factors. In areas where the growing season may be interrupted by periods of drought or where yields have to increase, irrigation needs generally increase too. When population increases, gross water demands may accelerate very rapidly.

This raises the question of how many people will the present African societies, within a few decades, manage to supply from each flow unit of water? This will of course depend on the potential to store water from the wet season for use during the dry season, on the technological capacity to redistribute water spatially to adapt the geographical pattern of availability to the pattern of water demands, on the administrative capacity to

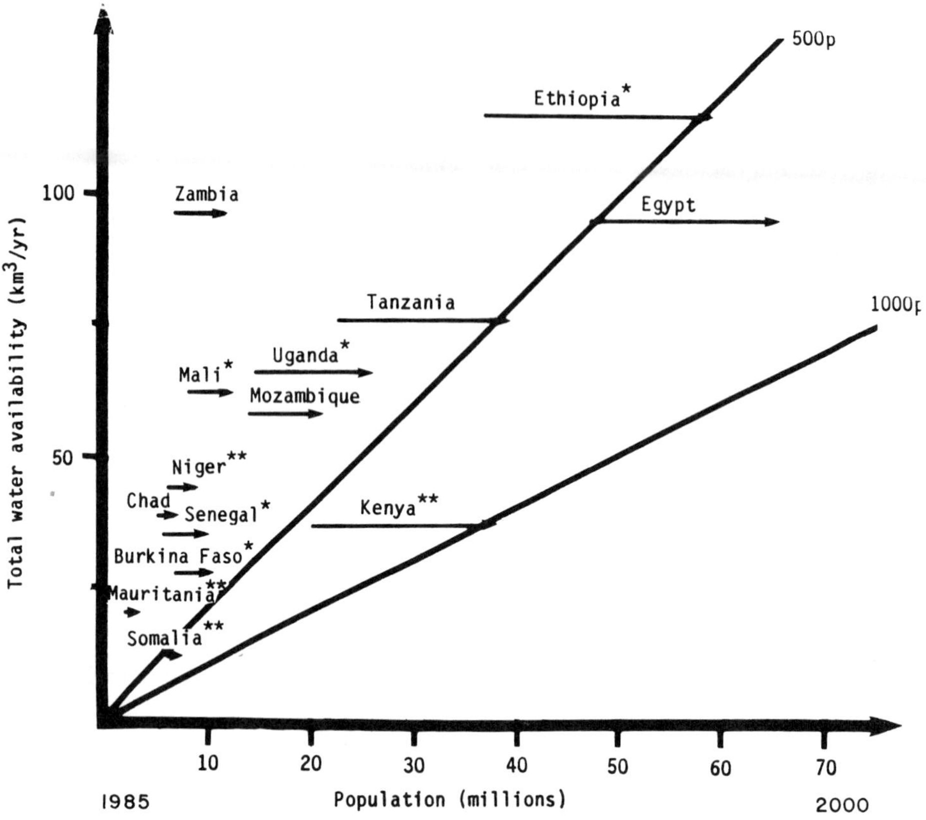

500p
1000p
Ethiopia*
Zambia
Egypt
Tanzania
Uganda*
Mali*
Mozambique
Niger**
Chad
Senegal*
Kenya**
Burkina Faso*
Mauritania**
Somalia**
100
50
Total water availability (km3/yr)
10
20
30
40
50
60
70
1985
Population (millions)
2000

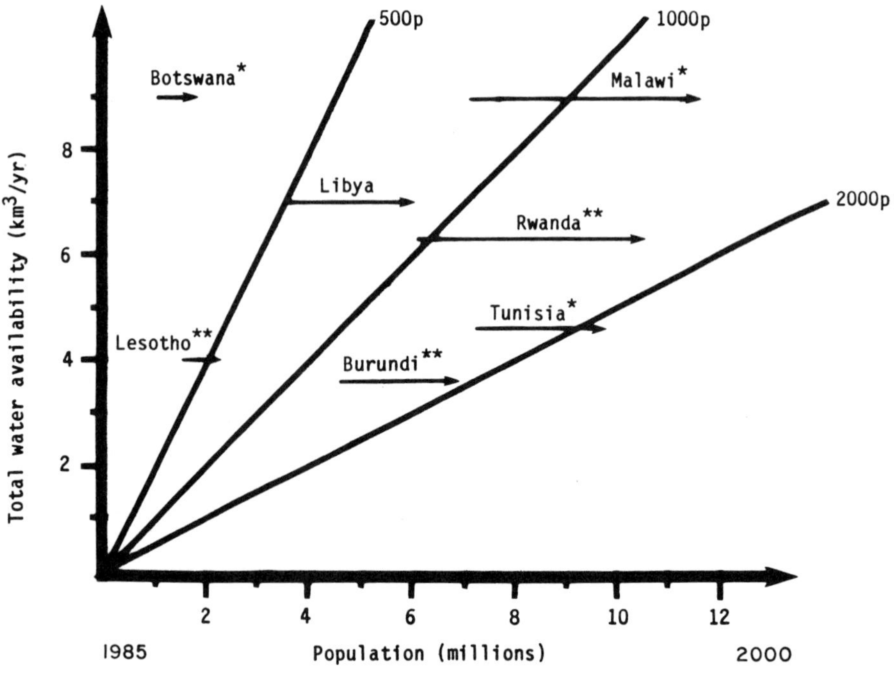

500p
1000p
2000p
Botswana*
Malawi*
Libya
Rwanda**
Tunisia*
Lesotho**
Burundi**
8
6
4
2
Total water availability (km3/yr)
2
4
6
8
10
12
1985
Population (millions)
2000

coordinate all the different sectors in society related to water, and on the overall need for irrigation water. Typically, irrigated agriculture needs 10–30 times the water amount needed for household supplies. The latter amount being about 35 m^3 per person per year—equivalent to 100 liters per person per day (abbreviated as H)—10–30 H would correspond to 350–1000 m^3 per person per year.

Judging from experience, one would expect that 2,000 p/unit constitutes the upper limit—the *water barrier*—under present conditions for a country with full access to advanced technology and sophisticated administrative capacity (Falkenmark, 1987). In a developing country, on the other hand, with poor data on the water availability and its interannual fluctuations, poor access to technology, lack of finance, and poor administrative capacity due to lack of manpower, the level of practical manageability may be much lower. In the long-term perspective things may change, but we are discussing short- to medium-term perspectives. If only 10–20% of the water availability is readily accessible, then only 200–400 p/unit could be supported on an overall demand level of 500 m^3 per person per year (about 15 H).

B. Rapid Approach of Large-Scale Water Crisis

Figure 7.8, which indicates the food security and water availability problems in Africa, includes code numbers that refer to both water for rainfed agriculture, and to the number of individuals jointly depending on each flow unit available to the country in the long branch of the water cycle. The first code number has been discussed in Section IV.A. The second code number indicates the water scarcity dimensions. By 2000, six countries will have to cope with problems of *absolute scarcity* (more than 1,000 p/unit), three in north Africa and three in east–central Africa. Another eight countries will be *water stressed* (500–1000 p/unit), dispersed over all the four African subregions. In all 150 million people will be living in countries in the first category, 350 million in the second. In all, around 60% of the African population will, in other words, be suffering from water scarcity problems, at least of the level today characterizing lower Colorado basin in United States.

Twenty-five years later, population growth will have driven close to 600 million people into conditions of absolute water scarcity, greatly com-

Figure 7.7 Water competition diagram shows relationship between total water availability (endogenous runoff plus exogenous inflow) and the population in 1985 and 2000 for some African countries: (*) critical population-supporting capacity by 2000 with low-level technological input, (**) capacity critical even with intermediate-level input. [From Falkenmark (1987); water data from L'vovich (1979), population data from World Resources Institute (1986).]

(a) 2000 AD

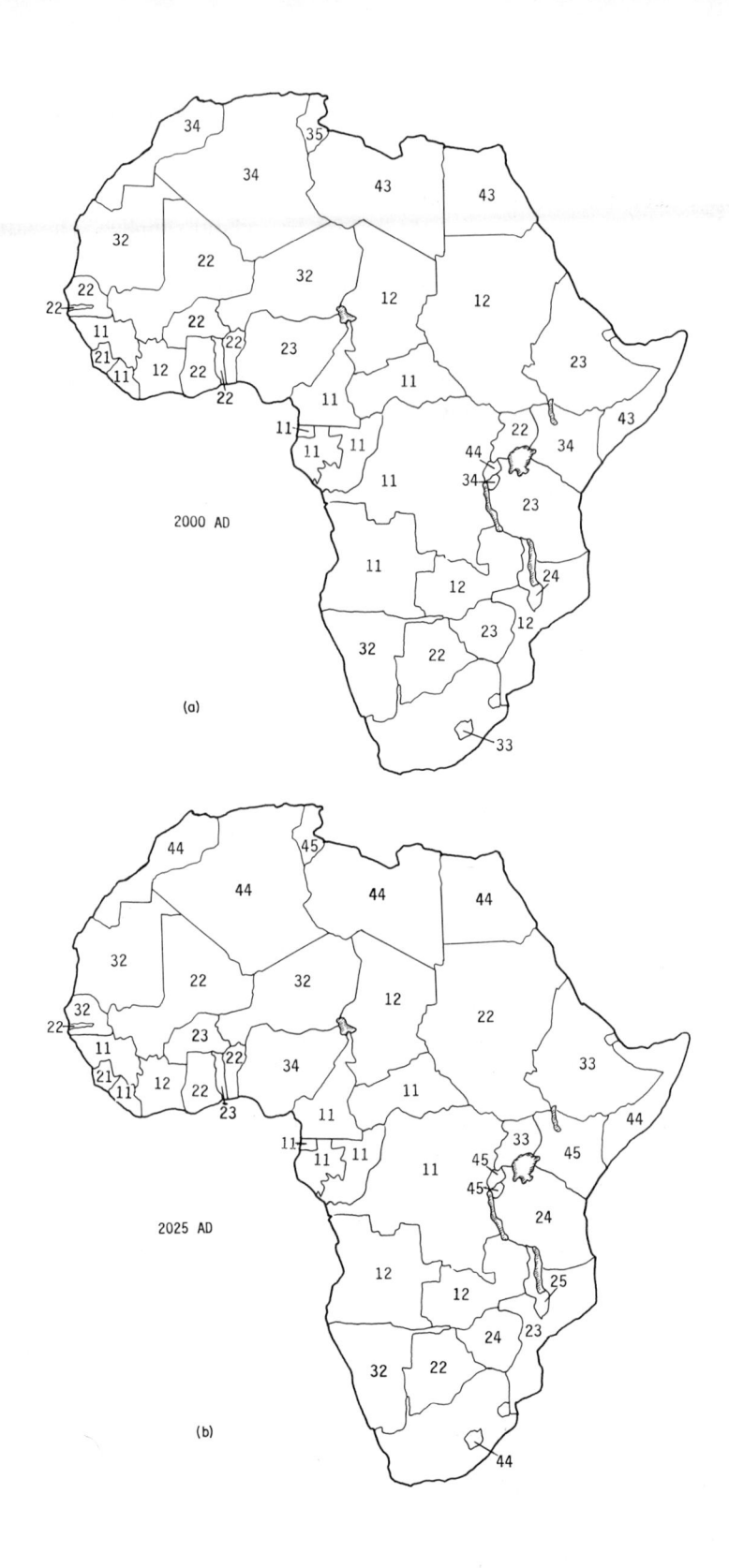

(b) 2025 AD

plicating the development of both water security in general and of food security in particular. In all, 14 countries will be in the group of absolute scarcity, and five countries in the water-stressed group.

Where does this information bring us? What are the implications from a development perspective? First of all we have to recall that in a modern society with a high level of water development, a great part of the potentially available water is also readily available, thanks to redistribution in time and space. When 100% of the potential availability can be used, a water competition level of 1,000 p/unit would allow an average supply level of 30 H. A level of 2,000 p/unit would allow a supply at the level of 15 H.

If, on the other hand, only dry season water availability can be counted on from rivers and groundwater, 1,000 p/unit would allow a supply level of only 6 H, and the level of 2,000 p/unit only 3 H. It is evident that absolute scarcity is indeed a severe constraint to water supply and food security unless massive storage possibilities exist. Even with a high level of water resources development, like that available in the Nile through the overyear storage in Lake Nasser, the average demand level of 1,000 m^3 per person per year, now typical for Egypt, will be forced downwards as the water competition rises into the interval of absolute scarcity.

V. THE COMBINED PICTURE

The two maps in Figure 7.8 provide us with a first idea of the crucial role of water scarcity in African development during the next few decades.

A. The Next Decade

During the 1990s massive water scarcity problems will complicate the road toward food security both in north and east–central Africa. Six countries have to cope with massive problems during this period: Morocco, Algeria and Tunisia in north Africa, and Kenya, Ruanda, and Burundi in east–

Figure 7.8 Water-related problems to be expected in societal development by (a) 2000 and (b) 2025. The first number in each two-digit series relates to self-sufficiency in rain-fed agriculture (see Figure 6). The second number refers to the level of water stress: (1) mainly dry season problems (water competition level >100 people per flow unit); (2) conditions comparable to central Europe (100–500 people per flow unit); (3) definite water stress (500–1,000 people per flow unit); (4) absolute water scarcity (1,000–2,000 people per flow unit); (5) present water barrier under advanced management conditions surpassed (>2,000 people per flow unit). [Data on water availability from L'vovich (1979); data on population from FAO (1986).]

central Africa. They are all dependent on far-reaching agricultural development to achieve self-sufficiency (at least yields corresponding to "high level inputs") and they all have absolute water scarcity. Tunisia is expected to be worst off by having reached the present "water barrier" for advanced management conditions (2,000 p/flow unit).

The next group of countries are those which will have moderately large water-stress problems (500–1,000 p/unit, i.e., conditions comparable to the lower Colorado basin in the United States). These countries are found in all four of the African subregions. The ones most dependent on water for food security (here referred to as "import dependent," i.e., code number 4 for yield-increasing technology) are Libya and Egypt in north Africa, and Somalia in east–central Africa. Countries needing far-reaching agricultural development are Lesotho in south Africa, and those depending on reaching intermediate-level agriculture are Ethiopia and Tanzania in east–central Africa, Nigeria in west Africa, and Zimbabwe in south Africa.

The third group of countries are those which will meet some water problems, i.e., dry season problems and other problems well known from the temperate zone. Water competition will be on a level comparable to many countries in central Europe (100–500 p/unit), and food security development will call for water development to increase the readily available amount of water. Dependent on yields corresponding to high technology agriculture are Mauretania and Niger in west Africa and Namibia in Southern. Dependent on intermediate-level agriculture are Uganda in east-central Africa, a number of countries in west Africa, and Botswana in south Africa.

B. First Decades of Next Century

If we extend our interest beyond the turn of the century, population growth will push most of these countries into higher- ranked categories of water-related development problems. The north and east–central African countries with massive water scarcity problems projected by 2000 will become import-dependent to feed their populations on a subsistence level. The already import-dependent countries in the same regions will advance from water-stressed conditions to absolute scarcity. In west Africa, Nigeria will advance from considerable to massive water scarcity problems.

A third group of countries will advance from some to considerable water scarcity problems (some west African countries, Mozambique in south Africa), from some to large problems (Uganda in east-central Africa), and from considerable to large (Ethiopia in east–central Africa). The Sahelian countries and Botswana in south Africa will remain in the category with some, i.e., more traditional, water development problems.

VI. CONCLUSIONS

In answering the initial question of whether water scarcity is a temporary or lasting problem in Africa, we might conclude our observations in this chapter with the following considerations. Many areas in Africa suffer from large-scale resource crises together with a degradation of fundamental life-support systems. One primary contributing factor is the poor understanding among both policy-makers and consultants of the particular African predicaments created by water-related limitations. Water availability limits both land productivity and development potential in general. Seasonal water deficiency in the root zone limits the overall crop yields, and interannual fluctuations add continuing risks for crop failures during semi-periodical droughts. The ability to manage water resources limits the population that can be supported by the water provided to a country by the global water cycle.

In a short-to-medium-term perspective, the poor economies of Africa force policy-makers to think in terms of food self-sufficiency simply because they lack the financial resources to purchase food abroad. The fundamental effort is, therefore, to strive for maximum increase of agricultural yields without turning to expensive large-scale solutions: in other words, to make the best possible use of local rain. This will primarily be a question of improving root-zone water security by small-scale water conservation systems. Protection of soil permeability is a key factor in increasing rainwater infiltration. Soil conservation has therefore to be integrated with water conservation, preferably on the watershed level. A number of success stories invite some optimism regarding the possibility of rapid results.

About a third of the African population lives in the countries of the drought-prone crescent of the semi-arid tropics, where part of the year is dry and the growing season too short for reliable rainfed production. There is a constant risk for crop failures during drought years. Only minor amounts of excess rainfall are available to recharge aquifers and local watercourses, and this excess is subject to the same interannual fluctuations, making this zone particularly sensitive to human interventions that disrupt soil permeability. Recurrent droughts, which are typical for the semi-arid climate, make contingency plans a crucial component of societal resource management.

Population growth requires the sacrifice of future economic opportunities because there exist clear limitations in the number of individuals that it is possible to sustain on each flow unit of water available from the global water cycle. This limitation can best be thought of as a *water barrier* to development, equivalent to the maximum providable gross water demand. The larger the population, the less water that can be allocated to each individual. The water barrier can be pushed forward with increased access to advanced management methods, up to a limit of about 2,000 p/

unit of water (1 million m³/yr). In the distant future, advanced recirculation and sequential reuse of water may push the limit further.

Increased agricultural yields based on better use of local rain will have to compete for water with other sectors of society. The rapidly increasing water scarcity, generated by population growth, therefore adds to problems with both food and water security. In a time perspective of a few decades, two-thirds of the African population will live in countries where the competition for water will be very large due to high population pressure on a constant resource (severe water stress or even absolute scarcity by 2000). Many of the hunger crescent countries belong to this category. By 2025 the populations subject to these constraints for development of self-reliance in basic food production will have increased beyond 1,000 million. This includes a number of countries that currently have absolute scarcity. Out of these 1,000 million people 700 million or half of the total population of Africa, will live in countries with absolute scarcity. In such countries not even 1,000 m³ per person per year (30 H) could be afforded even under sophisticated management methods.

The severity of the approaching water scarcity crisis calls for a variety of measures that are needed to create the necessary awareness and preparedness among planners and policy makers, both in the countries themselves and in the international community that bring advice, loans, and technical assistance. Among the most urgent measures is to develop the skills needed to meet the problems of water deficiency.

In fact, water scarcity can be foreseen to grow into a problem of alarming dimensions in Africa. This fact has gone unnoticed until now, even in the report on sustainable development by the recent World Commission on Environment and Development (1987). The fact that committees dominated by experts, living or educated in the temperate zone, have not been capable of describing the African water problems in an adequate way in fact adds to the present problems. It is of fundamental importance that the past undermining of problems descriptions, caused by climatic biases of temperate zone experts, be rapidly remedied by stimulating the development of indigenous expertise and research. Education and training are measures of highest priority and great urgency in order to reverse the present resource crisis. In particular, a hydrologically knowledgeable staff is crucial to provide planners and decision-makers with the basic information needed on the water resources, characterizing the country. Today, reliable information on the water resource available in the different African countries is extremely poor, calling for urgent development observation networks to make possible reliable water resources assessments. National water balances are necessary bases to support development plans, seeking realistic ways to reach optimum use of limited water resources.

African countries will be forced to look at water from a much broader perspective than the one practiced in the industrialized temperate zone.

The water circulation in the hydrological cycle is to be seen as a basic life-support system, supporting both plant production and societal development. In a time perspective of a few decades, small-scale socio-technical systems for the best possible use of local rainfall are much more realistic and promising solutions than traditional, large-scale irrigation systems based on exogenous water available from remote humid regions. Exceptions are countries crossed by large rivers such as the Nile, the Niger, the Zambesi, or the Limpopo. The rich water resources of the Congo River provide a long-term water reserve, but they will scarcely become accessible to water-deficient countries during the next few decades—the focus of this chapter—due to the vast scale of the river diversions needed.

Without such fundamental changes in overall perspectives and action, major perturbations will be unavoidable, and drought-trigged collapses of life-support systems are highly probable. In the wake will follow large-scale migration of environmental refugees and a considerable risk for political instability.

REFERENCES

Ahmad, Y. J. (1986). Special series of articles: The African Crisis. (1) It's a total breakdown in resources management. (2) Three sub-regions which need different priorities. (3) Recovery impossible till the emergency is ended. (4) Resource protection must be the first commandment. (5) Continental Harambee essential for survival. (6) A project is doomed if it doesn't involve the people. (7) It's water everywhere and not a drop to drink. (8) Steps that must be taken to maintain water safety. (9) UNEP's central place in ensuring water security. *Daily Nation* (Kenya) Autumn (1986).

Balek, J. (1977). "Hydrology and Water Resources in Tropical Africa." Elsevier Scientific, Amsterdam and New York.

Barnett, T. P. (1985). Long-term changes in precipitation patterns. *In* "Detecting the Climatic Effects of Increasing Carbon Dioxide" (M. C. MacCracken and F. M. Luther, eds), DOE/ER-0235, pp. 151–162. U.S. Dep. Energy, Washington, D.C.

Bentley, C. F., Holowaychuk, H., Leskiw, L., and Toogood, J. A. (1979). Soils. *Conf. Agric. Prod.: Res. Dev. Strategies 1980's, Bonn* Dtsch. Stiftung Int. Entwicklung, Dtsch. Ges. Tech. Zusammanarbeit, Bundesminist. Wirtsch. Zusammenarbeit and Rockefeller Found., New York.

Bolin, B., Doos, B. R., Jager, J., and Warrick, R. A., eds. (1986). "The Greenhouse Effect, Climatic Change and Ecosystems," Chap. 1. SCOPE, Wiley, Paris and New York.

Brown, L. R., and Jacobson, J. L. (1986). "Our Demographically Divided World," Worldwatch Pap. No. 74. Worldwatch Inst., Washington, D.C.

Falkenmark, M. (1984). New ecological approach to the water cycle: Ticket to the future. *Ambio* 13(3), 157–160.

Falkenmark, M. (1986). Fresh water—Time for a modified approach. *Ambio* 15(4), 192–200.

Falkenmark, M. (1987). Water-related constraints to African development in the next few decades. *IAHS Publ.* No. 164, 439–453.

FAO. (1986). "Need and Justification of Irrigation Development. Consultation on Irrigation in Africa, Lomé, Togo," AGL:IA/86/Doc I-D. FAO, Rome.

Hyams, E. (1976). "Soil and Civilisation." Harper, New York.
Korzun, V. I., ed. in chief. (1978). "World Water Balance and Water Resources of the Earth." Studies and Reports in Hydrology, No. 25. UNESCO, Paris.
Levine, G., Oram, P., and Zapata, J. A. (1979). Water. *Conf. Agric. Prod.: Res. Dev. Strategies 1980's, Bonn* Dtsch. Stiftung Int. Entwickling, Dtsch. Ges. Tech. Zusammanarbeit, Bundesminist. Wirtsch. Zusammenarbeit and Rockefeller Found., New York.
L'vovich, M. I. (1979). "World Water Resources and Their Future" (Am. Geophys. Union, transl.), p. 250. LithoCrafters, Chelsea, Michigan.
Nicholson, S. E., and Chervin, R. M. (1983). Recent rainfall fluctuations in Africa—interhemispheric teleconnections. *In* "Variations in the Global Water Budget" (A. Street-Perrott, M. Beran, and R. Ratcliffe, eds.), pp. 221–238. Reidel, Dordrecht, Netherlands.
Salunke, V. B. (1983). "Pani Panchayat, Dividing Line between Poverty and Prosperity." Gram Gourav Pratishthan, Shetkarinagar- Khalad Taluka-Purandhar, District, Pune, Maharashtra, India.
Swedish Red Cross. (1984). "Prevention Better than Cure," Report on Human and Environmental Disaster in the Third World. Stockholm and Geneva.
United Nations (U. N.). (1987). "Water Resources: Progress in the Implementation of the Mar de Plata Action Plan. Review of the Situation with Regard to the Development of Water Resources in the Drought-Stricken Countries of the African Region," Committee on National Resources, Tenth Session. Economic and Social Council E/C.7 1987/6.
World Commission on Environment and Development. (1987). "Our Common Future." Oxford Univ. Press, London and New York.
World Resources Institute. (1986). "World Resources 1986. An Assessment of the Resource Base that Supports the Global Economy." Basic Books, New York.
World Resources Institute. (1987). "World Resources 1987. An Assessment of the Resource Base that Supports the Global Economy." Basic Books, New York.

8

Agricultural Chemicals: Food and Environment

David A. Andow
David P. Davis

Department of Entomology
University of Minnesota
St. Paul, Minnesota

    I. Introduction
   II. Use of Agricultural Chemicals
  III. Nitrogen in Agroecosystems
      A. Flow of Nitrogen in Agroecosystems
      B. Management of Nitrogen in Agroecosystems
      C. Summary of Nitrogen in Agroecosystems
  IV. Pesticides
      A. Manufacture
      B. Use in Developing Countries
      C. Human Exposure
      D. Effects in Habitat of Application
      E. Movement Out of Habitat
      F. Summary of Pesticides in Agroecosystems
   V. Rational Use of Agricultural Chemicals
      A. Distribution of Costs and Benefits
      B. Interactions of Inputs and Cycles
      C. Toward More Rational Use
References

I. INTRODUCTION

Intensification of agricultural production has become essential to provide food and fiber as society has urbanized. In urbanized societies, less labor is available for agricultural production and large quantities of food and fiber are transported off-farm for consumption. Intensification is required to increase labor productivity on farms and to replace nutrients and energy leaving the farm. Modern industrial agriculture solves these problems with reduced variety of crops, increased animal stocking rates, increased genetic homogeneity of crops and livestock, and reduced fallowing of cropland and pasture. Concomitant with these practices is increased reliance on machines (Chapter 10), energy (Chapter 9), and agricultural chemicals.

Agricultural chemicals are now used in all areas of food production. Synthetic fertilizers and pesticides are integral to contemporary crop production. Feed additives, pesticides, and hormones are widely used in livestock production. Pesticides, ripeners, and preservatives are used to increase the storage life of agricultural products. It seems likely that industrial agriculture could not survive without them.

The dual character of agricultural chemicals provides the grist for a continued and frequently strident disagreement over their social utility. On the one hand, use of agricultural chemicals has increased and stabilized agricultural yields, thereby providing more food and fiber, alleviating some portion of world malnutrition and starvation, and enhancing economic stability in unstable environments. These benefits, however, have not been without environmental costs. Indeed, the darker side of agricultural chemicals has been unavoidable; inherent in their widespread use has been impaired human health and contamination of the environment. In addition, the rational pursuit of increased production on the farm has resulted in ecological irrationalities in the entire agroecosystem.

II. USE OF AGRICULTURAL CHEMICALS

Consumption of synthetic nitrogen has increased over six-fold in the last 40 years in the United States (Figure 8.1), and consumption of nitrogen fertilizer worldwide has increased over 17-fold in the last 30 years. Production of pesticides has increased nearly 20-fold since 1947 in the United States (Figure 8.2), and some agricultural chemicals, such as livestock antibiotic feed supplements and livestock hormones, were not commercially available until the 1950s. Indeed, livestock antibiotic feed supplements have become necessary because of the increased incidence of certain diseases associated with intensive livestock production (Harbourne, 1966). The intensification of agriculture has been extremely rapid since the Sec-

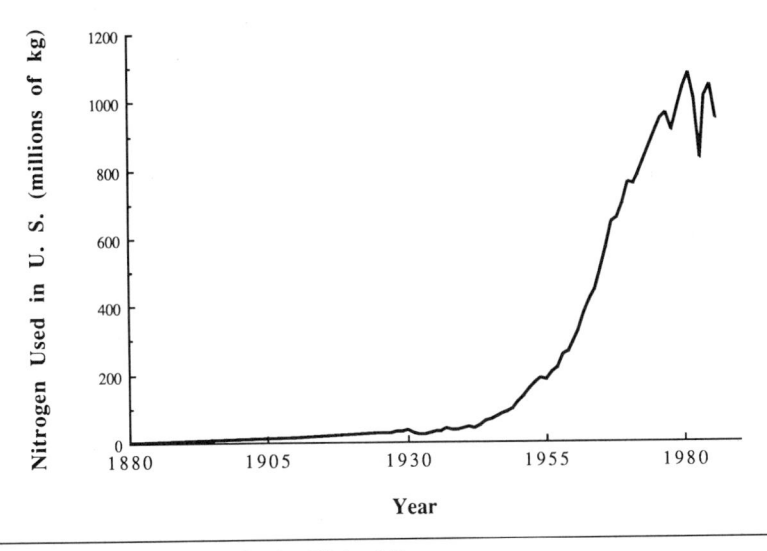

![Figure 8.1]

Figure 8.1 Nitrogen use in the United States.

ond World War; from that time, agriculture has been radically transformed to have an increased reliance on inputs that are external to the farm.

The amount of agricultural chemicals currently used in the United States is phenomenal. The United States consumes about 15.6% of the nitrogen, 13.0% of the phosphorus, 20% of the potassium, 45% of the

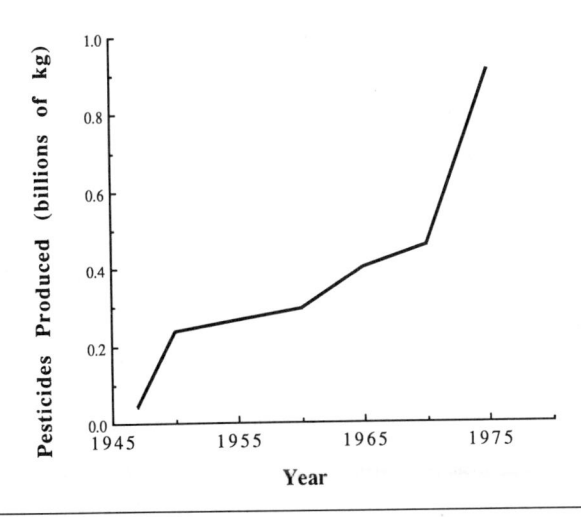

Figure 8.2 Pesticide production in the United States. [Redrawn from Pimentel *et al.* (1980).]

pesticides and the majority of livestock feed additives and hormones of the world total (Table 8.1). Over 300 kg/ha of synthetic fertilizer and 1,500 g/ha of active pesticide ingredients are applied to all arable lands, and livestock receive over 300 ppm of agricultural chemicals on average each year. Some lands and some animals receive substantially more than these average amounts of chemicals.

Use of agricultural chemicals has been less in developing countries than in developed countries. Developing countries, however, provide the main growth market for the agricultural chemical industry, and rates of increase of use of agricultural chemicals have been faster in developing countries than in developed countries. Consumption of synthetic fertilizer nitrogen in developing countries was only 32% of the world total in 1976

Table 8.1 Contemporary Use of Agricultural Chemicals in the United States and the World

Agricultural chemical	United States Year	Quantity	Units[a]	World Year	Quantity	Units[a]
Synthetic fertilizer	1985	44,640,000	t[b]			
Nitrogen (N)	1985	10,440,000	t[b]	1983–1984	66,907,000	t[c]
Phosphorus (P_2O_5)	1985	4,285,000	t[b]	1983–1984	32,856,000	t[c]
Potasasium (K_2O)	1985	5,088,000	t[b]	1983–1984	25,408,000	t[c]
Synthetic pesticides	—	217,160,000	kg a.i.[d]	—	482,580,000	kg a.i.[e]
Insecticides						
Crops[f]	1976	74,140,000	kg a.i.[g]			
Livestock	1976	4,815,000	kg a.i.[g]			
Herbicides (crops)	1976	101,800,000	kg a.i.[g]			
Fungicides (crops)	1976	19,640,000	kg a.i.[g]			
Defoliants and dessicants	1976	3,923,00	kg a.i.[g]			
Plant growth regulators	1976	2,850,000	kg a.i.[g]			
Fumigants						
Crops	1976	8,813,000	kg a.i.[g]			
Storage	1984?	1,180,000	kg a.i.[h]			
Botanical pesticides						
Insecticides						
Crops	1976	n.a.[i]				
Livestock	1976	72,300	kg a.i.[g]			
Livestock feed additives						
Antibiotics	1978	5,591,000	kg[j]			
NPN (area)	1980	359,000	t[k]			

[a] t = metric tons, kg = kilograms, a.i. = active ingredient.
[b] ERS (1987). [c] FAO (1985). [d] Sum of individual categories.
[e] The United States accounted for 45% of world pesticide consumption in 1970 (Furtick, 1976b).
[f] Includes miticides. [g] Eichers et al. (1978). [h] USDA (1986).
[i] Less than 22,700 kg; Eichers et al. (1978). [j] Burbee (1980). [k] USDA (1982).

(FAO, 1980), but had increased to 43% by 1983 (FAO, 1984). Pesticide use in developing countries was only 7% of the world total in 1970 (Furtick, 1976a), but had increased to 10% by 1974 (Furtick, 1976b).

In this chapter, we explore the food and environmental consequences of this intensive reliance on agricultural chemicals in industrial agriculture. Because of their large volumes and importance to agriculture, ecosystem function, and environmental health, we focus on the consequences of the use of synthetic nitrogen fertilizers and synthetic pesticides.

III. NITROGEN IN AGROECOSYSTEMS

The dual nature of nitrogen hinges on its ephemerality; biologically available nitrogen is transitory. Plants require specific inorganic forms of nitrogen for growth, but these forms are often limiting. Therefore, nitrogen has to be added regularly to maintain crop productivity. However, because of its transient nature, available nitrogen is easily lost from the agroecosystem to the atmosphere, groundwater, and surface waters.

The flow of nitrogen from soil to crops can deplete reserves of soil nitrogen in 5–20 years (Cooke, 1982). To maintain productivity, the input of nitrogen to a field has to be large enough to balance harvest removal. While sources of nitrogen are diverse, in modern industrial agricultural systems synthetic fertilizer is the primary source.

Synthetic nitrogen fertilizers were developed in the 1920s following the discovery by Habre in 1913 of a technique to convert atmospheric nitrogen to ammonia. Initially, synthetic nitrogen was inexpensive. The production of synthetic nitrogen, however, requires tremendous amounts of energy (1 ton of anhydrous ammonia requires 49 million Btu; Sherff, 1975), so the cost of production is tied to the cost of energy. This has resulted in greatly increased costs in recent times.

There are many benefits of synthetic nitrogen use. Forages have higher levels of total and digestable nitrogen, grains have higher protein content (Olsen *et al.*, 1976), fruit yield and vegetable leaf growth increase, carbohydrate and protein synthesis increase, grasslands can support more livestock, and plants are more drought and frost resistant (CEA, IFA, and IPI, 1983). The FAO has conducted over 200,000 fertilizer trials worldwide since 1961 (Cooke, 1982). The economic returns from fertilizer use (ratio of value of increased crop production to cost of fertilizer) are estimated to be 6.0 in Africa, 5.6 in Asia, 4.9 in Latin America and 3.2 in the Near East (overall average 4.8, Von Peter, 1980).

The role of synthetic nitrogen in providing food and fiber is extremely important. The challenge is to use our nitrogen resources wisely. In the United States during the 1960s and early 1970s farmers typically applied synthetic nitrogen in great excess because it was inexpensive. Today, however, the cost has increased greatly and environmental contamination, principally nitrates in ground water, has caused concern worldwide.

A. Flow of Nitrogen in Agroecosystems

Synthetic nitrogen is not the only source of nitrogen in agroecosystems. There are at least six other major sources of nitrogen inputs (Figure 8.3). To understand the ecological problems with current management of nitrogen in contemporary agriculture and to arrive at solutions, it is necessary to examine the flows of nitrogen in agroecosystems (Andow and Davis, 1988). A rational approach to nitrogen use suggests that on-farm resources and wastes are maximally used and synthetic nitrogen is supplementally used.

1. Synthetic Nitrogen Fertilizers

Effective use of synthetic nitrogen is complex. Usually, crops respond rapidly to initial additions of synthetic nitrogen, but returns diminish at

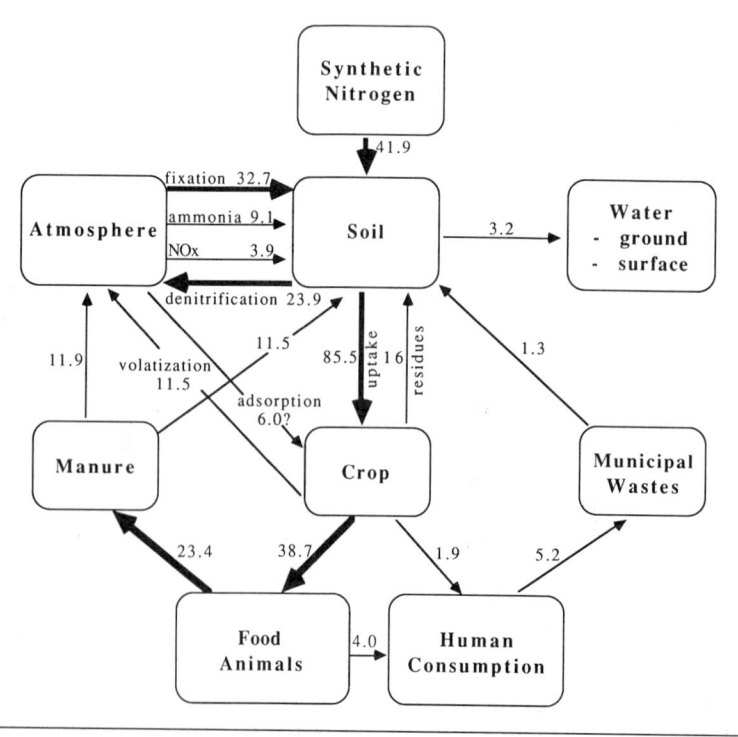

Figure 8.3 Flow of nitrogen in the United States. All values are in kg/person/year.

higher levels of input (Cooke, 1982). Crop response to nitrogen is affected by the balance of other nutrients, such as phosphorus and potassium (concentration of nitrogen, phosphorus, and potassium for synthetic fertilizers is given in Table 8.2). Other factors that affect crop response include soil nitrogen content, soil type, type of fertilizer and method of application, precipitation, and time of application. If synthetic nitrogen is added to soils already sufficient in nitrogen, no increase in production occurs (Westerman, 1987), and porous soils in humid regions require greater amounts of nitrogen because soil nitrogen leaches out of the root zone and into groundwater (Sherwood, 1985). Synthetic nitrogen may be added in liquid (e.g., urea, anhydrous ammonia) or solid forms (e.g., ammonium nitrate). Because liquid forms volatilize quickly if sprayed on the surface of soils, the type of fertilizer could affect availability to the plant. This can be minimized by injecting liquid nitrogen into the soil. Precipitation can have variable affects on synthetic nitrogen applications. While volatilization may be minimized, heavy rains may cause excessive leaching of nitrate out of the soil and into ground water. Furthermore, if synthetic nitrogen is applied in spring versus winter, the efficiency of uptake by the crop is increased by 48% (Pearson *et al.*, 1961), because nitrogen leaches out of the upper soil during winter when plants are not present and microorganisms are not active to use it.

Current use of synthetic fertilizer nitrogen is 9.5×10^9 kg N/yr in the United States (1975–1984 average; USDA, 1985a) or 41.9 kg/person/yr (USDC, 1983). This intensive use of nitrogen dominates soil nutrient status and could place agroecosystems on a unique successional pathway. Vitousek *et al.* (1987) showed that high levels of nitrogen input originating from fixation by a leguminous tree in traditionally nitrogen-poor volcanic soils altered successional dynamics, allowing many species to invade and displace naturally occurring species. Thus, the large nitrogen inputs from

Table 8.2 Composition of Widely Used Forms of Nitrogen, Phosphorus, and Potassium Fertilizers

Fertilizer	Nitrogen (%)	Phosphorous (%)	Potassium (%)
Anhydrous ammonia	82	0	0
Urea	46	0	0
Ammonium nitrate	33.5	0	0
Urea–ammonium nitrate	28	0	0
Concentrated superphosphate	0	45	0
Diammonium phosphate	18	46	0
Monoammonium phosphate	11	55	0
Potassium chloride	0	0	60

human sources may provide unique opportunities for ruderal plants and mobile pests.

2. Atmospheric Inputs

Additions of nitrogen to soil from the atmosphere occur through precipitation, dry deposition, fixation by free living bacteria (*Azotobacter* and *Clostridium*) and blue-green algae *(Anabaena)*, and fixation by plant-symbiotic bacteria *(Rhizobium)*. Only modest amounts of nitrogen enter with precipitation except in regions where industrial deposition may be involved (Smith *et al.*, 1987). Rain is estimated to carry 5.1 kg N/ha and snow 1.7 kg N/ha in humid temperate climates (Brady, 1974, p. 441). Thus, about 2.9×10^9 kg N/yr, or about 13.0 kg N/person/yr enters agricultural lands from precipitation; of this, approximately 9.1 kg N/person/yr is ammonium and 3.9 kg N/person/yr is nitrogen oxides, NO_x (Andow and Davis, 1988)

Dry deposition of nitrogen from the atmosphere can occur when suspended particulate matter settles on soil and plants, and when nitrogenous gases are directly adsorbed onto soil or plants. The importance of dry deposition as a nitrogen input to agriculture is unknown. In terrestrial ecosystems, it may occur at the same rate as precipitation inputs (Söderlund and Rosswall 1982), and near livestock facilities it can be considerable (Hutchinson et al. 1972). Overall, it is expected that dry deposition to agriculture is on the order of 6 kg N/person/yr (Andow and Davis, 1988).

Legumes fix varying amounts of nitrogen (e.g., alfalfa 79–224 kg/ha/yr, clover 49–183 kg/ha/yr, and soybeans 3–102 kg/ha/yr; Heichel, 1987). Fixed nitrogen can be used by the legume for growth or if in excess it may move into the soil or leach to groundwater (CEA, IFA, and IPI, 1983). Usually, a significant amount of leguminous nitrogen is removed with the crop at harvest, but crop residues and roots left in the soil provide some nitrogen for the next crop. The current level of biological nitrogen fixation on agricultural lands is about 7.4×10^9 kg N/yr, or 32.7 kg N/person/yr (Andow and Davis, 1988).

3. Volatilization and Denitrification

A significant amount of nitrogen is released to the atmosphere from the soil in the form of ammonia (NH_3) and nitrogen gas (N_2 and N_2O). Loss of ammonia has greatest potential when large amounts of nitrate-rich plant residues are decomposed (Dam Kofoed, 1985) or when fertilizers and manure are added to the surface of the soil. Ammonia also volatilizes directly from plant leaves (Farquhar *et al.*, 1983) at rates of 2.6×10^9 kg N/yr or 11.5 kg N/person/yr (Andow and Davis, 1988).

Several soil bacteria cause denitrification of nitrate to nitrogen gas.

Denitrification is most rapid in anaerobic soils containing adequate amounts of organic carbon, and the products of denitrification are N_2 and N_2O (Knowles, 1982). We estimated that biological denitrification results in about 5.4×10^9 kg N/yr, or 23.9 kg N/person/yr (Andow and Davis, 1988).

4. Residue and Compost

An emerging practice in row crop production is to leave plant residue behind after harvest. This results in less erosion and returns plant nutrient elements from non-harvested plant parts to the soil. This practice includes using cover crops, and minimum or no tillage. In the United States, an estimated 4×10^9 kg N/yr of plant residue is returned to the soil (Larsen et al., 1978), or about 16 kg N/person/yr. How much is used by plants is unknown.

Some of the plant residue is composted. Composting involves piling organic matter and allowing it to decay aerobically. The source of organic matter for the compost is varied and typically includes crop residues, manures, food processing residues, and even table scraps.

5. Municipal Sludge

Many communities and countries use sludges recovered from sewage plants as soil amendments. Milwaukee, Wisconsin, produces a sludge product called Milogranite composed of 6% nitrogen and 1% phosphorus. Only 15 cities in the United States are or have produced a similar product and it is estimated that less than 25% of the human waste in the United States is returned to the land, or 1.3 kg N/person/yr (Jones, 1982; NRC, 1978; USDC, 1983). In Denmark, as much as 48% of non-stabilized sludge is deposited on agricultural lands (Dam Kofoed, 1985). The use of municipal sludges as plant fertilizers is controversial because they may contain industrial sources of heavy metals that can accumulate in plant and animal tissues (CAST, 1985).

6. Manure Production, Volatilization, and Use

On an annual basis, farm animals in the United States produce about 5.3×10^9 kg N/yr, of manure or 23.4 kg N/person/yr (average of Frere, 1976; NRC, 1972; Van Dyne and Gilbertson, 1978; USDC, 1983). Cattle produce 82% of this total (Lauer, 1975), and 30.2% is deposited directly onto land in pastures (Van Dyne and Gilbertson, 1978). About 12% of the nitrogen voided on pastures and 50% of the nitrogen voided in confined conditions volatilize as ammonium directly to the atmosphere (Bouldin et al., 1984; Denmead et al., 1974; NRC, 1978). When manures and slurries are applied to soil, an additional 35% of nitrogen may be lost to the atmosphere (Sherwood, 1985). If slurry is injected into the soil, however, this value may be reduced by 11 or 12 times (Hall and Ryden, 1986). Thus, about 11.5 kg N/person/yr of all manure nitrogen makes its way back to the soil and the rest is lost to the atmosphere.

7. Crop Uptake

Accumulation of nitrogen in crops is one of the major annual flows of nitrogen. The amounts of N, P, and K removed for selected crops is given in Table 8.3. The National Research Council (1972) estimated crop uptake at harvest to be about 16.8×10^9 kg N/yr, or 74 kg N/person/yr in 1975. This does not count the nitrogen that is taken up and volatilized from plant leaves as ammonia. Adding in this 11.5 kg N/person/yr, we calculate total crop uptake to be about 85.5 kg N/person/yr.

8. Animal Feed

Most nitrogen in crops is passed on to food animals in the form of high-protein feed supplements (soybean meal), grain concentrates (corn), roughage, or pasture. We used feed consumption statistics and the nitrogen content of various feeds to estimate that about 38.7 kg N/person/yr is consumed by animals (Andow and Davis, 1988).

9. Human Consumption of Food

Nitrogen is a key element in proteins. Humans consume 100.4 gm protein per day (10-year average; USDA, 1985a) of which 67.5% is from animals

Table 8.3 Total Amounts of Nutrients Contained in Selected Crops[a]

Crop	Yield/acre[b]	Nitrogen (N)	Phosphorous (P_2O_5)	Potassium (K_2O)
Alfalfa	5 t	250[c]	60	225
Corn	150 bu	220	80	195
Cotton	1.5 bales	95	50	60
Coastal		150		
Bermuda grass	6 t		60	180
Soybean	40 bu	145[c]	40	75
Rice	6,500 lb	135	51	18
Wheat	60 bu	125	50	110
Oats	100 bu	100	40	120
Potatoes	400 bu	200	55	310
Peanuts	3,000 lb	220[c]	45	120

[a] Source: CAST (1985); International Minerals and Chemical Corporation (undated).
[b] Pounds per bushel of wheat, 60; corn, 56; soybeans, 60; oats, 32; potatoes, 60; rice, 45.
[c] Considerable nitrogen is obtained from atmosphere symbiotic fixation.

(USDA, 1985a). This amounts to 4.0 kg of animal nitrogen per person per year, and 1.9 kg of plant nitrogen per person per year.

10. Human Health Risk

Drinking nitrate in groundwater is a human health risk. In 1945, infant methemoglobinemia was attributed to ingestion of water containing high levels of nitrate (Shuval and Gruener, 1974). This results in cyanosis (bluish skin) and possibly death. Nitrate is converted to nitrite in the gastrointestinal tract and then moves into the blood stream. The nitrite then binds to hemoglobin, the oxygen carrying pigment of red blood cells, forming methemoglobin. Methemoglobin cannot transport oxygen and therefore oxygen transport is diminished. Normally, methemoglobin comprises less than 2% of total hemoglobin and is of no concern. The first signs of cyanosis can be seen at levels between 5–10% methemoglobin. The condition is most severe in infants less than 6 months of age; in 146 cases in Minnesota, 90% occurred in infants less than 8 weeks old (Rosenfield and Huston, 1950). In livestock, animals that are exposed to high levels of nitrate show symptoms in 3–5 days. Cyanosis, rapid breathing, extreme respiratory movements and nervousness are usually observed (Shuval and Gruener, 1974).

Because of these health problems, safety levels have been set at 10 parts nitrate-nitrogen per million for human consumption and 100 ppm for livestock (NAS, 1972). In addition, these health risks have incurred significant economic costs. Many wells in rural regions are being tested for nitrate contamination and some people have been forced to find other sources of drinking water.

11. Water Contamination

Nitrogen leaches from soils into groundwater at levels of about 1 ppm in forests and undisturbed grasslands. Normally, 99% of soil nitrogen is in an organic form which is largely unavailable to the crop (Johnson, 1987). Plants require mineral nitrogen and use mostly ammonium (NH_4OH) and nitrate (NO_3^-). Because nitrates have a negative charge, they are repelled by soil colloids and thus move freely with water. For leaching of nitrate to occur, the following conditions must hold: (1) nitrate is available in excess of the needs of soil microorganisms and the crop, (2) precipitation is in excess of evapotranspiration providing soil water for downward percolation of nitrates, and (3) soil must be porous to allow water and nitrate to percolate. Groundwater contamination has become an international concern with reports from New Zealand (Adams et al., 1979) Western Europe (Winteringham, 1984), India (Shinde and Vamadevan, 1974), Peru (Arca, 1974), and North America (CAST, 1985).

a. Synthetic nitrogen fertilizers. Groundwater contamination by synthetic nitrogen occurs on corn fields in Illinois where nitrogen has been overused since 1965 (Figure 8.4). Nitrate concentrations in groundwater are elevated in every European country and are correlated with plant and animal production and fertilizer use (Ahtlainen, 1986). In Denmark, nitrate in groundwater has increased three-fold in the last 20–30 years. In Sweden, the present nitrate levels reflect the fertilization levels of the 1960s because of the slow rate of infiltration of water to groundwater. Throughout most of Europe, this trend is expected to continue.

In the United States, groundwater supplies half of all drinking water and replenishes much of the surface water (Smith *et al.*, 1987). During 1974–1981 total nitrate load to Atlantic coast estuaries, the Gulf of Mexico, and the Great Lakes increased 20–50%. Further, of 383 surface water reporting stations, nitrate increased in 116 instances with decreases in only 27. Increases were associated with agricultural activities, including fertilized acreage as a percentage of basin area, livestock density and feed lot activity (atmospheric deposition may play a role in midwestern and eastern basins).

The range of agricultural systems contributing to contamination of groundwater is quite diverse, including orange production in California (45% losses of applied synthetic nitrogen to groundwater; Bingham *et al.*, 1971) and potato production in Long Island, New York (25–70% losses of applied synthetic nitrogen to groundwater; Baier and Rykbost, 1976).

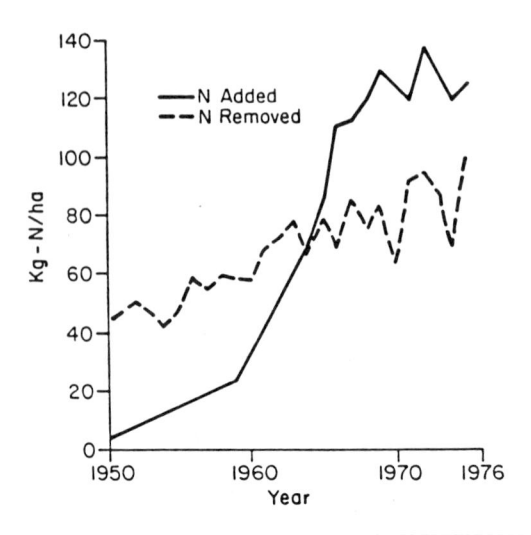

Figure 8.4 Annual average addition of fertilizer nitrogen and removal of crop nitrogen for corn in central Illinois (NRC, 1978).

8. Agricultural Chemicals: Food and Environment

The Big Spring Basin in northeastern Iowa is particularly well understood. The Big Spring basin is wholly agricultural with 91% of the land used for corn, pasture or hay. Here, the increase in groundwater nitrate directly parallels synthetic fertilizer application rates (Figure 8.5; Hallberg *et al.*, 1984). This aquifer, however, is very shallow, and deeper bedrock aquifers in Iowa show slower infiltration of nitrate (Libra *et al.*, 1987).

b. Manures. As soon as manure leaves the animal, it can contaminate water sources. Most fresh water ecosystems have evolved under limitations of nitrogen and phosphorus. When nitrogen and phosphorous are added, algae rapidly proliferate, causing a "bloom" which drives the oxygen level down. When manure runoff occurs in large quantities, the available oxygen in a body of water may go from 4% down to 0% in a matter of hours. The result is a fish kill. In 1967, three of eight major fish kills were due to manure runoff (USFWPCA, 1968).

Problems of manure management occur both in small farms where cattle have access to a stream and in cattle feedlots. Today, about 31.1%

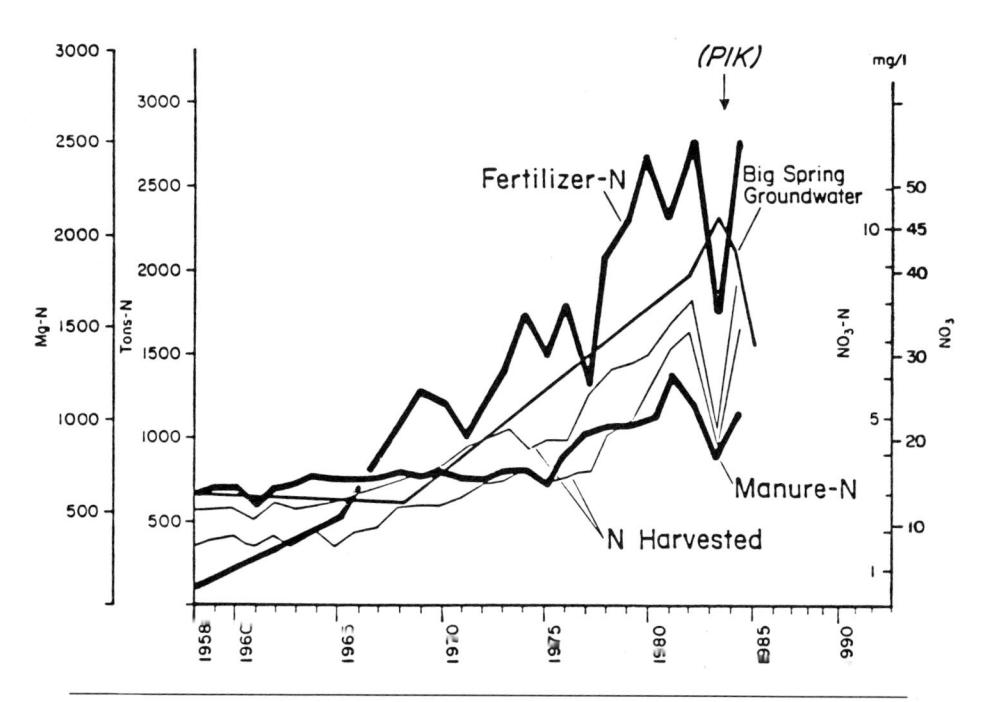

Figure 8.5 Mass of fertilizer and manure nitrogen applied in the Big Spring Basin and annual average nitrate concentration (right axis) in groundwater, and nitrogen removed in harvested corn (Hallberg *et al.*, 1984).

of beef cattle occur in concentrations exceeding 500 head per farm (Van Arsdall and Nelson, 1983), producing large quantities of manure in small areas and posing disposal problems (Manges et al., 1975). Nitrate levels in feedlots can be excessively high.

Leaching of nitrate in soils from manure is similar to leaching from synthetic nitrogen (Harris, 1987). However, manure nitrogen differs from synthetic nitrogen because the nitrogen is in many different forms that are not immediately available for leaching (being positively charged or organically bound). Thus, the nitrogen becomes available to the plant, or to leaching, gradually over the growing period. The amount of leaching depends on: (1) the amount of water moving downwards (leachate), (2) the amount and type of manure nitrogen applied and (3) the soil type (Table 8.4, Sherwood, 1985). Sherwood's data show that with higher rates of pig slurry more percent nitrogen moves with leachate and that soil type drastically affects percent nitrogen leaching; the relation between amount leachate and percent nitrogen leaching is less well established. Also, in four of the six comparisons, synthetic nitrogen exceeded all other inputs in percent leaching. This can be partly explained by an estimated 35% of

Table 8.4 Amount of Nitrogen in Leachate when Pig Slurry, Cattle Slurry, and Synthetic Nitrogen are Applied to Two Soil Types.[a]

Soil type and treatment	1976–1977 N in (Leachate = 637 mm)	1976–1977 % out	1977–1978 N in (Leachate = 546 mm)	1977–1978 % out	1978–1979 N in (Leachate = 399 mm)	1978–1979 % out
Moderately drained loam soil						
Control	—	—	—	—	—	—
Low pig slurry	469	2.4	284	1.7	483	0.3
Medium pig slurry	924	5.9	500	11.5	695	12.3
High pig slurry	1698	8.4	1026	22.3	1413	15.7
Cattle slurry	500	4.8	454	0.6	532	0.5
Synthetic nitrogen	360	27.0	360	26.5	360	8.5
	(Leachate = 426 mm)		(Leachate = 364 mm)		(Leachate = 266 mm)	
Impermeable gley soil						
Control	—	—	—	—	—	—
Low pig slurry	420	1.5	350	0.3	440	1.1
Medium pig slurry	756	1.1	618	0.4	923	2.0
High pig slurry	1480	0.3	1246	0.7	1669	4.3
Cattle slurry	295	0.9	436	0.5	566	0.8
Synthetic nitrogen	360	9.5	360	1.3	360	3.8

[a] Source: Sherwood (1985).

manure nitrogen lost by volatilization and 20% of manure nitrogen bound organically (Sherwood, 1985).

Thus, management of manure nitrogen poses many of the same problems as synthetic nitrogen. The major difference is that manure is an internal farm resource, requiring no additional expenditures other than proper storage and application. This would include covering manure stockpiles to protect from runoff, applying at rates that can be assimilated, and applying when crops are present and runoff potential is at a minimum. The inequitable spatial distribution of farm animals hinders proper utilization as some areas become manure deficient and others become inundated by the manure produced, resulting in a waste management problem.

B. Management of Nitrogen in Agroecosystems

Nitrogen moves through the agroecosystem in mass quantities. Its transient nature requires that we manage nitrogen flows better lest we contaminate our environment and ourselves. The goal of nitrogen management is to develop an agroecosystem that maintains soil fertility, minimizes environmental hazards, and maximizes use of internal farm resources (such as livestock wastes). For example, increased use of crop rotations with legumes would fix more atmospheric nitrogen. Manures could be applied at rates that the soil can assimilate. To calculate the amount of manure a particular soil can accomodate, the manure should be tested for macro and micro nutrients and the assimilative capacity of the soil could be determined (Overcash et al., 1983). This allows determination of the minimum number of acres a given amount of manure can be applied to.

An example is found in Delaware, where over 204 million broilers produced over 2.28×10^8 kg of manure annually (Harris, 1987). Because of a rise in nitrates in groundwaters (24% of all wells tested in 1982 had levels exceeding the maximum contamination level of 10 ppm), better manure handling and application practices were encouraged (Harris, 1987). Over 48% of farmers surveyed were applying manure in quantities of at least 539 kg N/ha (broiler manure at 4% nitrogen). Even though farmers recognized manure as a source of nitrogen, they may have continued synthetic nitrogen use because it was inexpensive and crop prices were high. The manure nitrogen alone represented an excess of 393 kg N/ha and resulted in crop damage under drought conditions. Excess nitrogen in corn silage resulted in deaths of some dairy cows.

The program MANURE (Managing Agricultural Nutrients Utilizing Resources Effectively) was started. Demonstrations showed that profitability was not affected when broiler manure was substituted for synthetic nitrogen. Further, 30% of farms had inadequate acreage to spread manure and were encouraged to sell their manure to local farms that could use it. Overall, more farmers began to calibrate their equipment (up 39%), analyze manures (up 7%; low due to being considered costly), spread at proper rates (up 38%), and cover their stockpiles with plastic to decrease volatilization, runoff, and leaching loss (up 15%). Programs like this encourage more rational use of nitrogen resources.

Because of snow and rain runoff, manures should not be applied in winter and spring. Table 8.5 gives equations to predict runoff potential of nitrogen during different seasons (Overcash et al., 1983). For example, a loading rate of 1020 kg N/ha results in runoff concentrations of 22.8 ppm for summer–fall and 83.5 ppm for winter–spring applications. Concomitantly, crop use efficiency of nitrogen is higher with smaller summer–fall runoff.

Synthetic nitrogen sources could be used if crop needs are not met by above sources. Better management of synthetic nitrogen would (1) time application to when crops need nitrogen, (2) place nitrogen in the soil to minimize volatilization and (3) ensure that crop and soil assimilation capacity is not surpassed.

C. Summary of Nitrogen in Agroecosystems

Alarming losses and gross inequalities in flow rates of nitrogen occur in modern agriculture (Figure 8.3). For example, the amount of crop nitrogen consumed by humans is only 4.9% of that consumed by food animals. This mass flow of crop nitrogen through food animals to humans (a conversion of only 10% of the nitrogen) is inefficient and produces by-products that must be handled wisely to avoid environmental contamination. This inefficiency is so great that the amount of nitrogen lost through manure volatilization is twice the total amount consumed by humans.

The major flow of nitrogen in modern industrial agriculture is application of synthetic nitrogen fertilizer to soils, uptake by crops, consumption by animals, and dissipation from manures. Humans intercept only a minor component of the nitrogen flows in industrial agroecosystems.

However, humans may intercept unhealthy amounts of nitrogen in drinking water. Nitrate contamination of water creates health risks and economic costs that are becoming more severe as nitrate contamination increases.

Table 8.5 Runoff Regressions to Predict Amount of Nitrogen and Phosphorus Lost When Manure Is Applied in Different Seasons[a]

Nutrient	Season	Equation	Correlation	Observations
Nitrogen	Winter–Spring	$Y = 2.45 + 0.18*X$	0.84	16
Nitrogen	Summer–Fall	$Y = -22.84 + 6.6 \,(\ln X)$	0.84	20

[a]Source: Overcash et al. (1983).

Agricultural systems will always have to balance the crop needs for enough nitrogen and nitrate–nitrogen loss to groundwater. Any of the flows of nitrogen into agricultural soils can result in excessive leaching of nitrate. Several types of solutions to these problems have been offered. The first is to modify the flows of nitrogen into and out of the soil. Some examples are genetically engineered crop varieties to fix more nitrogen, nitrification inhibitors, split applications of nitrogen, foliar applications of fertilizers, slow release formulations of fertilizers, and crop varieties that can produce at lower soil nitrogen levels.

The more progressive solutions suggest modifications in the structure of nitrogen flow. Some examples are to increase human consumption of crops, use more efficient animals (e.g., poultry is 2.7 times more efficient at converting nitrogen than cattle), restrict domestic ruminants to pasture, disaggregate animal production, and completely use all human and animal waste.

Today's agricultural system is unsurpassed in yield output. Yet, we must shift from a "maximization of yield" focus, to a "conservation of resources" focus. This does not mean a decrease in profitability to the farmer; overall, if less output were available, the farmer would see a better price for his or her commodity. This does mean, however, less environmental contamination and a healthier planet.

IV. PESTICIDES

The political controversy surrounding pesticides stems from their dual nature; they are valuable because they are lethal. Crop losses to pests have been estimated by Cramer (1967), USDA (various dates), and Pimentel et al., (1979). Cramer (1967) estimated total worldwide pre-harvest crop losses to be about 13.8% from arthropods, 11.6% from disease, and 9.5% from weeds, or a total of 34.9%. Pimentel (1976) and the USDA (1965) estimated total United States pre-harvest crop losses at 13% from arthropods, 12% from diseases, and 8% from weeds. The total dollar value (1974 dollars) for losses from pests were $25.4 billion in the United States (Pimentel et al., 1979) and $168.3 billion in the world (calculated from Cramer, 1967). Pre-harvest crop losses in the United States without the use of pesticides would be about 18% from arthropods, 15% from diseases, and 9% from weeds, so the benefits from pesticide use were valued at $8.7 billion (Pimentel et al., 1979).

The benefits of pesticide use in agriculture include increased crop and livestock production and greater longevity of food products. In nonagricultural uses, pesticides have been important in reducing the incidence of malaria and schistosomiasis and saving untold thousands of human lives. Few estimates of the economic benefits of the agricultural use of pesticides exist. Those that do, suggest dollar returns of from $3 to $5 for every dollar invested in pesticides in the United States (Headley, 1971; Pimentel, 1973; PSAC, 1965; Pimentel *et al.*, 1980). No estimates for the economic benefits of world use exist.

A wide variety of pesticides are used in modern industrialized agriculture. Over 45,000 formulations and 600 active ingredients are registered pesticides in the United States. Only 39 active ingredients, however, accounted for 79% of use in 1976 (Eichers *et al.*, 1978). Several classes of insecticides include inorganics (arsenicals), chlorinated hydrocarbons (DDT), organophosphates (parathion), carbamates (carbaryl), and pyrethroids (rotenone); several newer possibilities include hormone mimics, insect mating pheromones, and insect diseases. Herbicides include arsenicals, phenoxys (2,4-D), phenyl ureas (diuron), amides (propachlor), carbamates (butylate), triazines (atrazine), benzoics (amiben), and several others. Fungicides include dithiocarbamates (zineb), phthalimides (captan), and inorganics (sulphur, zinc, copper). Fumigants, which are used to sterilize soils and preserve stored grains, include ethylene dibromide (EDB), phosphine, and methyl bromide. We aggregate these classes for simplicity; conclusions about the social utility of pesticides, however, should be made separately on each chemical because each chemical has different effects on humans, non-target organisms, and the environment.

While pesticides have been used since antiquity, the modern era of chemical pesticides began in the mid-1800s with the discovery that an arsenic compound, Paris green, controlled Colorado potato beetle on potatoes (Boyce, 1976). Following this, many other inorganic pesticides were found and used, including other arsenicals, and copper and sulphur pesticides. Research on chemical warfare during the world wars provided the fundamental knowledge that ushered in the current era of synthetic organic pesticides. The first chemicals developed and marketed were the chlorinated hydrocarbons, followed by organophosphates and carbamates. In general, the evolutionary path in the use of pesticides in developed industrialized countries has been toward lower persistence in the environment and greater target specificity. In developing countries, however, many of the persistent, broad-spectrum pesticides are still commonly used (Weir and Shapiro, 1981).

Several factors mitigate the riskiness of pesticides in the environment. The first involves the persistence of the pesticide. Pesticides are degraded by light, chemicals, and microorganisms. Persistence is measured by the time it takes for a certain percentage of a given pesticide to degrade (Edwards, 1973). Inorganic pesticides and organochlorine insecticides are the most persistent with half-lives often of greater than one year; most herbicides, fungicides and the organophosphate insecticides degrade with half-lives of less than a year. The greater the persistence, the greater is the possibility of exposure to the pesticide. Thus, the historical evolution away from persistent pesticides has generally reduced environmental exposure to pesticides.

The second major factor is the spectrum of toxicity of the pesticide to other organisms (Table 8.6). For example, chlorinated hydrocarbons and pyrethroids tend to have low mammalian and avian toxicity, but high piscine toxicity. The organophosphates and carbamates, however, have some members with opposite toxicity spectra. Most herbicides and fungicides, and some organophosphates and carbamates, are relatively nontoxic to vertebrates and have a lower environmental hazard.

Other mitigating factors involve the formulation of the pesticide and its method of application. Granular formulations are safer than liquid formulations, and viscous liquids are safer than non-viscous liquids to humans. Placing pesticides into soils or injecting them into animals is safer than spraying pesticides on the target, and spraying from airplanes results in the greatest amount of pesticide drift and environmental exposure.

Pesticides are manufactured, transported and applied by humans; they disperse, degrade, and sometimes bioaccumulate. Unlike nitrogen, there is no global pesticide cycle; pesticides flow through ecosystems (Figure 8.6). As a result, we focus on minimizing the adverse impact of pesticides.

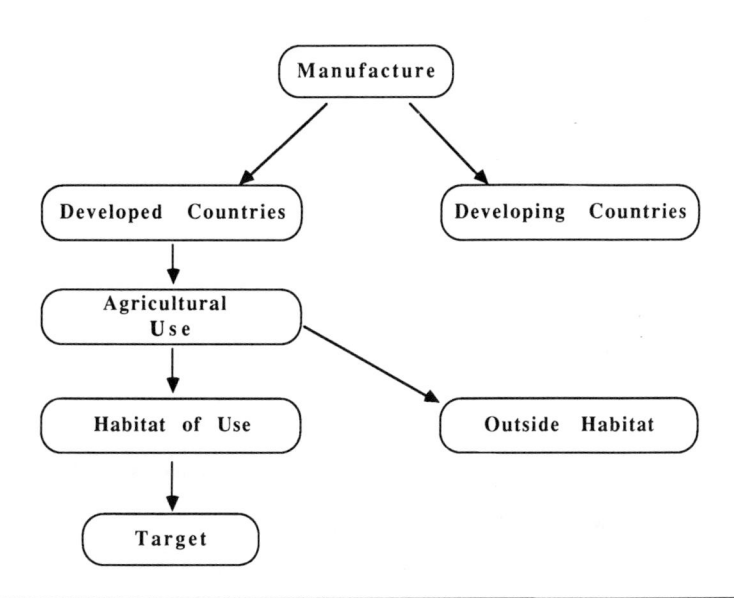

Figure 8.6 Flows of pesticides from production to final degradation.

Table 8.6 Spectra of Acute Toxicity of Selected Pesticides to Selected Taxa[a]

Compound	Class[b]	Rats LD$_{50}$ (mg/kg)	Young mallards LD$_{50}$ (mg/kg)	Rainbow trout LC$_{50}$ (ppm)	Chorus frog tadpole LC$_{50}$ (ppm)	Honeybee LD$_{50}$[c] (g/bee)
Paris green	I	22	>5000	—[d]	10	27.15
Sodium arsenite	IH	10–50	973[e]	36.5–100	—	—
Bordeaux mix	IF	High	2000	0.15–1.5[f]	—	High[g]
DDT	CH	420–800	>2240	0.007	1.4	6.19
Methoxychlor	CH	5000–6000	>2000	0.0072–0.052	0.44	High
Parathion	OP	4–30	1.9–2.1	2–2.7	1.6	0.177
Malathion	OP	480–1500	1485	0.17	0.56	0.726
Carbaryl	C	540	>2179	4.38	—	1.54
Carbofuran	C	11	0.40	—	—	0.149
Pyrethrum	P	820–1870	>10000	0.054	—	High
Rotenone	P	132	>2000	0.00015[h]	—	High
2,4-D	H	666	>>1000	250	100	High
Paraquat	H	150	—	>1	43–54	High
2,4,5-T	H	300	>5000[i]	1.3–12	—	High
Captan	F	9000	>5000[i]	—	—	High

[a]Source: Pimentel (1971).

[b]I = inorganic, CH = chlorinated hydrocarbon insecticide, OP = organophosphate insecticide, C = carbamate insecticide, P = pyrethroid insecticide, H = herbicide, F = fungicide.

[c]From NRCC (1981).   [d]Dash means not available.

[e]Maximum tolerated body concentration.   [f]Bluegill and striped bass.

[g]"High" in bee column greater than 11 g/bee.   [h]Coho salmon.   [i]LC$_{50}$ (ppm).

A. Manufacture

Manufacturing accidents are some of the most dramatic of the detrimental environmental and social effects of pesticides. These are typically high cost–low probability events, but the costs associated with any one of them are difficult to determine. For example, release of endrin from a manufacturing plant in Memphis, Tennessee, killed about 10–15 million fish, and while the value of these fish, and the impact on the fishery industries in Louisiana were large, the costs were not estimated (Graham, 1966). The James River is the principle fishery and seed oyster source in Virginia. Following the release of kepone these were closed, with over $10.8 million in direct loss. Even higher costs were incurred from forgone recreational use, and assessments against the manufacturer exceeded $13.3 million. The costs to clean the river were estimated as upwards of $7 billion, but these were not actually incurred (Pimentel *et al.*, 1980). The explosion at Seveso released 0.5 lb of dioxin on the surrounding area. The greatest health cost was an increased incidence of chloracne (20% of children in the zone of highest exposure), but major monetary costs were borne by the manufacturer, including purchasing all the homes in the zone of highest exposure and constructing special land fills to contain contaminated soil (Gough, 1986). The largest industrial accident occurred at Bhopal on December 2, 1984. Around 3,000 fatalities, 250,000 injuries, and increased miscarriages, stillbirths, and birth defects occurred; business losses totalled from $9–65 million plus the $25 million investment in the chemical plant; social losses included 2,800 jobs, distortions in grain and labor prices, and the tragic disruption of families, bankrupting some and transforming some into beggars. Lawsuits pending against the parent company total more than $100 billion, and estimates for minimal compensation to victims are $1.3–2.0 million (Shrivastava, 1987).

These accidents have complex causes. An industrial plant is not just machinery; it is a complex socio-technical system requiring constant human intervention. Human errors are a function of technical design, organizational factors, and the quality of plant personnel, which is affected by morale, number of people, training level, and quality of management. Accidents frequently relate to strategic pressures on the plant. For example, the low importance of the Bhopal plant to the parent company meant that the plant had fewer resources and poorer management, which in turn led to poor staffing, lax safety, and low morale (Shrivastava, 1987). The extent of damage depends on the responses of the groups affected and the local infrastructure. At Bhopal, the parent company responded slowly, and local resources could not treat all of the injured, could not bury all of the dead, could not house all of the homeless, and could not feed all of the hungry. In contrast, at Seveso, the parent company responded quickly and decisively, and adequate medical and social support minimized social impact.

B. Use in Developing Countries

In developing countries the contradictions between the productive and destructive nature of pesticides are most acute. Since food supplies are frequently short, the use of pesticides to increase food production can be a boon to the local population. However, the bane of pesticide use can be severe. Pesticide use in developing countries is much less than in industrialized countries, and while little quantitative data exist, human poisonings and environmental contamination are disproportionately high compared to industrialized countries.

Pesticides are more readily available to the general populace (Weir and Shapiro, 1981). Often they can be purchased over the counter in drug stores, and can be applied by anyone whether or not they are trained applicators. Some pesticide containers lack adequate warning labels, and some applicators cannot read the labels that exist. Sometimes, extra pesticide is stored in old soda pop bottles, where it has been mistaken for juice, and empty pesticide containers have been used to collect and store drinking water. Pesticides are often applied too frequently, in too large a dose, or in places they should not be applied. For example, some pesticides substitute for traditional fishing techniques. Pesticides are added to a stream and the fish, which are killed by the pesticide, are collected down stream. Frequently these fish contain large quantities of pesticide residues. All of these factors would lead to greater human and environmental exposure to pesticides than in industrialized countries. The rate of pesticide poisoning in developing countries is more than 13 times that in the United States (Weir and Shapiro, 1981).

The first major agricultural problem with pesticides occurred on cotton in the Cañete Valley in Peru (Boza-Barducci, 1972). In 1956, after intensive use of insecticides for several years, the cotton bollworm evolved resistance, that is, it was no longer killed by the insecticide. What ensued was an escalation of insecticide use, increased resistance in the pest, failure of the cotton industry, and environmental contamination.

Currently many pesticides that have been banned by the United States, European Economic Community, and Japan are widely used in developing countries. These include DDT, dieldrin, endrin, lindane, and aldrin. Pesticide companies have been accused of dumping dangerous chemicals on a poorly educated populace (Weir and Shapiro, 1981), or have been commended for supplying the only economical method to control many crop pests because these banned chemicals are significantly cheaper than their alternatives (Adam, 1976). This issue is both controversial and complicated, and is a clear case where the ethical concerns related to restricting the use of pesticides conflict with economic concerns of optimizing food production.

A major complication in this controversy is the many subsidies for pesticides in developing countries and the political motivation for these subsidies (Repetto, 1985). These subsidies range from 15–90% of what the total retail cost would be without a subsidy. Mechanisms include (1) subsidies from foreign governments, such as a reduced price, easy credit, and agricultural aid designated for the purchase of pesticides; and (2) subsidies by the local government, such as tax exemptions, easy credit, and sales below cost by government-controlled distribution networks. These subsidies encourage farmers to use more pesticides than they would if they had to pay the full cost.

C. Human Exposure

In the United States, about 45,000 people are treated and about 50 people die from pesticide poisonings each year (Pimentel *et al.*, 1980). In the entire world, about 500,000 pesticide poisonings occur each year (WHO, 1973), and between 5,000 (WHO, 1973) and 20,640 (Copplestone, 1977) deaths occur. Many of these deaths are suicides (Hayes and Vaughn, 1977; Copplestone, 1977) caused by the easy access to pesticides, and many of the poisonings are due to accidental acute exposure rather than chronic exposure (Peoples and Knaak, 1982). However, these estimates of exposure, morbidity, and mortality are very uncertain.

The groups of people at greatest risk of pesticide exposure are farm workers (Table 8.7), which includes applicators, farmers, and field workers. For example, aerial applicators and field workers are at special risk. Aerial application is stressful, requiring quick reactions when flying at very low altitudes, avoiding electric power lines, and making hundreds of turns each day. Pesticides can alter reaction time and fog mental judgement and may increase the chance of an airplane crash. In 1976, there were 174 airplane crashes involving pesticide applicators of which 11 were fatal (NTSB, 1977). Other factors, such as fatigue and heat stress could also cause these accidents (Richter *et al.*, 1980). A greater problem is that a large number of poisoned workers do not receive medical treatment and therefore are never counted in morbidity statistics (see references in Pimentel *et al.*, 1980). This is particularly true among field workers. Kahn (1976) surveyed field workers in Tulare County, California and found that fewer than 6% of the workers showing symptoms of pesticide poisoning had notified health officials, and Coye (cited in Wasserstrom and Wiles, 1985, p. 3) suggested that as many as 313,000 farm workers in the United States may suffer the effects of pesticide-related illness each year.

Table 8.7 Illness Due to Exposure to Pesticides[a]

Occupation	Number	%
Ground applicator (mixer, loader, applicator)	270	20.1
Mixer, loader (aircraft)	131	9.8
Field worker exposed to residues	165	12.3
Gardener	107	8.0
Nursery or greenhouse worker	100	7.4
Formulation plant worker	56	4.2
Warehouse worker, truck loader	45	3.4
Structural pest control worker	35	2.6
Fumigator of fields	22	1.6
Cleaner or repairer of machinery	35	2.6
Fireman exposed to pesticide fires	37	2.8
Tractor driver or irrigator	23	1.7
Worker exposed to drift	31	2.3
Flagman for aircraft application	16	1.2
Pilot of agricultural aircrafts	8	0.6
Indoor worker	79	5.9
Other type of pesticide user	183	13.6

[a] Reported to the California Health Department, 1975 (Coutts, 1980).

D. Effects in Habitat of Application

More than 99.9% of all pesticide applied to control pest insects misses the target pest and enters the environment (Metcalf, 1986), where it can cause significant adverse environmental effects. Pesticides can adversely affect all biotic components in the habitat of application, including microbes in soils, host plant physiology, and pollinating honeybees.

1. Resurgence

Pest resurgence after application of pesticides is one such problem. Resurgence occurs when the pesticide kills the target pest, but shortly afterwards either the target pest or another pest increases and causes damage. There are several mechanisms by which resurgence occurs. First, insecticides can kill predators and parasites of the target pest, which allows the pest to increase faster than usual. Applications of guthion or methoxychlor to potatoes frequently allows populations of green peach aphid to outbreak because lady bird beetles and lacewing larvae are killed. Second, insecticides can kill predators and parasites of other pests, causing secondary pest outbreaks. The destruction of predators and parasites after application of insecticides to control boll weevil in cotton released other insects, such as cotton budworm and cotton bollworm, and the destruction of fungal pathogens by a fungicide in soybean caused outbreaks of velvet bean caterpillars and cabbage loopers (see references in Pimental et al., 1980, p. 113).

Third, insecticides can alter host plant physiology, causing pest outbreaks (trophobiosis). Application of the insecticide, DDT, increased concentrations of sugar and nitrogen in leaves of field bean and soybean, making it a better food for mites (Rodriguez et al., 1960). Similarly, application of the herbicide, 2,4-D, to corn increased susceptibility to corn leaf aphid and corn smut (Oka and Pimentel, 1974).

Finally, resurgence can occur because the pesticide stimulates the target pest to reproduce more (hormoligosis; Luckey, 1968). Low, nonlethal doses of some insecticides on mites stimulate the fecundity of those mites (Dittrich *et al.,* 1974).

The cost to crop production of pest resurgence is substantial. At lease $153 million in additional control costs are spent in the United States annually to suppress these pests (Pimentel *et al.,* 1980). This does not include the additional crop losses sustained from pest resurgence that could not be prevented by additional pesticides.

2. Resistance

Any strong mortality agent that acts on a population of organisms is also a strong selective agent. Pesticides are strong mortality agents that select for pesticide-resistance in the target populations. In 1914 the first case of resistance to a pesticide was reported (Melander, 1914). Since that time, at least 428 species of arthropods, 91 species of plant pathogens, 5 species of weeds, and 2 species of plant parasitic nematodes have evolved resistance to pesticides (Georghiou and Mellon, 1983). The costs of increased resistance include the use of more or more expensive pesticides and are estimated to be at least $133 million annually for arthropods and weeds (Pimentel *et al.,* 1980). This does not include such adverse effects as major crop shifts and the collapse of crop industry. For example, the cotton industry in northeastern Mexico collapsed when resistance occurred, causing tremendous economic and social dislocations (Adkisson, 1972).

3. Crop and Animal Losses

Of the 99.9% of pesticide that misses the target pest, a significant portion contaminates crop and animal products, which can be condemned. At least $3.1 million of meat, $200,000 of milk, and $2.5 million of crops are condemned in the United States annually by FDA inspectors (Pimentel *et al.,* 1980). Additional losses include emergency recalls, such as the recall of milk and milk products that were contaminated by heptachlor in Hawaii (Smith, 1982).

Additional crop losses are sustained when herbicides drift onto sensitive crops or when herbicide residues in the soil inhibit crop growth or prevent rotation to a sensitive crop. These losses are more frequent when nearby plants are aerially sprayed, or when herbicide degradation is slow.

Based on extrapolations, Pimentel *et al.* (1980) suggested that these crop losses were nearly $60 million in direct loss and $7.9 million in indirect loss as insurance premiums.

Finally, due to human carelessness or animal curiosity, many animals are poisoned by pesticides. Animal poisonings are reported only when a veterinarian is involved, so estimates of animal poisoning are usually underestimated. Pimentel *et al.* (1980) extrapolated that about 18,400 cattle and cows, 1,400 swine and 1,300 poultry are poisoned annually in the United States. These losses were valued at $3.4 million.

4. Weed Community Shifts

Herbicide use causes changes in the weed species composition in a field. Herbicide-tolerant species rapidly replace susceptible species (Day, 1978) and expenditures for weed control and losses to weeds may be increasing. Commonly, perennial weeds replace annuals. The extent of economic loss, and the effects of these changes on the food web are not known.

5. Pollination

Many pesticides are toxic to insects, so it is not surprising that insect pollinators are frequently injured. Reduction in pollinator populations causes significant economic losses in reduced crop production, increased mortality of honey bee colonies and reduced honey production. These losses are estimated to be about $135 million per year (Pimentel *et al.*, 1980).

While many crops are self-pollinating or self-fertile, some crops, such as apples, almonds, alfalfa seed, and blueberries, require pollination, and others, such as melons and various fruits, benefit substantially from pollination. A few crops, such as cotton, soybeans, flax, and some varieties of citrus, can benefit from increased pollination (McGregor, 1976).

Honey and honey bee losses can be severe. Martin (1978) suggested that 5% of all honey bee colonies are killed outright by pesticides, and about 20% are affected. In California in 1978, 60,000 (11.9%) honey bee colonies were killed by pesticides (NRCC, 1981). In addition, because of heavy pesticide use, beekeepers are often unwilling to expose their bees to many crops with good apiary potential (Martin, 1978). This results in foregone honey production and sometimes reduced crop yield. For example, the 1977 blueberry crop in New Jersey was small partly because a cool spring set back insecticide schedules. That year, insecticides were applied closer to flowering time, and beekeepers were reluctant to risk exposing their bees to pesticides. Many early varieties were not adequately pollinated, and yields were reduced (Stricker, 1977).

Pesticides also reduce populations of wild pollinators, which include solitary bees, bumblebees, and hawkmoths. While some crops (red clover) rely on these wild pollinators, most of the effects of pesticides are unknown. Reduction of wild pollinators can reduce fruit set in some wild plants by as much as 50% (Thaler and Plowright, 1980), so frugivorous birds and mammals could be affected. In addition, the genetic structure of plant populations could be affected (NRCC, 1981), and it is possible that some sparse species may go locally extinct.

6. Soil Organisms

The ecological effects of pesticides on soil organisms in agricultural fields are poorly understood. Earthworms accelerate decomposition of organic matter by comminuting and moving leaf letter from soil surfaces to within the soil matrix where microorganisms can complete decomposition. Populations of earthworms are drastically reduced by some pesticides, particularly chlordane, endrin, parathion, phorate, and most nematicides and carbamate pesticides (Edwards, 1980a). The use of carbamate fungicides caused outbreaks of apple scab because earthworms, which had removed most of the infected apple leaves from the soil surface, had been killed (Edwards and Lofty, 1977).

Rates of decomposition and mineralization have been reduced by pesticides (Sobieszczanski, 1969; Ward and Wilson, 1973), rates of nitrification have been reduced (Debona and Audus, 1970; Pimentel, 1971) or enhanced (Balicka and Sobieszczanski, 1969) by certain herbicides. In some cases, pesticide application can reduce soil fertility (Dubey, 1970; Perfect *et al.*, 1979). The extent and overall significance of these effects are not known.

Clearly, many agricultural ecosystems have been under tremendous stress from pesticide application, and some orchards after a long history of pesticide treatments have very poor soil structure (Edwards, 1965). In addition, intensive use of some herbicides has so thoroughly eliminated extraneous plant organic matter, that normally harmless saprophytic collembola (springtails) have become a severe pest on the crop because there is nothing else to eat (Edwards, 1980b).

E. Movement Out of Habitat

About 65% of all agricultural pesticides are applied aerially (USDA, 1976), and between 20–80% of that misses the target area (Yates and Akesson, 1973), entering terrestrial and aquatic habitats. Substantial amounts of pesticide that is sprayed from ground equipment can also drift off of the target area, but there is little drift from granular pesticides and pesticides applied in the soil. The forms that are not aerially mobile, however, can percolate into groundwater or runoff into surface waters. Thus, many areas outside of the target habitat will be exposed to pesticides.

1. Water Contamination

Percolation to ground water and runoff to surface waters are the primary routes that pesticides enter water (Steenhuis and Walter, 1979; Canter, 1987; Cheng and Koskinen, 1986). Each pesticide has a dominant route, which is determined by the strength it adsorbs to soil particles (Table 8.8), and its persistance. Only persistent pesticides are of concern. A pesticide that strongly adsorbs to soil particles is difficult to remove from soil, while a weakly adsorbed pesticide readily dissociates. Organic matter is the most important sorptive surface in soil for all but a few highly ionic pesticides (Kenaga, 1975). Strongly adsorbed pesticides move primarily when the soil moves, and enter surface waters with eroded soils. Conservation tillage practices which reduce soil erosion also reduce water contamination from these pesticides. It is the moderately adsorbed pesticides that percolate into ground water. This is particularly true on porous soils, and in climates with high precipitation, but is also enhanced by some conservation tillage practices that increase soil permeability to ground water. Most weakly adsorbed pesticides degrade quickly.

Aldicarb, a moderately absorbed pesticide, illustrates many of the important environmental issues. Aldicarb is a systemic insecticide that is applied in the soil and was widely used to control Colorado potato beetle on potatoes throughout the United States. Potatoes are usually grown on sandy, porous soils and are often irrigated. Aldicarb was first found in the groundwater of the Long Island aquifer in 1979 near the major potato producing area. Many dispute whether or not observed concentrations present a health risk to humans, but this will be conclusively demonstrated only after many humans have been exposed and followed through epidemiological studies. If there is a risk, it will have already been incurred by the time it can be proved. Thus, the problem is more complex than simple risk analysis: The benefits of pesticide use must be compared to the costs of gaining or foregoing information about risk. Pesticide contamination of groundwater is now believed to be a widespread and critical environmental problem (Fairchild, 1987).

2. Aquatic and Terrestrial

Persistent pesticides are ubiquitous in the environment, occurring in the arctic and tropics, in sharks, earthworms, crayfish and walleye pike (Edwards, 1973). The persistent chlorinated hydrocarbon compounds probably constitute the most serious threat to fish and wildlife (Menzie, 1972). These pesticides can persist in the environment for 10–30 years (Edwards, 1973), but since their use has declined since the late 1960s, fewer residues are observed in soils (Carey, 1979), some fish (Frank *et al.*, 1978), and some aquatic birds (Blus *et al.*, 1979). There are, however, some localities where residue concentrations have increased (Fleming *et al.*, 1983).

Table 8.8 Classification of Pesticides and Plant Nutrients Based on Adsorption to Soil Particles[a]

	Strongly adsorbed	Moderately absorbed	Weakly adsorbed
Pesticides	Paraquat DDT Toxaphene	Trifluralin Atrazine Cyanazine 2,4-D Diphenamid Alachlor Dichlobenil Fluometuron	Chloramben Dicamba
Nutrients	Organic nitrogen Ammonium Phosphorus (solid phase)	Phosphorus (soluble inorganic)	Nitrate

[a]Source: Steenhuis and Walter (1979).

Bioconcentration of pesticides is the accumulation of pesticide residue in organisms (Kenaga, 1972, 1975). In aquatic environments, during the initial period of exposure, bioaccumulation occurs primarily through adsorption, the rate of which is determined by the affinity of the pesticide to solids. High rates of adsorption are often related to high surface area to mass ratios in organisms (Kenaga, 1975). For example, Reinert (1972) showed that daphnia and guppies accumulated more dieldrin residues directly from the water than from their food. This presumably occurs through adsorption on the filtering structures of daphnia and on the gills of guppies. Long-term bioaccumulation can attain steady state distributions, and is primarily related to the relative lipid-to-water solubility of the pesticide, not to adsorption (Kenaga, 1975). The more lipid soluble pesticides will, reach higher concentrations in the fatty tissues of organisms than the more water soluble pesticides.

In terrestrial environments, bioaccumulation occurs primarily through ingestion of contaminated food (Kenaga, 1972). Concentrations of pesticide residues can increase up the food chain if excretion of residue is slower than consumption, which is particularly true for the more lipid soluble pesticides.

The effects of pesticides on the natural biota are understood in very few cases. Raptors and fish-eating birds, such as peregrine falcons (Ratcliffe, 1967) and brown pelicans, suffer reproductive problems related to egg shell thinning caused by DDE, a common degradation product of DDT. Egg shell thinning reduced population densities of these birds, and was clearly linked to DDE concentrations (Hall, 1987). Direct kills of birds can be attributed to a small class of 13 organophosphate and one carbamate insecticide; there are about 20 reported die-offs each year (Hall, 1987). Other effects of pesticides to birds include changes in behavior (Peakall, 1985), cold tolerance, salt gland function, growth, and embryonic development (Hall, 1987).

Pesticides accounted for 18% of all reported fish kills in the United States and 3.7% of all major fish kills (those involving more than 100,000 fish). These kills are the manifestations of acute toxicity to fish, but the chronic or long-term effects are largely unknown (Nimmo *et al.*, 1987). The economic impact of bird and fish kills might be significant, since fishermen and hunters spent about $7 billion in 1970 pursuing their sport (USDI, 1970).

Many effects of pesticides applied at normal doses on ecosystem structure and function have been documented (Pimentel and Edwards, 1982). Populations of some organisms were reduced, predators and parasitoids were more likely to be seriously affected than herbivores, some pesticides retarded organic matter decomposition and turnover rates, and monocultures of crops were more likely to show signs of pesticide stress than ecosystems with a rich flora and fauna. At low doses, which would be more typical of pesticide drift, carbaryl had some opposite effects. Herbivores were more seriously affected than predators and parasitoids, and species-rich, old-field ecosystems were more sensitive to pesticide stress than corn monocultures (Risch *et al.*, 1986).

3. Food Contamination

People are exposed to minute quantities of pesticide by consuming residues in food and water. In the early 1970s about 50% of foods sampled by the U.S. FDA contained detectable levels of pesticide (Duggan and Duggan, 1973). As a result of this chronic exposure, pesticide residues are commonly found in human tissues. Virtually everyone in the United States harbors some pesticide residue (usually organochlorines), averaging 6 ppm in fatty tissues (Kutz *et al.*, 1977a). Even the very young have detectable residues because human milk and some cow milk contain residues (Jensen, 1983) and pesticides can cross the placental barrier (O'Leary *et al.*, 1970). In general, pesticide residue concentrations in humans have declined (Table 8.9; Davies and Doon, 1987).

While exposure to pesticide residues in foods and water clearly occurs, the hazards and risks associated with this exposure are more obscure. Some of the potential hazards from chronic exposure to pesticide residues are teratogenesis, which is the occurence of developmental abnormalities (Clegg and Khera, 1973), mutagenesis and oncogenesis. There is no direct epidemiological evidence that pesticides cause developmental abnormalities, mutations or cancer in humans, but most workers agree that some pesticides have these potentials (Pimentel *et al.*, 1980).

Table 8.9 Total DDT Equivalent Residues in Human Adipose Tissue in General U.S. Population, by Race and Age[a]

Race and age	Residues (ppm lipid weight)				
	FY 1970	FY 1971	FY 1972	FY 1973	FY 1974
Caucasians					
0–14	4.16	3.32	2.79	2.59	2.15
15–44	6.89	6.56	6.01	5.71	4.91
45 and above	8.01	7.50	7.00	6.63	6.55
Negroes					
0–14	5.54	7.30	4.68	3.16	4.02
15–44	10.88	13.92	11.32	9.97	9.18
45 and above	16.56	19.57	15.91	14.11	11.91

[a] Source: Kutz et al. (1976).

The main problem in estimating the hazard from exposure to small amounts of pesticide residues is one of extrapolation by hazard estimates to low doses. If the relation were like Figure 8.7(a), which indicates a synergism at very low doses, then extrapolation from higher doses would underestimate risk. If, on the other hand, the relation were like Figure 8.7(c), which indicates a threshold response, then extrapolation from higher doses would overestimate risk. Only if the relation were like Figure 8.7(b) would extrapolation from higher doses be accurate. While experimental approaches toward answering this question are fraught with difficulties, they are critical for more accurate risk assessment (Wilkenson, 1987).

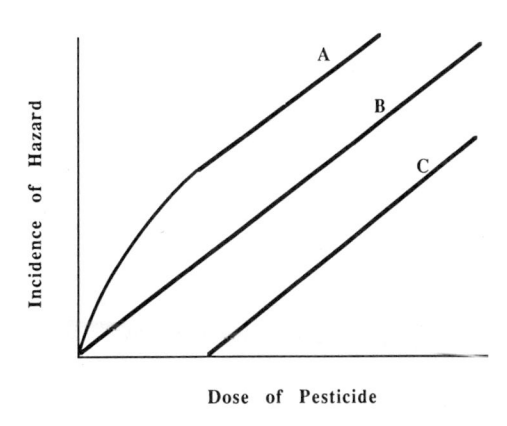

Figure 8.7 Possible dose–response curves from chronic exposure to low levels of pesticides.

A recent analysis of oncogenetic risk of residue exposure to humans in the United States suggested that the potential cancer risk following a lifetime's exposure to pesticide residues is one additional cancer case per 1,000 people (NAS, 1987). This was calculated by estimating the average food consumption and the average pesticide residue concentration on all foods in the United States. The product of these averages was used to estimate average exposure to pesticide residues. Hazard was estimated by assuming Figure 8.7(b) was an appropriate extrapolation of the dose-response relation from animal data. Total risk equals exposure times hazard, and was summed over all pesticide and food combinations. The cancer risk from pesticide residues is about 0.4% of the total cancer risk to people in the United States (NAS, 1987).

Some groups of people, however, are exposed to greater risk. U.S. negroes have higher concentrations of pesticide residues in body tissues than U.S. caucasians (Table 8.9). Other groups of people with diets that contain larger than average pesticide residues may be at greater risk. Some groups of people are accidentally exposed to abnormally high residue concentrations, and this could also increase risk. One example is the accidental contamination of milk with heptachlor in Hawaii, which exposed many people to abnormally high concentrations of heptachlor (Smith, 1982).

No estimates of teratogenic or mutagenic risk exist. These are also serious risks from chronic exposure to pesticide residues and need to be estimated.

F. Summary of Pesticides in Agroecosystems

Agricultural systems will always have to protect food resources from pest attack. Development and proliferation of pesticides as a primary tactic of pest control, however, have heightened awareness of the dual nature of pesticides. On the one hand, pesticides have preserved large quantities of food, over $8.7 billion annually in the United States, thereby preserving many lives. On the other hand, pesticides have poisoned over 500,000 people, killed over 5,000 people annually, may have increased cancer rates by 0.1%, incurred at least $2.2 billion in direct control costs and at least $498 million in indirect production costs, and done unquantified harm to soil fertility and structure, and ecosystem structure and function (Table 8.10).

While some people suggest that the choice is starvation or pesticides, several options exist. First, many environmental and social risks stem from the persistence and toxicity spectra of the pesticides. Efforts at developing more transitory, more specific pesticides have been underway for many years. Unfortunately, there is a limit to how transitory and specific future pesticides can be in a free-market economy. With development costs for a single pesticide exceeding $10 million in 1972 (Green, 1976),

Table 8.10 Summary of Environmental and Social Risks of Pesticide Use

Problems	Characteristics and magnitude
Manufacturing accidents (world) (Humans, economy, social structure, environment, legal)	Sporadic, localized Magnitude unknown
Developing countries (Humans, environment, resurgence, resistance, subsidies)	Continuous, dispersed Magnitude unknown
Human exposure, agricultural use	Continuous, localized
Poisonings (world)	500,000
Deaths (world)	5,000–20,640
Habitat of application	Continuous, dispersed
Resurgence (U.S. only)	$153 million extra control costs Additional yield loss
Resistance (U.S. only)	$133 million extra control costs Major crop shifts
Crop and animal loss (U.S. only)	$77.1 million "emergency" losses
Weed community shifts	Extra control costs
Pollination (U.S. only)	$135 million Effect on natural communities
Soil organisms	Pest outbreaks, soil fertility, Soil structure
Movement out of habitat	Continuous, dispersal
Water contamination	Environmental and human exposure,
Aquatic and terrestrial	Bird die-offs, fish-kills, Recreational sport economy, ecosystem structure and function
Food contamination	0.1% oncogenic risk (U.S.) Teratogenesis, mutagenesis

pesticide companies must be guaranteed a large enough market for the products under development to recoup their investment. This can be done by targeting large markets, such as cotton, or by not making the product too specific so that it can be used on several crops and several pests. Furthermore, because most pests attack crops over a period of several days, if a pesticide were too ephemeral, several applications might be necessary, requiring greater cost to the user. In addition, control of important pests on "minor" crops, which have small acreages, with species specific pesticides is unlikely because such pesticides are unlikely to be profitable. Finally even if a transitory, specific pesticide were developed, the pest could evolve resistance, which would require additional research and development costs. Thus, more transitory, more specific pesticides will continue to be developed, but there is a limit to how transitory and specific a pesticide can be because of economic constraints on the production process.

A more progressive alternative would be to rationalize the entire production system. Many of the problems with pesticides stem from their use as external inputs to the production process. Because they are external, they must be manufactured, transported, and applied. These processes result in accidents and drift. Efforts to internalize plant protection into the production process could reduce the need for these external inputs. Integrated pest management, which integrates diverse control tactics, such as biological control, cultural control, host plant resistance, and physical control, within an economic framework, will contribute toward this change (Andow and Rosset, 1989). In some cases, the agricultural system can be designed to eliminate the need for pesticides. For example, in the corn belt of the United States, profitability of the farm was maintained even without the use of pesticides (Lockeretz *et al.*, 1981).

V. RATIONAL USE OF AGRICULTURAL CHEMICALS

Rationale use of agricultural chemicals requires examination of the entire system of agricultural production. E. B. White stated, "I am pessimistic about the human race because it is too ingenious for its own good. Our approach to nature is to beat it into submission. We would stand a better chance of survival if we accomodated ourselves to this planet and viewed it appreciatively instead of skeptically and dictatorially." Our analysis of agricultural nitrogen and pesticide flows suggests that the goal of maximizing food production within the framework of microeconomic profit maximization results in system level irrationalities. Flows of nitrogen are wasted and extraordinary efforts incurred to dispose of potentially valuable fertilizer. Pesticides are spread widely, compromising life as they preserve it. While many solutions to each of these problems have been suggested, two aspects of these problems have been relatively ignored. We consider how the costs and benefits of pesticides and fertilizers are distributed in society and how pesticides and fertilizers interact.

A. Distribution of Costs and Benefits

All analyses of the relative risks and benefits of use of agrochemicals have aggregated all groups in society. Any social change, however, is likely to benefit one group in a society more than some other group. Thus, the distribution of risks and benefits among social groups is essential to know for evaluating solutions to environmental problems. For example, many of the acute health risks in pesticide use are incurred by unskilled groups in the working class, and significant benefits from synthetic nitrogen fertilizer use are received by grain merchants. Changes that cause unskilled workers to incur greater costs, even while beneficial in the aggregate, may be socially undesirable because they create a more inequitable society.

B. Interactions of Inputs and Cycles

The effects of pesticides, excess nitrogen, antibiotics, hormones and other inputs to industrial agriculture are not independent of each other. Intensification of animal production using confinement facilities increases the rate of encounter between animals, which facilitates spread of disease and requires use of antibiotics. In some crops, irrigation enhances spread of plant pathogenic fungi, which requires greater use of fungicide. As a consequence of these interactions, management practices form loosely coordinated management *syndromes,* where certain management practices co-occur. Industrial agriculture is one such management syndrome, where intensive application of agricultural chemicals is used to control natural variation to produce large quantities of a relatively homogeneous product in a timely manner. Two major interactions among agricultural chemicals are the effects of pesticides on the nitrogen cycle and the effects of fertilizer nitrogen on pest damage.

1. Nitrogen and Pests

Despite the nearly 20-fold increase in pesticide production since 1947, crop losses to pests have remained relatively constant (31.4–33%, Figure 8.8). Crop losses from weeds have decreased from 13.8% in the 1940s to 8.0% in 1974, but losses from insects have increased from 7.1 to 13.0% in the same time period (Figure 8.8).

Weed losses have decreased, probably because of greater use of more effective herbicides and better mechanical control, but why have insect losses increased? While part of the cause is pest resurgence from pesticide use, and increased cosmetic standards for some food products, increased use of synthetic nitrogen fertilizers have had a great effect. Fertilization of plants with nitrogen invariably increases the content of nitrogen and water in the various plant parts, and insects are frequently limited by the amount of available nitrogen and water in their diet (Mattson, 1980; Scriber and Slansky, 1981). The increase in nutritional quality of the plants increases the potential for the insects to cause damage.

One striking example was described by Dodd (1936). The moth *Cactoblastis cactorum* was imported into Australia to control prickly-pear cactus. When the cactus was deficient in nitrogen, the larvae of the moth were unable to complete development and were unable to damage the plant. After the cactus was fertilized, the moth prospered and decimated the plants.

Apparently, the commitment to use high levels of synthetic nitrogen

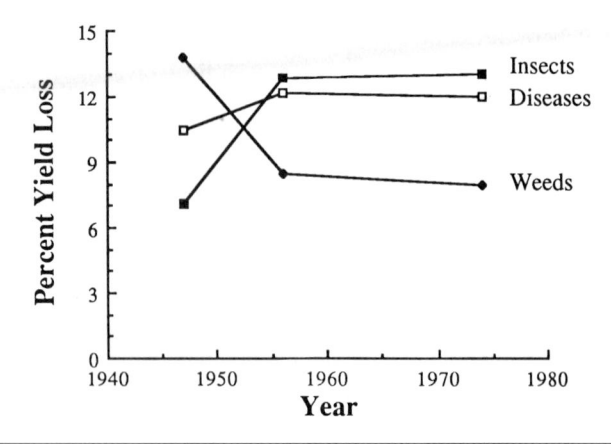

Figure 8.8 Pre-harvest losses to insects, diseases, and weeds since World War II in the United States. [Data from USDA (1954, 1965,) and Pimentel (1976).]

fertilizers has exposed modern industrialized agriculture to a greater potential for insect pest attack. This, in turn, has engendered a commitment to increased use of insecticides.

2. Pesticides and Nitrogen Cycle

Pesticides affect the nitrogen cycle in soils, the availability of nitrogen to plants, and the nutrient content of plants. Most blue-green algae are inhibited by most pesticides (Padhy, 1985), and several insecticides inhibit nodulation between legumes and *Rhizobium* (Agnihotri et al., 1981). Many insecticides inhibit nitrification and stimulate ammonification, which would benefit ammonia preferring plants, such as rice, pineapple, sugar cane, and coniferous seedlings (Agnihotri et al., 1981).

These changes in the nitrogen cycle can affect the availability of nitrogen to plants. There are, however, no generalizable effects; different pesticides have different effects (Martin, 1972; Greaves et al., 1976). Similarly, pesticides have unique effects on plant nutrition (Hance, 1981).

While the details of these effects have yet to be worked out, clearly, use of pesticides can change nitrogen flows in agroecosystems.

C. Toward More Rational Use

The transformation of U.S. agriculture has been incredibly rapid. In little more than three to four decades, it has become an intensive, industrial agriculture with great reliance on agricultural chemicals. Accompanying this increased dependence on agricultural chemicals has been increased realization that current practices can harm the environment and impair human health. We have suggested that while current practice is rational on the individual level—that is, all actors involved are making rational microeconomic decisions—it is not rational in the larger picture.

The rapid recent change and the deepening dissatisfaction with current agricultural practice portend significant future change. But how will agriculture change, and will agricultural chemicals be used rationally? It seems clear to us that agricultural chemicals are a necessary component in current U.S. agriculture. It is impossible to visualize the entire U.S. agricultural system in the near future as one that does not use any agricultural chemicals. Yet it seems necessary to visualize and strive to create an agricultural system for the United States that does not use any agricultural chemicals. The development of socially rational use of agricultural chemicals in the future will be enhanced by challenging their necessity today.

REFERENCES

Adam, A. V. (1976). The importance of pesticides in developing countries. In "Pesticides and Human Welfare" (D. L. Gunn and J. G. R. Stevens, eds.), pp. 115–130. Oxford Univ. Press, London and New York.

Adams, J. A., Campbell, A. S., McKeegan, W. R., McPherson, R. J., and Toukin, P. J. (1979). Nitrate and chloride in ground-water, surface water and deep soil profiles of central Canterbury, New Zealand. *Prog. Water Technol.* **11**, 351–360.

Adkisson, P. L. (1972). The integrated control of the insect pests of cotton. *Proc. Tall Timbers Conf. Ecol. Anim. Cont. Hab. Manag.* **4**, 175–188.

Agnihotri, V. P., Sinha, A. P., and Singh, K. (1981). Influence of insecticides on soil microorganisms and their biochemical activity. *Pesticides* **15**(9), 16–24.

Ahtiainen, M. (1986). Groundwater quality in relation to land application of sewage sludge. In "Factors Influencing Sludge Utilization Practices in Europe" (R. D. Davis, H. Haeni, and P. L'Hermite, eds.), pp. 51–62. Elsevier, New York.

Andow, D. A., and Davis, D. P. (1988). Nitrogen flows in contemporary United States agriculture. Unpublished manuscript.

Andow, D. A., and Rosset, P. M. (1989). Integrated pest management. In "Ecological Agriculture" (J. Vandermeer, C. R. Carroll, and P. M. Rosset, eds.). Macmillan, New York, in press.

Arca, M. (1974). Some aspects of nitrogen fertilization in Peru. In "Effects of Agricultural Production on Nitrates in Food and Water with particular References to Isotope Studies," pp. 53–62. Int. At. Energy Agency, Vienna.

Baier, J. H., and Rykbost, K. A. (1976). The contribution of fertilizer to the ground water of Long Island. *Groundwater* **14**, 439–447.

Balicka, N., and Sobieszczanski, J. (1969). The effect of herbicides on soil microflora. III. The effect of herbicides on ammonification and nitrification in the soil. *Acta Microbiol. Pol.* **18**, 7–10.

Bingham, F. T., Davis, S., and Shade, E. (1971). Water relations, salt balance, and nitrate leaching losses of a 960-acre citrus watershed. *Soil Sci.* **112**(6), 410–417.

Blus, L. J., Lamont, T. G., and Neely, B. S., Jr. (1979). Effects of organochlorine residues

on eggshell thickness, reproduction, and population status of brown pelicans *(Pelecanus occidentalis)* in South Carolina and Florida, 1969–76. *Pestic. Monit. J.* **12,** 172–184.
Bouldin, D. R., Klausner, S. D., and Reid, W. S. (1984). Use of nitrogen from manure. *In* "Nitrogen in Crop Production" (R. D. Hauck, ed.), pp. 221–245. Am. Soc. Agron.—Crop Sci. Soc. Am.—Soil Sci. Soc. Am., Madison, Wisconsin.
Boyce, A. M. (1976). Historical aspects of insecticide development. *In* "The Future for Insecticides: Needs and Prospects" (R. L. Metcalf and J. J. McKelvey, Jr. eds.), pp. 469–488. Wiley, New York.
Boza-Barducci, T. (1972). Ecological consequences of pesticides used for the control of cotton insects in Cañete Valley, Peru. *In* "The Careless Technology" (M. T. Farvar and J. P. Milton, eds.), pp. 423–438. Nat. Hist. Press, Garden City, New York.
Brady, N. C. (1974). "The Nature and Properties of Soils," 8th Ed. Macmillan, New York.
Burbee, C. (1980). Antibiotic feed additives. *Farmline* **1,** 9.
Burford, J. R., and Stefanson, R. C. (1973). Measurement of gaseous losses of nitrogen from soils. *Soil Biol. Biochem.* **5,** 133–141.
Canter, L. W. (1987). Nitrates and pesticides in ground water: An analysis of a computer-based literature search. *In* "Ground Water Quality and Agricultural Practices" (D. M. Fairchild, ed.), pp. 153–174. Lewis, Chelsea, Michigan.
Carey, A. E. (1979). Monitoring pesticides in agricultural and urban soils of the United States. *Pestic. Monit. J.* **13,** 23–27.
CAST. (1985). "Agriculture and Groundwater Quality," Rep. 103. Counc. Agric. Sci. Technol., Ames, Iowa.
CEA, IFA, and IPI. (1983). "Handbook on Environmental Aspects of Fertilizer Use." Nijhoff, Junk The Hague.
Cheng, H. H., and Koskinen, W. C. (1986). Processes and factors affecting transport of pesticides to ground water. *In* "Evaluation of Pesticides in Ground Water" (W. Y. Garner, R. C. Honeycutt, and H. N. Nigg, eds.), pp. 2–13. Am. Chem. Soc., Washington, D.C.
Clegg, D. J., and Khera, K. S. (1973). The teratogenicity of pesticides, their metabolites and contaminants. *In* "Pesticides and the Environment: A Continuing Controversy" (W. B. Deichmann, ed.), pp. 267–276. Intercontinental Med. Book Corp., New York.
Cooke, G. W. (1982). "Fertilizing For Maximum Yield." Macmillan, New York.
Copplestone, J. F. (1977). *In* "Global View of Pesticides" (D. Watson and A. W. A. Brown, eds.), pp. 147–157. Academic Press, New York.
Coutts, H. H. (1980). Field worker exposure during pesticide application. *In* "Field Worker Exposure During Pesticide Application" (W. F. Tordior and E. A. H. van Heemstra-Lequin, eds.), pp. 39–45. Elsevier, Amsterdam.
Cramer, H. H. (1967). Plant protection and world crop production. *Pflanzenshutz-Nachr.* **20,** 1–524.
Dam Kofoed, A. (1985). Pathways of nitrate and phosphate to ground and surface waters. *In* "Environment and Chemicals in Agriculture" (F. P. W. Winteringham, ed.), pp. 27–69. Elsevier, New York.
Davies, J. E., and Doon, R. (1987). Human health effects of pesticides. *In* "Silent Spring Revisited" (G. J. Marco, R. M. Hollingworth, and W. Durham, eds.), pp. 113–124. Am. Chem. Soc., Washington, D.C.
Day, B. E. (1978). The status and future of chemical weed control. *In* "Pest Control Strategies" (E. H. Smith and D. Pimentel, eds.), pp. 203–213. Academic Press, New York.
Debona, A. C., and Audus, L. J. (1970). Studies on the effects of herbicides on soil nitrification. *Weed Res.* **10,** 250–263.
Denmead, O. T., Simpson, J. R., and Freney, J. R. (1974). Ammonia flux into the atmosphere from a grazed pasture. *Science* **185,** 609–610.
Dittrich, V., Streibert, P., and Bathe, P. A. (1974). An old case reopened: Mite stimulation by insecticide residues. *Environ. Ent.* **3,** 534–540.

Dodd, A. P. (1936). The control and eradication of prickly pear in Australia. *Bull. Entomol. Res.* **26**, 503–522.
Dubey, D. (1970). Nitrogen deficiency disease of sugar cane probably caused by repeated pesticide application. *Phytopathology* **60**, 485–487.
Duggan, R. E., and Duggan, M. B. (1973). Pesticide residues in food. *In* "Environmental Pollution by Pesticides" (C. A. Edwards, ed.), pp. 334–364. Plenum, London.
Edwards, C. A. (1965). Effects of pesticide residues on soil invertebrates and plants. *In* "Ecology and the Industrial Society" (G. T. Goodman, R. W. Edwards, and J. M. Lambert, eds.), pp. 239–261. Blackwell, Oxford.
Edwards, C. A. (1973). "Persistant Pesticides in the Environment," 2nd Ed. CRC Press, Boca Raton, Florida.
Edwards, C. A. (1980a). Interactions between agricultural practice and earthworms. *In* "Soil Biology as Related to Land Use Practices" (D. L. Dindal, ed.), pp. 3–11. EPA/OPTS, Washington, D.C.
Edwards, C. A. (1980b). Investigations into the practicability of integrated control of beets. *In* "Integrated Control in Agriculture and Forestry" (D. L. Dindal, ed.), pp. 199–204. OILB/SROP, Washington, D.C.
Edwards, C. A., and Lofty, J. R. (1977). "The Biology of Earthworms." Chapman & Hall, London.
Eichers, T. R., Andrilenas, P. A., and Anderson, T. W. (1978). "Farmers' Use of Pesticides in 1976," Agricultural Economic Rep. No. 418. U.S. Dep. Agric., Econ., Stat., Coop. Serv., Washington, D.C.
ERS. (1987). "Agricultural Resources, Inputs, Situation and Outlook Report," Economic Research Service, AR-5. U.S. Dept. Agric., Washington, D.C.
Fairchild, D. M. (1987). A national assessment of ground water contamination from pesticides and fertilizers. *In* "Ground Water Quality and Agricultural Practices" (D. M. Fairchild, ed.), pp. 273–294. Lewis, Chelsea, Michigan.
FAO. (1980). "FAO Fertilizer Yearbook," Vol. 29. Food Agric. Organ. U.N., Rome.
FAO. (1984). "FAO Fertilizer Yearbook," Vol. 33. Food Agric. Organ. U.N., Rome.
FAO. (1985). "FAO Fertilizer Yearbook," Vol. 34. Food Agric. Organ. U.N., Rome.
Farquhar, G. D., Wetselaar, R., and Weir, B. (1983). Gaseous nitrogen losses from plants. *In* "Gaseous Loss of Nitrogen from Plant-Soil Systems" (J. R. Freney and J. R. Simpson, eds.), pp. 159–180. Martinus Nijhoff, The Hague.
Fleming, W. J., Clark, D. R., Jr., and Henny, C. J. (1983). Organochlorine pesticides and PCB's-A continuing problem for the 1980s. *Trans. North Am. Wildl. Nat. Conf.* **48**, 186–199.
Frank, R., Holdrinet, M., Braun, H. E., Dodge, D. P., and Sprangler, G. E. (1978). Residues of organochlorine insecticides and polychlorinated biphenyls in fish from Lakes Huron and Superior, Canada—1969–76. *Pestic. Monit. J.* **12**, 60–68.
Frere, M. H. (1976). "Nutrient Aspects of Pollution from Cropland. Control of Water Pollution from Cropland, Vol. II—An Overview," EPA-600/2-75-026b. U.S. Environ. Prot. Agency, Washington, D.C.
Furtick, W. R. (1976a). Insecticides in food production. *In* "The Future for Insecticides: Needs and Prospects" (R. L. Metcalf and J. J. McKelvey, Jr., eds.), *Adv. Environ. Sci* **6**, 1–15.
Furtick, W. R. (1976b). Uncontrolled pests or adequate food? *In* "Pesticides and Human Welfare" (D. L. Gunn and J. G. R. Stevens, eds.), pp. 3–12. Oxford Univ. Press, London and New York.
Georghiou, G. P., and Mellon, R. B. (1983). Pesticide resistance in time and space. *In* "Pest Resistance to Pesticides" (G. P. Georghiou and T. Saito, eds.), pp. 1–46. Plenum, New York.
Gough, M. (1986). "Dioxin, Agent Orange: The Facts," Chap. 9. Plenum, New York.
Graham, F., Jr. (1966). "Disaster by Default," Chap. 5. Evans, New York.

Greaves, M. P., Davies, H. A., Marsh, J. A. P., and Wingfield, G. F. (1976). Herbicides and soil microorganisms. *CRC Crit. Rev. Microbiol.* **5**, 1–38.

Green, M. B. (1976). "Pesticides, Boon or Bane?" Elek Books, London.

Hall, J. E., and Ryden, J. C. (1986). Current UK research into ammonia losses from sludges and slurries. *In* "Efficient Land Use of Sludges and Manure" (A. Dam Kofoed, J. H. Williams, and P. L'Hermite, eds.), pp. 180–192. Elsevier, New York.

Hall, R. J. (1987). Impact of pesticides on bird populations. *In* "Silent Spring Revisited" (G. J. Marco, R. M. Hollingworth, and W. Durham, eds.), pp. 85–112. Am. Chem. Soc., Washington, D.C.

Hallberg, G. R., Libra, R. D., Bettis, E. A., III, and Hoyer, B. E. (1984). "Hydrogeologic and water quality investigations in the Big Springs Basin, Clayton County Iowa. 1983 water year. *Iowa Geol. Surv., Open-File Rep.* No. 84-4.

Hance, R. J. (1981). Effects of pesticides on plant nutrition. *Residue Rev.* **78**, 13–41.

Harris, J. R., Jr. (1987). Poultry manure management and groundwater quality: the Delaware solution. *In* "Ground Water Quality and Agricultural Practices" (D. M. Fairchild, ed.), pp. 357–365. Lewis, Chelsea, Michigan.

Hayes, W. J., Jr., and Vaughn, W. K. (1977). Mortality from pesticides in the United States in 1973 and 1974. *Toxicol. Appl. Pharmacol.* **42**, 235–252.

Headley, J. C. (1971). Productivity of agricultural pesticides. *Econ. Res. Pestic. Policy Decis. Making, Proc. Symp. Econ. Res. Serv. (USDA)* pp. 80–88.

Heichel, G. H. (1987). Legume nitrogen: Symbiotic fixation and removal by subsequent crops. *In* "Energy in Plant Nutrition and Pest Control" (Z. R. Helsel, ed.), pp. 63–80. Elsevier Science Pub., Amsterdam.

Hutchinson, G. L., Millington, R. J., and Peters, D. B. (1972). Atmospheric ammonia: Absorption by plant leaves. *Science* **175**, 771–772.

Jensen, A. A. (1983). Chemical contaminants in human milk. *Residue Rev.* **89**, 2–128.

Johnson, G. V. (1987). Soil testing as a guide to prudent use of nitrogen fertilizers in Oklahoma agriculture. *In* "Ground Water Quality and Agricultural Practicse" (D. M. Fairchild, ed.) pp. 127–135. Lewis, Chelsea, Michigan.

Jones, U.S. (1982). "Fertilizers and Soil Fertility." Reston Publ. Co., Reston, Virginia.

Kahn, E. (1976). Pesticide related illness in California farmworkers. *J. Occup. Med.* **18**, 693–696.

Kenaga, E. E. (1972). Factors related to bioconcentration of pesticides. *In* "Environmental Toxicology of Pesticides" (F. Matsummura, G. M. Boush, and T. Misato, eds.), pp. 193–228. Academic Press, New York.

Kenaga, E. E. (1975). Partitioning and uptake of pesticides in biological systems. *In* "Environmental Dynamics of Pesticides" (R. Haque and V. H. Freed, eds.), pp. 217–273. Plenum, New York.

Knowles, R. (1982). Denitrification. *Microbiological Reviews* **46**, 43–70. *Occup. Med.* **18**, 693–696.

Kutz, F. W., Strassman, S. C., and Yobs, A. R. (1977a). Survey of pesticide residues and their metabolites in humans. *In* "Pesticide Management and Insecticide Resistance" (D. L. Watson and A. W. A. Brown, eds.), pp. 523–539. Academic Press, New York.

Kutz, F. W., Yobs, A. R., Strassman, S. C., and Viar, J. F., Jr. (1977b). Effects of reducing DDT usage on total DDT storage in humans. *Pestic. Monit. J.* **11**, 61–63.

Larsen, W. E., Holt, R. F., and Carlson, S. D. (1978). Residues for soil conservation. *In* "Crop Residue Management Systems" (W. R. Oschwald, ed.), ASA Spec. Publ. No. 31, pp. 1–16. Madison, Wisconsin.

Lauer, D. A. (1975). Limitations of animal waste replacement for inorganic fertilizers. *In* "Energy, Agriculture and Waste Management" (W. J. Jewell, ed.), pp. 409–432. Ann Arbor Sci. Publ., Ann Arbor, Michigan.

Libra, R. D., Hallberg, G. R., and Hoyer, B. E. (1987). Impacts of agricultural chemicals on ground water in Iowa. *In* "Ground Water Quality and Agricultural Practices" (D. M. Fairchild, ed.), pp. 185–215. Lewis, Chelsea, Michigan.

Lockeretz, W., Shearer, G., and Kohl, D. H. (1981). Organic farming in the Corn Belt. *Science* **211**, 540–547.

Luckey, T. D. (1968). Insecticide hormoligosis. *J. Econ. Ent.* **61**, 7–12.

McGregor, S. E. (1976). Insect pollination of cultivated crop plants. *U.S. Dep. Agric., Agric. Handb.* No. 496.

Manges, H. L., Lipper, R. I., Murphy, L. S., Powers, W. L., and Schmid, L. A. (1975). "Treatment and Ultimate Disposal of Cattle Feedlot Wastes," EPA 6602-75-013. U.S. Environ. Prot. Agency, Washington, D.C.

Martin, E. C. (1978). Impact of pesticides on honey bees. *Gleanings Bee Cult.* **106**, 318–320, 346.

Martin, J. P. (1972). Side effects of organic chemicals on soil properties and plant growth. *In* "Organic Chemicals in the Soil Environment" (C.A.I. Goring and J. W. Hamaker, eds.), pp. 733–000. Dekker, New York.

Mattson, W. J., Jr. (1980). Herbivory in relation to plant nitrogen content. *Annu. Rev. Ecol. Syst.* **11**, 119–161.

Melander, A. L. (1914). Can insects become resistant to sprays? *J. Econ. Entomol.* **7**, 167–173.

Menzie, C. M. (1972). Effects of pesticides on fish and wildlife. *In* "Environmental Toxicology of Pesticides" (F. Matsumura, G. M. Boush, and T. Misato, eds.), pp. 487–500, Academic Press, New York.

Metcalf, R. L. (1986). The ecology of insecticides and the chemical control of insects. *In* "Ecological Theory and Integrated Pest Management Practice" (M. Kogan, ed.), pp. 251–297. Wiley, New York.

NAS. (1972). "Water Quality Criteria 1972," EPA-R3-73-033. Natl. Acad. Sci., Washington, D.C.

NAS. (1987). "Regulating Pesticides in Food. The Delaney Paradox." Natl. Acad. Press, Washington, D.C.

Nimmo, D. R., Coppage, D. L., Pickering, Q. H., and Hansen, D. J. (1987). Assessing the toxicity of pesticides to aquatic organisms. *In* "Silent Spring Revisited" (G. J. Marco, R. M. Hollingworth, and W. Durham, eds.), pp. 49–70. Am. Chem. Soc., Washington, D.C.

NRC. (1972). "Accumulation of Nitrate," Committee on Nitrate Accumulation, Agricultural Board. Natl. Acad. Sci., Washington, D.C.

NRC. (1978). "Nitrates: An Environmental Assessment." Natl. Acad. Sci., Washington, D.C.

NRCC. (1981). "Pesticide–Pollinator Interactions," NRCC No. 18471. Natl. Res. Counc. Can., Assoc. Comm. Sci. Criteria Environ. Qual., Subcomm. Pestic. Ind. Org. Chem., Ottawa.

NTSB. (1977). "Aircraft Accident Reports," Vols. 1–4 for 1976 accidents, National Transportation Safety Board. U.S. Gov. Print. Off., Washington, D.C.

Oka, I. N., and Pimentel, D. (1974). Corn susceptibility to corn leaf aphids and common corn smut after herbicide treatment. *Environ. Entomol.* **3**, 911–915.

O'Leary, J. A., Davies, J. E., Edmundson, W. F., and Reich, G. A. (1970). Transplacental passage of pesticides. *Am. J. Obstet. Gynecol.* **107**, 65–68.

Olsen, R. A., Frank, K. D., Deibert, E. J., Dreier, A. F., Sander, D. H., and Johnson, V. A. (1976). Impact of residual mineral N in soil on grain protein yields of winter wheat and corn. *Agron. J.* **68**, 769–772.

Overcash, M. R., Humenik, F. J., and Miner, J. R. (1983). "Livestock Waste Management," Vol. II. CRC Press, Boca Raton, Florida.

Padhy, R. N. (1985). Cyanobacteria and pesticides. *Residue Rev.* **95**, 1–44.

Peakall, D. B. (1985). Behavioral responses of birds to pesticides and other contaminants. *Residue Rev.* **96**, 45–77.

Pearson, R. W., *et al.*, (1961). Residual effects of fall- and spring-applied nitrogen fertilizers on crop yields in the southeastern United States. *U.S. Dep. Agric. Tech. Bull.* No. 1254.

Peoples, S. A., and Knaak, J. B. (1982). Monitoring pesticide safety programs by measuring blood cholinesterase and analyzing blood and urine for pesticides and their metabolites. *In* "Pesticide Residues and Exposure" (J. R. Plimmer, ed.), pp. 410–457. Am. Chem. Soc., Washington, D.C.

Perfect, T. J., Cook, A. G., Critchley, B. R., Critchley, U., Davies, A. L., Swift, M. J., Russell-Smith, A., and Yeadon, R. (1979). The effect of DDT contamination on the productivity of a cultivate forest soil in the sub-humid tropics. *J. Appl. Ecol.* **16**, 705–719.

Pimentel, D. (1971). "Ecological Effects of Pesticides on Non- Target Species." Exec. Off. Pres., Off. Sci. Technol., Washington, D.C.

Pimentel, D. (1973). Extent of pesticide use, food supply, and pollution. *J. N.Y. Entomol. Soc.* **81**, 13–33.

Pimentel, D. (1976). World food crisis: energy and pests. *Bull. Entomol. Soc. Am.* **22**, 20–26.

Pimentel, D., and Edwards, C. A. (1982). Pesticides and ecosystems. *BioScience* **32**, 595–600.

Pimentel, D., Krummel, J., Gallahan, D., Hough, J., Merrill, A., Schreiner, I., Vittum, P., Koziol, F., Back, E., Yen, D., and Fiance, S. (1979). A cost–benefit analysis of pesticide use in U.S. food production. *In* "Pesticides: Contemporary Roles in Agriculture, Health, and Environment" (T. J. Sheets and D. Pimentel, eds.), pp. 97–150. Humana, Press, Clifton, New Jersey.

Pimentel, D., Andow, D. A., Gallahan, D., Schreiner, I., Thompson, T. E., Dyson-Hudson, R., Jacobson, S. N., Irish, M. A., Kroop, S. F., Moss, A. M., Shephard, M. D., and Vinzant, W. G. (1980). Pesticides: Environmental and social costs. *In* "Pest Control: Cultural and Environmental Aspects" (D. Pimentel and J. H. Perkins, eds.), pp. 99–158. Westview Press, Boulder, Colorado.

PSAC. (1965). "Restoring the Quality of our Environment," Report. Environ. Pollut. Panel, Press. Sci. Advis. Comm., Washington, D.C.

Ratcliffe, D. A. (1967). Decrease in eggshell weight in certain birds of prey. *Nature (London)* **215**, 208–210.

Reinert, R. E. (1972). Accumulation of dieldrin in an alga *(Scenedesmus obliquus)*, *Daphnia magma*, and the guppy *(Poecilia reticulata)*. *J. Fish. Res. Board Can.* **29**, 1413–1418.

Repetto, R. (1985). "Paying the Price: Pesticide Subsidies in Developing Countries." World Resour. Inst., Washington, D.C.

Richter, E. D., Gribetz, B., Krasna, M., and Gordon, M. (1980). Heat stress in aerial spray pilots. *In* "Field Worker Exposure During Pesticide Application" (W. F. Tordoir and E. A. H. van Heemstra-Lequin, eds.), pp. 129–136. Elsevier, Amsterdam.

Risch, S. J., Pimentel, D., and Grover, H. (1986). Corn monoculture versus old field: Effects of low levels of insecticides. *Ecology* **67**, 505–515.

Rodriguez, J. G., Maynard, D. E., and Smith, W. T., Jr. (1960). Effects of soil insectides and absorbents on plant sugars and resulting effects on mite nutrition. *J. Econ. Ent.* **53**, 491–495.

Rosenfield, A. B., and Huston, R. (1950). Infant methemoglobinemia in Minnesota due to nitrates in wellwater. *Minn. Med.* **33**, 787–796.

Scriber, J. M., and Slansky, F., Jr. (1981). The nutritional ecology of immature insects. *Annu. Rev. Entomol.* **26**, 183–211.

Sherff, J. L. (1975). Energy use and economics in the manufacture of fertilizers. *In* "Energy,

Agriculture and Waste Management" (W. J. Jewell, ed.), pp. 433–441. Ann Arbor Sci. Publ., Ann Arbor, Michigan.

Sherwood, M. (1985). Nitrate leaching following application of slurry and urine to field plots. *In* "Efficient Land Use of Sludge and Manure" (A. Dam Kofoed, J. H. Williams, and P. L'Hermite, eds.), pp. 150–157. Elsevier, New York.

Shinde, J. E., and Vamadevan, V. K. (1974). Fate and significance of nitrogen fertilizers and animal waste residues in relation to soil, water, climate and agronomic practices. *In* "Effects of Agricultural Production on Nitrates in Food and Water With Particular Reference to Isotope Studies," pp. 45–52. Int. At. Energy Agency, Vienna.

Shrivastava, P. (1987). "Bhophal: Anatomy of a Crisis." Ballinger, Cambridge, Massachusetts.

Shuval, H. I., and Gruener, N. (1974). Effects on man and animals of ingesting nitrates and nitrites in water and food. *In* "Effects of Agricultural Production on Nitrates in Food and Water with Particular Reference to Isotope Studies," pp. 117–130. Int. At. Energy Agency, Vienna.

Smith, R. A., Alexander, R. B., and Wolman, M. G. (1987). Water quality trends in the nations rivers. *Science* **235**, 1607–1615.

Smith, R. J. (1982). Hawaiian milk contamination creates alarm. *Science* **217**, 137–140.

Sobieszczanski, J. (1969). Herbicides as factors changing the biological equilibrium of soil. *Natl. Congr. Soil Sci., Soc., 1st, Sofia.*

Söderlund, R., and Rosswall, T. (1982). The nitrogen cycle. *In* "The Natural Environment and the Biogeochemical Cycles, Vol. 1, Part B," pp. 61–81. Springer-Verlag, Berlin.

Steenhuis, T. S., and Walter, M. F. (1979). Definitions and qualitative evaluation of soil and water conservation practices. *In* "Effectiveness of Soil and Water Conservation Practices for Pollution Control" (D. A. Haith and R. C. Loehr, eds.), EPA-600/3-79-106, pp. 14–38. Ecol. Res. Ser., EPA/OPER, Athens, Georgia.

Stevenson, F. J. (1982). Origin and distribution of nitrogen in soil. *In* "Nitrogen in Agricultural Soils" (F. J. Stevenson, ed.), Agron. Ser., No. 22. Am. Soc. Agron., Madison, Wisconsin.

Stricker, M. H. (1977). Blueberry pollination, 1977. *Gleanings Bee Cult.* **105**, 49–50.

Svensson, B. H., and Söderlund, R. (1976). "Nitrogen Phosphorous and Sulphur-Global Cycles, Scope 7 Report," Ecol. Bull., No. 22. Berlingska Boktryckeriet, Lund, Sweden.

Thaler, G. R., and Plowright, R. C. (1980). The effect of aerial insecticide spraying for spruce budworm control on the fecundity of entomophilous plants in New Brunswick. *Can. J. Bot.* **58**(3), 2022–2027.

USDA. (1954). "Losses in Agriculture," Agric. Res. Serv., 20–1 U.S. Dep. Agric., Washington, D.C.

USDA. (1965). Losses in agriculture. *U.S. Dep. Agric., Agric. Handb.* No. 291.

USDA. (1976). "The Pesticide Review (1975)." U.S. Dep. Agric., Agric. Stabil. Conserv. Serv., Washington, D.C.

USDA. (1982). "Feed Situation." U.S. Dep. Agric., Washington, D.C.

USDA. (1985a). "Agricultural Statistics." U.S. Dep. Agric., Washington, D.C.

USDA. (1985b). "Major Use of Land in the United States: 1982," Agric. Econ. Rep., No. 535. Econ. Res. Ser., U.S. Dep. Agric., Washington, D.C.

USDA. (1986). "Biologic and Economic Assessment of Stored Corn, Wheat, and Peanut Fumigants," Pestic. Assess. Lab., Agric. Res. Serv., U.S. Dep. Agric., Washington, D.C.

USDA. (1987). "Agricultural Resources Inputs Situation and Outlook Report," Econ. Res. Serv., AR-5. U.S. Dep. Agric., Washington, D.C.

USDC. (1983). "1980 Census of Population, Vol. I, Characteristics of the Population, Chapter A, Number of Inhabitants." Bur. Census, U.S. Dep. Commer., Washington, D.C.

USDI. (1970). "National Survey of Fishing and Hunting," Res. Publ. No. 95. Bur. Sport Fish Wild., Fish Wildl. Serv., U.S. Dep. Inter., Washington, D.C.

USFWPCA. (1968). "Pollution Caused Fish Kills in 1967," CWA-7. U.S. Fed. Water Pollut. Control Admin., Dep. Inter., Washington, D.C.

Van Arsdall, R. N., and Nelson, K. E. (1983). "Characteristics of Farmer Cattle Feeding," Agric. Econ. Rep., No. 503. Econ. Res. Serv., U.S. Dep. Agric., Washington, D.C.

Van Dyne, D. L., and Gilbertson, C. B. (1978). Estimating U.S. livestock and poultry manure and nutrient production. Handbook No. 619, U.S. Dep. Agric.

Vitousek, P. M., Walker, L. R., Whiteaker, L. D., Mueller- Dombois, D., and Matson, P. A. (1987). Biological invasion by *Myrica faya* alters ecosystem development in Hawaii. *Science* **238,** 802–804.

Von Peter, A. (1980). Fertilizer requirement in developing countries. *Proc. Fert. Soc.* No. 188.

Ward, D. J., and Wilson, R. E. (1973). Pesticide effects on decomposition and recycling of *Avena* litter in a monoculture ecosystem. *Am. Midl. Nat.* **90,** 266–276.

Wasserstrom, R. F., and Wiles, R. (1985). "Field Duty: U.S. Farm Workers and Pesticide Safety." World Resour. Inst., Washington, D.C.

Weir, D., and Shapiro, M. (1981). "Circle of Poison." Inst. Food Dev. Policy, San Francisco, California.

Westerman, R. L. (1987). Efficient nitrogen fertilization in agricultural production systems. *In* "Ground Water Quality and Agricultural Practices" (D. M. Farichild, ed.), pp. 137–151. Lewis, Chelsea, Michigan.

WHO. (1973). Safe use of pesticides, 20th report, World Health Organization to Expert Committee on Pesticides. *W. H. O. Tech. Rep. Ser.* No. 513, pp. 42–43.

Wilkinson, C. F. (1987). The science and politics of pesticides. *In* "Silent Spring Revisited" (G. J. Marco, R. M. Hollingworth, and W. Durham, eds.), pp. 25–46. Am. Chem. Soc., Washington, D.C.

Winteringham, F. P. W. (1984). "Environment and Chemicals in Agriculture." Elsevier, New York.

Yates, W. E., and Akesson, N. B. (1973). Reducing pesticide chemical drift. *In* "Pesticide Formulations," pp. 275–341. Dekker, New York.

9

Natural Gas as a Resource and Catalyst for Agroindustrial Development

Walter Vergara

Asia Technical Department
The World Bank
Washington, D.C.

I. Background
II. Monteagudo Agroindustrial Project
   A. Area of the Project
   B. Structure of the Project
III. Food Processing Plants
   A. Slaughterhouse and Meat Processing Plant
   B. Citrus Juice Concentrate Plant
   C. Vegetable Oil Plant
   D. Feed Mixing and Bagging Plant
IV. Energy Supply
   A. Cogeneration System
   B. Size Requirements of the Energy Supply System
   C. Electricity and Steam Generation Costs
V. Alternatives for the Supply of Electricity and Steam
   A. Low-Speed Reciprocating Engines
   B. Hydroelectric Supply
VI. Water Supply
VII. Gas Pipeline
VIII. Transportation System
IX. Impact of the Project
   A. Investment and Economic Performance
   B. Sensitivity Analysis
X. Conclusions
References

Note: The data and opinions included in this document are the sole responsibility of the author.

I. BACKGROUND

As a direct consequence of the fast pace of oil exploration in the last 15 years, world natural gas reserves have increased dramatically. The world's gas reserves/production ratio has also steadily increased, and significant gas reserves are now distributed among approximately 100 countries (supply, 1986). A number of agriculture-based developing countries in Latin America, Asia, and Africa have also seen dramatic increases in reserves and availability of natural gas (Table 9.1), and in many cases consumption of gas has likewise increased.

Although petroleum prices have decreased in real terms since 1982, the cost of petroleum fuels continues to be a critical factor for industrial activities in developing countries. If, as projected, over half of the world's oil reserves are used by the year 2000, the long-term outlook for crude oil costs is for further increases. Such increases would continue to strain the abilities of many nations to secure the fuels required for industrial development.

According to projected future increases in crude prices, prices of petroleum derivatives should continuously increase in real economic terms during the next decade, while natural gas (tied to opportunity costs) will gradually widen its competitive edge as a fuel substitute (Table 9.2). This

Table 9.1 Gas and Oil Reserves in Some Gas-Rich Developing Countries[a]

Country	Agriculture sector GDP (as % of total)[b]	Gas (10^6 toe)[c]	Petroleum (10^6 toe)[c]
Argentina	12	538	317
Bolivia	23	133	28
Brazil	12	34	317
Ecuador	14	96	193
Peru	8	44	124
Mexico	8	1,790	7,550
Venezuela	7	1,380	3,450
India	36	410	580
Indonesia	26	1,555	1,145
Thailand	23	173	13
Malaysia	21	1,158	386
Bangladesh	47	164	—
China	37	702	2,540
Algeria	6	2,480	1,214
Nigeria	26	1,100	2,210
Egypt	20	208	496

[a] Source: *Oil and Gas Journal* (1987).

[b] World Resources (1986).

[c] toe = ton of oil equivalent.

Table 9.2 Current and Projected Prices for Petroleum Derivatives and Natural Gas

	$US, 1986		
	1987	1990	1995
Fuel oil ($/ton)[a]	100	100	130
Naphtha ($/ton)[a]	160	150	220
Natural gas ($/MMBTU)			
U.S. wellhead	1.4	1.8	1.9
Middle East	0.5	0.5	0.5
Gas surplus ($/MMBTU)[b]			
Agricultural	0.5	0.5	0.5
Countries	1.0	1.0	1.0

[a]Based on World Bank estimates of future crude oil prices.
[b]In the lower range: Venezuela, Indonesia, Algeria, Bolivia.

chapter argues that the relative low cost and the wide availability of natural gas are major advantages for its use as primary energy source for supplying steam and electricity to agroindustrial projects located in rural and/or isolated areas in developing countries, and that the use of natural gas should be encouraged for such applications whenever economically justifiable.

Compared to other fossil fuels, natural gas has already proven to be a flexible, versatile, and economic fuel alternative. Because gas has customarily been treated as a byproduct of oil production in a number of developing countries, its pricing has allowed the displacement of other fuels in standard applications. It is being extensively used in the generation of electricity and supply of steam and heat to industry (as a substitute for diesel and fuel oil), as a transportation fuel (in substitution of gasoline and diesel), and as an attractive fuel in the domestic sector for cooking and heating.

Unfortunately for many gas-rich developing countries, the requirements for oil production and the lack of transport and distribution infrastructure have simultaneously forced a high rate of flaring, venting, and reinjection of associated gas. Therefore, in the short term, many of these countries will continue to have large amounts of gas reserves associated with very low long-run marginal costs of production (Mashayekhi, 1982), but lack the required infrastructure for storage, transportation, and marketing for optimal utilization. At the same time, in many of these areas where gas resources have been identified and developed, there remain vast rural, and otherwise richly endowed agricultural lands at the margin of industrial development. Progress is limited by lack of access to electricity and steam and because of the obstacles imposed by poor infrastructure and absence of adequate transportation.

The supply of electricity to rural areas in developing countries has always been faced by exceedingly high capital and distribution costs resulting from the relatively low and dispersed nature of the demand. The supply of liquid fuels for the domestic and commercial sectors of rural communities is also affected by high transport costs and unreliable availability. On the other hand, the use of wood, crop residues, and other renewable materials to meet energy requirements—although seemingly an attractive solution in agricultural areas—also faces increasing environmental and economic limitations as the long-term costs of indiscriminate deforestation and soil erosion become more and more apparent to developing countries (Pimentel *et al.*, 1987).

In summary, energy supply to rural communities is limited and deficient because demand is below an economic threshold to justify the required investments, and demand is low because it is limited by serious restrictions in supply. One of the effects is that development, particularly in the agroindustrial sector, is slow.

Industrial cogeneration of steam and electricity is particularly suitable as an economic energy supply alternative to the extension of the electricity grid and erection of steam generators, particularly in rural areas with prospects for agroindustrial development and ready access to pipeline gas. Among cogeneration systems, gas-based turbines are attractive because of the relatively low investment (Figure 9.1) and high efficiency. However,

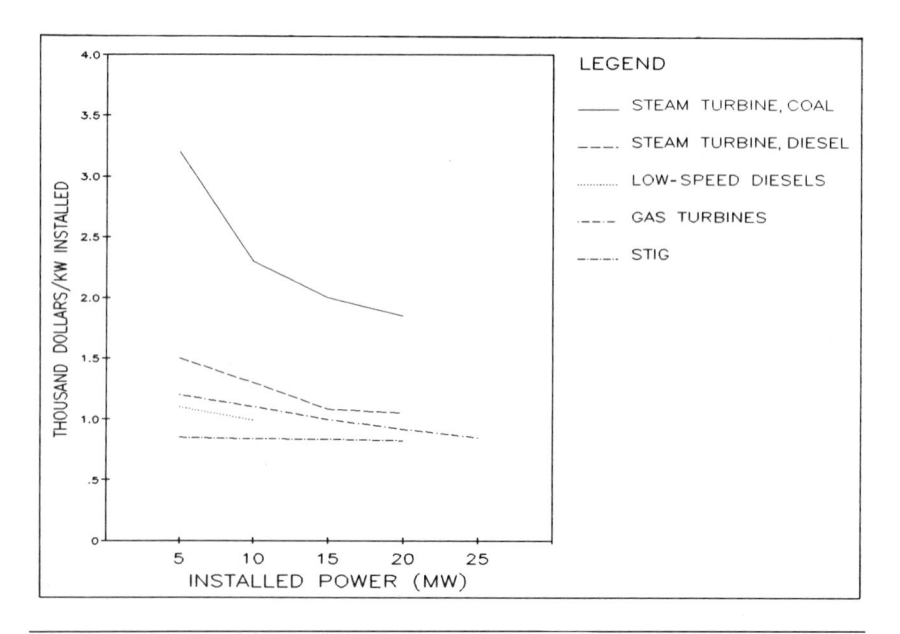

Figure 9.1 Comparison of installed capital costs of medium-capacity cogeneration systems.

9. Natural Gas and Agroindustrial Development

when the specific application requires variable loads, as is the case in most agroindustries, the economics of conventional gas turbines is often unfavorable due to the resulting low effective capacity in operation.

Steam-injected gas turbine (STIG) cogeneration systems solve the problem by allowing excess steam to be used as feedback to the turbine, thereby allowing additional electrical output and increased efficiency. Thus, steam injection leads to both increased flexibility and higher output and efficiency (Figure 9.2) (Larson and Williams, 1985).

The novel approaches to efficient use of stand-alone systems for generation of steam and electricity have large potential for development of industry in isolated areas where the cost of transport and storage difficulties make attractive the conversion of agricultural raw materials into high value added agroindustrial products. The value of production and other economic benefits obtained through agroindustry may in turn promote additional *in situ* development. This results in long term increased opportunities for business and employment and encourages the rural population to remain in the area as an attractive alternative to rural migration. In no place is this technology more promising than in those rural areas of developing countries where extensive pipeline networks have been laid over the last 15 years to transport gas to urban and traditional industrial centers for use as a domestic and industrial fuel.

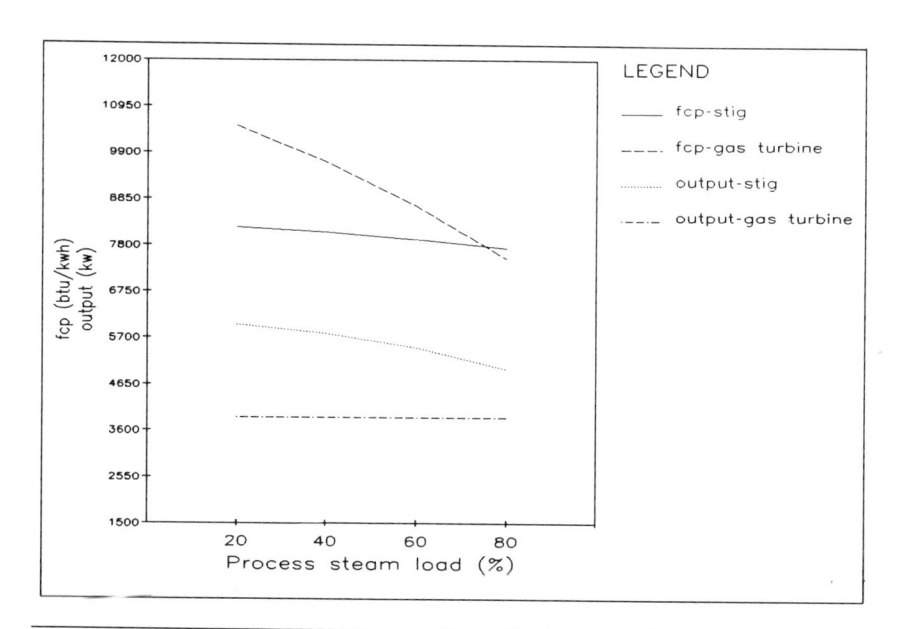

Figure 9.2 Fuel chargeable to power (FCP) and electrical output for steam-injected (STIG) and gas turbines. Based on 3.5 MW-rated turbine (Larson and Williams, 1985).

A quick glance through some of the specific situations reveals a number of sites with the required characteristics to advantageously benefit from the newly acquired availability of gas (Table 9.3). It is an extraordinary coincidence that most gas fields described in Table 9.3 are separated from the industrial and urban end users by some of the potentially richest, and in some cases, more isolated agricultural areas.

II. MONTEAGUDO AGROINDUSTRIAL PROJECT

As an example of the potential synergism of gas-based cogeneration and agroindustry, a specific case is analyzed. The Monteagudo Agroindustrial Project in Bolivia has recently been examined on its merits (as part of a development project jointly sponsored by the Government of Bolivia and the Organization of American States; Vergara, 1986), and some of the major conclusions and supporting data are hereby briefly presented.

Bolivia (population 9 million, with per capita GNP of US$470) is a poor agricultural country, with most of its population concentrated in the mountainous eastern side of the country and with sizeable expanses of agricultural land and significant reserves of natural gas in the less populated south west. As do many other countries in the region, Bolivia relies on agriculture for a large fraction of the domestic economic activity (27% of GDP in 1985; World Resources, 1986).

Table 9.3 Some Agricultural Areas Crossed by Pipeline from Remote Gas Fields

Country	Gas field	End users	Agricultural areas crossed by pipeline
Bolivia	Monteagudo Voltaredonda	Industry, electric sector	Santa Cruz, Chaco, Chuquisaca
Colombia	Guajira	Domestic, Industry, Transport, Electric sector	Lower Magdalena Valley
Ecuador	Suchufindi	Industry	East Andean valleys
Venezuela	Zulia	Industry, electric sector	Coastal plains
India	Bombay	Industry	Guyarat, Madhya, Uttar Pradesh
Indonesia	Kalimantan	Exports, chemical feedstock	—
Malaysia	Saba offshore	Industry	Saba

A. Area of the Project

The area selected for the project is the town of Monteagudo (population 3,000) in the Department of Chuquisaca (see Figure 9.3). This is an agricultural area with a long farming tradition that is geared to meet the needs of the local and provincial markets and that is relatively specialized in the production of corn, peanuts, citrus fruits, hogs, and cattle. Also, a significant fraction of the total production of hogs in Bolivia has its origin in the area around Monteagudo. Similarly, the Monteagudo area is the site of two natural gas fields currently under exploitation, and the Monteagudo–Sucre pipeline runs about 30 km south and west of the town. Regrettably, in contrast to the relatively rich endowment in soils, productivity, and fossil fuels, the area has remained devoid of infrastructure, with no paved roads or electric grids, and only a very primitive water supply system. As a consequence, the rich agricultural potential has not developed, no value-added operations are performed in the area, and most

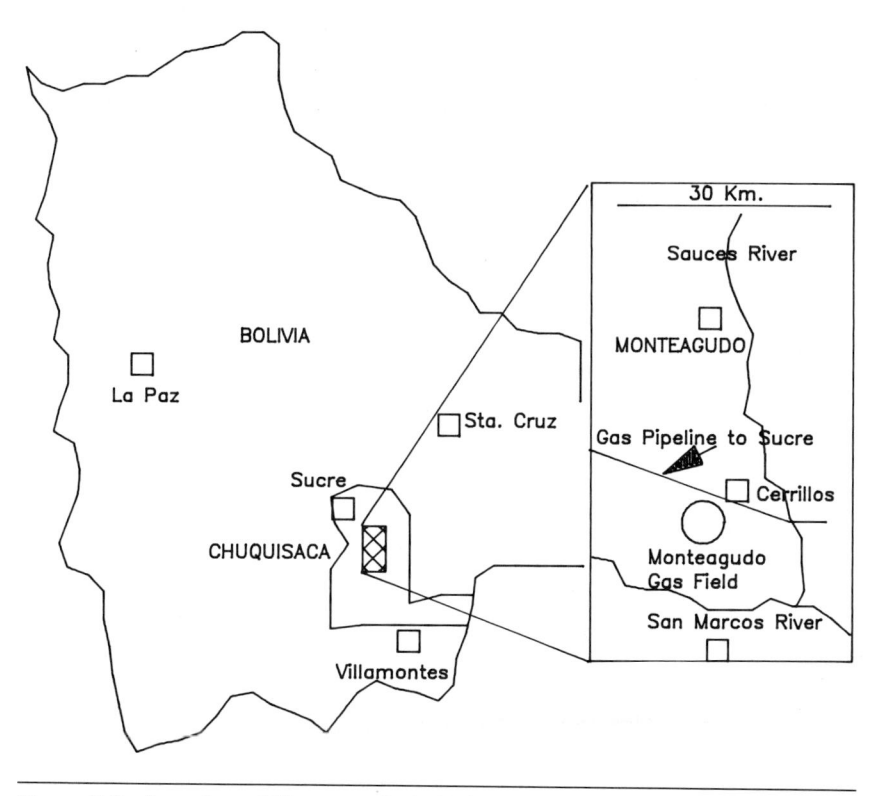

Figure 9.3 Location of Monteagudo in Bolivia.

of the production is exported as raw materials with little if any impact in the industrial development of the zone. Additionally, high transportation costs limit the area of influence and available markets for products originating in Monteagudo. Compounding the problem, the lack of development inhibits the availability of manpower and contributes to rural migration to the nearby urban centers.

B. Structure of the Project

The suggested project aims to provide development opportunities through the installation of an agroindustrial complex, in turn made feasible by the availability of gas in the area. The project consists of a number of closely related food processing facilities and the required infrastructure for steam, electricity, natural gas, water supply, and transportation hardware (Figure 9.4).

Specifically, in terms of food processing, the project consists of a slaughterhouse and meat packing plant, a citrus juice concentrate plant, a crude vegetable oil plant, and a feed mixing and bagging plant. The structure of the project allows for a strong linkage between the different processing facilities that supply raw materials and by-products to one another. For example, a crude vegetable oil plant produces also as a by-product, oilseed meal with a high protein content which can be utilized

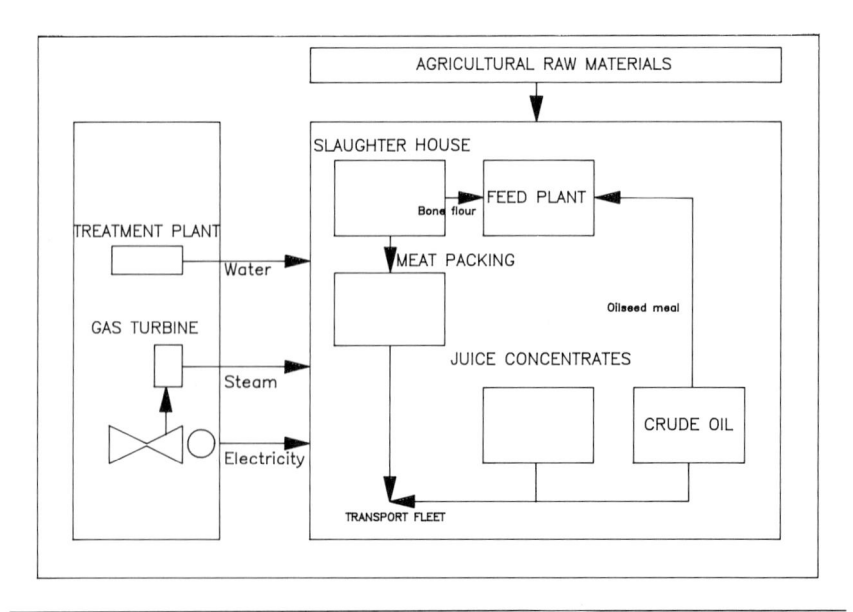

Figure 9.4 Structure of the Monteagudo project.

in the formulation of feed rations. Similarly, a slaughterhouse produces bone flour as a residue that could also be utilized at a feed mixing and bagging plant. In turn, all plants will share in the use of infrastructure specifically designed to meet the requirements of the project. Refrigerated transport for example, may not be justified for one product, but the combined production of the project will more likely achieve the necessary economies of scale.

The proposed infrastructure includes a cogeneration system, based on natural gas, sized to meet the thermal and electrical requirement of all the food processing installations; a water treatment plant; and a gas pipeline to bring the gas to the site from nearby gas fields and a refrigerated transport system.

III. FOOD PROCESSING PLANTS

A. Slaughterhouse and Meat Processing Plant

The area around Monteagudo ships for slaughtering about 40,000 hogs/year out of an estimated stock of 160,000 hogs. The establishment of a slaughter house in the area will (1) allow for greater economic margins through reduced costs and increased quality of raw materials; (2) greatly reduce the transportation costs of raw materials and allow for savings in the shipment of end products; and (3) benefit from the access in the area to a market for bone flour and other by-products from the plant to be used as feed rations for livestock.

The slaughtering and meat processing plant is sized to meet about 60% of the slaughtering needs of the local production, which is equivalent to 2,300 t (metric tons) of end products (cured and processed meat products). A unit of this size would supply a sizeable fraction of the Bolivian domestic market of processed meats. The plant would also produce 500 t/year of bone flour to be used as feed for livestock.

The major share of production costs are raw materials and energy. Raw materials costs are estimated at $3.5 million/yr ($1.3/kg). Energy requirements are met by steam and electricity. Steam is used in cleaning, sterilization, cooking and scalding operations. Total demand is estimated at about 6,700 t/yr (metric tons per year). Electricity is used for the operation of the processing, handling, and packing machinery. Total consumption has been estimated at 12.6 MWhr/day, or about 3.7 GWhr/yr. At the rates estimated in the section on energy supply, energy costs amount to US$0.5 million/year. The total processing cost is summarized in Table 9.4. The annual production cost is estimated at US$4.6 million, equivalent to $2/kg of meat products, out of which raw materials amount to about 75% and energy to about 11%.

Table 9.4 Slaughterhouse and Meat Packing Plant Costs

Feature	Amount
Size	90 animals/day
Operation	300 days/yr
Production	2,320 t/yr
Fixed capital cost[a]	$1.9 million

Production cost requirements	Annual cost US$ × 10³	%
Raw materials	3,462	76
All utilities[b]	519	(12)
Steam	(200)	(5)
Electricity	(262)	(6)
Other	(57)	(1)
Manpower[c]	83	2
Capital[d]	501	10
	4,565	100

[a] Fixed capital costs were estimated at US$1.9 million, out of which equipment is US$1.1 million and construction and engineering US$0.3 million.

[b] Based on $0.07/kWh and $30/ton of steam.

[c] Based on 52 person-days/day at rates about twice the existing salaries in urban areas.

[d] Includes maintenance charges (4% of equipment costs), overhead (4% of fixed capital cost), interest charges 14% of working capital) and depreciation (11% on fixed capital cost).

B. Citrus Juice Concentrate Plant

The total area dedicated to orange orchards around Monteagudo is estimated at 550 ha and the total commercial production is around 4,000 t/yr. But, because of (1) the high transport costs; (2) the lack of storage infrastructure; and (3) the seasonal nature of the production, a significant fraction of the total is lost. The net effect is a restrictive supply situation that discourages demand and local producers from further exploiting this crop. On the other hand, recent market studies have indicated a high potential demand for fresh juice and concentrate in many of the urban centers out of reach to local production.

The juice concentrate plant is sized to process all the fruit it can collect from the area (a minimum of 6,000 t/yr) and feed with concentrate a juice reconstitution plant located on or near the major consumption centers. The total production would be 530 t of concentrate, equivalent to about 2,750 t of fresh juice. The plant would also produce about 700 tons of saleable by-products. The installation of the plant in the area will benefit from (1) the direct access to the production zone with the implied reduction in transport costs and a feedback effect on the quality and volume of production in the area; (2) a drastic reduction in transportation costs (8.8 t of concentrate equal the juice of 100 t of fresh fruit), and (3) a market in the area for the pulp and peel produced as by-products.

Processing costs consist mostly of the cost of the oranges (72%), energy (10%), and capital (14%). Steam and electricity costs are the components of energy costs. Steam is required in the concentration and pasteurization stages of juice processing and for drying the pulp and peel for further processing at the feed rations plant. Total steam requirements at the plant are sized at about 4,400 t/year. Electricity requirements are sized at 210 MWhr/yr. Total processing costs are estimated at about US$8,000/t ($2.8/kg of concentrate) (Table 9.5).

C. Vegetable Oil Plant

Before 1984, Bolivia was largely dependent on imports of vegetable oils to meet local requirements. Since then, substantial increase in the local

Table 9.5 Orange Juice Concentrate Plant Costs

Feature	Amount
Size	40 t/day of fresh fruit
Operation	150 days/yr
Production	530 t/yr juice concentrate
Fixed capital cost[a]	$1.1 million
Working capital	$0.2 million

Production cost requirements	Annual cost US$ × 10³	%
Raw materials[b]	1,070	72
All utilities[c]	167	12
Steam	(132)	(9)
Electricity	(15)	(1)
Water and other	(20)	(2)
Manpower[d]	32	2
Capital[e]	203	14
	1,492	100

[a]Includes $0.6 million for equipment and $0.3 million for construction and engineering. Other changes and contingency make up the balance.

[b]6,000 t/season at $179/ton.

[c]Based on $0.07/kWh and $30/ton of steam.

[d]Based on 11.5 person-days/day.

[e]Includes maintenance charges (3% of equipment cost, 4% of fixed capital cost), interest charges (14% of working capital), and depreciation (10% on fixed capital cost).

production of soybeans and an expansion in processing capacity have enabled an improvement in the local supply of oils. But localized shortages of raw materials are still prevalent in the country's southwest areas, where prohibitive transportation costs restrict the supply of soybean seed to the southern complex of Villamontes. One way to solve the problem is by manufacturing crude oil in small factories located in or around the producing areas for subsequent transport and refining at Villamontes at considerable savings in transportation costs. This is, in summary, the rationale for establishing one such unit around Monteagudo. The crude oil plant is sized to process up to 20 t/day of oilseeds and produce up to 1,400 t of crude oil and 4,500 t of oilseed meal per year. The installation of a vegetable oil plant in the area is a logic component of the agroindustrial project given the following reasons:

1. The direct access to a rich, productive corn producing area which provides a plentiful supply of starch to the feed rations plants and a rich, high-quality raw material for vegetable oil production.
2. The availability of soybeans in the area, a raw material for oil extraction and for the production of a protein rich byproduct also suitable for use at the feed rations plant for livestock.
3. The relative location of a major vegetable oil refinery in southern Bolivia that will benefit from the supply of crude oil at Monteagudo.

Processing costs at the crude vegetable oil plant are summarized in Table 9.6. Raw material costs (corn kernels, soybean seeds) account for 68%, energy costs 15%, and capital 14%. Energy costs are made up of steam and electricity costs. Steam is required for the operation of the hot presses (oil extractors) and the distillation during solvent recovery. Steam is also utilized for detoxification of the soybean seeds and other minor operations. Total steam demand has been estimated at 4,400 t/yr which, at the estimated steam generation costs, is equivalent to 11% of total production costs. Similarly, electricity requirements sized at 0.9 GWhr/yr represent about 4% of production costs.

D. Feed Mixing and Bagging Plant

The livestock feed mixing and bagging plant already being installed in the area is an important component of the future project. The plant is a convenient outlet for by-products from the meat packing plant (bone flour), from the vegetable oil plant (oilseed meal), and the juice plant (pulp), thereby reducing transportation costs, eliminating environmental problems, and increasing profitability. The plant is sized to produce 10,000 t/year of livestock feed rations and with slight modifications could separate the kernels from the corn for further processing and oil extraction at the

Table 9.6 Vegetable Oil Plant Costs

Feature	Amount
Size	25 t/day of oilseeds
Operation	250 days/yr
Production	1,440 t/yr of crude oil; 4,640 t/yr of oilseed meal
Fixed capital cost[a]	$1.05 million
Working capital	$0.20 million

Production cost requirements	Annual cost US$ × 10³	%
Raw materials[b]	868	68
All utilities[c]	206	16
Steam	(133)	(11)
Electricity	(63)	(4)
Water	(10)	(1)
Manpower[d]	30	2
Capital[e]	172	14
	1,275	100

[a] Total fixed capital cost includes $0.6 million in equipment cost and $0.3 million in construction and engineering. Other changes and contingencies make up the balance.

[b] The raw materials include about 800 t of corn kernels and 5,000 tons of soybeans. The kernels are assumed to be obtained in exchange for about 2,000 tons of cornmeal for the feed mixing and bagging plant.

[c] Based on $0.07/kWh and $30/t of steam.

[d] Based on 13 person-days/day.

[e] Includes maintenance (3% of equipment cost), insurance (0.6% of fixed capital cost), overhead (4% of fixed capital cost), interest charges (14% of working capital), and depreciation (8% of fixed capital cost).

vegetable oil plant. No significant additional investments are required to integrate this plant with the rest of the project. On the contrary, it will benefit from savings in the supply of electricity and steam through the cogeneration system.

IV. ENERGY SUPPLY

The proposed food processing plants, as shown in the previous paragraphs, have sizeable thermal and electrical requirements which may amount to a significant fraction of total production costs. The need for an adequate energy supply has been identified repeatedly as one of the deterrents for agroindustrial activities in rural areas. Because the area under consideration lacks electricity supply, but has a rich endowment of natural gas, the system proposed consists of a gas turbine-based cogeneration system for the supply of steam, heat, and electricity.

Cogeneration gas turbines are in use in many industrialized countries and are generally recognized as an efficient, competitive, small- to medium-size system for the joint supply of steam and electricity to industry and commerce. The advantages for selecting such a system for the project include:

1. The relatively smaller capital costs associated with the gas turbine when compared with the investments associated with small-scale hydroelectric development or reciprocating engines;
2. The high thermodynamic efficiency achieved through the joint generation of electricity and heat (steam, waste heat) for energy-intensive industries such as agroindustry where significant requirements of both are typically needed; and
3. The flexibility of operation obtained through the use of steam injection under the expected variable load conditions.

A. Cogeneration System

Figure 9.5 depicts the proposed cogeneration system. Gas from nearby gas fields is transported by gas pipeline to the site and is burned in the combustion chamber. The heat released operates the turbine, generating

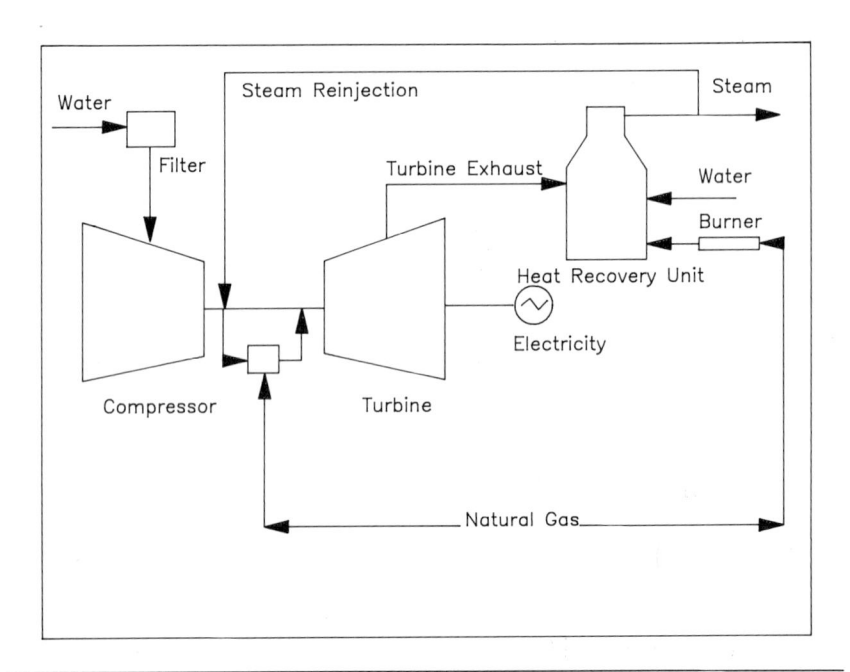

Figure 9.5 Flow chart of the cogeneration system.

electricity. The hot gases released through the exhaust are sent to the Heat Recovery Steam Generator (HRSG), where steam is produced at the required pressure. Flexible steam/electricity supply is achieved through the use of steam injection and gas burners.

B. Size of the Energy Supply System

The power requirements at the project are of the order of 1.7 MW, associated to a demand equivalent to 5.4 GWhr/year. The installed power in the domestic and commercial sectors is about 0.6 MW, these and other power requirements add to 3.0 MW and a total demand of 8.6 GWhr (see Table 9.7). Similarly, steam requirements are estimated at 17,500 tons. To meet the energy requirements, the nominal power of the system should be at least 3 MW which at the scheduled electricity demand implies a load factor of about 46% over 7,000 hr of operation per year. An Allison 501 KH turbine (Stuart and Stevenson, 1986) or equivalent will meet the power requirements and also generate up to 7.5 tons steam/hr under the expected operating conditions. But, given the relative isolation of the project, standby equipment is required. The estimated cost, performance, and other characteristics of the proposed system are summarized in Table 9.8.

C. Electricity and Steam Generation Costs

Cogeneration costs are essentially made up of the cost of natural gas and the capital costs. The total demand for electricity was estimated at 8.6

Table 9.7 Estimated Total Energy Requirements

Plant	Annual electricity demand (GWhr)	Annual steam demand (10^3 t)
Food processing		
Slaughterhouse and meat processing plant	3.74	6.7
Orange juice concentrate plant	0.21	4.4
Vegetable oil plant	0.90	4.4
Feed plant	0.54	0.5
Infrastructure		
Water treatment plant	0.60	—
GLP plant	0.15	—
Others	0.31	—
Current domestic and industrial sector	1.19	—
Other demands	0.40	1.5
Short-term growth of domestic sector[a]	0.60	—

[a] Assumes that availability of electricity in the area will contribute to a sharp short-term increase in consumption by the domestic sector.

Gwhr/yr. If an average electrical efficiency of 27% is achieved, the total requirements for gas are about 109 billion BTU/yr:

$$\text{Gas requirements} = \frac{8.64 \times 10^6 \text{ kWhr} \times 3{,}412 \text{ BTU/kWhr}}{0.27}$$

The generation system will also supply around 3.7 tons steam/hr (4.7 t/hr peak supply) out of a maximum steam availability estimated at 6.5 t/hr. Under these conditions, the fuel charged to power (FCP), defined as the fuel required in the cogeneration system in excess of the requirements to produce steam in an independent boiler, is described by:

$$\text{FCP} = K(1 - S/N_b)/E = 6{,}890 \text{ BTU/kWhh},$$

where S and E are the gas equivalent fractions of the amounts of steam (0.40) and electricity (0.27) generated by the system, N_b is the assumed

Table 9.8 Cogeneration System Costs

Feature	Amount
Electricity	
Rating	3.1 MW
Effective power	2.7 MW
Electricity demand	8.6 GWhr/yr
Steam demand	17,500 t/yr
Load factors	46% electrical, 38% steam
Operation	7,000 h/yr
Fixed capital cost[a]	US$3.5 million

Generation costs	Quantity	Unit costs (US$/kWhr)
Natural gas[b]	109 MCF	0.016
Other supplies		0.001
Water		0.001
Manpower[c]		0.004
Capital costs		
Maintenance (3% of equipment cost)		(0.010)
Insurance (0.6% of fixed capital cost)		(0.002)
Overhead (4% of fixed capital cost)		(0.016)
Capital recovery factory[d]		(0.080)
Total		0.130
Electricity generation costs (54% of total)[d]		(0.070)

[a]Total fixed capital cost is $2.9 million in equipment costs, which includes a spare gas turbine, and $0.6 million in installation and contingencies.
[b]Based on $1.3 per million BTU.
[c]Based on 9 person-days/day.
[d]Corresponds to the fuel charged to power (FCP).

efficiency of the boiler (0.88), and K is the thermal equivalent of electricity (3,412 BTU/kWhr). This means that under the expected average requirements of the project, about 54% of the gas used can be charged to power generation,

$$\text{FCP} = \frac{6{,}890 \times 0.27}{K} = .54,$$

or about $0.070/kWh based on the generation costs described in Table 9.8. Consequently, the steam production costs (ST) are equivalent to about US$30/t:

$$\text{ST} = \frac{[(1 - \text{FCP}) \times \$0.130]/\text{kWhr} \times 8.6 \times 10^6 \text{ kWhr}}{17{,}500 \text{ t}}$$

This calculation is based on 46% of the total generation costs assigned to 17,500 t of steam/year.

The capital recovery factor allowed for the cogeneration system (Table 9.8) will permit a reasonable return of the investment and at the same time very attractive rates for steam and electricity, comparable to power services from centralized systems in urban areas. This is largely achieved through the benefits of cogeneration.

As can be seen in Table 9.8, only about 12% of total generation costs are due to the cost of natural gas which has been charged at about the ongoing tariff for industrial purposes, which in turn is sufficient to cover cash costs and return on investment for the gas distributor.

A sensitivity analysis of electricity costs and food processing costs to the ability of selling steam, and to the cost of natural gas, clearly indicates that the key factor in keeping energy costs low is the volume of steam demand by the food processing plants (Figure 9.6). The nature of energy requirements at food processing plants in general, and of the specific plants considered in this project in particular, with high steam requirements due to sterilization, pasturization and concentration make them specially suitable for cogeneration systems.

V. ALTERNATIVES FOR THE SUPPLY OF ELECTRICITY AND STEAM

For comparison purposes, two alternative systems for the supply of energy to the project have been investigated: (1) the supply of steam and electricity through the use of low-speed reciprocating engines, and (2) the development of a hydroelectric site.

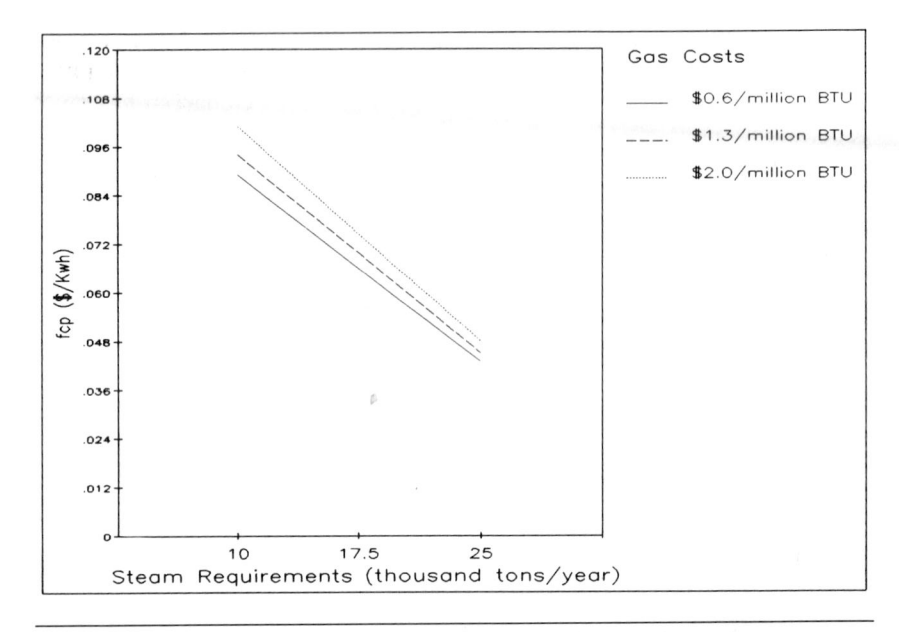

Figure 9.6 Fuel charged to power, as a function of steam demand and gas costs.

A. Low-Speed Reciprocating Engines

The advantages of low-speed reciprocating engines are related to (1) the generally lower rating of these engines which allows for a tighter fit to the electricity demand curves with the associated savings and increased energy efficiency, and (2) the lower speed which requires less stringent maintenance routines and therefore becomes more amenable to the technical capabilities to be found in rural areas.

Nevertheless, the use of low-speed reciprocating engines is also associated with a number of drawbacks: (1) steam cogenerated through this system is of low pressure and limited in quantity, which limits its applications to low steam demand situations and therefore may not be suitable for agroindustrial activities; (2) engine longevity is still under question as these machines were originally designed for discontinuous loads and have not yet been in operation long enough to prove its resilience under continuous load regimes; and (3) the capital cost of the resulting system is generally higher (1.3–1.5 times) per installed kW than the capital cost of comparably sized gas turbines.

For comparison purposes, a system of reciprocating engines has been sized to meet the expected power and steam requirements using as its base a continuous cycle, turbocharged Caterpillar engine (Caterpillar, 1986). The cost of the engine is $3.6 million, to which allowances have to be made for the steam generators ($0.5 million) and installation, engineering, and transport costs for a total of $4.5 million. This is considerably higher than the investment associated with the gas turbine.

B. Hydroelectric Supply

As an alternative to cogeneration systems, a hydroelectric development in the area could meet the power requirements of the project and be complemented with gas-based boilers to supply the steam demand. Hydroelectricity is based on a renewable non-polluting source of energy for which there are a number of attractive sites in the area. Also one of the sites (Rio Azero) has been extensively surveyed and included as an option in the development program of the region. This site is located about 30 km northwest of Monteagudo and could support an installed power of 1 MW (Table 9.9). The associated investment, including the costs of a transmission line to Monteagudo, were estimated at $3.3 million (Rio Azero, 1985).

From the viewpoint of the project, the option of electricity supply from Rio Azero presents the following characteristics: (1) both its generation cost and start-up period are considerably higher when compared with the cogeneration system; (2) the power rating at the site will not be enough to meet the requirements; and (3) even if costs and size were appropriate, this option does not solve the issue of steam supply at the project.

VI. WATER SUPPLY

Next to energy, the reliable supply of quality water is the most important utility requirement for agroindustrial activities in rural areas. Water is required for a variety of purposes in the food processing plants, including

Table 9.9 Alternatives for Electricity Supply to the Agroindustrial Project

System	Rating (MW)	Fixed capital cost (US$ million)	Cost per installed kW (US$ thousand)	Generation cost (US$/kWhr)
Gas turbine (STIG)[a]	3	3.5	1.2	0.07
Low-speed reciprocating engine[a]	3	4.5	1.5	0.16[b]
Hydroelectric development at Rio Azero	1	3.3	3.3	0.20

[a]Includes the steam generation equipment.
[b]Does not account for partial steam supply to the project.

steam generation, cooling requirements, and cleaning and other housekeeping operations. Water for steam generation corresponds to the actual production of steam minus any condensates recovered. Although the food processing plants are located together, it is doubtful that all condensates will be recovered. Further, the water plant should be sized to meet 100% water make-up on the peak steam requirements of 7.5 m^3/hr. Cooling and cleaning water are assumed to be of the order of 7.5 m^3/hr. Aside from industrial water requirements, the installation of a water treatment system enables the supply of treated water for the domestic sector of the surrounding area. These requirements were sized at 10.5 m^3/hr. The total demand is 25 m^3/hr. The equipment should be sturdy and reliable, and it should require the minimum maintenance and supervision. A modular water treatment plant which can be expanded as needed is advised, and the cost is estimated at US$0.83/m^3 of water treated (Table 9.10).

VII. GAS PIPELINE

To be made available to the project, gas will have to be piped from the nearby compression station at Cerrillos (see Figure 9.3). Total gas demand, as previously discussed, is equivalent to about 109 MMft^3/year. But, because the processing season is limited by the harvesting periods, actual demand will be concentrated over a period of 250 days. The peak gas supply under these conditions is about 0.7 MMft^3/day while the average gas demand is only about 0.4 MMft^3/day.

While the variations expected in gas requirements are relatively wide, the effect on the volume of gas being shipped on the Cerrillos–Monteagudo

Table 9.10 Water Treatment Plant Costs

Feature	Amount
Capacity	25m^3/hr
Operation	300 days/yr
Fixed capital cost[a]	$0.2 million
Working capital	$50,000
Treatment cost item	Unit costs ($/m^3)
Utilities	0.05
Manpower	0.16
Capital costs	0.62
	0.83

[a]Based on a total equipment cost of $0.16 million and $0.04 million in engineering and transport costs.

pipeline will be relatively minor because (1) the actual volume sales to the project on a yearly basis are only about 3% of the total volume sent via the main pipeline, and (2) the main gas consumer at the project has also the longest processing season, which helps to dampen somewhat the expected variations in gas requirements.

The pipeline required to meet the expected volumes under the existing pressure conditions at Cerillos and over a distance of 30 km has a calculated diameter of 3 inches (7.5 cm). The cost of the pipeline, including accessories, is estimated at US$1.0 million and should last 20 years or more. At 150 psi entry pressure, this pipe will be able to handle up to six times the current peak requirements.

VIII. TRANSPORTATION SYSTEM

As mentioned above, a major obstacle for agroindustrial activity in the area is the absence or poor quality of the road infrastructure which makes access to markets difficult, time consuming, and costly. By the same token, any effort geared to reduce transportation requirements by weight and volume is bound to improve the competitiveness of local agroindustrial activities.

Short of major investments in improvement of roads and bridges, transportation of processed agroindustrial products (high value-added) is the most efficient way to reduce transportation costs. Transportation costs from Monteagudo to the surrounding markets are a function of the type of vehicle utilized, the preservation requirements of the product to be moved, and the markets targeted for distribution of the materials. In particular, transportation of meat, its by-products, and orange juice concentrate require refrigeration.

The daily transport capacity required was estimated at about 45 t/day, for which a fleet of ten vehicles with 15 t of capcity are required (six in route, three on load, and one spare). The transportation cost was estimated as about US$0.89/km (see Table 9.11). Transportation costs to the main consuming centers have been estimated based on actual road distances. These costs are summarized in Table 9.12. As can be seen, transport costs to La Paz may be as high as 15–20% of production costs. Because the final products weigh only a fraction of the raw materials, the actual savings accrued through the processing of the agricultural products are substantial.

IX. IMPACT OF THE PROJECT

A. Investment and Economic Performance

The total investment required for the agroindustrial project is close to $10.1 million but the agroindustrial activities by themselves will only account for 45% of the total investment (Figure 9.7). The energy system is

Table 9.11 Refrigerated Transport Cost[a]

Item	US$/km
Fuel[b]	0.375
Maintenance and materials[c]	0.160
Manpower[d]	0.100
Capital[e]	0.255
	0.890

[a] Equipment: diesel truck with 15-t capacity, 3 axles, and refrigeration chamber for operation at 0°C. Fixed capital cost: US$0.1 million.

[b] Based on fuel requirements of 1.5 l/km and fuel costs of $0.25/liter.

[c] Includes tires, oil, and supplies.

[d] Salaries and expenses.

[e] Based on a capital recovery factor reflecting 20% return on investment/year and depreciation over 1 million km.

the second most expensive item, in part because being a critical activity for the entire operation a back-up turbine has been included in the estimate.

The key to the success of the agroindustrial project is the competitiveness achieved in the market through use of the right economies of scale, the reasonable cost of energy and other supplies, the direct access to the raw materials and savings in transportation costs. The estimated economic performance of the project has been measured estimating the capital recovery factors (Table 9.13). The expected results are quite satisfactory.

The entire impact is only partially gauged through the economic performance of the plants. In addition, (a) the number of jobs created and

Table 9.12 Estimate of Transport Costs from Project Site to Major Consumer Points[a]

	Cost ($/t)	
City	Refrigerated	Ambient
Sucre	37	32
Cochabamba	79	70
La Paz	123	108
Santa Cruz	51	45
Villamontes		42

[a] Based on actual load distances and estimated transport costs per km-t. Assumes trucks return empty to project site.

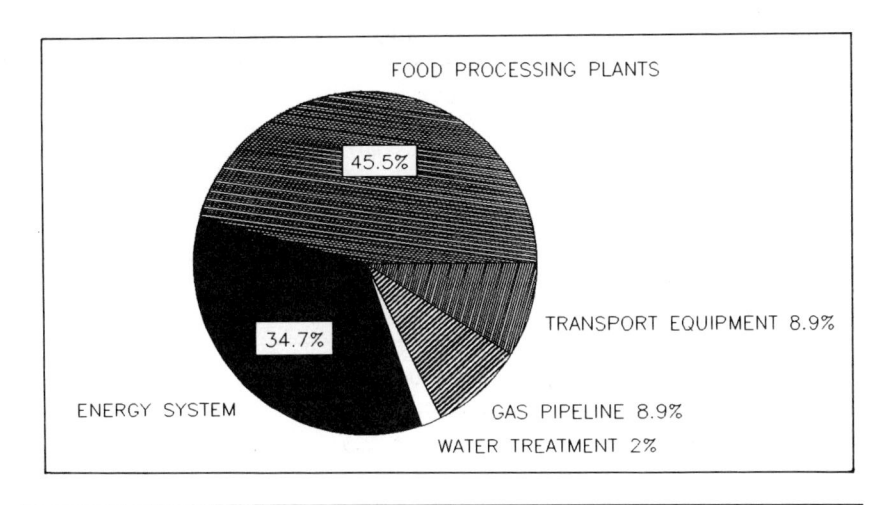

Figure 9.7 Composition of total investment for Monteagudo project.

the volume of sales of agricultural materials to the project should also spur a flurry of activities in the area (Table 9.14); (b) the improvements in electricity and water supply will significantly raise the quality of life of the local population; and (c) sound investments in productive activities will go a long way in restoring confidence in the viability of rural livelihood in agricultural areas.

B. Sensitivity Analysis

Energy is second only to raw materials as a component of food processing costs. Increases in the cost of steam and electricity have a substantial impact in processing costs and will consequently affect the expected return on investments. Figure 9.8 shows the effect on processing costs of higher energy prices and the estimated effect on capital recovery factors for each of the food processing units involved. A doubling of energy costs will render at least some of the projects economically unattractive.

Table 9.13 Estimated Capital Recovery Factors

Plant	Capital recovery factor
Slaughterhouse and meat processing	0.42
Orange juice concentrate	0.36
Vegetable oil	0.28
Energy supply	0.20
Water supply	0.19
Transport system	0.30

Table 9.14 Jobs Created and Sales Volume of Raw Materials

Plant	Jobs created	Investment per job ($ thousand)	Sales volume of raw materials thousand t/yr	$ million/yr
Slaughterhouse and meat processing	52	48	2.6	3.4
Orange juice concentrate	14	79	6.0	1.1
Vegetable oil	9	111	5.9	0.9
Energy system	9	390	109.0[a]	0.1
Water supply	3	67	—	—
Transport system	10	90	—	—
	97	104		5.5

[a] Million ft^3/yr.

X. CONCLUSIONS

Natural gas is an abundant, versatile, and increasingly competitive fuel now available to some 100 countries worldwide. A number of these countries with agriculture-based economies also possess extensive gas reserves associated with very low long-run marginal costs.

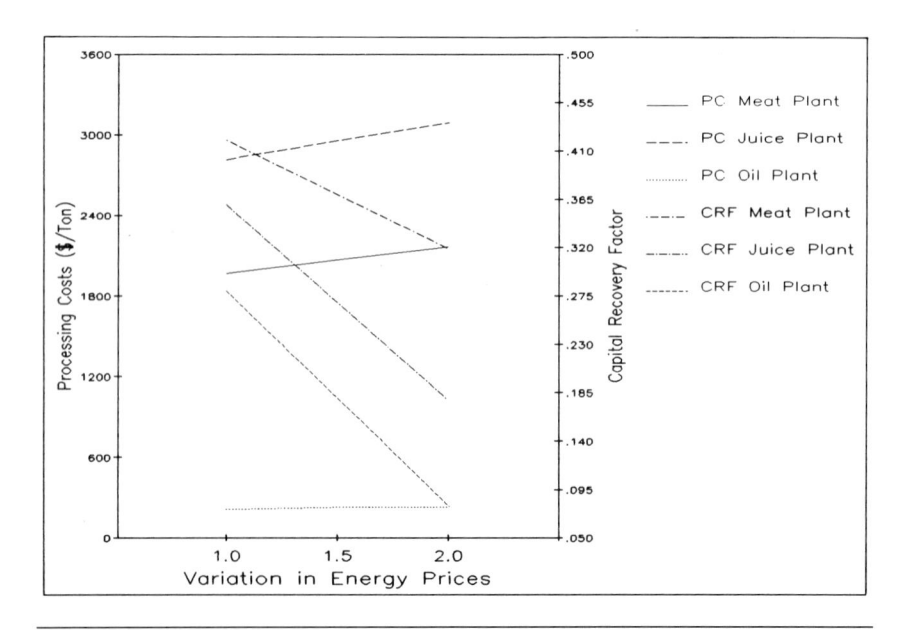

Figure 9.8 Impact of energy prices on processing costs (PC) and capital recovery factors (CRF).

Gas-field development has, in many instances, taken place in relatively remote locations and its transport to end users has required extensive pipeline networking. As a result, pipeline gas is in fact available to some of the major agricultural areas in Venezuela, Colombia, Ecuador, Bolivia, Argentina, India, Indonesia, Bangladesh, Thailand, and China, among others.

Gas-based congeneration of steam and electricity is a competitive alternative to extension of the grid and construction of stand-alone boiler systems for the supply of energy to industrial users in rural areas. The relative ratio of requirements of steam to electricity by food processing operations makes this industry particularly suitable for cogeneration systems with substantial benefits in processing costs. Even though energy costs in food processing operations are second to raw materials costs, the availability of steam is a key factor in the development of agroindustry in rural areas.

Monteagudo, an agriculture-based community in southwest Bolivia, exemplifies an area in which the availability of a nearby gas pipeline permits the establishment of a gas-based cogeneration system to supply the energy needs of an agroindustrial project at very competitive costs, enabling the development of the whole area and acting as a true catalyst to agroindustrial development.

REFERENCES

Caterpillar (1986). Price quote.
Larson, E., and Williams, R. (1985). Technical and economic analysis of steam injected gas turbine cogeneration. *In* "Energy Sources: Conservation and Renewables. Am. Inst. Phys., New York.
Mashayekhi, A. (1982). "Marginal Cost of Natural Gas in Developing Countries. Concepts and Applications," Energy Dep. Pap. No. 10. World Bank, Washington, D.C.
Oil and Gas Journal. (1987). Gas report. **85**(28), 33–80.
Pimentel, D., *et al.* (1987). World agriculture and soil erosion. *Bioscience* **37**, 277–283.
Rio Azero Feasibility Report. (1985). Report to the Corporación de Desarollo de Choqulsaca, Cordech Sucre, Bolivia. (In Span.).
Stuart and Stevenson. (1986). Personal communication concerning technical and cost data for the Allison 501.KH.
Supply seen no constraint to growth of world gas market. (1986). *Oil Gas J.* **84** (39), 25–26.
Vergara, W. (1986). "Monteagudo Agroindustrial–Energy Project. Prefeasibility Analysis." Ministerio de Energia e Hidro Carburos and Organization of American States, La Paz, Bolivia. (In Span.)
World Resources. (1986). A report by the World Resources Institute and the International Institute for Environment and Development, New York.

10

Mechanization and Food Availability

Carl W. Hall

Directorate for Engineering
National Science Foundation
Washington, D.C.

I. Introduction
II. Farming and Agriculture
III. Mechanization, Tractorization, and Electrification
    A. Mechanization
    B. Decrease in Food Availability with Mechanization
    C. Small-Scale versus Large-Scale Mechanization
    D. Labor Resources
    E. Other Characteristics of Mechanization
IV. Summary
References

I. INTRODUCTION

The mechanization of agriculture and food production must be considered in terms of physical, economic, and human resources. An industrial nation cannot expect to have a highly developed economy and an advanced, technological society with a primitive agricultural system. Wages, technological developments, and degree of automation are as important to agriculture as to other segments of the economy.

To provide power for agriculture there has been an evolution from manual labor to the use of tools, draft animals, naturally powered engines, and now fossil fuel engines and electricity (Table 10.1). As controls, sensing, automation, and the application of computers developed in industry, they have been successfully used in agriculture. Adoption of advanced technology in agriculture systems, including production and processing, has often preceeded other segments of the economy such as manufacturing.

Food availability is generally increased by mechanization, in which there is increased productivity of land and labor. However, food availability can be decreased as a result of negative impacts on natural resources, including soil and water loss or degradation. Increased dependence on fossil fuel and petroleum has occurred as a result of industrialized agriculture. Of particular interest is the opportunity to replace fossil fuels, which consist primarily of hydrocarbons, with materials produced on farms, which consist primarily of carbohydrate and lignocellulose materials.

Table 10.1 Comparison of Energy Inputs in Tilling 1 ha of Soil by Manpower, Oxen, 6-HP and 50-HP Tractor[a]

Tilling unit	Worker (hr)	Machinery	Petroleum	Manpower	Oxen power	Total
Manpower	400	6,000	—	194,000	260,000	200,000
Oxen (pair)	65	6,000	—	31,525	260,000[b]	297,525
Tractor						
6-hp	25	191,631[c]	237,562[d]	12,125	—	441,318
50-hp	4	245,288[e]	306,303[f]	2,400	—	553,991

[a]Source: Pimentel and Pimentel (1979).
[b]Each ox is assumed to consume 20,000 kcal/day of feed.
[c]An estimated 191,631 kcal machinery was used in the tillage operation.
[d]An estimated 23.5 liters of gasoline used.
[e]An estimated 245,288 kcal machinery was used in the tillage operation.
[f]An estimated 30.3 liters of gasoline used.

II. FARMING AND AGRICULTURE

Farming refers to working the land to produce products used for feed and food. Agriculture refers to a broad spectrum of activities ranging from farming to processing farm products into food. In primitive agriculture most of the activities for food processing take place on the farm. In market-oriented production and processing agriculture, farming is a small segment of the total activity (Figure 10.1). Increased mechanization causes a shift from farming as the major activity to farming as a small part of the food system, with much of the investment, processing, and other activities occurring off the farm.

III. MECHANIZATION, TRACTORIZATION, AND ELECTRIFICATION

Mechanization, as a first step, involves the machinery and devices to till and cultivate the land, usually with animal power. Animals are then replaced with engines and tractors as the evolution of mechanization occurs

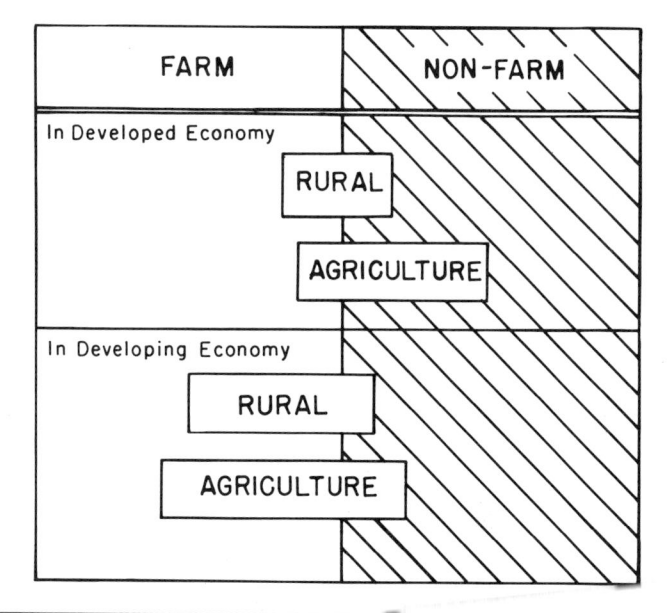

Figure 10.1 Terminology diagram: farm, rural, and agriculture (Esmay and Hall, 1973).

264 Carl W. Hall

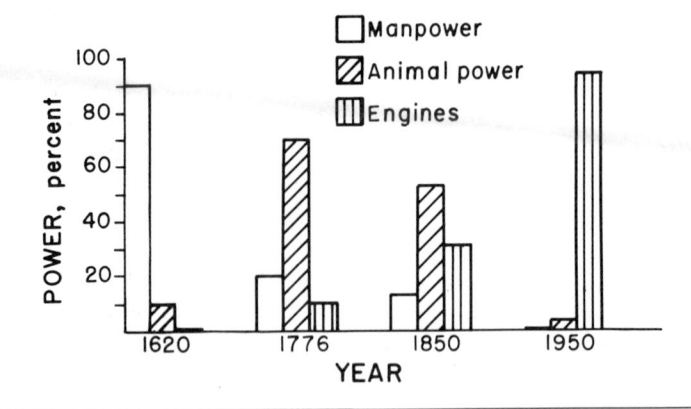

Figure 10.2 Percentage of power provided by manpower, animals, and engines in U. S. history (Pimentel and Pimentel, 1979).

(Figure 10.2). Mechanization and tractorization are often improperly considered as being the same. In the initial stage mechanization could consist of providing mechanical tools to the workers. Often mechanization involves a stationary engine or motor for pumping, irrigating, grinding, chopping, and handling materials. Electrification is another means of supporting mechanization, principally for stationary tasks. The degree of mechanization is often measured in terms of horsepower per hectare, and is illustrated for many countries of the world in Figure 10.3).

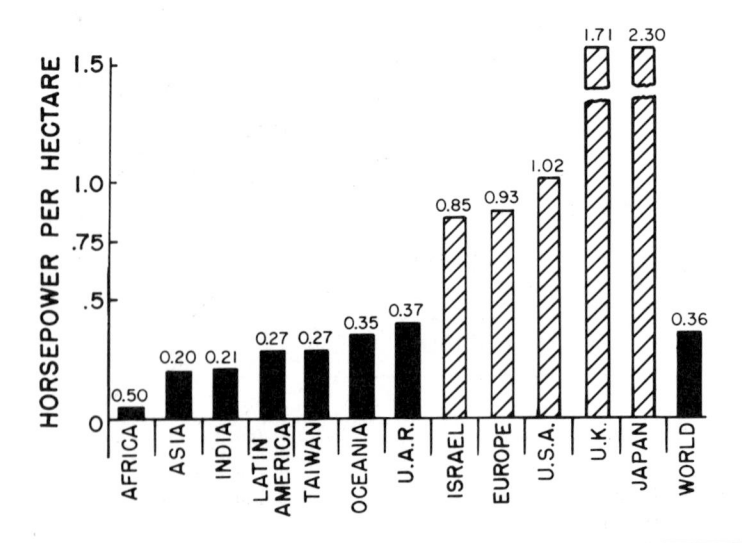

Figure 10.3 Power available for field production, worldwide, 1964–1965 (Esmay and Hall, 1973).

A major criticism of the shift from animal power to tractors, engines, and electric power is the dependence on commercial fossil fuels, a nonrenewable resource. An alternative is to burn the agricultural products or convert them to liquid and gaseous fuels. When draft animals provide the power for production, they consume about 20% of the farm production. As conversion from animal to tractor and electric power occurred, much of the grain production was used to feed a hungry population, and more recently has provided surpluses. The forages used by draft animals were used for food producing animals, principally cattle for milk and meat. With decreasing world demand for U.S. agricultural products, consideration is being given to converting a portion of that production to fuel for mechanization and electrification.

A. Mechanization

In general, mechanization is initially undertaken to promote intensified land use and magnification of worker output (Table 10.2). It is desirable that mechanization be achieved without unnecessary worker displacement, that it improve the environment in which the worker operates, and that it be done on an economic basis with the investment justified on the basis of returns. Mechanization must be considered in connection with irrigation, seeding, cultivating, and harvesting to obtain the benefit of high-producing varieties. In developing countries where 0.2–0.3 HP/ha (horsepower per hectare) are normally used, about twice that amount (0.5–0.8 HP/ha; Figure 10.4) is required to handle the increased yields of new varieties. Additionally, irrigation, fertilizer, and plant protection must be considered. (Chou et al., 1977). In the United States, 40–50 HP are available per worker (Table 10.3) to meet the needs of increased yields, use of chemicals, and high labor productivity. Mechanization is closely related to irrigation, fertilization, and plant protection.

Mechanization is used to carry out operations on a timely basis.

Table 10.2 Worker Productivity by Alternative Harvesting Methods[a]

Year	Method	Time to harvest 1 ha of wheat
1800	Sickle	2.5 hr
1850	Scythe	37 min
1900	Binder	5 min
1920	Tractor-binder	1 ⅔ min
1945	Combine-harvester	1 ⅓ min (harvest and thresh)

[a]Source: FAO (1965).

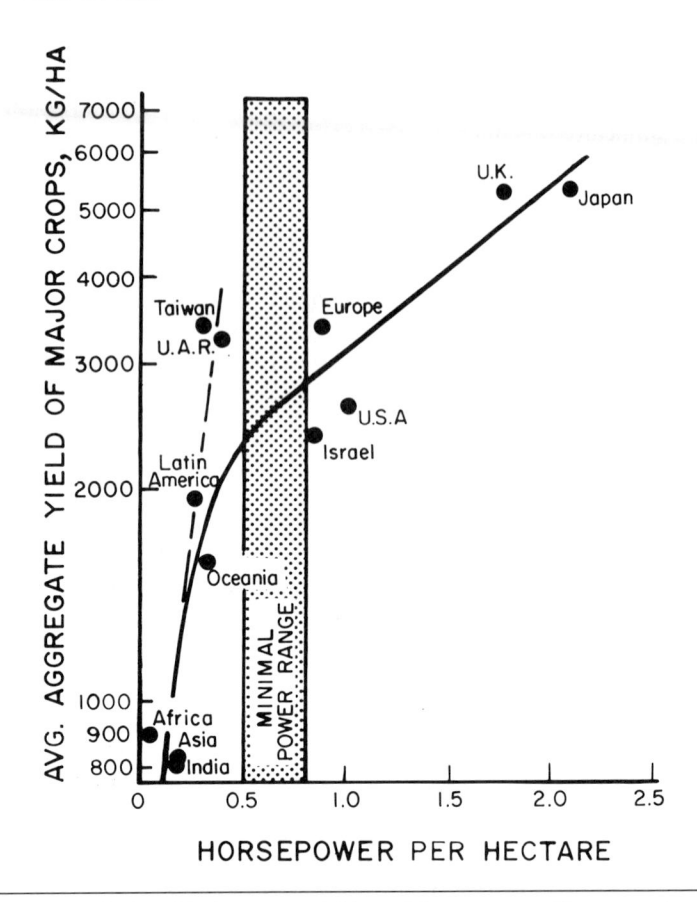

Figure 10.4 Relationship between yields and horsepower (Esmay and Hall, 1973). Major food crops include cereals, pulses, oilseeds, sugar crops (raw sugar), potatoes, cassava, onions, and tomatoes. Data for Asia exclude mainland China.

Table 10.3 Horsepower Available per Farm Worker in United States, Including Electric Motors and Trucks

Year	Hp	Year	Hp
1890	1.0	1950	27
1920	3.3	1960	35
1930	12	1970	73[a]
1940	16	1980	130[a]

[a]Calculated by author based on data in USDC (1970, 1986).

Mechanization can be a factor in optimizing production with the weather and growing season, thereby making most appropriate use of water resources. Mechanization and irrigation, where appropriate, and water control or drainage can contribute to increasing worker productivity (production per hour of labor) (Table 10.4) and land productivity (production per hectare of land). The increase in labor productivity can lead to a reduction in land productivity as a result of careless operations, particularly from losses occurring at harvest. Generally the cost and benefits of these mechanized activities can be measured. The value of some advantages, such as reducing drudgery, getting workers out of dusty and hazardous environments, and eliminating undesirable work conditions, may be difficult to measure. Workers might be discouraged because of the monotony of some mechanized operations, but that is probably no worse than the monotony of manually planting, hoeing, picking, and sorting.

The relations between labor input and energy input in terms of fuels and chemicals and the food energy outputs are reasonably well documented (Figures 10.5 and 10.6) for the United States. Of significance is that the increase in yields previously obtained by increasing energy output has reached a point of diminishing returns. Although the term kilocalorie provides a useful and convenient name for an energy unit, a kilocalorie of energy in a fossil fuel does not have the same food value as a kilocalorie of energy in a food. A kilocalorie of coal, although inedible, provides the same heat as a kilocalorie of food.

If multicropping is feasible, mechanization may be needed to complete the planting or harvesting within the time available. The productivity of the worker as well as the productivity of land are also increased under these conditions. With mechanization and multicropping the work load can be distributed more uniformly throughout the year. With increased use of high-yielding crop varieties, mechanization may be needed to quickly harvest the crop with minimum loss from shattering, cracking, and mold-

Table 10.4 Comparison of Average Worker Productivity in United States, per Man-Hour[a]

	Year		
	1930	1965	1980
Corn (bushel)	1	15	30
Silage (ton)	0.1	1.3	3.0
Hay (bales)	29	160	300
Plowing (acre)	0.25	1.1	3.0
Eggs (dozen)	4	15	30
Milk (cwt)	29	102	200

[a] Source: Esmay and Hall (1973); updated by author.

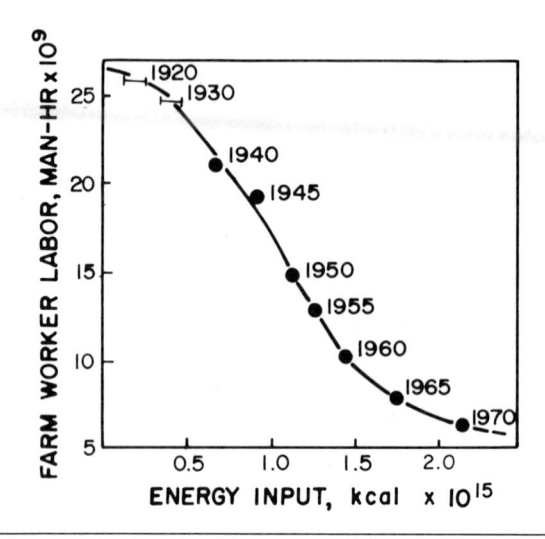

Figure 10.5 Labor use on farms as a function of energy use in the U. S. food system (Steinhart and Steinhart, 1975).

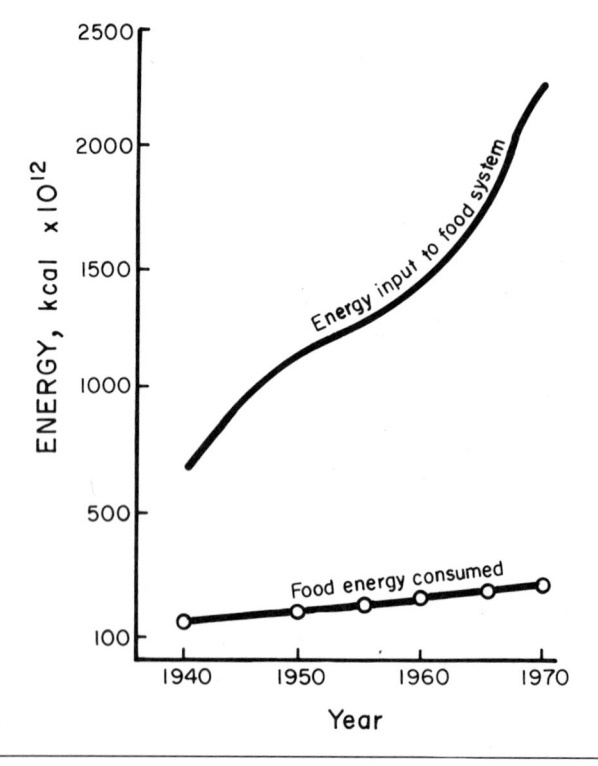

Figure 10.6 Energy use in food system compared to the calorie content of the food consumed (Steinhart and Steinhart, 1975).

ing. Harvest often comes at a time when there is a high demand for labor for harvesting other crops.

A characteristic of mechanization is uniformity of treatment of soil, plants, and products. That treatment by mechanization could be favorable or unfavorable. Location of the seed, placement of the fertilizer, height of cutting stalks, or the distance between rows can be accurately attained. On the other hand, an inaccurate meter or a broken belt could cause skipping of a needed treatment. A malfunction of a cutting, separating, or handling device could cause damage to the product. These malfunctions can continue for some time in a rapidly moving mechanized operation without the knowledge of the operator.

B. Decrease in Food Availability with Mechanization

A decrease in food availability can result from the actual loss of the product from overripeness, weather impacts, and improperly operating equipment. On a longer term basis, improper mechanizing of agriculture and use of land unsuited for mechanization can result in a gradual decrease in food production and availability, due to factors such as wind and water erosion and nutrient and organic matter loss. Improper use of heavy implements and tractors on some soils can cause soil compaction and formation of hardpan in the root zone. These could result in a reduction of production or an ability to grow certain crops. Harvesting of some crops, particularly fruits and vegetables, is often done before they are fully mature to serve the processing market. Harvesting before maturity may increase availability of food. Handling of fruits and vegetables mechanically may cause cuts, bruises, cracking, and other damage to some of the products, making them less available or unavailable particularly for fresh food markets. Many successful efforts are being made to develop devices to improve harvesting and handling.

C. Small-Scale versus Large-Scale Mechanization

Small-scale mechanization refers to utilizing appropriate mechanization for small farms, usually under 10 ha. In Asia, the average farm size varies from 1–4 ha, with 1.0 ha/person average area. (Kitani, 1985). Whereas the usual prerequisites for farm mechanization are considered to be ample land, scarce labor, and abundance of capital, this is opposite to the situation for small farms. With small farms, fields are too small to utilize large equipment. Where capital is limited and money is not available for fertilizer, mechanization (including tractorization and electrification) does not provide the soil with nutrients that might otherwise be available if animals were used for power. The change from hand labor to animal power to engine power from 1920 to 1980 in Japan, where small farms predominate, is shown in Figure 10.7).

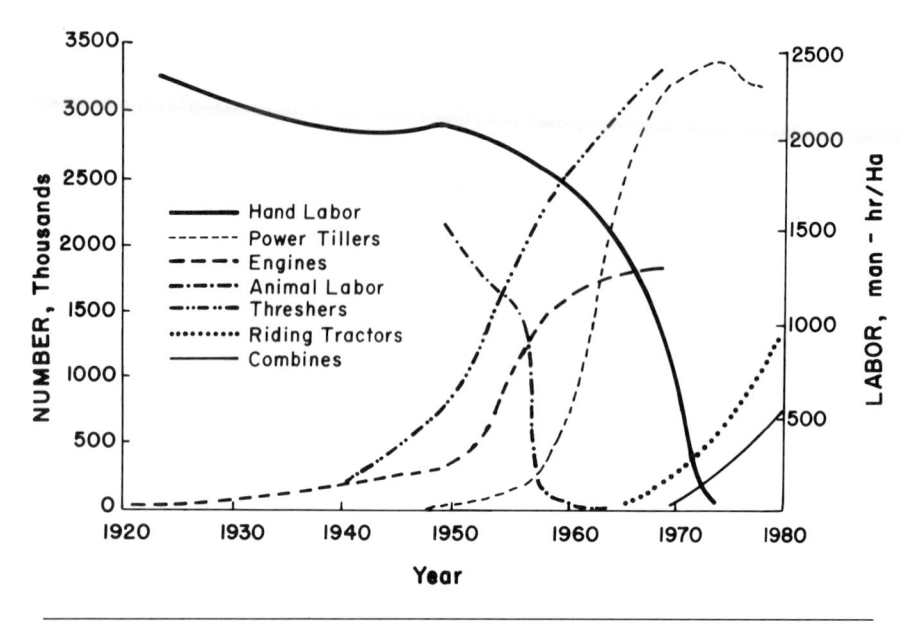

Figure 10.7 Labor requirements and farm machines for rice production in Japan 1920–1980 (Kitani, 1985).

In large-scale mechanization many inputs such as fuel, fertilizer, and seeds come from outside of the farming system. In developed countries there is a large infrastructure known as agribusiness that includes the production, flow, and utilization of these many inputs external to the farm. Large-scale mechanization involves creating large parcels of land, eliminating ditches and fencerows, and providing paved roads.

D. Labor Resources

A shortage of labor due to the demands of industrialization will pull labor away from the farms, justifying mechanization (including electrification) of many tasks. The degree of mechanization needs to be commensurate with the farming activity, size of land area, crop and animal production, and economy. Mechanization may permit adding livestock enterprises to crop production, particularly for feed grains and forages. Mechanization could stimulate the development of the farm machinery and tractor industry which will require and utilize some of the labor released. Mechanization may free workers for agricultural, processing, and distribution industries. Although the number of farms in the United States has decreased dramatically, the number of people involved in the agriculture enterprise, including transportation, processing, and distribution, has remained approximately constant at about 30 million people over the past 40 years.

Farm workers are often pushed out of their jobs or replaced as a result of mechanization. Without appropriate skills these people may find it difficult to find new jobs. So it is incumbent that these workers take advantage of local, state, and federal programs and industry opportunities to maintain and develop new skills needed in the economy.

Generally, labor costs increase more rapidly than the value of agricultural products. The farmer has three major options with which to respond: mechanize, intensify, and specialize (Knorr and Watkins, 1984). As American farmers compete in the international markets for the sale of many products, mechanization provides a means of competing with products from other parts of the world, particularly those areas where hourly labor costs are one-tenth to one-fourth those in the United States. The production of rice and wheat in the United States can compete with many parts of the world because these crops can be raised under optimal conditions with the use of mechanization. A classic case with respect to labor and mechanization is the development of the cotton picker during and following World War II. The cotton picker was developed and utilized to fill the need for timely harvest because workers were not available. On the other hand, considerable controversy has arisen with respect to the development of the tomato harvester. The controversy revolves around the subject of potential displacement of organized workers, both a social and economic issue, as a result of mechanization of harvest. In general, mechanization of farming will develop parallel to industrialization of a country. To do otherwise would require the country to support an entirely different culture and economy for the farm labor force, an almost impossible task in a democratic society.

E. Other Characteristics of Mechanization

Some aspects of mechanization are considered as an advantage by some while being a disadvantage to others. Several of these characteristics are listed below.

1. Mechanization leads to more specialization and less diversification.

2. The wastes of combustion from the tractor, engines, or power plants pollute the atmosphere.

3. Mechanization by itself may have little effect on yield unless other changes, such as selecting crop variety, providing fertilizer, and controlling moisture are made.

4. Mechanization is related to manpower and energy inputs. The farm output per unit of energy input has reached a plateau.

5. The costs of maintenance, parts, and storage for machinery are often not included in economic considerations.

6. Whether increased particulates due to dust and dirt in the atmosphere as a result of stirring the soil can be attributable to mechanization per se is questionable. Increased particulates could cause a cooling of the atmosphere. However, the time of residence of soil particulates in the atmosphere is considered short-lived. For land without cover in a dry climate with wind, more dust and fine particles would be added to the atmosphere. For particulates, the cultural practice rather than the degree of mechanization could be considered as a major controlling factor. With more timely operations mechanization could help minimize the particulate load in the atmosphere.

7. With deep tillage large amounts of power are needed that are almost impossible to obtain from draft animals.

8. Methods of plowing and tilling the soil are being modified through no-till or minimum tillage practices to reduce the tractor-fuel required and minimize the disturbance of the top soil to reduce runoff and erosion from wind and water.

IV. SUMMARY

In its broadest sense mechanization of agriculture reflects the industrial development in which power and machines are used to provide more production in a timely manner on the land. Some of the advantages of mechanization include reducing drudgery and decreasing heavy work in a potentially contaminated environment. Some workers are negatively impacted by loss of jobs and thereby their livelihood as a result of mechanization. These displaced workers, if properly trained, can often find productive work in agribusiness but usually have to relocate to take advantage of those job opportunities. With improper use of mechanization, soil can be compacted, products damaged, soil and water lost, and erosion increased. A major impact from tractorization and electrification is the shift from dependence on energy from plants and animals, a part of the agricultural system, to energy from fossil fuels for engines and generators, providing a dependence on a non-renewable resource external to the agricultural system.

REFERENCES

Chou, M., Harmon, D. P., Jr., Kahn, H., and Wittwer, S. H. (1977). "World Food Prospects and Agricultural Potential." Praeger, New York.
Esmay, M. L., and Hall, C. W. (1973). "Agricultural Mechanization in Developing Countries." Shin-Norinsha Co., Tokyo.
FAO (1965). "Report of the World Food Congress," Washington, D.C., 1963, Vol. II. Rome.

Kitani, O. (1985). "Small-Scale Mechanization in Asia," ASAE Pap. No. 85-5048. Michigan State Univ., East Lansing.
Knorr, D., and Watkins, T. R. (1984). "Alternatives in Food Production." Van Nostrand Reinhold, New York.
NAS (1974). "World Hunger—Approaches to Engineering Actions," Report of a Seminar. Washington, D.C.
Pimentel, D., and Pimentel, M. (1979). "Food, Energy and Society." Arnold, London.
Steinhart, J. S., and Steinhart C. E. (1975). Energy use in the food system. In "Food: Politics, Economics, Nutrition, and Research" (P. H. Abelson, ed.), pp. 33–43. Am Assoc. Adv. Sci., Washington, D.C.
USDC (1970, 1986). "Statistical Abstracts." U.S. Gov. Print. Off., Washington, D.C.

11

Population, Food, and the Economy of Nations

William J. Hudson
The Andersons Management Corp.
Maumee, Ohio

I. Is Population Limited by Food?
II. Is Food Driven by Population or by the Economy of Nations?
III. Can the Worst Fears of Environmentalists Be Substantiated?
IV. Vision 2020
    References

I. IS POPULATION LIMITED BY FOOD?

From the 15th through the 18th centuries world population doubled; however, within that period, from decade to decade, population rose and fell like a series of tides—depending on the prevalence of epidemics and plagues, and on the adequacy of food supplies, which were determined almost solely by variations in climate. However, with the beginning of the Industrial Revolution, this rhythm changed: Now for nearly two centuries, population growth has registered a continuous, steady rise—more or less rapid according to different observers but always (Braudel, 1985).

Mathematical analysis of population trends shows that present rates of growth will lead to a doubling of world population every 30–40 years. Most observers intuitively feel that such growth will not be sustainable, that it will run into natural limits, and that the precursors of such limits are already apparent in pollution, erosion, and other environmental worries. Perhaps the most influential book to advocate this view was *The Limits to Growth,* in 1972; another was *Global 2000,* in 1980. These books have been read by millions of people. The two books to take the other side, *The Ultimate Resource* (Julian Simon, 1981) and *The Resourceful Earth* (Julian Simon and Herman Kahn, 1984) are perhaps more carefully argued and documented, but certainly not nearly so well circulated. Human intuition favors limits. Popular economics is based on scarcity, not abundance. Pessimists far outnumber optimists in any sample of humanity. Plenty of scientific studies have supported the majority; but one has to wonder what the research results might be if an even distribution of predispositions could somehow have been implanted in the investigators.

In the present chapter, let us consider a middle course. Let us concede and discuss some limits, and then try our imagination ("the ultimate resource") on what might be done, or what might happen in the economy of nations, to improve the global food system ahead of its supposed decline.

Dr. Melvin Calvin (1986) of the University of California at Berkeley has made an interesting calculation on the maximum amount of food that could be produced on earth, and the maximum population which that would support, 22×10^9 persons. Dr. Calvin began with an assumption of total solar energy reaching the earth (5.5×10^{16} kcal/day) multiplied that by the fraction of the earth's surface which is arable land (20%), multiplied that by an assumed photosynthetic efficiency for crops (0.1%), and then multiplied that by an upper limit of food production efficiency (50%, vs. 10% at present). He arrived at a total amount of solar energy available for food production, 5.5×10^{13} kcal/day. He then divided this available energy by 2,500 kcal/day, a common measure of the average food energy intake required by human beings in "normal" activities.

Another approach to this "limit" of 22 billion is shown in Table 11.1, calculated by the author based on USDA world grain production data.

11. Population, Food, and the Economy of Nations

Table 11.1 Population Capacity of Earth, Estimated by "Maximum" Grain Area and Yield[a,b]

		Increase yield		Add emergency areas
Area	As of 1986	To U.S. level	With biotechnology	
United States				
Area (million ha)	67	67	67	100
Yield (Mt/ha)	4.7	4.7	9.4	9.4
Production (Mt)	314	314	628	936
Population (millions)	241	241	241	241
Lb/capita	2,873	2,873	5,747	8,564
Foreign				
Area (million ha)	643	643	497	600
Yield (Mt/ha)	2.1	4.7	9.4	9.4
Production (Mt)	1,369	3,008	4,652	5,616
Population (millions)	4,719	4,719	4,719	4,719
Lb/capita	639	1,405	2,174	2,624
World				
Area (million ha)	710	710	564	700
Yield (Mt/ha)	2.4	4.7	9.4	9.4
Production (Mt)	1,683	3,322	5,280	6,552
Population (millions)	4,960	4,960	4,960	4,960
Lb/capita	748	1,477	2,347	2,913
Potential sustainable world population @ 748 lb/capita (millions)	—	9,793	15,565	19,315

[a]Source: USDA (1987).
[b]Total grain production is sum of wheat, coarse grains, and rice.

World grain production in 1986 was 1,683 million metric tons (Mt), and population was 4,960 million. Pounds per capita of grain production averaged 748. If all countries could learn to produce grain yields similar to those of the United States, then a population of 9.8 billion could be supported at similar diets to today. If world yields were doubled from the U.S. levels, through biotechnology discoveries, then a population of 16 billion could be supported. It should be noted that crop yields which are *quadruple* the national averages are already obtained routinely in test and contest plots, so the assumption of doubling is in one sense conservative. Finally, if all semi-arable land were pressed into emergency service and if yields could be kept high, then a population of 20 billion could still have an average of 748 pounds of grain each. This agrees roughly with the solar energy calculations made by Calvin.

So by two approaches we arrive at a figure in the range of 20–25 billion as a limit or ceiling—about five times today's population. The approaches are reasonably straightforward. They both assume that energy and other resources used in the production of food will be adequate, at least in the purely quantitative sense. In Calvin's method, the assumption is that the process of plant biology will capture solar radiation and convert it to nutritious biomass at rates in advance of today's average technologies, but not in advance of present experimental capability. In my own method, the assumption is two-fold: (1) that fossil fuels and fertilizers are adequate to continue present production trends for at least a decade or two, and (2) that advances in biotechnology and plant science will replace physical resources with knowledge, without loss of yield. Such advances are already well under way in the laboratory, and have begun movement to the field.

Neither calculation of population limit includes any outright "science fiction"—e.g., no artificial space colonies, no importation of nitrogen from a moon of Saturn, no entirely new extraplanetary sources of energy, no breakthroughs in human metabolism (to be able to live on fewer pounds or calories), no "easy" conversion of non-photosynthetic energy to food via massive hydroponics.

The purpose of the two calculations is not to quantify precisely the world's remaining resources—fossil fuels, fertilizer, water, and so on. The calculations are meant to concede that a limit to population may be likely—one which may indeed be imposed eventually by resources—and to estimate in a very rough way what the magnitude might be. This in turn may set the stage for less emotional discussion of how best to conduct ourselves in the middle of the two extremes of human outcome—imminent decline and eternal rapid growth.

Suppose then that we have room for about 20–25 billion people, and that today there some five billion in the "room." Is this about how crowded the room feels? Subjectively, we might acknowledge that a few people in the room already feel very crowded and are concerned about the prospect of the room's filling up further—but they can't really get the attention of others. Perhaps it is simply too soon. No decline in food availability has begun. In fact, and quite objectively, just the opposite is true. According to the USDA (1987):

	World population (millions)	World grain usage (Mt)	World grain usage (lb/capita
1960	3,063	832	599
1986	4,960	1,646	731

Thus, if official data are correct, average world grain usage per person has increased from 599 lb/yr in 1960 to 731 lb/yr in 1986—an improvement of nearly one-fourth and a trend which shows no early signs of stopping.

Furthermore, and still objectively, the principal food problem in the

United States, Western Europe, Canada, Australia, Argentina, and a few other countries is too much food rather than too little. As we will discuss later, the United States pays its farmers not to grow grain on some 20–40 million acres per year, which at typical yields would produce some three billion bushels of additional, unsaleable grains—enough to add three-fifths of a bushel to world per capita consumption, or about 35 lb/yr to the above average of 731 lb/yr. Alternatively, this three billion bushels of "full-cost non-production" would provide (at 731 lb each) the complete annual requirements of 250 million people—a figure much greater than anyone's estimate of the acutely starving and malnourished people in the entire world.

This contradiction, of heart-breaking hunger amidst vast unwanted food surplus, staggers a few people of the five billion in the not-yet-crowded "room," but these observers seem powerless to act. The predominant opinion remains that of perpetual scarcity. Specific cases of hunger are viewed as evidence of general scarcity. Cases of surplus are seen as separate—as exceptions—and bound not to last long. Few people indeed match the current shortage with the current surplus, and wonder why a solution can't be had. The majority seem to be saying that we're all aboard the Titanic, and that there's no point in trying to make all the cabins comfortable because the ship will ultimately sink. But what if the iceberg is ten years away, or twenty, or more? Would we not be well advised to make maximum, wise use of our resources now—in the hope that passenger comfort (perhaps even peace) will aid in finding ways to avoid impact? Is the only way to avoid impact to return to port (cancel all human growth)? Is it in fact possible to return to port?

II. IS FOOD DRIVEN BY POPULATION OR BY THE ECONOMY OF NATIONS?

At one level, this question is completely rhetorical, for it is certainly true that the final demand for food is by people; in other words, it is certainly the world's population which consumes its food. At another level, however, it is very important to distinguish between the pressure put upon food supplies by population growth itself, versus the pressure from the *economic* capability and growth of that population.

Figure 11.1 shows the smooth and continuous growth of population in contrast with the upward but volatile curve of world grain trade. It can be seen from this graph, prima facie, that annual variations in population growth have virtually nothing to do with annual variations in world grain trade. Instead, grain trade depends more immediately on such variables as economic growth (see Figure 11.2, which give world GNP per capita)

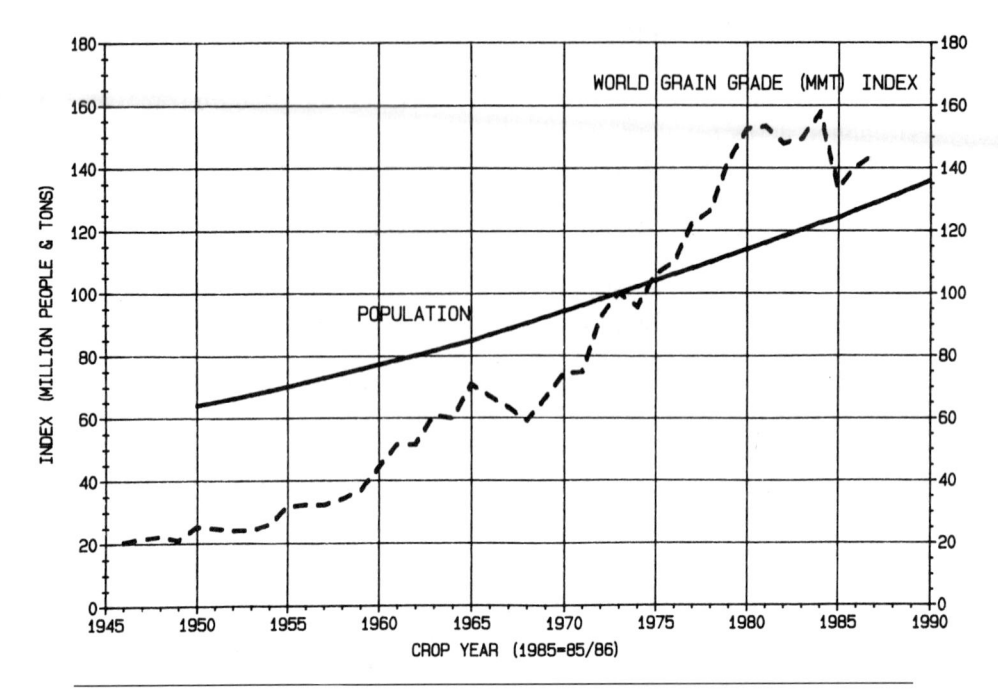

Figure 11.1 Population versus world grain trade. Population data from World Indices (1986). World grain trade index from *World Development* (1987), adjusted to 100 in 1973.

and the creation of money and credit (see Figure 11.3, which shows the parallel between world trade of all goods, including food, and the creation of Eurocredit).

It can also be seen from Figure 11.1 that grain trade, in the most recent four decades, has been growing at a steeper slope than population itself. What this means is that the desire to improve diets by the existing population is more important (to trade) than the amount of new food required for each year's additional 75 million people. This can also be seen in straightforward calculation: If 75 million additional people each consume 773 lb/yr of grains, then the total required to support population growth per se is 25 Mt. World trade of grains has been about 250 Mt/yr in recent years, and thus the requirement by population growth itself would only be 10% of trade, *even if the entire diet of the 75 million were supplied by imports*. This would not be the case, because most food—both for new and existing people—is produced and consumed domestically, and not imported from world markets. The demand for over 90% of the 250 Mt of grain in international trade is being driven by the economic ability of people to make marginal improvements in their diets.

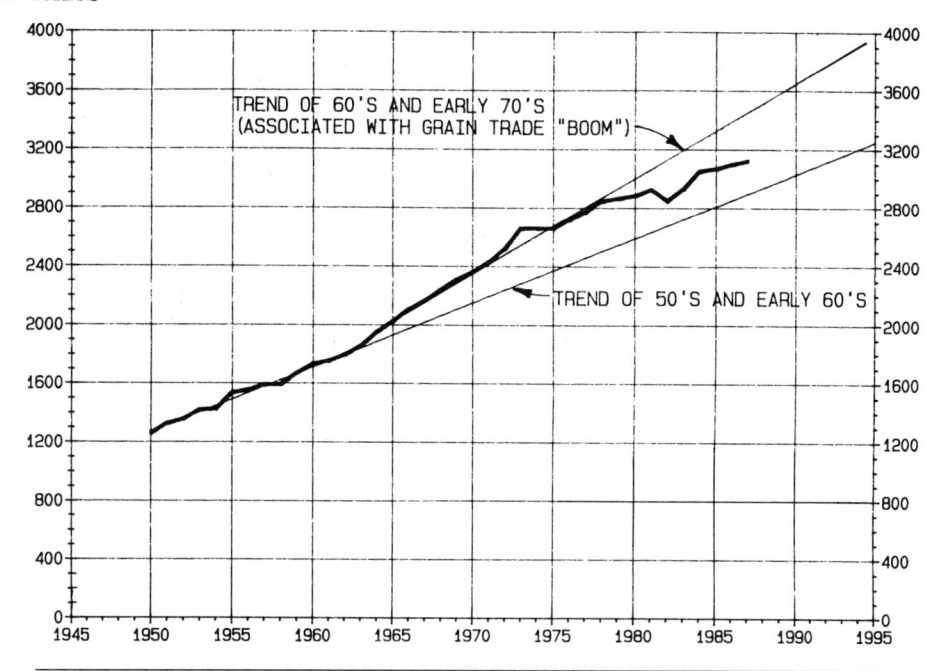

11. Population, Food, and the Economy of Nations   281

Figure 11.2 Per capita world gross national product (heavy line), with trend of early 1960s and 1970s (upper limit) and trend of 1950s and early 1960s (lower limit). World GNP data from Calvin (1986) and Insel (1987), adjusted to 1986 dollars with U.S. GNP export deflator from *Economic Indicators,* (1987) and *Economic Report* (1987). Population data from *World Indices* (1986) year per-capita figures.

As mentioned earlier, food consumption per capita, world-wide, is in a rising trend. Figure 11.4 shows this using a USDA series of food production (or consumption) in dollar terms, adjusted for inflation. The increase has been from about $200/person in 1955 to about $235/person in 1985.

In the meantime, real food prices per unit of volume or weight have declined. Note Figure 11.5, which charts the average price of wheat, coarse grains, and soybeans in trade since 1950—both in actual and deflated dollars. Figure 11.6 shows the downward trend of this real price of grains.

If the deviations in annual food production and the deviations in traded food price are both stated in terms of percentages and plotted together, the result is Figure 11.7. It becomes clear that high price deviations (or low) are not the result of equally large declines (or rises) in food production. In other words, the principal and popular index for food "scare," namely

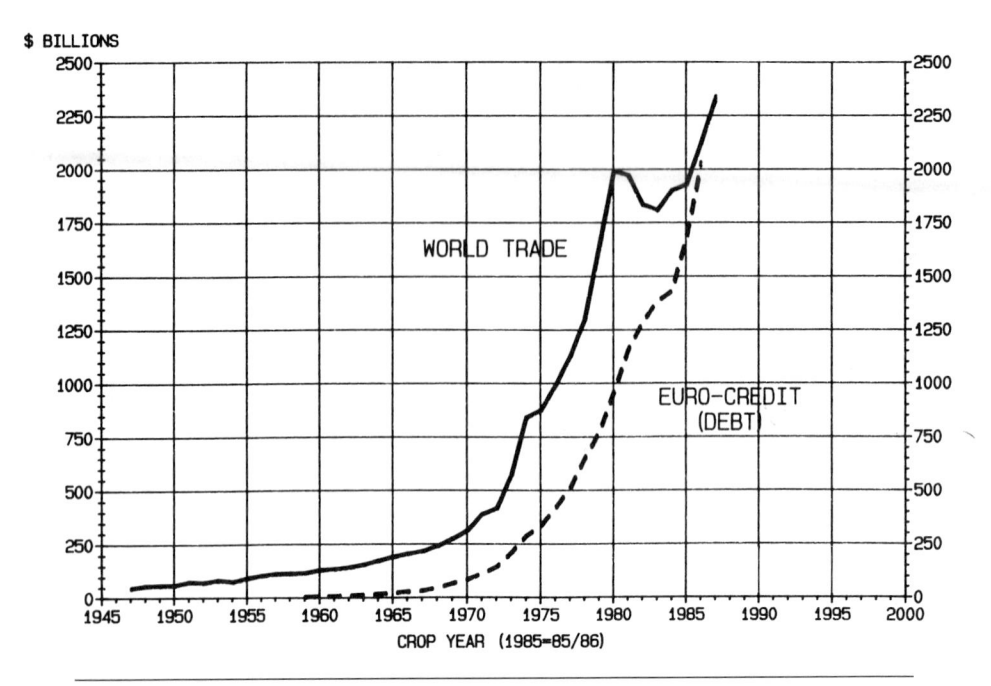

Figure 11.3 World trade (solid line) versus Eurocredit (debt) (broken line). World trade data from Insel (1987). Eurocredit data from *World Population* (1986).

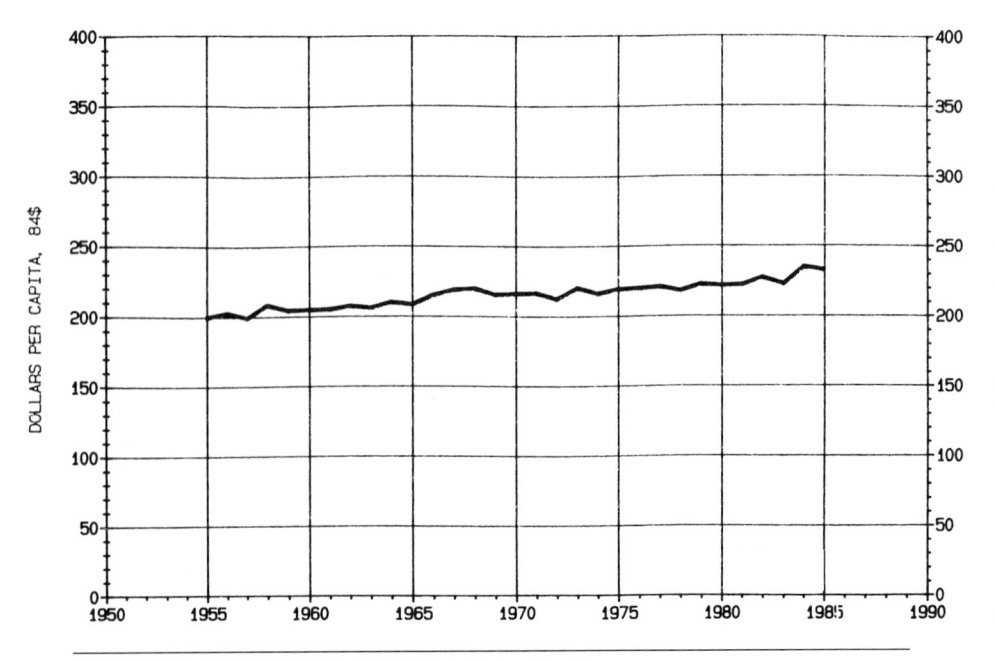

Figure 11.4 Per capita world food production (*World Agriculture*, 1987).

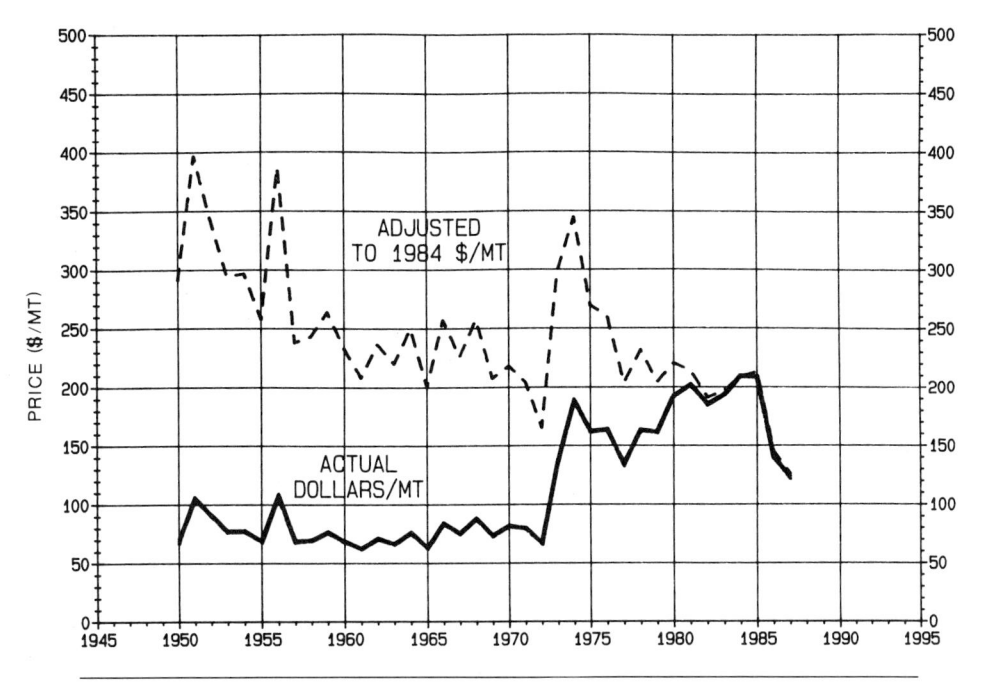

Figure 11.5 Actual grain prices (solid line) versus prices adjusted to 1984 U.S. dollars (broken line) for wheat, coarse grains, and soybeans in trade. Actual prices from Thompson (1987), *World Development* (1987), and *World Financial* (1987), adjusted by tonnages in trade. Adjusted to 1984 U.S. dollars by U.S. GNP export deflator in *Economic Report* (1987).

price, does not parallel changes in food production on a world scale. Price scares concern groups of affluent purchasers, affected certainly by regional harvests, but more affected by global economic and political factors (for example, the OPEC oil price increase together with the recycling of oil credit and the subsequent monetary inflation).

Figure 11.8 plots world food production per capita versus GNP per capita, in common dollar units. Clearly, food production has not maintained pace as a share of man's economic activity in the past thirty years. The percent of world GNP occupied by food production has dropped from 13% in the late 1950s to about 7% at the present time (Figure 11.9).

The data in this section are arguing as follows:

1. Population growth has little direct effect, in any given year, on food trade.
2. Food trade is affected, over the long run, by the desire and the economic ability of peoples to improve their diets. Food trade, in any given year, is strongly affected by economic growth and

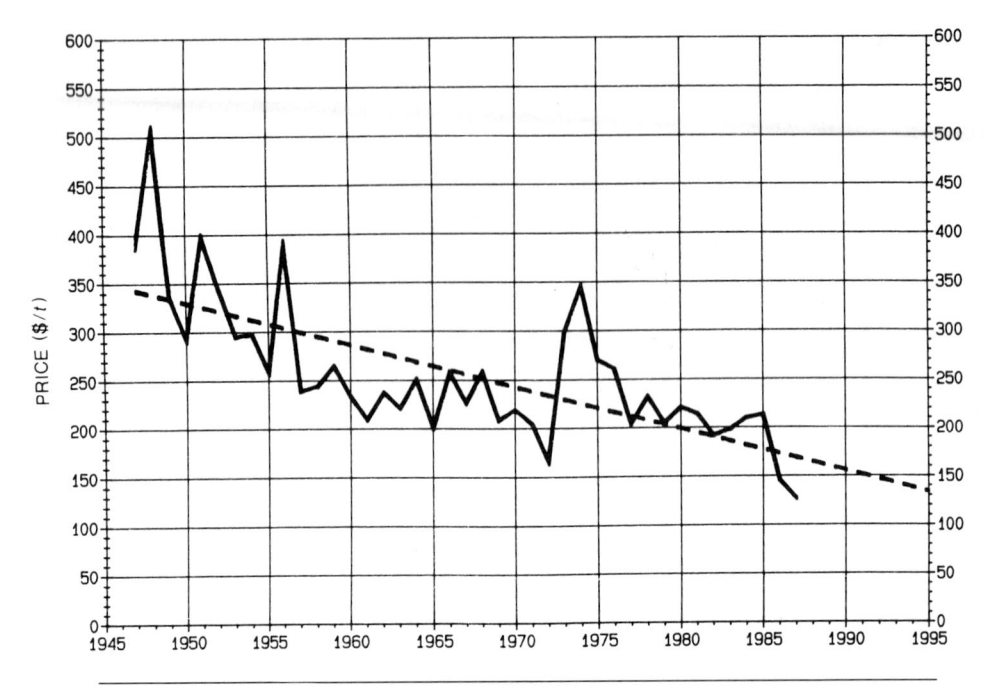

Figure 11.6 Trend for real prices for wheat, coarse grains, and soybeans in trade in 1984 U.S. dollars. Data sources same as Figure 11.5.

sometimes dramatically by financial conditions and political events—especially those concerning the most affluent nations.
3. Food price scares are unreliable guides to how many people the earth will accommodate. Real food prices, as judged by bushels being traded, are steadily declining.
4. Food production, in global aggregate, is becoming less and less a share of man's total economic activities—i.e., it is getting easier to produce food, even though population is growing and even though the average consumption per person is steadily rising.

III. CAN THE WORST FEARS OF ENVIRONMENTALISTS BE SUBSTANTIATED?

Whether or not population is limited by potential food supply, and whether or not that population is on the order of 20–25 billion people, the world in total is currently in a period of improved ease at securing its food. The

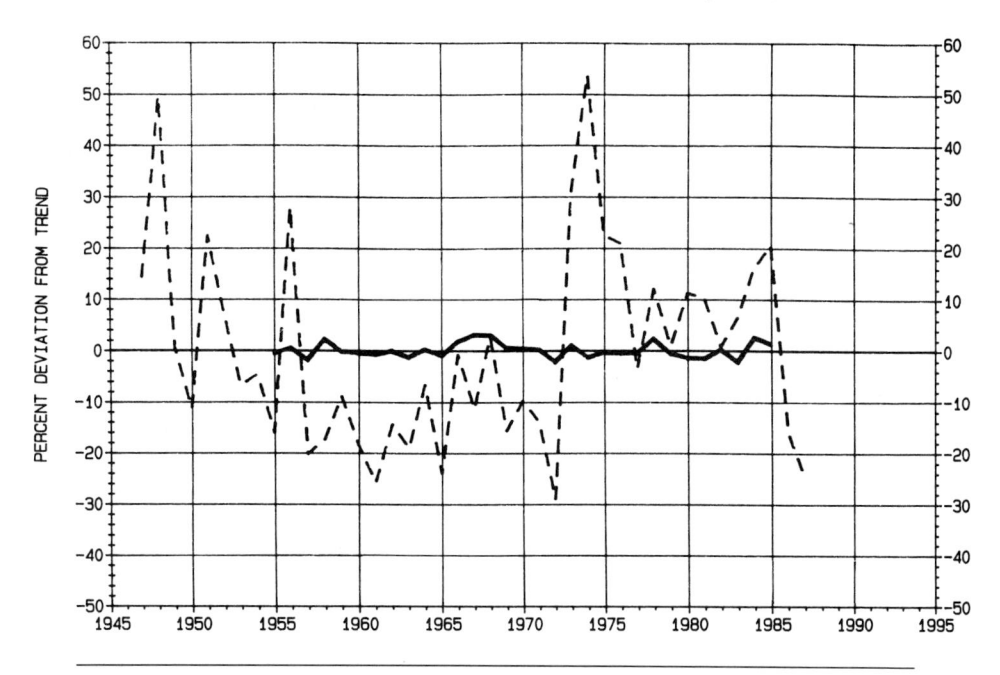

11. Population, Food, and the Economy of Nations 285

Figure 11.7 Deviation in world per capita food production (solid line) versus deviation in traded grain prices in 1984 U.S. dollars (broken line). Data sources same as in Figures 11.4 and 11.5. Percentage is actual minus (linear) trend divided by (linear) trend.

situation is summed up by the title of an article in *Foreign Affairs:* "A World Awash in Grain" (Insel, 1987).

Several questions come to mind simultaneously. Is the surplus being caused primarily by governmental distortion of food production incentives? Do we have enough energy and other resources to maintain this production? And, is this condition of ease and surplus coming at the cost of irrevocable damage to the environment?

Let us here recall that we have conceded a limit to population, on the order of 20–25 billion people, associated with such basic resource constraints as sunlight and photosynthesis. What we want to address is the potential for comfort on the route between the present five and the hypothetical 20–25 billion people. What actions by man and his technology, commerce, and governments might improve our quality of life between now and later, and produce a global culture in which all of us take pride in sustaining?

Earlier in this chapter we argued against the simple syllogism that "population growth causes food shortages." What we found was that there

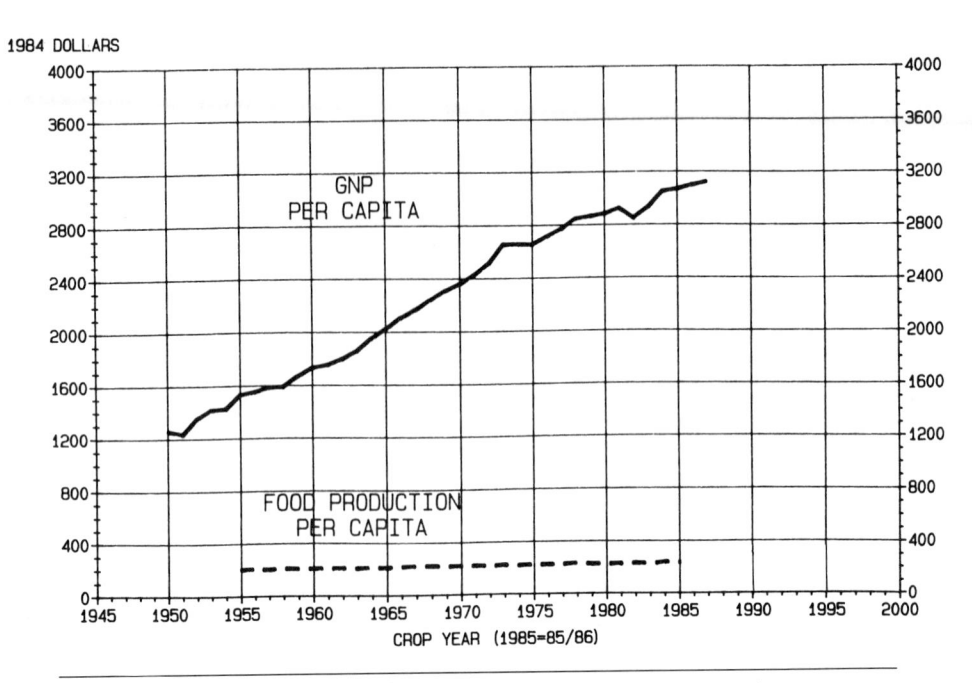

Figure 11.8 Per capita world gross national product (solid line) and per capita world food production (broken line). Data sources same as in Figures 11.2 and 11.4.

is no present shortage, globally, and that economics and politics are more directly associated with food price, production, and trade—in the near term—than population growth per se. Let us take the same approach to breaking unjustified "logic" in the area of environmentalism. Let's keep our eye not so much on the sheer presence of such problems as erosion, but on their historical context and whether or not change is occurring.

Perhaps the leading voice of environmental concern is the Worldwatch Institute of Washington, D.C. In its report, *State of the World 1987*, the Institute said that "human use of the air, water, land, forests and other systems that support life on earth were pushing those systems over 'thresholds' beyond which they cannot absorb such use without permanent change and damage. The result has been declining food and fuel production in many parts of the world and, for the world as a whole, contamination of the atmosphere, climatic change, a mass extinction of plant and animal species and the long-term prospect of a decline in the quality of life" (cited in Shabecoff, 1987).

This conclusion seems to have great emotional appeal to the majority of people. It is an update on the voyage of the Titanic that they are already

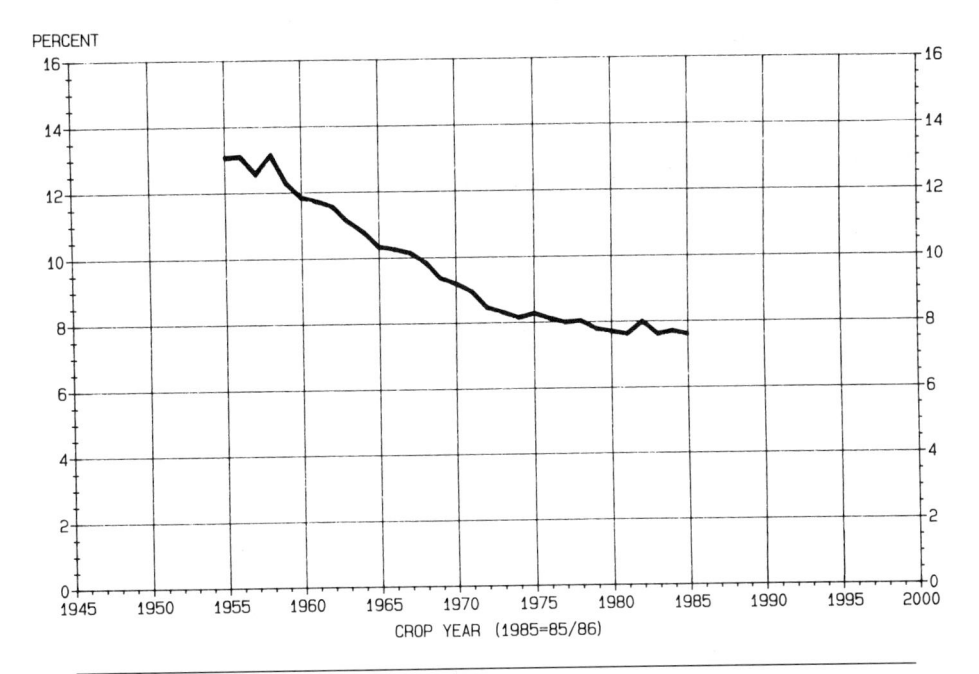

11. Population, Food, and the Economy of Nations   287

Figure 11.9 Per capita world food production as a percentage of per capita world gross national product (both in 1984 U.S. dollars). Data sources same as in Figures 11.2 and 11.4.

convinced we are on. Few of them study the report and weigh its evidence. It is assumed to be reputable and "scientific," and some may even assume that a multitude of careful observers around the world have sent reams of objective data to the Institute.

In actuality, the Institute employs a handful of researchers and relies, as all of us must, upon the publication of data by the governments and other principal public agencies of the world. The Institute's function is to interpret in the way it chooses the same data that all the rest of us are also interpreting. There is no way from this, it can be no different—the reader is the final judge.

Suppose you believe that the data presented so far in this chapter have indeed been taken accurately from the original government reports and assembled here in good order. If so, the conclusion is clear and simple: world food production is *growing*, not declining. Note that the Worldwatch Institute says that food production is *declining* "in many parts of the world." This too is certainly true, particularly for sub-Saharan Africa. But what of the whole? Why doesn't the Institute acknowledge that the world total is growing, and that some areas have massive surpluses which could easily make up for the declines elsewhere? By omission, the Institute tells you its bias—that the majority prefers to believe in decline, and that evidence of decline in a few places is as good as decline in general, because apparently the majority also believes in the concept that all regions must be strictly self-sufficient. Furthermore, the Institute seems to say that if *parts* are in decline, they must not be blamed directly, but rather we must blame the activities of the whole. Might not the *parts* have political systems or other indigenous inadequacies which do not indict the capacity of the entire earth?

The worst fears of environmentalists concerning food (especially food produced with "energy-intensive" methods) and population are:

1. Uncontrolled erosion of cropland
2. Pollution of water and air by agri-chemicals
3. Loss of biological species needed for future
4. Alteration of climate because of deforestation fossil fuel consumption, and ozone depletion

Cases demonstrating erosion, pollution, and loss of species undeniably exist. In fact, there are many hundreds and thousands of such cases. But, in general, globally, what is the overall impact and what is the trend? Are these problems getting better or worse—and and by what means can you tell? The truth is that:

1. Cases are difficult to quantify and group together for mathematical treatment;
2. Standardized data are reported for only a handful of countries, and principally just the United States; and thus
3. Global conclusions about environment, and the progress or lack thereof, are heavily, if not primarily anecdotal interpretations.

Today's sophisticated reader is not very patient with the assertion that we lack data; he or she is used to "instant coverage." But coverage which is "instant" is that which is journalistic or anecdotal and should not be confused with that which is scientific.

The concern about climate change is not quite the same as erosion, pollution, and species loss; these latter three we can see with our own eyes, if we visit the scene of a specific case. But who can see a change in (global) climate? The interesting thing about the concern with climate is that reputable scientists are coming forth with correlations of weather data and declaring alarm; how can this be other than amusing to us, when the same "scientific" methods of weather prediction cannot predict rain or shine better than 50–50 next week, or even day after tomorrow? The only objective thing to do with climate change, at this stage of our ability, is to label it an aspect of Fate (with a capital "F"). Today's reader does not like this either—he or she prefers that all questions have answers. He

11. Population, Food, and the Economy of Nations   289

or she further prefers that the answers be "scientific" and clear. But climate has simply not yet yielded to science, nor does it appear the least bit ready.

The recent "history of climate," as captured by crop yields, is given in Figure 11.10. It is seen that world grain yield is steadily upward, with small fluctuations. The grain yield of an individual country, such as the United States, may fluctuate dramatically, as shown, but not the world in total. The figure is comforting, but only on the short scale of time shown. We do not know what may be happening on the scale of centuries or millenia.

Anecdotal evidence of erosion, especially in underdeveloped countries, is, I will concede, alarming. But is the news from the erosion front all bad? Can any *trend* be established?

The United States government, virtually alone among nations, has conducted some five major surveys of erosion in this country—in 1934, 1958, 1967, 1977, and 1982. The surveys are difficult to compare, because with advancing knowledge, different items were chosen for measurement and different language was used to describe the results. Table 11.2 is my attempt to interpret this very important time-series of disparate technical information. It would appear that, at a very general level, erosion in the United States has gotten better, not worse. Not even the most worried

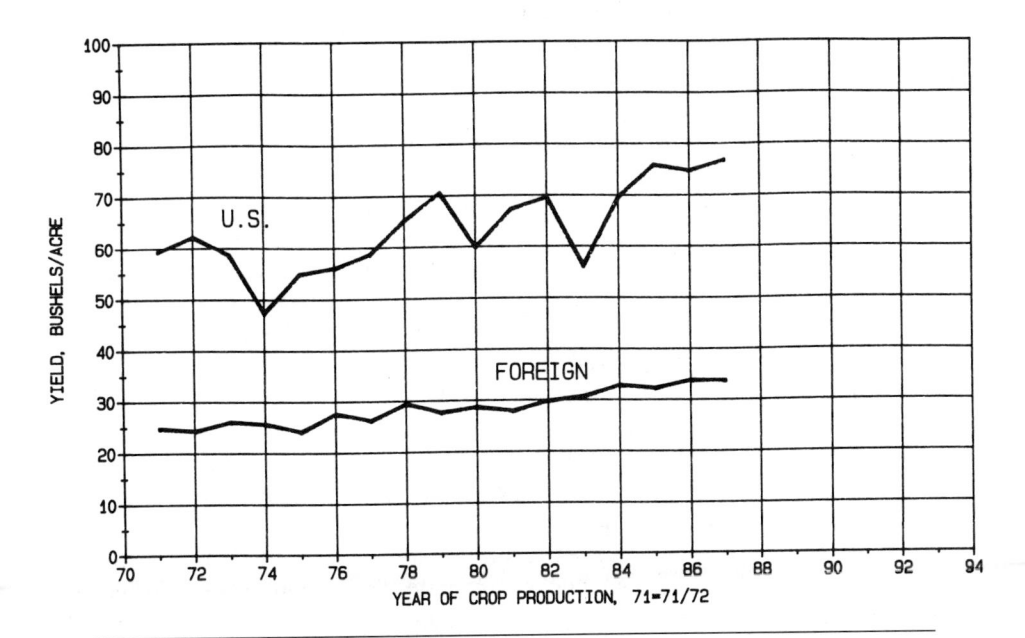

Figure 11.10 Foreign grain yield versus U. S. grain yield (wheat plus coarse grains). [From *World Development* (1987).]

Table 11.2 Comparison of U.S. Cropland Erosion Studies

Date and source	Finding	%
1934, USDA[a]	Slight erosion	56
	Moderate erosion	32
	Severe erosion (essentially destroyed for tillage)	12
1958, USDA	Conservation treatment adequate	36
	Conservation treatment needed	64
1967, USDA	Conservation treatment adequate	36
	Conservation treatment needed	64
1977, USDA	Erosion <5 t/acre/yr	66
	Erosion 5–13.9 t/acre/yr	23
	Erosion >14 t/acre/yr	12
1982, USDA[b]	Erosion <5 t/acre/yr	66
	Erosion >5 t/acre/yr	34
1987, estimate[c]	Erosion <5 t/acre/yr	75
	Erosion >5 t/acre/yr	25

[a] Covered only east and west regions of north-central United States.

[b] Includes new measurement for "erosion tolerance" beyond 5 t/acre/yr. Some cropland is tolerant of more, some of less than 5 t.

[c] Estimated by adding 20 million acres set aside for conservation use and 10 million acres in long-term conservation reserve to the category "<5 t/acre/yr."

of environmentalists suggests that this *data* shows a worsening; environmentalists need only assert that erosion exists, which it does, and the majority of listeners, by reflex, assume that it is very bad and getting worse.

But what if the facts suggest that the trend in erosion is getting better? What if, with government encouragement, including a 40-million-acre Conservation Reserve Program in the 1985 Farm Bill, the American farmer is making slow but general progress against erosion? The farmer is certainly being pushed in this direction, gently, by the ironic fact that higher yields in case cash grain crops, especially corn, are associated with such practices as "no-till" and with such "conservation tillage" techniques as chisel plow incorporation of fall residue.

If the evidence indicates that the world's most productive agricultural country, the United States, is not achieving its abundance through an indiscrimate, high-energy "rape" of the land, would we not assume (in the absence of foreign data) that other countries, perhaps now proceeding unwisely, will eventually follow the same path?

One or two final generalities about erosion—in the interest of seeing the plainest of all connections between climate, crops, and erosion. Where it rains, we grow crops. Where it rains, there are rivers to return water to the sea. Where it rains, there is also erosion. A rill is a small river. It

11. Population, Food, and the Economy of Nations 291

is normal. Not even a saint could produce crops with zero erosion. It is not an absolute question nor a moral question.

The "tolerable" amount of erosion cannot be calculated, especially not in terms of the "natural" process of soil formation—for the soil will *never* be turned backed over to natural processes. The soil will always be tended, and repaired (if need be), by human processes. The repair will require knowledge, technology, and energy—and without predicting the future success of technology and the future price of energy, we cannot predict the ease of repairing soil, nor thus the tolerable erosion today. Neither side of this double-edged prediction can be reliably made.

The continental United States receives about 2,300 km^3/yr of rain (*Global 2000*, 1980). About one-third of this falls on that part of the Mississippi River drainage basin called the "cornbelt." About half of this falls during the crop production season. So about 400 km^3 of rain are involved in the production of a corn crop, let's say of recent average size, of 8 billion bushels. The energy required for the sun to evaporate a cubic kilometer of water from the ocean is calculated by scientists at 2.3 quadrillion BTU (Quads) (Foley, 1981). Let's say that atmospheric circulation, to deliver the water from sea to land, is "free." Then the 8 billion bushels have required about 400 km^3 × 2.3 Quad/km = 920 Quads. This is an average input from climate of 100,000,000 BTU/bushel. Pimentel and Pimentel (1979) and others have estimated the farmer's input energy (including fossil fuel, fertilizers, pesticides, labor, etc.) at about 100,000 BTU/bushel.

Thus the "climate lever," or the role of nature in modern crop production, is 1,000-to-1 over man. It so happens that this enormous lever does two things: it produces food, and it produces erosion. As the saying goes, "there are no effects without side effects." But in terms of food production, the investment of 1 BTU of fossil fuel at the tip of a 1,000-to-1 lever looks like a very good risk, a good business to be in. What we see today, in the global surplus of food, is the constant progress of technology at the tip of this mighty, natural lever. What we see also today, in the widespread concern about erosion, is that the natural lever may so magnify the "side-effect" as to imperil the effect. It is a worthwhile worry, but apparently (though still arguably) the trend is more positive than negative.

An objective appraisal of the environmental impact of agrichemicals runs into the same problems as erosion—lack of data and a surplus of anecdotal evidence. With agrichemicals the data problem is much worse, because even in the United States, widespread pesticide usage is a recent phenomenon and effects have not been extensively surveyed through time. But the case-by-case evidence of pesticide damage tends to be much more acute and sensational: Thousands of people were killed in the 1984

Bhopal, India, pesticide gas release, and the entire Rhine River seemed to be threatened by the Swiss chemical spill of 1986.

These serious industrial accidents overshadow both the trend toward safer, more biodegradable chemicals and the many positive results of pesticides for more comfortable living. In the United States, the Environmental Protection Agency has caused the removal from the market of the most hazardous pesticides and has established a market incentive for chemicals which are (1) more target specific, (2) more efficacious with much less volume of active ingredient, and (3) readily biodegradable. Farming methods are also evolving toward application of active materials (pesticides and fertilizers both) on the areas immediately contiguous to plant roots rather than broadcast to the entire field. Furthermore, interest in "regenerative farming," which might be called the substitution of wise management practices for unthinking abundance of chemicals, has come for real to the American cornbelt—as farmers seek *profitable* volume rather than volume itself.

The progress reported earlier in food production per capita could hardly be happening without agrichemicals. Fertilizer and pesticides are both crucial to current yields. Biotechnology may provide an alternative way to reach crop genetic potentials, but the science is still young.

To stress the good results of pesticides and to highlight the fact that good news loses out to bad, let me mention the success recently reported by the World Health Organization over river blindness (onchocerciasis) in Burkino Faso (Brooke, 1987). For centuries, farmers in this west African country avoided fertile lowlands because of black flies which gave them "oncho" in 20% of the male population. New larvacides (pesticides), approved by independent ecologists, have been developed and applied in a $200 million program over the past five years. Five hundred thousand square miles are under treatment, and now 1,000 square miles of the best cropland available to the country have been returned to cultivation. Even at African yields (which are low compared to the world average), this is enough to feed well over a million people—or to increase the average grain protein diet of 6 million Burkino Fasoans by over 20% (*World Indices*, 1986).

What I have tried to argue in the above paragraphs is that the worst fears of environmentalists (erosion, pollution, loss of species, and climate change) are easy to call up and emotionalize in almost any readership—but that scientific trend analysis is lacking. Good news is not reported equally with bad. In a few areas where multi-decade data do exist, as with U.S. erosion surveys, it is not accurate to conclude that things are getting worse. The action of enlightened legislation and agencies may be quite adequate to interrupt the fate which Garrett Hardin named, "The Failure of the Commons."

IV. VISION 2020

On September 4, 1987, leaders of the People's Republic of China revealed to a visiting team from the World Bank that Chinese leadership had plans to "shift the emphasis away from grain production in the relatively prosperous coastal provinces and rely far more on grain imports" (Thomson, 1987). The Chinese said that this overturning of traditional policy was "in the interests of making agricultural production more efficient . . ., and to allow coastal provinces, which are major grain producers, to concentrate on cash crops and rural industry, while covering the grain deficit through a marked increase in imports."

It must be observed that these plans are long range, and that not all members of the leadership concur in them—that in fact, there will probably be a major struggle over whether to implement them or not. Nonetheless, for the leadership of the world's most populace nation to argue publicly for "agricultural production efficiency" even at the cost of greater grain imports is something of a major watershed in rational thinking about world food policy.

"Self-sufficiency" has been the dominant issue in global agricultural policy for many years. Food self-sufficiency is pursued regardless of a country's natural endowments. One major exception to this is Japan. Japan is completely surrounded by the sea, a very inhospitable surface for producing food. But isn't Japan thus similar to an arid land (of which many exist in Africa and Asia), surrounded by desert or by jungle or by poor soils of whatever kind? Japan's food security rests on producing economic value added rather than converting its inhospitable surroundings (the sea) directly to protein. Why can't other countries follow this example, and rely on the advantage that various continental countries have in producing a surplus of food?

Three kinds of advantage need to be distinguished—"comparative advantage," "absolute advantage," and "competitive advantage." To illustrate the differences, suppose we consider a hypothetical case of a lawyer and his or her secretary. Suppose the lawyer, in working his way through law school, became a competent typist—in fact, so good that he could type 150 words per minute, and won the city championship in typing. But now he is a competent lawyer as well, and commands fees of $150/hr. His secretary is a good typist, capable of 80 words per minute. The question in this case is "Who has the comparative advantage in typing, the lawyer or the secretary?" The answer is the secretary. Compared to her other skills, the secretary can type much better than she can lawyer. Compared to his other skills, the lawyer can practice law much better than he can type. He has an absolute advantage in typing, but it would not help his earnings to use it; his typing does not constitute a competitive advantage either, most of which would come from his knowledge of the law, his ability to argue to juries, his skill in counseling clients, etc.

In Ricardo's system of comparative advantage, nations would specialize according to their comparative advantages amongst all goods and services, and by so doing the economic output of the whole would be greater. The system is roughly at work today, but with certain inefficiencies. If in France, the price of wheat is about $5 per bushel, and in the United States, the price for a similar bushel of wheat is about $3, it would seem impossible for France to export wheat. But from the point of view of France, it has a comparative advantage with wheat—it can grow and export wheat better than most of the other goods it is capable of producing. So the government subsidizes the sales of wheat, to bring the French price in line with the United States and other suppliers who have an absolute advantage.

China has a comparative advantage in corn; China can produce and export corn better than it can produce and export, say, automobiles or steel or other goods that might be needed by its Pacific Rim neighbors. So China exports corn, depriving its own population of this food, in order to earn exchange for what the population apparently wants worse, such items as televisions and appliances. What China is realizing, in its proposed plan to reduce grain production in favor of cash crops (vegetables, fruit, etc.) and rural industry is that the *absolute advantage* by other countries in grain production may be so great that the price will be much lower than their own, and quite affordable—especially if rural industries blossom and add economic value to the people's labor that could not be added with food commodities. If basic food commodities can be imported cheaply, it is more efficient to buy them and to devote the labor to manufactured goods which command much greater added value.

Self-sufficiency need not be taken on an item-by-item basis. It can be approached in terms of general economic success, as is the case with Japan. Once the criterion of *food* self-sufficiency is removed, the potential for expanded diets worldwide emerges strongly, because certain countries have enormous absolute advantages in food production that are presently lying idle.

Figure 11.11 shows the percent share of the United States in exports of all goods and of coarse grains (including corn). The U.S. presently has about a 10% share of world trade in all goods, but about a 60% share of world trade in coarse grains. It can be said, therefore, that the United States is six times better at exporting corn and other coarse grains than it is at exporting the average item that it exports. The United States has a "revealed comparative advantage" in corn of 6. The term "revealed" means that despite everything, including (unwise) farm programs, the record shows that the U.S. can export corn five times better than most other items. Figures 11.12 and 11.13 show that the revealed comparative ad-

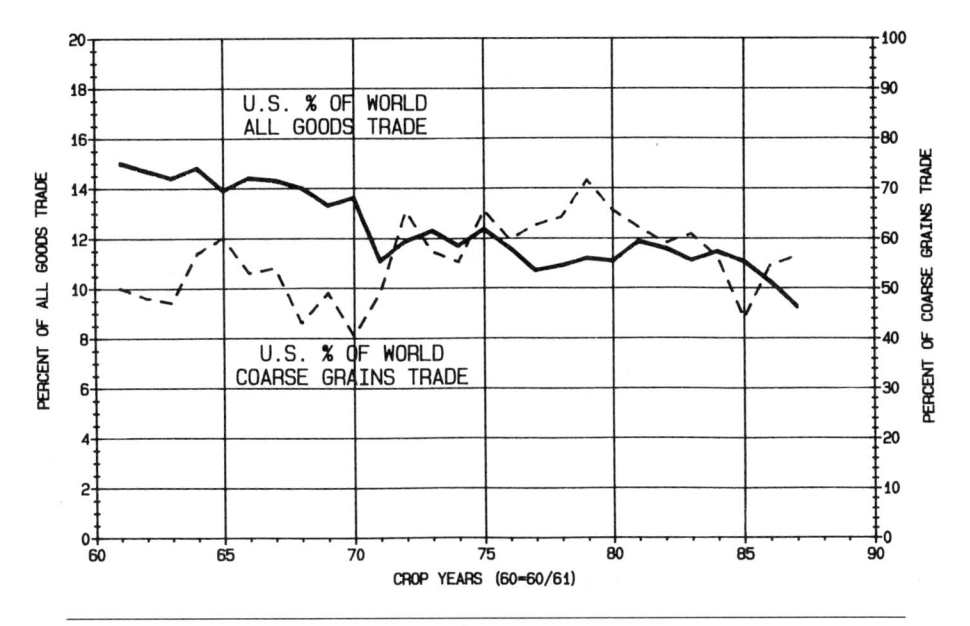

Figure 11.11 United States' share of world trade for all goods (solid line) and coarse grains (broken line). Data for U.S. all goods trade from *Economic Report* (1987); data for world trade from Insel (1987); data for trade of coarse grains from *World Development* (1987).

vantage in coarse grains is greater than in wheat or soybeans and that it has remained strong even in the last few years of tough economic times for U.S. farmers. In other words, despite large subsidies by the U.S. government, bent on *raising* the price of corn and preserving the income of farmers, U.S. exports of corn continue to outshine other goods from America on world markets. This testifies to an enormous absolute advantage in agriculture, in the form of fertile, rainfed cropland (the corn and wheat belts) together with a very small population to feed domestically—compared certainly with China, India, or indeed most other countries.

Other grain exporters have larger comparative advantages than the U.S. for exporting grain—for instance, Argentina (whose coarse grain revealed comparative advantage may be about 20), Brazil, Canada, Australia, South Africa, Thailand, and others. But all the other exporting countries, in total, have only about one-half the scope, or total production capacity, as the United States.

What we must consider is this: If the "room" is limited to 20 or 25 billion people, won't efficiency and absolute advantage become the crucial determinants of policy—rather than the kind of "revealed comparative

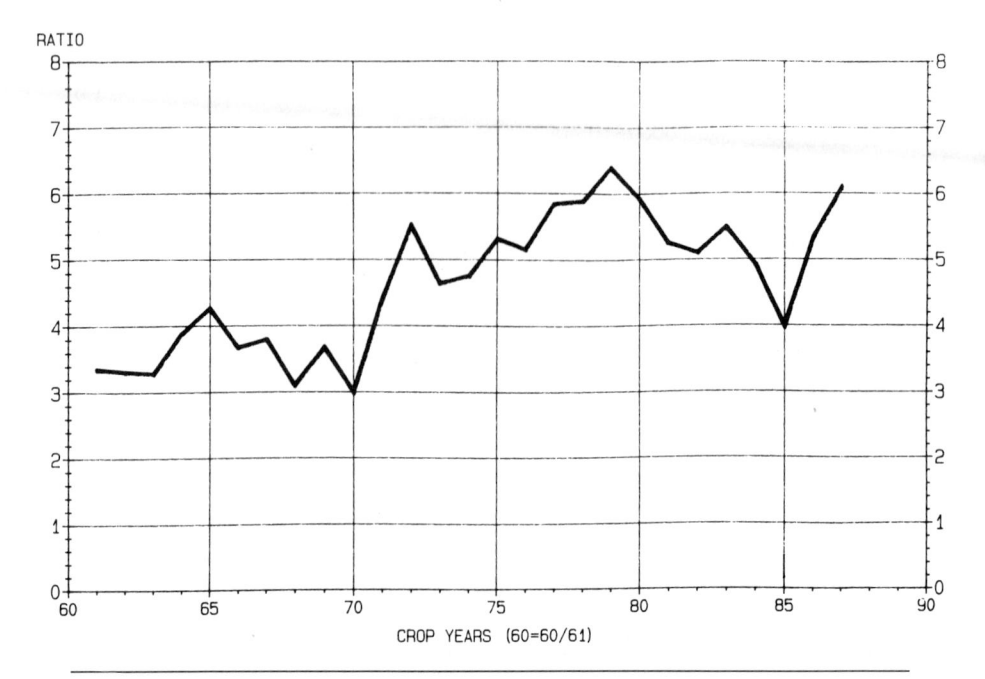

Figure 11.12 "Revealed" comparative advantage coefficient for U.S. coarse grains versus all goods. Data from *Economic Report* (1987) and *World Development* (1987); ratio calculated.

advantage" we have now, in which the insistence on item-by-item food self-sufficiency brings government into play to preserve inefficient arrangements? As the room gets crowded, must not all the occupants become more efficient, rely on the specialists with greatest advantage, and seek greater exchange with each other, in order for the whole to be most comfortable?

Table 11.3 indicates what could be accomplished within the present bounds of agricultural knowledge and practice. World grain production is about 1,600 Mt per year, which is 62 billion bushels. This is an average of 12.5 bushels per person. The acutely hungry people on earth, estimated at 50 million, average only about 5 bushels each—compared to a minimum nutritional need of about 6 or 7 bushels each. The malnourished, estimated at 300 million people, average 6 bushels each. The surplus countries, the present grain exporters, estimated at 800 million, average 19.5 bushels each. This leaves the rest of humanity, 3,850 million people, at 11.5 bushels each.

To cure hunger means supplying 50 million people with 6.5 bushels each—a total of 300 million bushels, which is one-tenth of the present

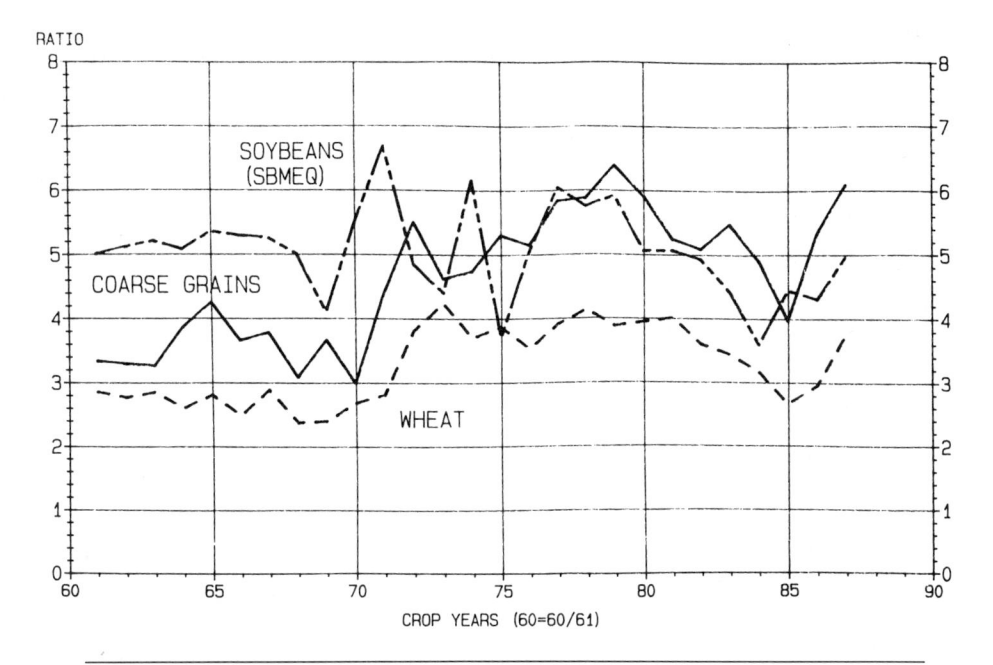

Figure 11.13 "Revealed" comparative advantage coefficient for U.S. wheat (broken line), coarse grains (solid line), and soybeans (dashed line) versus coefficient for all goods. Data sources same as in Figure 11.12.

surplus in grain exporting countries. To cure malnutrition would require 1.6 billion bushels, or about half the amount that is set-aside and not grown in the U.S. every year. To feed the 75 million additional people every year, at the 11.5 bushel per year average, requires 900 million bushels, which is also small versus the surplus, and which is below the average annual growth of grain production outside the surplus countries.

To raise the average world citizen's diet to that of the grain surplus areas would mean adding 8 bushels to 3,850 people, for a total of 30 billion new bushels. This could be done by doubling the yields of the present grain surplus countries, enabling them to use their geoclimatic absolute advantages. As mentioned earlier, yields of quadruple the present country averages are reported routinely by test stations and contest winners. The genetic potential of crops may hardly have been approached.

By the time we reach the year 2020, the world's population may have reached 10 billion, if present trends continue, and if nuclear quarrels desist. If the limit of carrying capacity is 20 to 25 billion people, the "room" by 2020 will be half full. At some point in those next three decades, the pres-

298   William J. Hudson

Table 11.3 World Food Demand[a]

1986 Status[c]	People (million)	Grain usage per person[b] (bu/yr)	Total (billion bu)	(million t)
World population	5,000	12.5	62	1,600
Acutely hungry	50	5.0	0.2	5
Malnourished	300	6.0	1.8	45
"Average"	3,850	11.5	44.0	1,130
Surplus countries	800	19.5	16.0	420
Demands				
Annual growth of population	75	11.5	0.9	25
Cure hunger	50	6.5	0.3	8
Cure malnutrition	300	5.5	1.6	40
Raise "average" to that of surplus countries	3,850	8.0	30.0	770
Supply				
Present surplus of surplus countries			3.0	75
Surplus if double yields of surplus countries			27.0	700
Trade				
Today			8.0	200
Potential			38.0	975

[a]Sources: USDA (1987); World Indices (1986).
[b]Includes wheat, coarse grains, and rice, but not oilseeds.
[c]Surplus countries are the United States, EEC, Canada, Brazil, Argentina, S. Africa, Australia, and Thailand.

sure to use the earth's natural resources more efficiently will no doubt greatly intensify. But why wait? Why not abandon the concept of item-by-item food self-sufficiency now, in lieu of the unused capacity of absolute food surplus countries? Is not the principal one of these, the United States, the one with the best demonstrable trend in environmental protection? Won't specialization help save rather than harm the world's geoclimatic systems? Might not the provision of good diet to all five billion present occupants make the "room" much more comfortable on the route to 2020?

REFERENCES

"Agricultural Outlook." (1987). U.S. Gov. Print. Off., Washington, D.C.
Braudel, F. (1985). "The Structures of Everyday Life." Harper, New York.
Brooke, J. (1987). Old scourge loses ground in West Africa. *New York Times* Sept. 3, p. 1.
Calvin, M. (1986). Letter to D. Pimentel, Dec. 18.
"Economic Indicators." (1987). Prepared for the Joint Economic Committee. U.S. Gov. Print. Off., Washington, D.C.

"Economic Report to the President." (1987). U.S. Gov. Print. Off., Washington, D.C.
Foley, K. (1981). Letter to W. J. Hudson, Oct. 17.
"Handbook of Economic Statistics 1986." (1987). Natl. Tech. Inf. Cent., Springfield, Virginia.
"The Global 2000 Report to the President: Entering the Twenty- First Century: A Report." (1980). Prepared by the Council on Environmental Quality and the Department of State, Vol. 2. U.S. Gov. Print. Off., Washington, D.C.
Insel, B. (1987). A world awash in grain. *Foreign Aff.* **63**, 892–912.
"Monthly Bulletin of Statistics." (1987). United Nations, New York.
1980 Appraisal, Part I: Soil, Water, and Related Resources in the United States: Status Condition and Trends." (1981). U.S. Dep. Agric., Washington, D.C.
Pimentel, D., and Pimentel, M. (1979). "Food, Energy, and Society." Wiley, New York.
"The Second RCA Appraisal: Soil, Water, and Related Resources on Nonfederal Land in the United States: Analysis of Condition and Trends." (1987). U.S. Dep. Agric., Washington, D.C.
Shabecoff, P. (1987). Man said to tax Earth's systems: Report says environment may have reached its limits for permanent change. *New York Times* Feb. 15, p. 4.
Thomson, R. (1987). Peking unveils wide changes in economic policy. *Financial Times* Sept. 4, p. 1.
"World Agricultural Supply and Demand Estimates." (1987). World Agric. Outlook Board, Washington, D.C.
"World Development Report." (1987). Oxford Univ. Press, New York.
"World Financial, Markets." (1987). Morgan Guaranty Trust Co., New York.
USDA (1987). "World Grain Situation/Outlook. Foreign Agriculture Circular. Grains." Foreign Agric. Serv., Washington, D.C.
"World Indices of Agricultural and Food Production 1976–1985." (1986). U.S. Gov. Print. Off., Washington, D.C.
"World Oilseed Situation and Market Highlights. Foreign Agriculture Circular. Oilseeds and Products." (1987). Foreign Agric. Serv., Washington, D.C.
"World Population Profile: 1985." (1986). U.S. Gov. Print. Off., Washington, D.C.

12

Ecological Resource Management for a Productive, Sustainable Agriculture

David Pimentel, Thomas W. Culliney, Imo W. Buttler, Douglas J. Reinemann, and Kenneth B. Beckman

College of Agriculture and Life Sciences
Cornell University
Ithaca, New York

- I. Introduction
- II. Principles for a Productive, Sustainable Agriculture
- III. Soil Nutrient and Water Resources
- IV. Pests and Their Control
- V. Importance of Biological Resources
- VI. Environmental and Economic Aspects of Ecological Agricultural Management
- VII. Conclusion
  References

I. INTRODUCTION

Because of high production costs and low commodity prices, U.S. farmers have financial problems and about 20% face bankruptcy (USDA, 1985a,b; NYEH, 1986). In part, the high production costs are due to environmental degradation, including soil and water losses, and loss of biological diversity. States like Iowa, which have some of the best soils in the nation, report losses of one-half of their topsoil after just a little over 100 years of farming (Risser, 1981). The nation as a whole has lost over one-third of its topsoil (Handler, 1970). With an average soil loss rate of 18 t/ha/yr (Lee, 1984), the United States is losing topsoil 18 times faster than it is being replaced (Swanson and Harshbarger, 1964; Hudson, 1981; Larson, 1981; McCormack et al., 1982; Lal, 1984; Elwell, 1985).

Not only do the sediments and water that run off agricultural land represent a specific loss to agriculture, they also contribute to the pollution of groundwater and surface waters (OTA, 1983). It has been estimated, in fact, that sediments and water runoff cause about $6 billion in off-site damages annually (Clark, 1985). Fertilizers and pesticides are common pollutants of water resources (Reilly, 1985; Thomas, 1985), and agriculture is reported to be the greatest nonpoint polluter of water resources in the nation (Chesters and Scheirow, 1985; Myers et al., 1985).

Two significant and costly inputs to U.S. agriculture are fertilizers and pesticides. United States agriculture uses about 49 million t (metric tons) of commercial fertilizers (USDA, 1983) and 350,000 t of pesticides per year (Pimentel and Levitan, 1986). The ecological effects of pesticides on the environment are a major concern. Direct losses to agriculture from pesticides include destruction of natural enemies (OTA, 1979), increased pesticide resistance (Georghiou and Saito, 1983), and destruction of honey bees and other nontarget organisms (Pimentel et al., 1980a). It has been estimated that the social and environmental losses due to pesticides are at least $1 billion annually (Pimentel et al., 1980a).

Another of the primary inputs to agriculture is fossil energy. An estimated 17% of the annual fossil energy consumption of the United States is used to supply the nation with its food and fiber needs, one-third of which is for crop and livestock production (Pimentel, 1984). Currently, about 1,100 liters of oil equivalents are required to produce a hectare of a crop like corn (Pimentel, 1984). The energy input for nitrogen fertilizer alone is now greater than the total energy inputs for raising corn in 1945— about a 20-fold increase in the amount of nitrogen fertilizer versus a 3-fold increase in corn yield. This intensive management of agroecosystems requires an investment of more than $500/ha/yr for corn excluding land and taxes (USDA, 1984a). Clearly, chemical use for high productivity and compensation for soil, water, and biological resource degradation contribute to the high production costs and other problems of U.S. agriculture.

We propose that high crop yields could be maintained and input costs reduced by the appropriate management of soil, water, energy, and biological resources. In this chapter we examine the ecological principles and alternative practices that might be employed to make agriculture environmentally and economically sound and sustainable in the long term. A clear need exists for reducing input costs and making agriculture more productive (Farm bill, 1985 [Public Law 99-205]; GAO, 1985; Buttel *et al.*, 1986).

II. PRINCIPLES FOR A PRODUCTIVE, SUSTAINABLE AGRICULTURE

Agricultural production depends on soil, water, air, energy, and biological resources. Clearly, for a productive, sustainable agriculture, the complex interactions among these resources must be understood so that they can be managed as an integrated system (Figure 12.1).

The major principles that underlie an agricultural system that will be productive while protecting the environment are outlined in Figure 12.1. These include:

1. Adapting and designing the agricultural system to the environment of the region. This means, for example, culturing crops and/or forages (livestock) that are ecologically adapted to the soil, water, climate, and biota present at the site.

2. Optimizing the use of biological resources in the agroecosystem. This includes making effective use of biological pest control, green manures, cover crops, rotations, agricultural wastes, and other biological resources (Edens *et al.*, 1985; Vietmeyer, 1986).

3. Developing strategies that induce minimal changes in the natural ecosystem to protect the environment and minimize the use of fossil energy in manipulating the agroecosystem.

Although this holistic approach is complex, this complexity may be overcome in part by focusing primarily on four factors that are commonly manipulated in an agroecosystem—soil nutrients, water, energy, and pests (Figure 12.1). The goal is to conserve soil nutrients and water, while at the same time encouraging beneficial organisms and discouraging pests. Soil nutrients (nitrogen, phosphorus, potassium, and others) and water are essential to a productive agriculture. Conserving soil and water resources reduces the inputs of commercial fertilizers and irrigation needed and thus decreases costs. Similarly, manipulations of the agroecosystem that encourage biological pest control and make the environment unfavorable for pests reduce the use of pesticides. Combined, these strategies

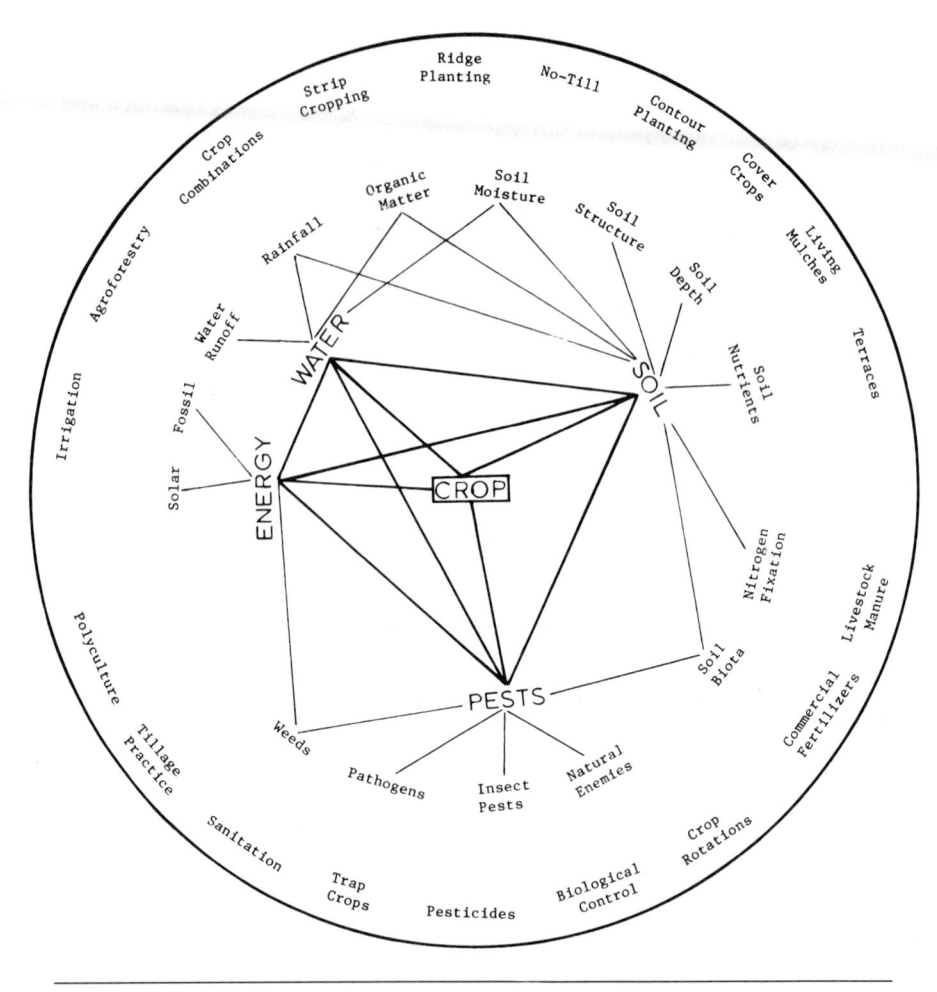

Figure 12.1 Some of the complex ecological interactions among soil, water, energy, and biological resources in crop ecosystems. Around the outer ring are several management practices that if appropriately employed can improve the productivity and sustainability of agriculture and at the same time achieve lower input costs.

will reduce input costs and help maintain a highly productive, ecologically sound agriculture (Figure 12.1).

III. SOIL NUTRIENT AND WATER RESOURCES

Soil erosion on U.S. croplands averages 18 t/ha/yr and ranges from about 0–300 t/ha/yr for both water and wind erosion combined (Lee, 1984). On some of the best agricultural lands of the nation, such as in Iowa and Missouri (Major Land Resource Area #10), soil erosion averages 36 t/ha/

yr (USDA, 1980; Lee, 1984). Average values suggest the seriousness of the problem overall, but some cropland resources are managed extremely well and have erosion rates at the acceptable level of 1 t/ha/yr or less (F. R. Troeh, 1987 unpublished data).

Erosion adversely affects crop productivity by reducing water availability, removing nutrients, reducing organic matter, and restricting rooting depth as the soil thins (OTA, 1982; Schertz et al., 1985). It is primarily the loss of water through runoff that is responsible for reducing productivity (NSESPRPC, 1981). When vegetation is absent, water runs off the land rapidly. For example, water runoff rates have been measured to be as much as 10- to 100-fold greater on cleared land than on vegetation-covered land (Charreau, 1972; USDA–ARS and EPA–ORD, 1976). Conventional tillage of corn was reported to allow nine times greater water runoff compared with no-till grown corn (Angle et al., 1984).

Both water and wind erosion also reduce the available water-holding capacity of soil by selectively removing organic matter and the finer soil particles (Buntley and Bell, 1976). Increasing soil organic matter by applying livestock manure increased the water infiltration rate by more than 90% (Meek and Donovan, 1982; Sweeten and Mathers, 1985), mainly by decreasing the rate of water runoff (Mueller et al., 1984).

Besides water, shortages of soil nutrients (nitrogen, phosphorus, potassium, calcium, etc.) are the most important factors limiting crop productivity. One metric ton of rich agricultural soil from the upper few centimeters may contain 4 kg of nitrogen, 1 kg of phosphorus, 20 kg of potassium, and 10 kg of calcium (Alexander, 1977; Bohn et al., 1979; Scheffer and Schachtschabel, 1979; Greenland and Hayes, 1981). Therefore, the loss of 18 t/ha/yr of soil represents a total of 72 kg/ha of nitrogen, which is almost half of the average of 152 kg/ha/yr of nitrogen fertilizer that is applied to U.S. corn (USDA, 1982) and involves a substantial loss of nutrients (Correll, 1983). The harvest of the corn crop itself removes from 25 to 50% of the nitrogen applied. Additional amounts (15 to 25%) of the nitrogen are lost by volatilization (Allison, 1973; Schroder, 1985) and 10 to 50% by leaching (Schroder, 1985).

Erosion does not remove all the components of soil equally. Several studies have demonstrated that the eroded material is usually 1.3 to 5 times richer in organic material than the remaining soil (Barrows and Kilmer, 1963; Allison, 1973). Organic matter is important to soil quality because of its positive effects on water retention, soil structure, and cation exchange capacity. Further, it is the major source of nutrients needed by plants (Allison, 1973; Volk and Loeppert, 1982). Ninety-five percent of the nitrogen in the surface soil and 15–80% of the phosphorus is found in soil organic matter (Allison, 1973). Reducing the amount of soil organic matter from 3.8–1.8% is reported to lower the yield of corn about 25% for some soils (Lucas et al., 1977).

Losses of water and soil nutrients and reduced organic matter are the major factors reducing crop productivity from erosion (Battiston et al., 1985; Schertz et al., 1985). However, some assessments of the effect of erosion on crop productivity are based only on reduced soil depth while holding these other factors constant (Craft et al., 1985; Crosson, 1985). In these kinds of studies, for example, corn yields are reported to decrease less than 1% per centimeter of soil depth reduction (Craft et al., 1985). Using figures such as this, a loss of 18 t/ha/yr of soil, which removes about 1.3 mm of soil depth, is reported to result in a reduced corn yield of less than 0.1%. Because this degree in reduction in rooting depth and productivity is relatively minor, several studies have concluded that the costs of implementing certain conservation technologies are greater than the annual benefits they would produce (Shrader et al., 1963; Berglund and Michalson, 1981; Crosson and Stout, 1983; Mueller et al., 1985).

If, however, the total effects of erosion are measured instead of the effects of reduced soil depth, then reductions in crop yields from 15–30% result from moderate to severe erosion (Battiston et al., 1985; McDaniel and Hajek, 1985; Schertz et al., 1985). Thus, the total benefits of soil conservation that prevent losses of water, nutrients, and organic matter are significant (Lee et al., 1974; Pollard et al., 1979; Pope et al., 1983; Wijewardene and Waidyanatha, 1984; Crowder et al., 1985; Mueller et al., 1985). For example, yields from corn grown on the contour were about 12% greater than from corn grown with the slope (Smith, 1946; Sauer and Case, 1954). On land with a 7% slope, yields from cotton grown in rotation were increased 30%, while erosion was reduced nearly one-half (Hendrickson et al., 1963). In tests using rotations, the yields of corn were about 10% larger than continuously grown corn, and weed control was improved (Ewing, 1978; Muhtar et al., 1982; Sundquist et al., 1982; Oldham and Odell, 1983–1984; Barker et al., 1984). Although yields of corn by no-till may be higher, especially under hot, dry conditions, overall corn yields with no-till average about the same as conventional (Van Doren et al., 1977; Wentzel and Robinson, 1983; Wiese, 1983; Bitzer et al., 1985; Mueller et al., 1985).

A cultural practice developed in China (Wan et al., 1959), ridge-planting (Figure 12.2), combines the advantages of contour planting and no-till for soil, water, and energy conservation while eliminating several of the disadvantages like heavy pesticide use and poor germination (Griffith et al., 1973, 1986a,b; Deszo, 1979; Comis and Howell, 1982; Campbell and Brown, 1983; Gebhardt et al., 1985; RAA, 1985; Griffith and Mannering, 1986). In ridge-planting, seeds are planted on top of contour ridges that are about 20 cm high and relatively dry and warm because the vegetation and crop residues are pushed to the bottom of trenches at time of planting. The ridges laid out along the contour and the presence of crop residues in the trenches enhance soil and water conservation.

Weeds in ridge-planting can be controlled without herbicides, first by removing about 5 cm of soil at the top of the ridge at time of planting and

Fall-Winter

Cover Crop
Crop Residue

Spring Planting

Seed

Mid-Summer

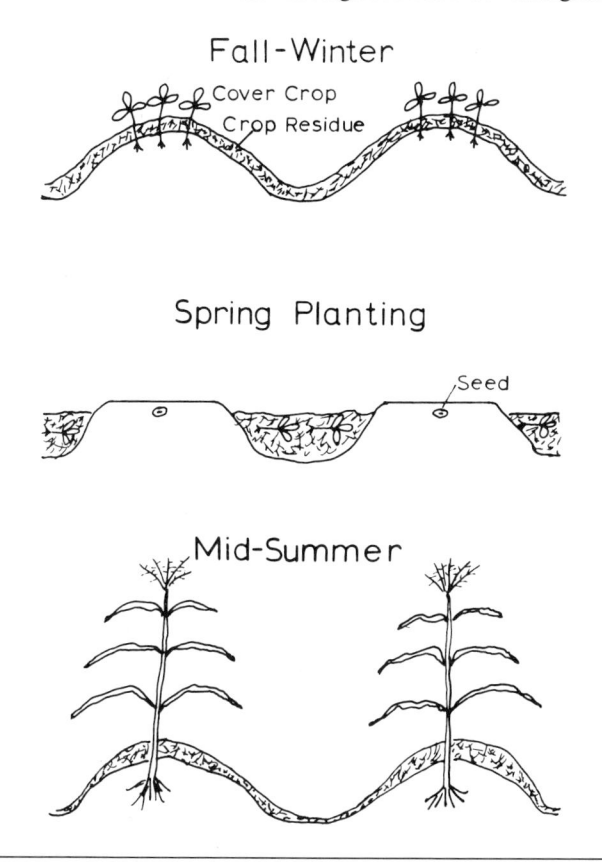

Figure 12.2 The ridge-planting system through fall–winter, spring, and mid-summer.

pushing it into the trench. Later weeds are controlled by cultivating and pulling soil and organic matter from the bottom of the trench to the base of the rapidly growing crop plants. Preliminary data indicate that yields from ridge-planting average about the same (Campbell and Brown, 1983) or higher than both no-till and conventional corn production because of better soil and water conservation. Moreover, the warmer soil temperatures improve germination and the stand of corn (Comis and Howell, 1982; L. M. Thompson, 1985 personal communication).

IV. PESTS AND THEIR CONTROL

Along with the careful management of soil and water resources for high crop yields, the farmer must also control pests. Currently, despite all pest controls, which include about 350,000 t of pesticide annually, about 37%

of total potential U.S. crop production or about $50 billion is lost to pests (Pimentel, 1986). Although heavy pesticide use has substantially reduced damage caused by some pests, no overall reduction in crop losses from pests has occurred. For example, since 1945 U.S. crop losses to pathogens and weeds have fluctuated but never declined (Pimentel et al., 1978). Rather surprisingly, crop losses due to insects have increased nearly two-fold (from 7% to about 13% of crop yields) from 1945 to the present. This has occurred in spite of a more than ten-fold increase in insecticide use during the same period (Pimentel, 1986).

The significant increase in insect damage to crops can be attributed to several major changes that have taken place in U.S. agricultural production practices since the 1940s. Specifically, reductions in crop rotations, field sanitation and crop diversity, plus increased monoculture have contributed to increased need and use of pesticides (Pimentel et al., 1978; OTA, 1979, 1982).

Insecticides have also reduced the number of natural enemies that are present. When this occurs, more insecticide has to be used, yet losses due to pests increase. With cotton, for instance, four to five additional sprays are applied to compensate for the destruction of natural enemies of the cotton bollworm and budworm (Pimentel et al., 1977).

The above examples and those mentioned earlier illustrate several of the problems associated with heavy dependence on pesticidal controls. Integrated pest management and bioenvironmental pest control suggest that pests can be reduced by a combination of controls instead of a single factor like pesticides (PSAC, 1965). Some of the bioenvironmental controls include natural enemies, rotations, host plant resistance, sanitation, timing of planting, tillage, and crop and genetic diversity (PSAC, 1965; OTA, 1979).

Although most pest control manipulations of the agroecosystem are independent of soil and water conservation practices, a few are complementary. For example, crop rotations and strip cropping helped control pests (e.g., corn insect pests, diseases, and weeds), while at the same time conserving soil and water (PSAC, 1965; NAS, 1968; OTA, 1979, 1982). Reports of insect pests in some no-till systems are also encouraging. Blumberg and Crossley (1983) and House et al. (1984) reported that insect damage to sorghum leaves was reduced more than two-fold in no-till sorghum compared with conventional-tillage sorghum. This reduction in damage is possibly due to increased numbers of carabid beetles, other predators, and parasites found associated with some no-till systems (House and All, 1981; Blumberg and Crossley, 1983; Ferguson and McPherson, 1985; House and Parmelee, 1985). Designing and managing agroecosystems based on concern for the environment and cropping system will help accomplish the goal of conserving soil and water and resources while making the crop environment less favorable for pests.

V. IMPORTANCE OF BIOLOGICAL RESOURCES

Crops and livestock are but a fraction of the total biological resources used in agriculture. Most of the 200,000 species of plants and animals that exist in the U.S. natural ecosystem are involved in agricultural production (Pimentel *et al.*, 1980b). These natural biota perform many essential functions in agriculture, such as: degrading wastes, recycling nutrients, protecting crops and livestock from pest attack, pollinating crops, conserving soil and water resources, and preserving genetic material for crop and livestock breeding.

The value of natural predators and parasites in biological pest control has already been discussed. Wild bees and honey bees pollinate about $20 billion worth of crops annually (Levin, 1984). Insecticide use kills bees and decreases their effectiveness for pollination. By reducing insecticide use and making judicious applications, opportunities exist to reduce the current $135 million annual honey bee and wild bee losses (Pimentel *et al.*, 1980a).

Nitrogen is second only to water as a limiting component in U.S. agricultural production (Delwiche, 1978). Nitrogen must be supplied in crop production systems to maintain high yields. In the United States, about 11 million t of nitrogen fertilizer, worth about $6 billion, are applied annually (USDA, 1983). Thus, the estimated 14 million t of nitrogen that are biologically fixed by microorganisms in the United States annually (Delwiche, 1970)—with a calculated value of about $7 billion—are of great economic value in U.S. crop production.

Although it may be impractical to plant a hectare to a legume for a season to produce nitrogen, because the price of land and other inputs is high and the legume might compete with the main crop for water, it may be feasible in some situations to interplant some row crops with a legume. For example, legumes can be planted between the rows of corn in July and August and then the legumes plowed under in early spring when the field is being prepared for crop planting. Winter vetch and other legumes planted in this manner contribute from 50 to 150 kg/ha of nitrogen depending on the growing season (Mitchell and Teel, 1977; Scott *et al.*, 1984). In addition, some legumes serve as a cover crop and living mulch that protect the soil from wind and water erosion and reduce weeds and insect pests (Vrabel *et al.*, 1980; Horwith, 1983; Palada *et al.*, 1983; Altieri *et al.*, 1985). These crops also collect and store soil nutrients during fall and winter and, of course, the residues add organic matter to the soil when plowed under.

A good quality soil that on average contains about 6,700 kg/ha in biomass is in large part living (Hole, 1981). For instance, the average biomass of biota per hectare in the upper 15 cm of rich soil is: insects and earthworms about 1000 kg/ha each (Wolcott, 1937; Edwards and Lofty, 1977);

protozoa and algae about 150 kg/ha each (Alexander, 1977); bacteria about 1,700 kg/ha (Alexander, 1977); and fungi 2,700 kg/ha (Alexander, 1977).

The level of organic matter in agricultural soils should be about 4% in moist-temperate environments to assure sufficient biota in the soil for degrading wastes and recycling nutrients, and sufficient earthworm and insect populations and tunnels in the soil for the effective percolation of water (T. W. Scott, 1986 unpublished data). The abundance of soil biota is directly related to the amount of organic matter in the soil. For example, raising soil organic matter from about 2% to 6% increased earthworm biomass about ten-fold or up to 1,200 kg/ha (Edwards and Lofty, 1982). Similarly, earthworm and microorganism biomass increased about five-fold when the quantity of manure applied to pastureland was about doubled (Ricou, 1979).

In addition to degrading wastes and recycling nutrients like nitrogen, phosphorus, and potassium, these organisms play an important role in soil formation. Earthworms commonly bring from underground to the soil surface from 10–50 t/ha/yr of soil (Lee, 1983), while insects bring about one-tenth this amount (Lyford, 1963; Hole, 1981; Kalisz and Stone, 1981; Beattie and Culver, 1983; Culver and Beattie, 1983; Davidson and Morton, 1983; Zacharias and Grube, 1984; Lockaby and Adams, 1985). Therefore, tillage systems such as no-till and ridge-planting, which increase soil organic matter by decreasing erosion, will foster soil biota and thus increase the productivity of soil (Hole, 1981; Edwards and Lofty, 1982). In addition to contributing to soil formation, these organisms increase water infiltration. For instance, one square meter may have more than 10,000 earthworm channels.

Another valuable biological resource available in the United States is livestock manure, which totals 1.6 billion t/yr (Anderson, 1972). This amount of manure contains about 80 million t of nitrogen, 20 million t of phosphorus, and 64 million t of potassium (Thompson and Troeh, 1978). These quantities of nutrients are significantly greater than the quantities of commercial fertilizer applied annually in the United States, which contain 11 million t of nitrogen, 5 million t of phosphorus, and 6 million t of potassium (USDA, 1983).

From these data, it appears that from five to seven times more nutrients are available in manure than are applied annually in commercial fertilizer. However, as Safley et al. (1983) calculated, only 2 million t of the total nitrogen in manure are economically recoverable and usable with present technology. This is due in part to the uneven distribution of livestock facilities and crop areas. In some cases, like feedlots, manure constitutes a serious waste problem. From 30–90% of the nitrogen in manure can be lost through ammonia volatilization when manure is exposed on crop and pasturelands (Vanderholm, 1975). But less than 5% is lost as NH^3 volatilization when manure is plowed under immediately, and about 15% is lost when disced into the soil.

Although livestock manure is produced year round, immediate application is often impractical because of cropping patterns and weather, necessitating the storage of manure. Composting is one means of stabilizing nitrogen during storage; however, composting manure and other organic matter may result in large nutrient losses—particularly nitrogen—if not managed properly. Nitrogen is lost during the composting process primarily through ammonia volatilization as aerobic microorganisms degrade the organic matter (Sikora and Sowers, 1985). Although nitrogen is lost during the composting process, compost material has advantageous characteristics including its structure and tendency not to immobilize soil nitrogen (Sikora and Sowers, 1985).

A system in which manure is stored over winter in anaerobic lagoons with minimal surface exposure and covered immediately with soil during spring application reduces nitrogen losses to about 20% (Vanderholm, 1975; Bezdicek et al., 1977).

VI. ENVIRONMENTAL AND ECONOMIC ASPECTS OF ECOLOGICAL AGRICULTURAL MANAGEMENT

Three major difficulties with conventional U.S. agriculture are the high costs of production, the serious problem of environmental resource degradation, and the instability of crop yields (Brown, 1984). In Table 12.1 the economic and environmental benefits of two soil and water conservation methods for cultivating corn are compared to conventional corn production. The two conservation practices—no-till and ridge-planting—considerably reduced current input costs of $523/ha. Included with no-till is the alternative practice of rotating corn with another appropriate crop. Both practices reduce erosion (as mentioned), and the rotation eliminates the need to use an insecticide treatment for the control of the corn rootworm complex, a typical pest problem in continuously grown, conventional corn (Pimentel et al., 1977). Selecting the appropriate crops for rotation with corn reduces corn diseases (Pearson, 1967; Mora and Moreno, 1984) and weed problems (NAS, 1968; Mulvaney and Paul, 1984). Although rotations offer many advantages, some disadvantages include inconvenience of producing multi-crops and sometimes less profit if the alternate crop produces less net return than corn.

For the ridge-planting system, several low-input alternative practices are added (Figure 12.2). These include livestock manure and use of cover crops with continuous corn. The advantages of including livestock manure were mentioned earlier. The use of legume cover crops is of value in reducing soil erosion and water runoff, reducing weed problems, and conserving soil nutrients—soil nutrients are picked up and stored by the cover

Table 12.1 Energy and Economic Inputs per Hectare for Conventional and Alternative Corn Production Systems

	Conventional			No-till and rotation			Low-input alternatives and ridge-planting		
	Qty.	10³ kcal	Economic	Qty.	10³ kcal	Economic	Qty.	10³ kcal	Economic
Labor (hr)	10[a]	7[f]	50[r]	7[cc]	6[f]	35[r]	12[jj]	9[f]	60[r]
Machinery	55[b]	1,485[g]	91[s]	45[dd]	1,215[g]	75[s]	45[dd]	1,215[g]	75[s]
Fuel (liter)	115[b]	1,255[b]	38[t]	70[ee]	764[h]	23[t]	70[ee]	764[h]	23[t]
N (kg)	152[b]	3,192[i]	81[u]	152[ff]	3,192[i]	81[u]	(27t)[kk]	559[qq]	17[rr]
P (kg)	75[b]	473[j]	53[v]	75[ff]	473[j]	53[v]	34[ll]	214[j]	17[u]
K (kg)	96[b]	240[k]	26[w]	96[ff]	240[h]	26[w]	15[mm]	38[k]	4[w]
Limestone (kg)	426[b]	134[l]	64[x]	426[ff]	134[l]	64[x]	426[ff]	134[l]	64[x]
Corn seeds (kg)	21[b]	520[m]	45[y]	24[gg]	594[m]	51[v]	21[b]	520[m]	45[y]
Cover crop seeds (kg)	—	—	—	—	—	—	10[oo]	120[oo]	10[xx]
Insecticdes (kg)	1.5[c]	150[n]	15[z]	0[hh]	0	0	1.5[c]	150[n]	15[z]
Herbicides (kg)	2[c]	200[n]	20[z]	4[ii]	400[n]	40[z]	0[pp]	0	0
Electricity (10³ kcal)	100[b]	100[o]	8[aa]	100[b]	100[o]	8[aa]	100[b]	100[o]	8[aa]
Transport (kg)	322[d]	89[p]	32[bb]	196[d]	54[p]	20[bb]	140[d]	39[p]	14[bb]
Total		7,845	$523		7,172	$476		3,862	$352
Yield (kg)	6,500[e]	26,000[q]		6,500	26,000[q]		6,500	26,000[q]	
Output/input ratio		3.31			3.63			6.73	

[a] Labor input was estimated to be 10 hr because of the extra time required for tillage and cultivation compared with no-till, which required 7 hr (USDA, 1984a).
[b] Pimentel and Wen (1987).
[c] Mueller et al. (1985).
[d] Transport of machinery, fuel, and nitrogen fertilizer (Pimentel and Wen, 1987).
[e] Three-year running average yield (USDA, 1982).
[f] Food energy consumed per laborer per day was assumed to be 3,500 kcal.
[g] The energy input per kilogram of steel in tools and other machinery was 18,500 kcal (Doering, 1980) plus 46% added input (Fluck and Baird, 1980) for repairs.
[h] Fuel includes a combination of gasoline and diesel. A liter of gasoline and diesel fuel was calculated to contain 10,000 and 11,400 kcal, respectively (Cervinka, 1980). Weighted average value of 10,900 used in calculations. These values include the energy input for mining and refining.
[i] Nitrogen = 21,000 kcal/kg (Dovring and McDowell, 1980).
[j] Phosphorus = 6,300 kcal/kg (Dovring and McDowell, 1980).

[k]Potassium = 2,500 kcal/kg (Dovring and McDowell, 1980).
[l]Limestone = 315 kcal/kg (Terhune, 1980).
[m]Hybrid seed = 24,750 kcal/kg (Heichel, 1980).
[n]Energy input for insecticides and herbicides was calculated to be 100,000 kcal/kg (Pimentel, 1980).
[o]Includes energy input required to produce the electricity.
[p]For the goods transported to the farm, an input of 275 kcal/kg was included (Pimentel, 1980).
[q]A kilogram of corn was calculated to have 4,000 kcal.
[r]Labor = $5/hr.
[s]USDA, 1984a.
[t]Liter = $0.33.
[u]N = $0.53.
[v]P = $0.51.
[w]K = $0.27.
[x]Limestone = $0.15.
[y]USDA (1984a).
[z]Insecticide and herbicide treatments = $10/kg for both the material and application costs.
[aa]kwh = 7¢.
[bb]Transport = 10¢/kg
[cc]No-till requires less labor than conventional because tillage and cultivation are reduced (Colvin et al., 1982; Mueller et al., 1985).
[dd]20% smaller machinery was used because less power is needed in no-till and ridge planting (Colvin et al., 1982; Muhtar and Rotz, 1982; Allen and Hollingsworth, 1983; Hamlett et al., 1983; USDA, 1984b).
[ee]Nearly 40% less fuel is required compared with conventional because the soil was not tilled, only lightly cultivated (Colvin et al., 1983; Mueller et al., 1985).
[ff]Assumed that same amount of N, P, K, and Ca required in no-till.
[gg]About 15% more seed was planted to offset poor germination in no-till (USDA, 1984b).
[hh]No insecticide was used because the corn was planted in rotation after soybeans.
[ii]Twice as much herbicide was used compared with conventional tillage to control weeds.
[jj]Five additional hours were necessary for collecting and spreading 27 t of manure (Pimentel et al., 1984).
[kk]A total of 27 t of cattle manure was applied to provide 152 kg of N.
[ll]A total of 41 kg of P was provided by the manure.
[mm]A total of 81 kg of K was provided by the manure.
[nn]Cultivation and cover crop used for weed control.
[oo]About 10 kg of cover crop seeds were used (Heichel, 1980).
[pp]No herbicide used, weed control carried out by cultivation and cover crop.
[qq]About 1.9 liters of fuel were required to collect and apply 1 t of manure (Pimentel et al., 1984).
[rr]The value of manure was given for the fuel required to transport and spread.
[ss]1 kg of cover crop seed = $1.

crop. Note, ridge-planting is not suitable for all soils, rainfall, and crops (Lal, 1977, 1985), which emphasizes the need for care in selecting appropriate technologies for ecological resource management.

Numerous other alternate technologies could have been considered for this example, including other cropping systems, green manures, and pest control practices, but the technologies we selected illustrate the potential of an alternative system to conserve soil and water resources, reduce the need for pesticides, and improve the sustainability of the agroecosystem.

Average input data for conventional corn production are listed in Table 12.1. It is assumed that this crop is grown in a region where rainfall averages 1,000 mm/yr, and on land with a slope of 3–5% and erosion rate of 18 t/ha/yr. Average U.S. corn yield is 6,500 kg/ha, and the energy input is calculated to be 8.0 million kcal/ha with 12 hr of human labor. The energy production ratio, i.e., the ratio of kcal output per kcal input, is 3.2 (Table 12.1). Total production costs are calculated to be $523/ha.

The no-till system is assumed to be planted in an environment similar to that of conventional corn. The major differences between no-till and conventional are (1) erosion is reduced from 18 t/ha/yr to about 1 t/ha/yr; (2) labor is reduced from 12 hr to 10 hr; (3) smaller tractors are employed; (4) less tractor fuel is used; (5) about twice as much herbicide is used to control weeds; and (6) no insecticide is used because the corn is planted in rotation after a nonhost crop such as a legume (Table 12.1). The total energy inputs and costs are about 10% less than those for conventional. Also, the yield of corn in no-till is assumed to be similar to that of conventional (Van Doren et al., 1977; Taylor et al., 1984; Hargrove, 1985).

As mentioned, several alternative practices are integrated in the ridge-planting system (Table 12.1). For this system the assumptions are (1) ridge-planting is carried out on the contour and crop residues are left on the surface, thus erosion is reduced from 18 t/ha/yr to a tolerable level of less than 1 t/ha/yr; (2) available livestock manure is substituted for all the nitrogen needs and most of the phosphorus and potassium needs; (3) labor input is raised to 15 hr/ha to include the time required for manure spreading compared with 12 hr/ha for the conventional system; (4) corn is planted in rotation after a nonhost plant like a legume, thus no insecticide is used; (5) because of the cover crop and well-designed tillage system, no herbicide is included; and (6) smaller tractors are used and less fuel is consumed (L. M. Thompson, 1985 personal communication). The total energy inputs for the ridge-planting system are reduced by nearly half, and production costs are reduced by one-third of the conventional system (Table 12.1). Results similar to those calculated for this low-input system have been obtained by farmers who have used a like low-input system for crop production (L. M. Thompson, 1985 personal communication). The production costs of these were about $100/ha less than conventional systems (L. M. Thompson, 1985 personal communication).

Although the corn yield for the low-input system is assumed to be equal to the conventional system, yields would probably be much higher in the long term. Using sound soil and water conservation measures will slow the loss of soil and decline of productivity. Over a 20-year period about 2.6 cm of soil can be expected to be lost in the conventional corn system, with a soil loss of 18 t/ha/yr. About 500 years would be required to replace this 2.6 cm of lost soil. If this soil degradation were offset with increased energy inputs like fertilizer and irrigation, then this would involve substituting a nonrenewable resource (fossil energy) for a renewable resource. With the cost of fuel to rise in the coming decades, the substitution of a nonrenewable resource for a renewable resource will become very costly to farmers and society. Thus, soil and water conservation can pay major dividends in the long term.

This analysis suggests that the use of ecologically sound practices will maintain high yields while reducing production costs and protecting the environment—especially soil, water, energy, and biological resources. For example, with the ridge-planting system, soil erosion and water runoff are controlled and pesticide use is reduced. All of this reduces costs by decreasing fertilizer, pesticide and machinery costs. Of major importance is the fact that the productivity of the soil and integrity of the entire agroecosystem is maintained for the future. Fortunately, numerous alternative practices for soil and water conservation and pest control are readily available for use in productive agriculture (PSAC, 1965; Troeh *et al.*, 1980). Each set of agricultural technologies, however, has to be selected and adapted to the particular environmental site of the region.

VII. CONCLUSION

Degradation of soil, water, and biological resources that are essential to agricultural production contributes to current high production costs. By employing various alternative practices that improve the environment and the use of resources, production costs can be significantly reduced as illustrated by our two models.

A wide array of soil and water conservation technologies exists that could be integrated into alternative crop management programs (Troeh *et al.*, 1980; Lockeretz, 1983; Pimentel *et al.*, 1987). Similarly, numerous bioenvironmental pest control technologies are available that could help reduce costly pesticide inputs (PSAC, 1965; OTA, 1979).

Selecting the particular combination of alternative practices depends on the conditions of soil, water, climate, and biota and the crop and/or livestock to be produced. The agroecosystem has to be designed and adapted for a particular environment. In addition to conserving soil and water, the improved use of biological resources for biological control and obtaining nutrients (nitrogen) from legumes and other technologies can help reduce production costs.

Clearly, this ecological approach is complex and requires detailed understanding of the resources, crops, livestock, and environment. Designing a holistic management scheme for agriculture for a particular site depends on a multidisciplinary effort by scientists to help farmers adapt this sophisticated approach to agricultural production. Applying the broad principles for an ecological management strategy will help develop a productive, environmentally sound agriculture with greatly reduced production costs.

Why does this ecological approach to agriculture have potential now? Economic problems and growing environmental concerns plus the challenge of producing more world food are encouraging agriculture to look to improved resource management practices if a profitable and environmentally sound agriculture is to be achieved. At the same time more sophisticated biological knowledge and technologies are now available, than ever before. These help us integrate basic information on soil, water, energy, and biological resources and enable us to adapt crop/livestock systems to a particular environment.

ACKNOWLEDGMENTS

We thank the following people for reading an earlier draft of this article and for their many helpful suggestions: M. Altieri, D. Andow, G. Berardi, D. F. Bezdicek, W. Dritschilo, D. Horn, B. Horwith, G. House, W. J. Hudson, J. Krummel, R. I. Papendick, W. Parham, F. R. Troeh, K. Watt, and K. Wilde. And at Cornell University we thank F. Buttel, B. Chabot, R. McNeil, W. Naegeli, and N. R. Scott.

REFERENCES

Alexander, M. (1977). "Introduction to Soil Microbiology," 2nd Ed. Wiley, New York.
Allen, R. R., and Hollingsworth, L. D. (1983). Limited tillage sorghum on wide beds. *ASAE Pap.* No. 83-1517.
Allison, F. E. (1973). "Soil Organic Matter and its Role in Crop Production." Elsevier, New York.
Altieri, M. A., Wilson, R. C., and Schmidt, L. L. (1985). The effects of living mulches and weed cover on the dynamics of foliage- and soil-arthropod communities in three crop systems. *Crop Prot.* **4**, 201–213.
Anderson, L. L. (1972). Energy potential from organic wastes: a review of the quantities and sources. *Inf. Circ. U.S. Bur. Mines* No. 8549.

Angle, J. S., McClung, G., McIntosh, M. S., Thomas, P. M., and Wolf, D. C. (1984). Nutrient losses in runoff from conventional and no-till corn watersheds. *J. Environ. Qual.* **13**, 431–435.

Barker, G. L., Sanford, J. O., and Reinschmeidt, L. L. (1984). Crop-rotations versus monocrop systems for the hill areas of Mississippi (cotton, soybeans, maize, wheat). *Res. Rep. Miss. Agric. For. Exp. Stn.* **9**(7).

Barrows, H. L., and Kilmer, V. J. (1963). Plant nutrient losses from soils by water erosion. *Adv. Agron.* **15**, 303–315.

Battiston, L. A., McBridge, R. A., Miller, M. H., and Brklacich, M. J. (1985). Soil erosion productivity research in southern Ontario. *In* "Erosion and Soil Productivity," ASAE Publ. No. 8-85, pp. 25–38. Am. Soc. Agric. Eng., St. Joseph, Michigan.

Beattie, A. J., and Culver, D. C. (1983). The nest chemistry of two seed-dispersing ant species. *Oecologia* **56**, 99–103.

Berglund, S. H., and Michalson, E. L. (1981). Soil erosion control in Idaho's Cow Creek watershed: an economic analysis. *J. Soil Water Conserv.* **36**, 158–161.

Bezdicek, D. F., Sims, J. M., Ehlers, M. H., Cronath, J., and Hermanson, R. H. (1977). Nutrient budget in a dairy anaerobic lagoon. *In* "Food Fertilizer and Agricultural Residue" (R. C. Loehr, ed.), pp. 681–692. Ann Arbor Sci Publ., Ann Arbor, Michigan.

Bitzer, M. J., Blevins, R. L., Aswad, M., Deaton, P., Childers, J., Henry, D., and Amos, H. (1985). Effect on tillage on soil loss and corn grain yields on sloping land. *Proc. 1985 South. Reg. No-Till Conf.* (W. L. Hargrove, F. C. Boswell, and G. W. Langdale, eds.), pp. 163–164. Univ. of Georgia, Athens.

Blumberg, A. Y., and Crossley, D. A., Jr. (1983). Comparison of soil surface arthropod populations in conventional tillage, no-tillage and old field systems. *Agro-Ecosystems* **8**, 247–253.

Bohn, H. L., McNeal, B. L., and O'Connor, G. A. (1979). "Soil Chemistry." Wiley, New York.

Brown, W. L. (1984). Some observations on changing trends in agricultural production systems. *Agric. Res. Inst. Conf., Changing Agric. Prod. Syst. Fate Agric. Chem., Chevy Chase, Md.*

Buntley, G. J., and Bell, F. F. (1976). Yield estimates for the major crops grown in the soils of west Tennessee. *Tenn. Agric. Exp. Stn. Bull.* No. 561.

Buttel, F. H., Gillespie, G. W., Jr., Janke, R., Caldwell, B., and Sarrantonio, M. (1986). Reduced-input agricultural systems: Rationale and prospects. *Annu. Meet. South. Rural Sociol. Assoc., Orlando, Fla.*

Campbell, J. K., and Brown, W. H. (1983). Ridge tillage for corn production on wet soils. *ASAE Pap.* No. NAR 83-102.

Cervinka, V. (1980). Fuel and energy efficiency. *In* "Handbook of Energy Utilization in Agriculture" (D. Pimentel, ed.), pp. 15–24. CRC Press, Boca Raton, Florida.

Charreau, C. (1972). Problèmes posés par l'utilisation agricole des sols tropicaux par des cultures annuelles. *Agron. Trop. (Paris)* **27**, 905–929.

Chesters, G., and Scheirow, L. (1985). A primer on nonpoint pollution. *J. Soil Water Conserv.* **40**, 9–13.

Clark, E. H., II. (1985). The off-site costs of soil erosion. *J. Soil Water Conserv.* **40**, 19–22.

Colvin, T. S., Hamlett, C. A., and Rodriguez, A. (1982). Effect of tillage system on farm machinery selection. *ASAE Pap.* No. 82-1029.

Colvin, T., Erbach, D., Marley, S., and Erickson, H. (1983). Large-scale evaluation of a till plant system. *ASAE Pap.* No. 83-1027.

Comis, D. L., and Howell, R. (1982). Ridge tillage (Northern Corn Belt, Indiana, Michigan). Soil and water conservation news. *USDA Soil Conserv. Serv.* **3**(8), 8–10.

Correll, D. L. (1983). N and P in soils and runoff of three coastal plain land uses. *In* "Nutrient Cycling in Agricultural Ecosystems" (R. R. Lowrance, R. L. Todd, L. E. Asmussen, and R. A. Leonard, eds.), Spec. Publ. No. 23, pp. 207–244. Univ. of Georgia Exp. Stn., Athens.

Craft, E. M., Carlson, S. A., and Cruse, R. M. (1985). A model of erosion and subsequent fertilization impacts on soil productivity. *In* "Erosion and Soil Productivity," ASAE Publ. 8-85, pp. 143–151. Am. Soc. Agric. Eng., St. Joseph, Michigan.

Crosson, P. (1985). National costs of erosion on productivity. *In* "Erosion and Soil Productivity," ASAE Publ. 8-85, pp. 254–265. Am. Soc. Agric. Eng., St. Joseph, Michigan.

Crosson, P. N., and Stout, A. T. (1983). "Productivity Effects of Cropland Erosion in the United States." Resources for the Future, Washington, D.C.

Crowder, B. M., Poinke, H. B., Epp, D. J., and Young, C. E. (1985). Using CREAMS and economic modeling to evaluate conservation practices: an application. *J. Environ. Qual.* **14**, 428–434.

Culver, D. C., and Beattie, A. J. (1983). Effects of ant mounds on soil chemistry and vegetation patterns in a Colorado mountain meadow. *Ecology* **64**, 485–492.

Davidson, D. W., and Morton, S. R. (1983). Myrmecochory in some plants *(F. chenopodiaceae)* of the Australian arid zone. *Oecologia* **50**, 357–366.

Delwiche, C. C. (1970). The nitrogen cycle. *Sci. Am.* **223**(3), 137–158.

Delwiche, C. C. (1978). Legumes—past, present and future. *BioScience* **28**, 565–570.

Deszo, J. (1979). [Possibilities for decreasing soil cultivation in maize growing.] A talamuveles csokkentesenek lehetosege a kukocricater-mesztesben. *In* "Tessedik Samuel" Tiszantuli Mezogazdasagi Tudomanyos Napok, Debrecen, Hungary; Agrartudomanyi Egyetem. 54–55 [Hu] DATE Mezogazdasag tudomanyi Egyetemi Kar, Debrecen, Hungary. *Field Crop Abstr.* **35**(1), 18 (1982).

Doering, O. C. (1980). Accounting for energy in farm machinery and buildings. *In* "Handbook of Energy Utilization in Agriculture" (D. Pimentel, ed.), pp. 9–14. CRC Press, Boca Raton, Florida.

Dovring, F., and McDowell, D. R. (1980). "Energy Use for Fertilizers." Dep. Agric. Econ. Staff Pap. No. 80 E-102. Univ. of Illinois, Urbana.

Edens, T. C., Fridgen, C., and Battenfield, S. L., eds. (1985). "Sustainable Agriculture and Integrated Farming Systems." Michigan State Univ. Press, East Lansing.

Edwards, C. A., and Lofty, J. R. (1977). "Biology of Earthworms." Chapman & Hall, London.

Edwards, C. A., and Lofty, J. R. (1982). Nitrogenous fertilizers and earthworm populations in agricultural soils. *Soil Biol. Biochem.* **14**(5), 515–521.

Elwell, H. A. (1985). An assessment of soil erosion in Zimbabwe. *Zimbabwe Sci. News* **19**, 27–31.

Ewing, L. (1978). Rotation [of corn and soybeans] cuts costs and boosts yields. *Soybean Dig.* **39**(1), 18.

Ferguson, H. J., and McPherson, R. M. (1985). Abundance and diversity of adult Carabidae in four soybean cropping systems in Virginia. *J. Entomol. Sci.* **20**(2), 163–171.

Fluck, R. C., and Baird, C. D. (1980). "Agricultural Energetics." AVI, Westport, Connecticut.

Gebhardt, M. R., Daniel, T. C., Schweizer, E. E., and Allmaras, R. R. (1985). Conservation tillage. *Science* **230**, 625–630.

General Accounting Office (GAO). (1985). "Agriculture Overview—U.S. Food/Agriculture in a Volatile World Economy," Briefing Report to the Congress, Nov. 1985. U.S. Gen. Accounting Off., Washington D.C.

Georghiou, G. P., and Saito, T., eds. (1983). "Pest Resistance to Pesticides." Plenum, New York.

Greenland, D. J., and Hayes, M. H. B. (1981). "The Chemistry of Soil Processes." Wiley, New York.

Griffith, D. R., and Mannering, J. V. (1986). "Differences in Crop Yields as a Function of Tillage System, Crop Management and Soil Characteristics," Mimeo. Dep. Agron., Purdue Univ., West Lafayette, Indiana.

Griffith, D. R., Mannering, J. V., Galloway, H. M., Parsons, S. D., and Richey, C. B. (1973). Effect of eight tillage-planting systems on soil temperature, percent stand, plant growth, and yield of corn on five Indiana soils. *Agron. J.* **65**, 321–326.

Griffith, D. R., Mannering, J. V., and Box, J. E. (1986a). Soil and moisture management with reduced tillage. *In* "No- Tillage and Surface-Tillage Agriculture: The Tillage Revolution" (M. A. Sprague and G. B. Triplett, eds.), pp. 19–57. Wiley, New York.

Griffith, D. R., Mannering, J. V., Mengel, D. B., Parsons, S. D., Bauman, T. T., Scott, D. H., Turpin, F. T., and Doster, D. H. (1986b). "A Guide to Till-Planting for Corn and Soybeans in Indiana (Tillage) ID-148." Coop. Ext. Serv., Purdue Univ., West Lafayette Indiana.

Hamlett, C. A., Colvin, T. S., and Musselman, A. (1983). Economic potential of conservation tillage in Iowa. *Trans. ASAE* **26**(3), 719–722.

Handler, P., ed. (1970). "Biology and the Future of Man." Oxford Univ. Press, New York.

Hargrove, W. L. (1985). Influence of tillage on nutrient uptake and yield of corn. *Agron. J.* **77**, 763–768.

Heichel, G. H. (1980). Assessing the fossil energy costs of propagating agricultural crops. *In* "Handbook of Energy Utilization in Agriculture" (D. Pimentel, ed.), pp. 27–33. CRC Press, Boca Raton, Florida.

Hendrickson, B. H., Barnett, A. P., Carreker, J. R., and Adams, W. E. (1963). Runoff and erosion control studies on Cecil soil in the southern Piedmont. *U.S. Dep. Agric. Tech. Bull.* No. 1281.

Hole, F. D. (1981). Effects of animals on soil. *Geoderma* **25**, 75–112.

Horwith, B. J. (1983). Ecological interactions of plant species following plowing of old-fields in Michigan. Ph.D. Thesis, Univ. of Michigan, Ann Arbor.

House, G. J., and All, J. N. (1981). Carabid beetles in soybean agroecosystems. *Environ. Entomol.* **10**(2), 194–196.

House, G. J., and Parmelee, R. W. (1985). Comparison of soil arthropods and earthworms from conventional and no-tillage agroecosystems. *Soil Tillage Res.* **5**, 351–360.

House, G. J., Stinner, B. R., Crossley, D. A., Jr., Odum, E. P., and Langdale, G. W. (1984). Nitrogen cycling in conventional and no-tillage agroecosystems in the Southern Piedmont. *J. Soil Water Conserv.* **39**, 194–200.

Hudson, N. (1981). "Soil Conservation," 2nd ed. Cornell Univ. Press, Ithaca, New York.

Kalisz, P. J., and Stone, E. L. (1981). Soil mixing by scarab beetles. *Agron. Abstr.* p. 227.

Lal, R. (1977). Soil-conserving versus soil-degrading crops and soil management for erosion control. *In* "Soil Conservation in the Tropics" (D. Greenland and R. Lal, eds.), pp. 81–86. Wiley, New York.

Lal, R. (1984). Productivity assessment of tropical soils and the effects of erosion. *In* "Quantification of the Effect of Erosion on Soil Productivity in an International Context" (F. R. Rijsberman and M. G. Wolman, eds.), pp. 70–94. Delft Hydraul. Lab., Delft, Netherlands.

Lal, R. (1985). A soil suitability guide for different tillage systems in the tropics. *Soil Tillage Res.* **5**, 179–196.

Larson, W. E. (1981). Protecting the soil resource base. *J. Soil Water Conserv.* **36**, 13–16.

Lee, K. E. (1983). The influence of earthworms and termites on soil nitrogen cycling. *In* "New Trends in Soil Biology" (P. Lebrun, H. M. André, A. de Medts, C. Gregoire-Wibo, and G. Wauthy, eds.), pp. 35–48. Univ. Catholique de Louvain, Louvain-La-Neuve, Belgium.

Lee, L. K. (1984). Land use and soil loss: a 1982 update. *J. Soil Water Conserv.* **39**, 226–228.

Lee, M. T., Narayanon, A. S., and Swanson, E. R. (1974). "Economic Analysis of Erosion and Sedimentation," Econ. Res. Rep. No. 130. Dep. Agric. Econ., Univ. of Illinois, Urbana.

Levin, M. D. (1984). Value of bee pollination to United States agriculture. *Am. Bee. J.* **124**(3), 184–186.

Lockaby, B. G., and Adams, J. C. (1985). Pedoturbation of a forest soil by fire ants. *Soil Sci. Soc. Am. J.* **49**(1), 220–223.

Lockeretz, W. (1983). "Environmentally Sound Agriculture." Praeger, New York.

Lucas, R. E., Holtman, J. B., and Connor, L. J. (1977). Soil carbon dynamics and cropping practices. *In* "Agriculture and Energy" (W. Lockeretz, ed.), pp. 333–351. Academic Press, New York.

Lyford, W. H. (1963). Importance of ants to brown podzolic soil genesis in New England. *Harvard For. Pap.* **7**, 1–18.

McCormack, D. E., Young, K. K., and Kimberlim, L. W. (1982). "Current Criteria for Determining Soil Loss Tolerance," ASA Spec. Publ. No. 45. Am. Soc. Agron., Madison, Wisconsin.

McDaniel, T. A., and Hajek, B. F. (1985). Soil erosion effects on crop productivity and soil properties in Alabama. *In* "Erosion and Soil Productivity," ASAE Publ. No. 8-85, pp. 48–58. Am. Soc. Agric. Eng., St. Joseph, Michigan.

Meek, B., and Donovan, T. (1982). Long term effects of manure on soil nitrogen, potassium, sodium, organic matter, and water infiltration rate. *Soil Sci. Soc. Am. J.* **46**(5), 1014–1019.

Mitchell, W. H., and Teel, M. R. (1977). Winter annual cover crops for no-tillage corn production. *Agron. J.* **69**, 569–573.

Mora, L. E., and Moreno, R. A. (1984). Cropping pattern and soil management influence on plant diseases: I. *Diplodia macrospora* leaf spot of maize. *Turrialbo* **34**(1), 35–40.

Mueller, D. H., Wendt, R. C., and Daniel, T. C. (1984). Soil and water losses as affected by tillage and manure applications. *Soil Sci. Soc. Am. J.* **48**(4), 896–900.

Mueller, D. H., Klemme, R. M., and Daniel, T. C. (1985). Short- and long-term cost comparisons of conventional and conservation tillage systems in corn production. *J. Soil Water Conserv.* **40**, 466–470.

Muhtar, H. A., and Rotz, C. A. (1982). A multi-crop machinery selection algorithm for different tillage systems. *ASAE Pap.* No. 82-1031.

Muhtar, H. A., Black, J. R., Burkhardt, T. H., and Christenson, D. (1982). Economic impact of conservation tillage in Michigan. *ASAE Pap.* No. 82-1033.

Mulvaney, D. L. and Paul, L. (1984). Rotating crops and tillage. Both sometimes better than just one. *Crops Soils Mag.* **36**(7), 18–19.

Myers, C. F., Meek, J., Tuller, S., and Weinberg, A. (1985). Nonpoint sources of water pollution. *J. Soil Water Conserv.* **40**, 14–18.

National Academy of Sciences (NAS). (1968). "Principles of Plant and Animal Pest Control," Vol. 2. Weed Control Publ. No. 1597. NAS, Washington, D.C.

NSESPRPC (National Soil Erosion–Soil Productivity Research Planning Committee). (1981). Soil erosion effects on soil productivity: a research perspective. *J. Soil Water Conserv.* **32**, 82–90.

New York Economic Handbook (NYEH). (1986). "Agricultural Situation and Outlook. New York Economic Handbook 1986," A. E. 85-29. Dep. Agric. Econ., Cornell Univ., Ithaca, New York.

Office of Technology Assessment (OTA). (1979). "Pest Management Strategies in Crop Protection," Vols. I and II. OTA, U.S. Gov. Print. Off., Washington, D.C.

Office of Technology Assessment (OTA). (1982). "Impacts of Technology on U.S. Cropland and Rangeland Productivity." OTA, U.S. Gov. Print. Off., Washington, D.C.

Office of Technology Assessment (OTA). (1983). "Water-Related Technologies for Sus-

12. Ecological Resource Management 321

tainable Agriculture in U.S. Arid/Semiarid Lands,'' OTA-F-212. U.S. Gov. Print. Off., Washington, D.C.
Oldham, M. G., and Odell, R. T. (1983-1984). The Morrow Plots—America's oldest experimental field (University of Illinois research plots laid out in 1876 to study the effect of crop rotation on yield, corn, oats, clover, soybeans). *Better Crops Plant Food* **68**(Winter), 12-14.
Palada, M. C., Gonser, S., Hofsetter, R., Volak, B., and Culik, M. (1983). Association of interseeded legume cover crops and annual row crops in year-round cropping systems. *In* "Environmentally Sound Agriculture" (W. Lockeretz, ed.), pp. 269-276. Praeger, New York.
Pearson, L. C. (1967). "Principles of Agronomy." Reinhold, New York.
Pimentel, D., ed. (1980). "Handbook of Energy Utilization in Agriculture." CRC Press, Boca Raton, Florida.
Pimentel, D. (1984). Energy flow in the food system. *In* "Food and Energy Resources" (D. Pimentel and C. W. Hall, eds.), pp. 1-24. Academic Press, New York.
Pimentel, D. (1986). Agroecology and economics. *In* "Ecological Theory and Integrated Pest Management Practice" (M. Kogan, ed.), pp. 299-319. Wiley, New York.
Pimentel, D., and Levitan, L. (1986). Pesticides: amounts applied and amounts reaching pests. *BioScience* **36**, 86-91.
Pimentel, D., and Wen, D. (1987). Technological changes in energy use in U.S. agricultural production. *In* "Research Approaches in Agricultural Ecology" (S. R. Gliessman, ed.). Springer-Verlag, New York.
Pimentel, D., Shoemaker, C., LaDue, E. L., Rovinsky, R. B., and Russell, N. P. (1977). "Alternatives for Reducing Insecticides on Cotton and Corn: Economic and Environmental Impact," Environ. Res. Lab., Off. Res. Dev., EPA, Athens, Georgia (issued in 1979).
Pimentel, D., Krummel, J., Gallahan, D., Hough, J., Merrill, A., Schreiner, I., Vittum, P., Koziol, F., Back, E., Yen, D., and Fiance, S. (1978). Benefits and costs of pesticide use in U.S. food production. *BioScience* **28**, 772, 778-784.
Pimentel, D., Andow, D., Dyson-Hudson, R., Gallahan, D., Jacobson, S., Irish, M., Kroop, S., Moss, A., Schreiner, I., Shepard, M., Thompson, T., and Vinzant, B. (1980a). Environmental and social costs of pesticides: a preliminary assessment. *Oikos* **34**, 127-140.
Pimentel, D., Garnick, E., Berkowitz, A., Jacobson, S., Napolitano, S., Black, P., Valdes-Cogliano, S., Vinzant, B., Hudes, E., and Littman, S. (1980b). Environmental quality and natural biota. *BioScience* **30**, 750-755.
Pimentel, D., Berardi, G., and Fast, S. (1984). Energy efficiencies of farming wheat, corn, and potatoes organically. *In* "Organic Farming: Current Technology and Its Role in a Sustainable Agriculture," ASA Spec. Publ. No. 46, pp. 151-161. Am. Soc. Agron., Madison, Wisconsin.
Pimentel, D., Allen, J., Beers, A., Guinand, L., Linder, R., McLaughlin, P., Meer, B., Musonda, D., Perdue, D., Poisson, S., Siebert, S., Stoner, K., Salazar, R., and Hawkins, A. (1987). World agriculture and soil erosion. *BioScience* **37**, 277-283.
Pollard, R. W., Sharp, B. M. H., and Madison, F. W. (1979). Farmers' experience with conservation tillage: a Wisconsin survey. *J. Soil Water Conserv.* **34**, 215-219.
Pope, A. P., III, Bhide, S., and Heady, E. O. (1983). Economics of conservation tillage in Iowa. *J. Soil Water Conserv.* **38**, 370-373.
President's Science Advisory Committee (PSAC). (1965). "Restoring the Quality of Our Environment." Report of the Environmental Pollution Panel, President's Science Advisory Committee, The White House, Washington, D.C.
Regenerative Agricultural Association (RAA). (1985). "The Thompson Farm, Nature's Ag School." Regenerative Agric. Assoc., Emmaus, Pennsylvania.

Reilly, W. K. (1985). Protecting groundwater. *J. Soil Water Conserv.* **40**, 280.
Ricou, G. A. E. (1979). Consumers in meadows and pastures: pastures. *In* "Grassland Ecosystems of the World: Analysis of Grasslands and Their Uses" (R. T. Coupland, ed.), pp. 147–153. Cambridge Univ. Press, London and New York.
Risser, J. (1981). A renewed threat of soil erosion: it's worse than the dust bowl. *Smithsonian* **11**, 121–131.
Safley, L. H., Nelson, D. W., and Westermann, P. W. (1983). Conserving manurial nitrogen. *Trans. ASAE* **26**, 1166–1170.
Sauer, E. L., and Case, H. C. M. (1954). Soil conservation pays off. Results of ten years of conservation farming in Illinois. *Univ. Ill. Agric. Exp. Stn. Bull.* No. 575.
Scheffer, F., and Schachtschabel, P. (1979). "Lehrbuch der Bodenkunde." Enke, Stuttgart.
Schertz, D. L., Moldenhauer, W. C., Franzmeier, D. P., and Sinclair, H. R., Jr. (1985). Field evaluation of the effect of soil erosion on crop productivity. *In* "Erosion and Soil Productivity," ASAE Publ. 8-85, pp. 9–17. Am. Soc. Agric. Eng., St. Joseph, Michigan.
Schroder, H. (1985). Nitrogen losses from Danish agriculture—trends and consequences. *Agric. Ecosyst. Environ.* **14**, 279–289.
Scott, T. W., Burt, R. F., and Otis, D. J. (1984). "Crop, Intercrop and Cover Crop Systems," Agronomy Mimeo, No. 84-5. Cornell Univ., Ithaca, New York.
Shrader, W. D., Johnson, H. P., and Timmons, J. F. (1963). Applying erosion control principles. *J. Soil Water Conserv.* **18**, 195–200.
Sikora, L. J., and Sowers, M. A. (1985). Effect of temperature control on the composting process. *J. Environ. Qual.* **14**, 434–439.
Smith, D. D. (1946). The effect of contour planting on crop yield and erosion losses in Missouri. *J. Am. Soc. Agron.* **38**, 810–819.
Sundquist, W. B., Menz, K. M., and Neumeyer, C. F. (1982). Technology assessment of commercial corn production in the United States. *Stn. Bull. Minn. Agric. Exp. Stn.* No. 546.
Swanson, E. R., and Harshbarger, C. E. (1964). An economic analysis of effects of soil loss on crop yields. *J. Soil Water Conserv.* **19**, 183–186.
Sweeten, J. M., and Mathers, A. C. (1985). Improving soils with livestock manure. *J. Soil Water Conserv.* **40**, 206–210.
Taylor, F., Raghaven, G. S. V., Negi, S. C., McKyes, E., Vigier, B., and Watson, A. K. (1984). Corn grown in a Ste. Rosalie clay under zero and traditional tillage. *Can. Agric. Eng.* **26**(2), 91–95.
Terhune, E. C. (1980). Energy used in the United States for agricultural liming materials. *In* "Handbook of Energy Utilization in Agriculture" (D. Pimentel, ed.), pp. 25–26. CRC Press, Boca Raton, Florida.
Thomas, L. M. (1985). Management of nonpoint-source pollution: what priority? *J. Soil Water Conserv.* **40**, 8.
Thompson, L. M., and Troeh, F. R. (1978). "Soils and Soil Fertility," 4th Ed. McGraw-Hill, New York.
Troeh, F. R., Hobbs, J. A., and Donahue, R. L. (1980). "Soil and Water Conservation for Productivity and Environmental Protection." Prentice-Hall, Englewood Cliffs, New Jersey.
USDA–ARS and EPA–ORD. (1976). "Control of Water Pollution from Cropland," EPA Rep. No. EPA-600/2-75-026A, ARS Rep. No. ARS-H-5-2, 2 vols. U.S. Gov. Print. Off., Washington, D.C.
U.S. Department of Agriculture (USDA). (1980). "America's Soil and Water: Conditions and Trends." USDA Soil Conserv. Serv., U.S. Gov. Print. Off., Washington, D.C.
U.S. Department of Agriculture (USDA). (1982). "Fertilizer: Outlook and Situation," Econ. Res. Serv., FS-13. USDA, Washington, D.C.

U.S. Department of Agriculture (USDA). (1983). "Agricultural Statistics 1983." U.S. Gov. Print. Off., Washington, D.C.
U.S. Department of Agriculture (USDA). (1984a). "Economic Indicators of the Farm Sector, Costs of Production," Econ. Res. Serv., ECIFS 4-1. USDA, Washington, D.C.
U.S. Department of Agriculture (USDA). (1984b). Returns to corn and soybean tillage practices. Econ. Res. Serv., Agric. Econ. Rept. No. 508. USDA, Washington, D.C.
U.S. Department of Agriculture (USDA). (1985a). Financial characteristics of U.S. farms, January 1985. *Agric. Inf. Bull. (U.S. Dep. Agric.)* No. 495.
U.S. Department of Agriculture (USDA). (1985b). The current financial condition of farmers and farm lenders. *Agric. Inf. Bull. (U.S. Dep. Agric.)* No. 490.
Vanderholm, D. H. (1975). Nutrient losses from livestock waste during storage, treatment and handling. *In* "Managing Livestock Waste," pp. 282–285. Am. Soc. Agric. Eng., St. Joseph, Michigan.
Van Doren, D. M., Triplett, G. B., Jr., and Henry, J. E. (1977). Influence of long-term tillage and crop rotation combinations on crop yields and selected soil parameters for an Aeric Ochraqualf soil [Maize]. *Ohio Agric. Res. Dev. Cent. Res. Bull.* No. 1091, Map. Ref. Sept.
Vietmeyer, N. D. (1986). Lesser-known plants of potential use in agriculture and forestry. *Science* **232,** 1379–1384.
Volk, B. G., and Loeppert, R. H. (1982). Soil organic matter. *In* "Handbook of Soils and Climate in Agriculture" (V. J. Kilmer, ed.), pp. 211–268. CRC Press, Boca Raton, Florida.
Vrabel, T. E., Minnotti, P. L., and Sweet, R. D. (1980). "Seeded Legumes as Living Mulches in Corn," Pap. No. 764. Dep. Veg. Crops, Cornell Univ., Ithaca, New York.
Wan, G., Gu, Y., and Li, C. (1959). The cultivating principles in the book, "Lu's Chinqiu." *In* "Agronomy History in China" (Zhonggue Nongxueshi) (First draft), Vol. 1, pp. 77–102. Kexue Press, Beijing.
Wentzel, R., and Robinson, K. L. (1983). "Farmer's Experience with No-Till Corn Production in Ontario County, N.Y.," A. E. Res. 83-8. Dep. Agric. Econ., Cornell Univ.,
Wiese, A. F. (1983). No-tillage crop production in temperate agriculture. *In* "No-Tillage Crop Production in the Tropics" (I. O. Akebundu and A. E. Deutsch, eds.), pp. 7–24. Published for West Afr. and Int. Weed Sci. Soc. by Int. Plant Prot. Cent., Oregon State Univ., Corvallis.
Wijewardene, R., and Waidyanatha, P. (1984). "Systems, Techniques and Tools. Conservation Farming for Small Farmers in the Humid Tropics." Dep. Agric., Sri Lanka and Commonw. Consultative Group Agric. Asia-Pac. Reg.
Wolcott, G. N. (1937). An animal census of two pastures and meadow in Northern New York. *Ecol. Monogr.* **7,** 1–90.
Zacharias, T. P., and Grube, A. H. (1984). An economic evaluation of weed control methods used in combination with crop rotation: A stochastic dominance approach. *North Cent. J. Agric. Econ.* **6**(1), 113–120.

13

Population Growth, Agrarian Structure, Food Production, and Food Distribution in the Third World

Frederick H. Buttel
Laura T. Raynolds

Department of Rural Sociology
Cornell University
Ithaca, NY

I. Introduction
II. The Malnutrition Debate
   A. Theoretical and Empirical Issues
   B. Nature and Consequences of Undernutrition and Malnutrition
III. Population Growth and Hunger
   A. Theoretical Issues
   B. Demographic Change in the Third World
   C. Relationships between Population Growth, Income, Food Production, and Food Consumption among Third World Nations
IV. The Green Revolution and the Alleviation of Hunger: Contribution and Controversy
   A. Brief History of the Green Revolution
   B. Green Revolution Controversy
   C. International Agricultural Research and the Green Revolution in Context
V. Agrarian Structure, Food Production, and Hunger
VI. Food Policy, Food Consumption, and Nutrition
VII. Discussion
Appendix: Data Sources and Operationalization of Variables for the Empirical Analysis of Food Access in Third World Countries
References

I. INTRODUCTION

There are probably no more ideologically charged issues in agriculture than those of how many hungry people there are in the world, why they have inadequate access to food, and what steps should be taken to reduce hunger and malnutrition. After decades of social science research, debates on these issues remain just as lively as they were 10 to 15 years ago.

These debates, however, are by no means confined to the social science community. Indeed, much of the intensity of these longstanding disagreements is caused by rivalries between the agricultural-biological and rural social sciences. Agricultural production scientists and scientific institutions have historically promulgated an image of their work as constituting a race between burgeoning populations and the development of new, productivity-increasing technologies. This view is shared by a significant share of the social science community, principally among economists. It has been argued that the most immediate contributions that can be made to reducing world hunger are to hasten the development of improved agricultural technology and to reduce the rate of population growth.[1] In the main, proponents of this perspective see that because of important developments in agricultural production technology over the past three decades, largely through the so-called Green Revolution, world hunger has slowly but surely declined, though the food situation in sub-Saharan Africa remains of critical concern and demands intensified efforts at developing new agricultural production technology for application there.

The majority of the social science community (and a substantial minority of agricultural scientists), however, argues that the level of food production per se tends not to be the most important factor that determines hunger and malnutrition. Instead, it is suggested that food supplies are currently adequate to provide subsistence for all people in the world and that hunger is principally a problem of the distribution of food supplies. In other words, hunger and malnutrition are principally reflections of household incomes and intrahousehold income and food distribution, of inequality of income distribution, and of forces that perpetuate poverty and inequality in the Third World. Furthermore, population growth, rather than being seen as an autonomous cause of hunger, is viewed as being yet another consequence of Third World underdevelopment and ine-

---

[1] This position can best be labeled the "productionist-Malthusian" perspective, since it combines a Malthusian view of population-food dynamics with the argument that increased food production (rather than increased income and economic security for the poor) is the most important factor in reducing world hunger (see Table 13.1). As we will point out below, however, Malthusian postures are not always shared among all of those who argue for the primacy of agricultural technology development and productivity increases in alleviating world hunger (see, e.g., Simon, 1981).

qualities of income and wealth; rapid population growth and hunger tend to occur together not so much because population growth causes hunger, but instead because population growth and hunger are common consequences of economic inequality. Finally, because there is no apparent trend toward higher and more equal incomes across Third World countries, proponents of this perspective see that hunger and malnutrition continue to worsen.

In this chapter we will devote principal attention to explicating, comparing, and assessing these rival interpretations of the causes of world hunger. We will do so in terms of five key issues in debates over access to food: (1) the statistical controversy over the prevalence of hunger and malnutrition, (2) the impact of population growth on hunger, (3) the role of the Green Revolution in combating hunger, (4) the effects of changes in agrarian structures on food production and food access, and (5) the types of food and food security policies that play a positive role in reducing hunger and malnutrition.

In addition to providing an overview of a set of issues pertaining to food distribution and consumption, we will present the results of a cross-national study of food access. These cross-national data, which consist of cross-sectional (circa 1984) data on and measures of change (circa 1974–1984) for average daily per capita calorie consumption and several explanatory variables, will enable us to shed some empirical light on issues of concern to this paper. More specifically, we will be in a position to assess the role of factors such as national income (gross national product), population growth, ratio of population to hectares of arable land, equality of income distribution, and food exports in shaping access to food among comtemporary Third World nations.

II. THE MALNUTRITION DEBATE

A. Theoretical and Empirical Issues

Reflecting on the process of preparing the monumental—though largely neglected—report of the Presidential Commission on World Hunger (1984), Walter Falcon (1984, p. 176) has noted that "[w]ith all of the previous studies on world hunger, it is truly amazing that such widely divergent views still exist on the number of people suffering moderate to severe protein-calorie malnutrition." Any person reviewing the published evidence cannot help but be similarly impressed.

The following is a representative survey of the range of estimates of world hunger (see also Poleman, 1981). There are currently somewhat over 5 billion human beings in the world. Of these 5 billion persons, Eberstadt (1981) has estimated that as few as 100 million people are affected by protein-calorie malnutrition, while Poleman (1981) has put the figure at less than 300 million. By contrast, Reutlinger and Selowsky (1976) have concluded that more than 1 billion people—more than a fifth of the world's population and about one-third of Third World people—are afflicted by protein-calorie malnutrition. Most estimates, however, fall between these two extremes. Recent studies by the World Bank have placed the number of persons afflicted by protein–calorie malnutrition at between 340 to 720 million, with the former number representing a caloric intake below which stunted growth among children and serious health risks would result, and the latter based on a caloric intake allowing for an active working life (*New York Times*, 1987). The most recent estimate by the UN Food and Agriculture Organization (FAO) is 512 million, roughly at the midpoint of the World Bank's range. The Presidential Commission on World Hunger (1980) concluded that between 500 million and 1 billion people suffer from protein-calorie malnutrition.

The estimates of the extent of malnutrition in the world appear to vary due to a combination of definitional, statistical, and ideological factors. There are differences of viewpoint about the daily requirements for various nutrients, the extent to which behavioral or biological adjustments compensate for low nutrient intake, the extent to which disease exacerbates low calorie and protein intake, and the degree to which hunger statistics should include those who are subclinically malnourished (e.g., malnutrition that inhibits an active working life) in addition to those who suffer from clinical malnutrition (e.g., stunted growth, kwashiorkor, marasmus).

Further, since the vast bulk of world hunger occurs in Third World countries in which social and health statistics are often of poor quality, many of these disagreements over the extent of hunger result from the lack of adequate data. Poleman (1981), for example, argues that official statistics underestimate both nonmarket income (in-kind remuneration) and noncommercial food production by households for their own use. Poleman thus argues that prevailing estimates of world hunger, which have been based primarily on such official statistics, tend to exaggerate the extent of world hunger. Poleman (1981) has also argued that world hunger estimates tend to ignore the role of strategies by the poor to compensate for low incomes or dietary intake (e.g., switching to inexpensive starchy staples, reducing the frequency of vigorous activity).

Finally, there is a clear ideological dimension to differences in estimates of the extent of world hunger and malnutrition. There is an ideological affinity for those who take relatively conservative positions on world hunger issues to make low estimates of the extent of malnutrition, while those who see hunger primarily in terms of income inequality and underdevelopment tend to make higher estimates.

Despite sharp disagreements among various observers of world hunger data, there is a general consensus on the fact that the extent of world hunger has increased over the past 15 years, especially since 1980. The

World Food Council of FAO estimated that during the 1970s, a decade in which there was moderately rapid economic expansion in the developed industrial and developing countries, the number of hungry people grew about 1.5 million per year on average. During the 1980s, however, with an intensification of global recession which was felt most sharply in the developing world, the number of hungry people has grown by approximately 8 million per year on average. The World Bank concurs that protein-calorie malnutrition has increased over the past decade (*New York Times*, 1987).

B. Nature and Consequences of Undernutrition and Malnutrition[2]

There are a number of frequent misperceptions about the incidence and nature of hunger. Perhaps the most frequent misperception is that hunger is most widespread in Africa. This perception has been caused by the high incidence of drought and famine in Africa during recent years. (See Sen, 1981, for an excellent overview of famines in the context of Third World poverty and hunger.) Perhaps as well an impression has been created that the successes of the Green Revolution in Asia have led to rapid expansion in food supplies and to many Asian nations becoming net food exporters. There is a certain degree of truth to these propositions, but they are also misleading. For example, it is estimated that as many as 300 million people in India remain malnourished even though India has had a successful Green Revolution, has become a net food exporter, and has severe problems of disposing of wheat and rice surpluses. These 300 million Indians lack sufficient income to purchase the grain surpluses that are exported or remain as carryover stocks.

It is generally agreed that two-thirds of the world's hungry are in the densely-populated countries in Asia, especially India, Pakistan, Bangladesh, Indonesia, the Philippines, and Kampuchea. These countries, along with Zaire and Ethiopia in Africa and Brazil in Latin America, are estimated to account for about 70% of all the world's hungry (Falcon, 1984, p. 176). Data on preschool child malnutrition also suggest that rates of chronic and acute malnutrition are generally lower in Africa in "average" (nonfamine) years than in South and Southeast Asia (Kumar, 1987).

A second frequent misperception about hunger is that the principal hunger problem is that of protein shortfalls. For two decades nutritionists had promulgated this notion, which is now generally recognized to be false (see, e.g., Lipton and Longhurst, 1985). For all but a few localized

---

[2] Undernutrition refers to inadequate caloric intake in which clinical symptoms are not generally present, whereas malnutrition refers to food intake that is deficient enough to cause clinical symptoms (Latham, 1984).

regions the principal malnutrition problem is the lack of calories. A third misperception, likely due to the increased attention in the popular press to the Sahelian and Ethiopian droughts and war-induced famines in Kampuchea and Somalia, is that hunger is principally due to these famine events, which are often perceived as occurring more frequently than in previous decades. In the main, however, global hunger is largely a phenomenon of chronic malnutrition. Also, the incidence and severity of famines are less than five decades ago, and the international donor community is now far better able to respond to famine events (Falcon, 1984).

The impacts of food intake shortfalls are not even across populations. The most vulnerable group is that of young children from roughly ages one to five for whom cereal-based diets are inadequate. Even mild malnutrition has been found to lead to stunted growth, while more severe malnutrition causes retardation in physical, psychological, and behavioral development. Among children under five years of age the interaction of undernutrition and disease commonly leads to death. Malnutrition typically is the primary or associated cause of 50% or more of the deaths of children in developing countries (Latham, 1984). Infant mortality rates of 140 per 1,000 children under one year of age are common in sub-Saharan Africa, by comparison with 25 per 1,000 in the U.S. and 13 per 1,000 in Sweden in 1983 (IBRD, 1985, pp. 218–219). It is often argued that high infant mortality rates tend to sustain high fertility rates, since having many children represents insurance that several children (especially boys) will survive to adulthood (Falcon, 1984).

Another vulnerable group is that of pregnant and lactating women, since the added stresses of childbearing exacerbate undernutrition. There is also evidence that in some countries women fare poorly in intrahousehold food allocation because of their subordinate position in the social structure (Sen, 1983).

III. POPULATION GROWTH AND HUNGER

A. Theoretical Issues

The causes and consequences of population growth have been controversial issues ever since Thomas Robert Malthus published the first edition of his *An Essay on the Principle of Population* in 1798. In particular, Karl Marx's criticisms of Malthus' theory of population in the 1850s remain widely read and of considerable relevance over 130 years later. Nonetheless, Malthus' theory of population and more recent neo-Malthusian versions have historically been the most widely accepted perspective on poopulation in development circles, though perhaps more so among agricultural scientists than social scientists. For example, there is scarcely an agricultural research and development planning document that does not call attention to burgeoning world population growth and to the need for further agricultural research to keep pace with the relentlessly growing number of hungry mouths (see, e.g., CGIAR, 1985; Plucknett and Smith, 1982).

Table 13.1 provides a typology of theoretical perspectives on food production and hunger. The typology demonstrates that the major theoretical arguments on hunger are shaped by positions on two key issues. The first issue is whether increased agricultural production is the principal factor in alleviating hunger, and the second is whether population or nonpopulation factors are major causal antecedents of hunger. The typology yields four categories, which we term (1) conservative non-Malthusianism, (2) non-Malthusian political economy, (3) productionist neo-Malthusianism, and (4) ecological neo-Malthusianism. Table 13.1 also lists illustrative exemplars for each theoretical perspective.

As noted earlier, the two variants of neo-Malthusianism have historically been the prominent views of population and hunger issues in development circles. But it should be stressed that there are some major differences between Malthus' theory and that of contemporary neo-Malthusianism. Malthus' theory of population had three major postulates: that population has a natural tendency to increase faster than subsistence, that population increases more or less rapidly according to the abundance of subsistence, and that population increase has a natural tendency to promote poverty (see Petersen, 1979, for a comprehensive summary of Malthus' population and economic theories). Of these, modern neo-Malthusians would tend to agree only with the third; indeed, the most common argument among contemporary neo-Malthusians is that rapid population

Table 13.1 A Typology of Theoretical Perspectives on Hunger in Developing Countries, with Contemporary Exemplars

Assumptions regarding the role of increased food production in alleviating hunger	Assumptions regarding the role of population growth	
	Non-Malthusian	Neo-Malthusian
"Productionist" (increased food production is central in reducing hunger)	Conservative non-Malthusianism: Simon, 1981; Simon and Kahn, 1984	Productionist-Malthusianism: Plucknett and Smith, 1982; CGIAR, 1985
"Nonproductionist" (nonproduction factors [e.g., reduction of inequality, increased economic security among the poor, reduction of economic dependency] are most central in reducing hunger)	Non-Malthusian political economy: Lappe, 1986; Mamdani, 1972; Murdock, 1980; de Janvry, 1981	Ecological Malthusianism: Brown, 1987; Brown et al., 1985; Meadows et al., 1972

growth dictates that income, land, natural resources, and food be divided among more and more persons, thereby reinforcing poverty and hunger. There are, however, two other distinctive components of neo-Malthusianism, both of which were absent in Malthus' own writings. One, stressed mainly by productionist-Malthusians, is that rapid population growth inhibits GNP and income growth by increasing the level of investments required to employ a growing labor force and by severely straining state budgets in providing education to a large number of children (see, e.g., IBRD, 1984). The second, stressed by ecological Malthusians, is that rapid population growth leads to poverty, underdevelopment, and malnutrition not only because of the logic of more persons among whom to share food and income, but also because population growth undermines the natural resource base. That is, growing population is seen to lead to deforestation, soil erosion, desertification, land degradation, and so on, which subsequently exacerbate poverty and lead to further population growth (see, e.g., Brown et al., 1985).

It is probably fair to say, though, that the persuasiveness of neo-Malthusianism has declined over the past decade. One reason, for example, is the arguably un-Malthusian nature of the population problem in Africa over the past three decades. For example, population growth rates in Africa in the 1960s were relatively modest—generally 1.5–2.0% annually (Eicher, 1986). New land was being brought into production by subsistence farmers, which enabled African countries to slowly increase their food output and to be self- sufficient in staple foods (while some countries, such as Senegal and Nigeria, were major food exporters). After independence, however, the initiation of foreign-aid-financed development projects and the pursuit of economic development by African states had led to population growth rates in the 2.5–4.4% range and to declining per capita food production and income (Office of Technology Assessment, 1984; Paulino, 1987). African development problems thus cannot be seen as a straightforward reflection of a long, autonomous dynamic of population growth, since rapid population growth is a relatively new phenomenon that seems to have been caused by, rather than having been the immediate cause of, poverty.

The prestige of neo-Malthusianism has also suffered with continual criticism of "limits to growth" (see, e.g., Meadows et al., 1972; *Global 2000 Report to the President,* 1980) type models and projections (see Humphrey and Buttel, 1982, for a summary). In particular, these models have generally projected that population growth and resource scarcities, particularly of land and fossil fuels, would result in declining per capita food production. Observers from across the theoretical and political spectrum (e.g., from Lappe, 1986, to Simon and Kahn, 1984) have presented theoretical arguments and convincing data that neo-Malthusian projections of declining per capita food production have proven to be misleading. Indeed, a further reason for the declining persuasiveness of neo-Malthu-

paths these countries have taken to exhibiting relatively comparable population growth rates (of roughly 2.3–2.5% annually). In 1960 the low-income and lower-middle-income countries had comparable crude birth rates of 44 and 43 per 1,000, respectively, while the crude death rate in low-income countries was 24 per 1,000, and that of the lower-middle-income countries 17 per 1,000 during the same period. By contrast, the 1960 average crude birth rate for the upper-middle-income countries was 40 and the crude death rate 13. From 1960 to 1982, the low-income countries (including China) exhibited the most rapid declines in the crude birth rate (from 44 to 30) and in the crude death rate (24 to 11). Again, however, the data for the low-income countries are heavily dominated by spectacular declines in the crude birth and death rates of China (from 39 to 19, and from 24 to 7, respectively). Less China, the low-income countries in 1982 had an average crude birth rate in excess of 40 per 1,000 and an average death rate of about 15 per 1,000, figures substantially greater than those of the lower- and upper-middle-income countries. The lower-middle-income and upper-middle-income countries exhibited significant, but smaller declines in their crude birth and death rates than did the low-income countries as a whole.

Particularly startling is the fact that the crude death rate for the upper-middle-income countries for 1982 was a mere eight per 1,000, which was even slightly lower than that of the advanced industrial countries (nine per 1,000). The very rapid decline in the upper-middle-income countries' crude death rates is consistent with the historical tendency for mortality rates (especially infant mortality rates) to decline in advance of fertility rates under conditions of economic growth. Thus, while the upper-middle-income countries have crude birth rates that are low by Third World standards, averaging 31 per 1,000 in 1982 (by contrast with 56 per 1,000 in Malawi, 55 per 1,000 in Kenya, and 14 per 1,000 in the advanced industrial nations), improvements in sanitation and health care have lowered the upper-middle-income countries' mortality rates sufficiently to maintain rates of population growth that are only slightly below those of the two other categories of Third World nations.

A final observation that can be made is that variations in per capita income levels are far more closely associated with fertility and mortality rates than they are with overall population growth rates. In other words, *among contemporary developing countries those with relatively low levels of per capita GNP tend to have high crude death and birth rates, while countries with higher per capita GNPs tend to have low crude death and birth rates, yielding roughly comparable high rates of population growth of 2.3 to 2.5% per annum.* As can be seen from Table 13.2, in which we report cross-national data on population growth and food consumption, there is only a modest association between per capita GNP in 1984 and the average annual rate of population growth from the mid-1970s to the

Table 13.2 Product-Moment Correlation Coefficients for the Relationships between Calories per Capita, 1983, and Selected Variables, circa 1984, Third World Nations[a]

	1	2	3	4	5	6	7	8	9
1. Per capita GNP, 1984	—								
2. Percentage of income, lowest 20th percentile	−042	—							
3. Population growth rate, 1973–1984	−373	−396	—						
4. Infant mortality rate, 1984	−650	004	303	—					
5. Food production per capita, 1984–1985	530	−141	−395	−606	—				
6. Total agricultural exports per capita, 1984	624	−105	−223	−471	570	—			
7. Cereal exports per capita, 1984	467	021	−308	−280	393	750	—		
8. Population per arable hectare, 1984–1985	627	113	−172	−259	089	784	647	—	
9. Calories per capita, 1983	558	231	−278	−628	643	259	205	135	—

[a]For most coefficients, $N = 93$, but some Ns are less than 93 when there are missing data. The coefficients are computed with pairwise deletion. Some variables are measures of change or pertain to years other than circa 1983–1984. See text and appendix for measurement details. Decimals are omitted.

mid-1980s among Third World nations. Furthermore, there is virtually no association between the population growth rate and growth in Gross Domestic Product (GDP) during this period of time (see Table 13.3).

C. Relationships between Population Growth, Income, Food Production, and Food Consumption among Third World Nations

With this background of major theoretical arguments and recent patterns of world population growth, it seems apparent that a number of arguments advanced thus far can be subjected to empirical test. In this section we will undertake an elementary empirical analysis of the impacts of several variables on mean daily calorie supply among 93 Third World nations. The nations included in the analysis are all those for which data are given in the *World Development Report 1986* (IBRD, 1986), less the developed industrial countries, European state-socialist countries, the high-income oil exporters, Cuba, and China. Two types of data will be examined. First, cross-sectional correlations among nine variables for circa 1984 have been computed. Second, measures of change for seven variables for the period of circa 1974–1984 will be presented.

While details on the measurement of variables in the study are given

Table 13.3 Product-Moment correlation Coefficients for the Relationships between Change in Calories per Capita, 1974–1983, and Selected Variables (Change from circa 1974–1984), Third World Nations[a]

	1	2	3	4	5	6	7
1. Gross domestic product growth, 1973–1984	—						
2. Agriculture growth rate, 1973–1984	304	—					
3. Change in food production per capita, 1974–1976 to 1982–1984	223	637	—				
4. Percentage of income, lowest 20th percentile	297	076	408	—			
5. Population growth rate, 1973–1984	077	014	−175	−396	—		
6. Change in population/ha arable land, 1975–1985	169	−087	−290	−056	357	—	
7. Change in calories per capita, 1974–1983	350	290	219	429	136	065	—

[a] For most coefficients, $N = 93$, but some Ns are less than 93 when there are missing data. The coefficients are computed with pairwise deletion. Some variables are measures of change or pertain to years other than circa 1983–1984. See text and appendix for measurement details. Decimals are omitted.

in the appendix, let us make a few comments on the dependent variable of the analysis, daily caloric supply per capita. These data were taken from World Bank calculations (IBRD, 1986) based on FAO data. The calculation was done by dividing the caloric equivalent of the food supplies in an economy by the population. Food supplies are taken to include domestic production, imports less exports, and changes in food stocks. Animal feed, seeds for use in agriculture, and food lost in processing and distribution are excluded.

A key issue in the use of this indicator, given our comments above about the limitations of per capita measures, is whether this variable measures the degree to which a country's population is well fed. Unlike many other per capita measures such as per capita income, daily calorie supply per capita has a practical upper bound, since well-fed people tend to consume only so many calories per day even if they are quite rich. Thus, daily calorie supply per capita should be a sensitive indicator of the degree to which the bulk of a country's population has adequate caloric intake. Undernutrition and malnutrition are conventionally understood to have their most direct reflection in high rates of infant mortality. As Table 13.2 shows, there is a relatively high correlation between daily calorie supply per capita and the rate of infant mortality ($r = -.628$), adding credence to our assumption that calorie supply data comprise a valid indicator of the adequacy of food consumption.

Table 13.2 reports the cross-sectional correlation coefficients for circa 1984. These data show that two variables—per capita GNP and food production per capita—have the closest relationships with daily calorie consumption per capita (other than infant mortality, as noted earlier). The strong positive relationship of daily calorie consumption with per capita food production would tend to support the productionist-Malthusian perspective discussed in the introduction of the paper (see also footnote 1). The strong positive relationship between daily calorie consumption and per capita GNP, however, is consistent with non-Malthusian notions.

The validity of neo-Malthusian postures can be gauged, in part, by examining the strength of relationships between population growth rate and the ratio of population to arable land, on one hand, with daily calorie supply per capita, on the other, among the sample of developing countries. Table 13.2 shows that the correlation between population growth rate and calorie supply is modest ($r = -.278$), though consistent with the neo-Malthusian expectation, while that for the ratio of population to arable land is negligible ($r = .135$) and inconsistent with the point of view.

In Table 13.1 and our previous discussion of population and hunger we suggested a more fine-grained portrayal of the debate over Malthusianism—having indicated, in particular, that there are two distinct variants of both neo-Malthusianism and non-Malthusianism. One of the major arguments of the non-Malthusian political economy perspective is that income inequality is a major factor shaping food access. Table 13.2 indicates that there is a modest correlation ($r = .231$) between percent of income accounted for by the lowest 20th percentile of income earners and daily calorie supply per capita, consistent with the non-Malthusian political economy (and, to a degree, the conservative non-Malthusian) perspective. A further argument that is frequently made among non-Malthusian political economy observers is that agricultural exports, an indicator of the economic dependency of Third World countries on the developed world, tend to be associated with insufficiency of food access. The relevant data in Table 13.2 (correlations between calorie supply and total agricultural and cereal exports) do not, however, lend support to this argument.

We have also computed comparable measures of change in the variables of this study, the correlations among which are reported in Table 13.3.[5] These data provide a somewhat different perspective on the issues

---

[5] There are some exceptions to this general procedure. One is that the cross-sectional measure of income equality is included in Table 13.3 because data over time are not available. The two exports variables are likewise cross-sectional variables. Finally, gross domestic product growth, 1973–1984, is used as the indicator of change in the level of aggregate economic activity.

of population, development, and hunger. In particular, these data show that percent of income accruing to the bottom 20th percentile of income earners has the largest correlation with change in calories per capita ($r = .429$) of all variables in the analysis, followed by change in Gross Domestic Product, 1973–84. These results could be said to be most consistent with the non-Malthusian political economy posture. The data in Table 13.3 also reveal only weak evidence that improvement in food consumption has any necessary connection with increases in agricultural productivity. The correlation between agriculture growth and change in per capita calorie supply is modest ($r = .290$), while that for change in food production per capita is even smaller ($r = .219$). These results would suggest that agricultural production and productivity improvements tend to lead to only small improvements in the adequacy of food access.

The data in Table 13.3 are particularly inconsistent with prevailing neo-Malthusian arguments. The correlation between the 1973–1984 rate of population growth and change in calories per capita is small ($r = .136$), and in fact the sign is positive—indicating that countries with the most rapid population growth tended (slightly) to have the highest rates of improvement in food adequacy. Likewise, the correlation between change in the ratio of population per hectare of arable land and food access is also small and positive ($r = .065$), inconsistent with neo-Malthusian views.[6]

IV. THE GREEN REVOLUTION AND THE ALLEVIATION OF HUNGER: CONTRIBUTION AND CONTROVERSY

A. Brief History of the Green Revolution

What is now commonly referred to as the Green Revolution had its origins in a joint food crop research program in the Mexican Ministry of Agriculture that was initiated in 1943 by the Rockefeller Foundation. When this research program was founded, Mexican wheat yields averaged 11 bushels/acre and Mexico imported half of the wheat it consumed. By the

---

[6] For both the cross sectional and longitudinal data analyses we have computed first-order partial correlation coefficients controlling for the GNP and GDP variables. The following are the first-order partials (controlling for per capita GNP) in the prediction of calories per capita, 1983; percentage of income accounted for by the lowest 20th percentile of households (.253); population growth rate ($-.170$), food production per capita (.375), population per arable hectare ($-.265$), agricultural exports per capita ($-.095$), and cereal exports per capita ($-.092$). The following are the first-order partial correlation coefficients (controlling for change in GDP, 1973–1984) in the prediction of change in calories per capita, 1974–1983: agricultural growth (.138), change in per capita food production (.207), percentage of income accounted for by the lowest 20th percentile of income earners (.366), population growth rate (.066), and change in population per arable hectare (.032).

end of the 1960s, only eight years after the first new wheat varieties had been released, Mexican wheat yields had more than tripled, to 39 bushels/acre, and Mexico had become self-sufficient in wheat despite substantial population growth in the intervening quarter century (Baum, 1986). In the interim, in 1963, the original Rockefeller-funded research program in the Mexican government was reorganized and renamed the International Center for the Improvement of Maize and Wheat (the Spanish acronym for which is CIMMYT).

Soon after the first modern wheat varieties were released in Mexico, comparable varieties were released in Pakistan and India. The results there were similar. Wheat production in Pakistan increased from 3.9 million t in 1966 to 7.3 million t in 1971. By 1980, the Pakistani wheat harvest had increased to 10.8 million t. The results were even more dramatic in India, where the release of modern varieties (MVs) in 1968 led to a 50% increase in the Indian wheat crop over the previous year (Baum, 1986, pp. 10–11).

Impressive gains in Asian rice yields also followed upon the establishment of the International Rice Research Institute (IRRI) in the Philippines in 1960 by the Rockefeller and Ford Foundations. By 1966, IRRI, only four years after it had begun operations, had released its first rice MV: IR8, a short-statured, sturdy-stemmed, day-length-insensitive variety. IR8 and subsequent rice varieties spread through Asia as fast as had CIMMYT-bred wheats. Two other agricultural research institutes, the International Institute for Tropical Agriculture (IITA) in Nigeria and Centro Internacional de Agricultura Tropical (CIAT) in Colombia, were jointly established by the Rockefeller and Ford Foundations in the 1960s. In 1970, Norman Borlaug, the well-known CIMMYT wheat breeder whose efforts are widely considered to have been essential in the South Asian wheat Green Revolution, was awarded the Nobel Peace Prize. Building on the momentum of Green Revolution success stories in Mexico and Asia, the Consultative Group on International Agricultural Research (CGIAR), the umbrella organization of the international agricultural research centers (IARCs), was founded in 1971 to facilitate raising funds for expansion of the system. Currently there are 13 IARCs with a total annual base operating budget in excess of $180 million.

Dalrymple (1985) has estimated that by 1982–1983, rice MVs were planted on 72.6 million ha in the major Third World continents, and wheat MVs on 50.7 million ha. Both rice and wheat MVs are planted on slightly more than 50% of the hectares of both crops grown in the Third World. The spread of wheat and rice MVs and MVs in general, though, has been uneven and appears to have slowed worldwide in recent years. Nearly three-quarters of wheat MV hectares are in only three countries (India, Argentina, and Pakistan), though outside of the Near East and Communist Asia large proportions of wheat hectares are now given over to MVs. In the case of rice, about 40% of MV hectares are in Communist Asia, and over 62% of the rest is in only two countries (India and Indonesia). Less than 10% of rice hectares in Africa and the Near East are planted to MVs, less than one-third in Latin America, and less than 45% of non-Communist Asia (Dalrymple, 1985, p. 1069).

Because Third World wheat production is heavily concentrated in Asia and two Latin American countries, while rice production is even more concentrated worldwide (in Asia), these data on the extent to which wheat and rice MVs have diffused worldwide may serve to exaggerate the global role of the Green Revolution. For all practical purposes, the Green Revolution has made virtually no inroads in Africa. The Green Revolution has penetrated Latin America to a greater extent, though still only a small proportion of its total arable land is planted in MVs.

The general lack of impact of MVs outside of Asia is accounted for by several factors. One is that Third World cultivators grow literally hundreds of crops, only some of which are the object of agricultural research. Also, technical progress has been slow in many crops in which the IARCs and national agricultural research institutes currently engage in research. Finally, many of the "technical packages" that have been created by agricultural researchers have led to varieties that are difficult for peasants to adopt for socioeconomic, cultural, or agronomic reasons. For example, in Mexico the bulk of maize is produced by smallholding peasants, in association with beans or squash, and largely for household consumption. The maize varieties produced by CIMMYT have tended not to be well suited to cultivation in association with other crops and have required expensive purchased inputs which are typically beyond the means of smallholder peasants. Accordingly, Mexican peasants have been reluctant to invest major sums in inputs for a crop that is largely produced for home consumption. As a result, maize production technology on small farms in Mexico has changed very little despite the fact that CIMMYT is located within its borders (DeWalt and Barkin, 1987).

For crops other than rice and wheat progress has been even slower. As noted earlier, maize, one of the original "big three" emphasized in the IARCs in the 1960s to the present, has never experienced a substantial "green revolution." Progress in the more than 30 other crops to which IARC research is now devoted has generally been even less because of the recentness of research efforts, technical problems, inadequate technological "packages," farmer resistance, and other factors.

To our knowledge there do not exist data on the additional Third World food supplies that the Green Revolution has made possible. The Green Revolution has no doubt made a major impact, albeit one confined to limited geographical zones. Major factors other than MVs that have contributed to increased Third World agricultural production include fertilizers, biocides, irrigation and water control, and (in some areas such as Brazil) expansion of the land area under cultivation.

B. Green Revolution Controversy

The record of the Green Revolution in leading to Third World agricultural production increases has been mixed. Accordingly, the Green Revolution has received mixed reviews on several other counts, particularly its socioeconomic—and, to a lesser extent, its environmental—impacts.

Norman Borlaug had yet to receive his Nobel Peace Prize when, in 1969, Clifton Wharton (1969), a well-known development economist, future President of Michigan State University, and future Chancellor of the State University of New York, raised the specter that the Green Revolution might open up a Pandora's box by increasing land concentration and landlessness. A few months later Johnston and Cownie (1969) made comparable arguments. Numerous other critiques would follow, among the most prominent of which were Frankel (1971), Byres (1972), Cleaver (1972, 1982), Griffin (1974), Oasa and Jennings (1982), and Pearse (1980).

While these critiques have set forth diverse arguments, the following points were among the most frequent and typical. First, it was argued that because MVs were made available in "packages" involving essentially obligatory use of expensive complementary inputs (fertilizers, biocides, irrigation), small farmers tended to have less access to MVs because of financial constraints than did more prosperous large farmers. Second, because MVs made agriculture more profitable—particularly when grain prices were generously supported by Third World governments—many large landlords tended to evict their sharecropppers and tenants, thereby increasing the extent of landlessness. Third, because large farmers were able to use MVs earlier and more effectively than smallholders, these farmers were able to consolidate the lands of their smaller neighbors who had been forced out of business. Fourth, it was argued that because of MV-induced increases in the size of already large farms, an incentive was created to mechanize planting and harvesting operations, resulting in decreased employment opportunities. Fifth, MVs (and associated irrigation, fertilizer, and grain price subsidies by governments) tended to result in increased land prices, benefiting large owners of land over smallholders and the landless. Finally, it has been argued that MVs were applicable largely only to favored agroecological zones, which has caused the Green Revolution to be highly unequal spatially. The empirical bases for these arguments generally were studies of the adoption of wheat MVs in Mexico and of rice and especially wheat MVs in South Asia.

Further, several observers of the Green Revolution, most prominently Perelman (1977), Ophuls (1977), and Mooney (1979), have been critical of the Green Revolution on ecological grounds. Their criticisms have generally focused on the tendency of MV technology to intensify monocultural cropping; decrease fallow periods; lead to waterlogging, salinization, and other problems because of expansion of irrigation; increase chemical runoff and soil erosion; and lead to the loss of the genetic diversity represented in traditional varieties and landraces and in wild relatives.

Proponents of the Green Revolution began to respond to the critics during the mid- to late-1970s. Their studies and publications have drawn disproportionately on the Asian rice MV experience (and virtually not at all on the Mexican wheat Green Revolution).[7] Important publications in this genre have included Ruttan (1977), Hayami (1981), and Barker and Hayami (1978). The response has included the following major arguments. Neither farm size nor land tenure has been a major constraint to MV adoption. Small farmers and tenants have been found to adopt MVs just as or more rapidly than large owner-operators, and where smallholders' adoption has come later they are soon to catch up. MVs also do not tend to increase disproportionately the productivity of large farmers over smaller ones; in most studies the yields obtained from MVs on fields of smallholders are comparable to those of large farmers. The contribution of MVs to mechanization has also been sharply disputed. Where mechanization has occurred it has been primarily attributed to distortions in the price of capital (e.g., overvalued exchange rates or subsidized credit) which have created an artificial incentive on the part of big farmers to purchase large machinery.

Whereas the critics of the Green Revolution have suggested that it has tended to increase economic inequality, Green Revolution proponents have argued that this assessment ignores the role of MVs in increasing labor demand. Various studies have indicated that MVs, especially because they often permit doublecropping, tend to increase annual labor input per hectare by 10–50% (see, e.g., Barker and Cordova, 1978; Barker et al., 1985). Many of the defenders of MVs, however, have acknowledged that MV adoption has often tended to result in a disproportionately small share of the increased output being allocated to labor rather than to land and capital (Ruttan, 1977; Mellor and Lele, 1973). They have also acknowledged that "the introduction of MV technology into a community in which resources are very inequitably distributed tends to reinforce the existing inequality" (Hayami and Ruttan, 1985, p. 338).

A final point raised in defense of MVs and the Green Revolution has

---

[7] See Bray (1986) for a discussion of some of the unique aspects of the Asian rice economies, especially the tendency toward small operational holdings (because of limited scale economies), toward highly labor-intensive cultivation practices, and toward rice being the only major crop produced. It is arguably the case that in rice areas the Green Revolution progressed so rapidly and had more modest socioeconomic impacts because of these characteristics. For example, the restricted range of inequality in the size of operational landholdings reduced the financial differences among cultivators. Also, unlike other zones where large and small farmers tend to produce different crops (de Janvry, 1981), in the rice zones the uniformity of cropping patterns created a stronger competitive compulsion for all farmers to adopt the MVs.

been that the productivity increases it has afforded have caused food prices to be lower than they otherwise would have been (CGIAR, 1985; Pinstrup-Andersen, 1982). Thus, it is argued that the Green Revolution has benefited poor consumers over their more privileged counterparts because food expenses account for a higher proportion of the income of the poor (often up to 80 percent).

Defenders of the Green Revolution not only launched formidable counterattack, but it could be said that 1970s pessimism about the impacts of the Green Revolution has been supplanted by a new enthusiasm rivaling by that of the late 1960s (see, e.g., Lipton and Longhurst, 1985). This has no doubt been due to a combination of factors, including (1) the fact that the Green Revolution has matured, such that smallerholders have caught up in their MV adoption and the major socioeconomic dislocations have occurred over a decade ago, (2) the increasingly conservative political tenor of the times, and (3) the fact that critics of the Green Revolution have not been very successful in offering plausible alternatives to MV technology.

This is not to suggest that criticism of the Green Revolution has had no effect on research in the IARCs of the CGIAR network (or on that in First World and Third World countries' agricultural research institutes). Criticism of the Green Revolution no doubt had several major impacts. One has been to increase the emphasis on integrated pest management and other nonchemical means of plant protection to enable peasants to reduce their use of expensive, purchased pesticides. Another has been the embracement of farming systems and cropping systems research programs by most of the IARCs. A third impact has been to increase the attention paid to genetic resource conservation within the CGIAR system. Criticism of the fact that MVs were primarily suitable for favored agroecological zones has led to more research on MVs suitable for less favored zones (e.g., on upland and deep-water rice). Finally, it can be said that IARC research is far more attuned than one to two decades ago to emphasize the development of varieties that require fewer purchased inputs.

Some have argued that these changes may be more apparent than real (see, e.g., Koppel and Oasa, 1987). For example, while the IARCs' technical and social science staff has developed an increasingly sophisticated understanding of peasant economies and the processes of MV adoption during the rice Green Revolution, this knowledge has been largely based on the relatively commercial peasantries in the high-productivity "rice-bowl" regions of Asia. Thus, what is known about technology adoption among rice farmers in Asia may be of limited applicability to the situations of the "limited resources" cultivators in unfavored zones and peripheral regions (Koppel, 1985). The fact remains, however, that the international agricultural research community has become increasingly more willing to respond to the concerns raised by the critics of MV technology.

Nonetheless, over the past five years there has been an ideological hardening in the views of the critics and defenders of the Green Revolution. As Lipton and Longhurst (1985) have put the matter, the pendulum of opinion has swung from naive optimism of the late 1960s, to excessively harsh criticism during the 1970s, to what arguably are exaggerated claims for the benefits of MVs in the 1980s (see also Buttel, 1986). In their words, "The pendulum has now swung too far. . . . The new 'MV euphoria' needs a critical review" (1985, pp. 2–3). They proceed to do such a review, which is among the more comprehensive, even-handed, and constructive that have yet to appear.

C. International Agricultural Research and the Green Revolution in Context

In conclusion to this section on the Green Revolution we would like to make some summary observations and comments on the past and future role of international agricultural research. First, we believe it is not particularly useful to enter into a lengthy examination of empirical evidence in order to pronounce one side or another in the Green Revolution controversy the winner. (Extensive reviews of the literature are provided by Lipton and Longhurst 1985, Ruttan, 1977, Hayami and Ruttan, 1985, Barker *et al.*, 1985, Farmer, 1977, Bayliss-Smith and Wanmali, 1984, and Pearse, 1980, with startling differences in the conclusions that are drawn.) We will merely note, much like Lipton and Longhurst (1985) have, that the impacts of the Green Revolution of the past, and the prospective impacts of MVs of the future, depend substantially on the socioeconomic context in which they are deployed and on the research priorities that are pursued. When deployed under conditions of relative equality in the distribution of resources, MVs tend to have relatively neutral impacts. When superimposed on highly unequal landholding systems and political structures, however, MVs of the sort that predominated during the early years of the Green Revolution are likely to exacerbate these inequalities.

Second, again following Lipton and Longhurst (1985), it is important to note that the more recent MVs are in many respects quite different from those that characterized the heady, but ultimately troubled, days of the Green Revolution in Mexico and Asia. Lipton and Longhurst, for example, characterize the early MVs such as IR8 as virtual "pest museums"—a reference to the fact that these varieties were highly vulnerable to pests and required extensive use of biocides. It is apparent, however, that IARC research strategies have slowly moved toward nonchemical means of pest and pathogen control. More recent varieties have greater pest and disease resistance and are, on balance, arguably more genetically diverse than the earliest ones. The Green Revolution is now being extended to nearly three dozen more food crops than was the case in the mid-1960s, expanding the long-term potential for more broad-based increases in food production.

Third, recalling our empirical analysis in the preceding section, it must be kept in mind that increased agricultural productivity is, in and of itself, merely a contributory factor to alleviating hunger and malnutrition. The advances in human welfare that can be made possible through agricultural research and production increases are not automatic; these gains are most problematic in countries with high levels of inequality and slow growth in income. Social reforms—especially income and land redistribution—are far more direct ways of reducing hunger. Yet there is now, probably quite justifiably so, a growing pessimism as to whether reforms of these sorts are possible in an era of global economic stagnation and pressure by international financial institutions, which have had the effect of reducing the ability of Third World governments to redistribute income and land (see, e.g., de Janvry, 1983). Neither supporters not critics of the Green Revolution should expect that increased agricultural productivity will rapidly or directly reduce hunger or enhance human well-being. There are many other factors that shape such outcomes.

By the same token, however, investments in agricultural research and productivity increase must be recognized as laying the foundation for long-term improvements in well-being, particularly if major institutional and social reforms become possible. Put somewhat differently, agricultural research generally has long lead times before major pay-offs are realized—the wheat and rice Green Revolutions being significant, and in some ways unfortunate, exceptions. The early successes of the wheat and rice Green Revolutions have probably misled the international donor community, Third World government officials, and others into thinking that rapid gains can be made within a few years. Such gains in the future, especially in crops other than the cereal grains in which major gains cannot be made through dwarfing and repartitioning of biomass from stalks and leaves into edible food, are, more than likely, impossible. Yet there will be major long-term costs if agricultural research is deemphasized or abandoned because it cannot yield dramatic results within five or fewer years.

Fourth, the force of the Green Revolution in affecting Third World food production capacity and agrarian structures should not cause us to see that international agricultural research is a relentless juggernaut. De Janvry and Dethier (1985), for example, while by no means entirely enthusiastic about the Green Revolution, have called attention to the weak support that agricultural research has in the developing world and the problems that would be caused by reduction in research activity. The IARCs have arguably been underfunded from the very outset. The first two IARCs, IRRI and CIMMYT, were, in the early years when they developed the first MVs, funded at a level smaller than that of any one of the five or so largest U.S. state agricultural experiment stations. With such a low level and the precarious nature of their funding, the IARCs have been forced into stressing "high-impact" research, i.e., to make the publicly visible "big hit" such as a rapid doubling of wheat yields in the Indian and Pakistan Punjabs. [The same could be said of the developing countries' national agricultural research institutes, which are essential in adapting IARC-developed materials to local conditions (de Janvry and Dethier, 1985).]

Fortunately, the CGIAR network with 13 IARCs is now in a position to broaden its research portfolio. But its $180 million of annual base funding—roughly equal to the combined annual budgets of the four largest U.S. state agricultural experiment stations—is very low when it is recognized that this amounts to barely over $1 million for each developing country. This funding is also said to be precarious in this era of state fiscal crises and concern that funding developing-country-oriented agricultural research might harm industrial-country farmers.

Fifth, future broad-based gains—from many crops, adapted to both favored and unfavored agroecological zones, and suitable for all classes of cultivators—will be difficult to achieve and will require long lead times. Even in rice, one of the glamour crops of the Green Revolution, it is now widely recognized that the yield potential of the newest varieties is no greater than that of the first IRRI variety, IR8 (Dalrymple, 1985). Each of the major research frontiers for boosting rice yields—hybridization and biotechnology—has potential problems. Hybrid rice seeds will be considerably more expensive than reproductively stable varieties, raising the possibility of a new rice Green Revolution that is more socially unequal than was the first. Biotechnology research, which is expensive and involves moving research more "upstream" to the basic or fundamental end than has been typical in the IARCs, is not likely to yield major results in rice for some time. Private sector dominance of biotechnology in the industrial countries also introduces the possibility of a socially unequal, private-sector-led second Green Revolution (Kenney and Buttel, 1985).

Sixth, the nature of Green Revolutions to come will depend largely on the research priorities that are pursued in the IARCs, the national agricultural research institutes, and quite possibly private sector research laboratories. MVs are neither inherently good nor inherently bad. While virtually any technological change can be expected to have some social impact, careful selection of research priorities can minimize these impacts, particularly on peasant smallholders.

Much of the vitriol of early criticisms of the Green Revolution came from a deeply held sense of social justice—that MVs, which were developed by a CGIAR network that presented its work as benefiting the Third World poor, had exhibited quite the opposite effects. For some this social justice conception remains persuasive, and for others not. But there can be little doubt that the problems of Third World poverty and hunger will be exacerbated if productivity-increasing technological changes in agriculture have the impact of uprooting smallholder peasantries, of increasing Third World landlessness, and of increasing rural unemployment. The vast bulk of the world's poor are rural if not agricultural people, though the nature of rural poverty—and hence the ability of technological change to affect rural poverty beneficially or adversely—varies greatly by continent (Lipton and Longhurst, 1985). Nonetheless, there are only a very small handful of countries that have the labor absorption capacity to employ gainfully agricultural and rural peoples uprooted through technological change, making retention of small farmers an important goal for the near future. These realities must bear heavily on the selection of future research priorities.

V. AGRARIAN STRUCTURE, FOOD PRODUCTION, AND HUNGER

World data sources, as sketchy as they are, do nonetheless permit us to generate a reasonably accurate picture of changes in demographic structure, health and social welfare indicators, and other indicators among a wide range of Third World nations. Unfortunately, there are essentially no comprehensive comparative data on agrarian structure and structural change for more than a handful of developing countries. The only such data source, FAO's (1970) *1970 World Census of Agriculture*, reported early 1960s (pre-Green Revolution) data that were largely limited to the size distribution of farms and land tenure. From these data we know that the pre-Green Revolution agrarian structures of the Latin American countries were characterized by the greatest degree of land concentration, followed by Africa and Asia (IBRD, 1982, p. 82). [It should be stressed that these data are reported in terms of operational (rather than ownership) units. These data therefore tend to understate the degree of land concentration in Asia where landlord-tenant relationships predominate in many countries.] This observation quite likely remains accurate today, but we have little sense of how the many changes in agricultural technology and the world and national economies have affected agrarian structures since that time. Nonetheless, comparative agrarian structure research remains very difficult to do because of the lack of data and conceptual difficulties in undertaking such comparative assessments (Ghai *et al.*, 1979).

Recognizing the severe limits of comparative data on the structure and dynamics of agrarian systems across the world, in this section of the paper we would like to identify what we feel are some of the major issues concerning agrarian structure and to provide some prospective assessments. We will begin with two separate, but interrelated issues: (1) can small farms and agricultural development programs aimed at small farmers be a meaningful engine of accumulation and a viable strategy for improving the lot of the rural poor, and (2) what are the relative efficiencies of small farms and large farms?

It should be noted at the outset that positions on these two issues transcend the left-right political-theoretical spectrum that we referred to earlier. It is widely acknowledged that the basic thrust of agricultural development efforts over the past three decades has led to few benefits to smallholders and the rural poor (see, e.g., Chambers, 1983; Johnston and Clark, 1982; de Janvry, 1981). It is also generally accepted that the relative position of smallholder peasants has deteriorated, especially in recent years as redistributionist policies have been abandoned in favor of increased production (Bailey, 1986; Grindle, 1985). Is the disappointing record of agricultural development on behalf of small farmers inherently the case, or largely the product of poorly designed agricultural development programs?

In beginning to answer this question, an initial consideration is that rural and agrarian structures differ greatly across and within the major Third World continents (see, e.g., Hyden, 1986; Ghai *et al.*, 1979; Lipton and Longhurst, 1985) so that making generalizations is particularly hazardous. Latin American peasantries, for example, are generally characterized by their subordination within latifundia-minifundia systems, high levels of agricultural and rural surplus labor, fragmentation of small landholdings, and the prevalence of semiproletarianization;[8] African peasantries by the general absence of private alienation of land, the high autonomy of peasants vis-a-vis elite classes and state policy, and lack of agricultural surplus labor; and Asian peasantries by the prevalence of tenancy relationships with landlords, intensive cultivation of small plots, increased fragmentation of landholdings, and persistent and growing problems of landlessness and surplus rural labor. These observations, of course, hold at only the most general level. They are, nonetheless, illustrative of the very wide range of circumstances in which peasant cultivators find themselves.

De Janvry (1981), in his influential account of Latin American agrarian structure, has argued that there are very formidable limits to agricultural development programs and "reformism" in improving the lot of the rural poor. He makes three key arguments in this regard. First, the Latin American peasantry is increasingly more a proletariat than a class of cultivators; accordingly, peasants can be benefited more by enhancing their status as

---

[8] Semiproletarianization refers to the process by which smallholder peasants, because of their increasingly fragmented landholdings, become increasingly dependent on wage labor for family income.

wage workers (e.g., minimum wage legislation) rather than as cultivators. Second, Latin American peasants are increasingly cultivating small, fragmented, sub-subsistence, "postage-stamp-sized" plots. Their landholdings are now too small to be viable base of accumulation. Third, there are formidable political-economic constraints to developing policy reforms that will make a major difference in the lot of peasants *qua* cultivators. De Janvry suggests that land reform is now a dead issue in Latin America because of the importance of large-scale farms in generating export income and due to the crucial role that subsistence peasants play in keeping wage and food prices low. Even less drastic reforms, such as major agricultural development programs that would increase the productivity of peasant agriculture and their security on small plots, are likely to be strongly resisted by powerful agrarian classes who depend on peasants as a cheap labor force. (It should be stressed that de Janvry is referring specifically to the Latin American situation. The dependence of large farmers on the labor of neighboring smallholder peasants, which characterizes the bulk of Latin America, is less common in Asia and Africa.)

One of the key arguments employed in support of the notion that small peasant agriculture can serve as a viable engine of accumulation and as a means for reducing rural poverty and malnutrition relates to the second question set forth above—that is, whether small farms are more efficient than large farms. The notion of the superiority of small farm productivity—particularly with respect to land and capital productivity—has been advanced by observers as diverse as Lappe and Collins (1979, Part V) and the World Bank (IBRD, 1982). The general argument is that smallholders farm their plots more intensively and carefully than larger farmers, leading to greater gross output per hectare than larger cultivators. The efficiency of small farmers has long been the basis for arguments in favor of land reform, i.e., that land reform would put redistributed lands into the hands of smaller-scale, more efficient farmers, resulting in greater output and employment opportunities (see, e.g., Dorner, 1973). Impressive compendia of data have been marshaled in support of this argument (see, especially, Berry and Cline, 1978).

While recognizing that there are many instances in which the generalization of small-farmers' superior productivity holds, two caveats should be mentioned. One is that the vast bulk of the data in support of the argument are relatively old, typically collected prior to the time when large, semifeudal, extensively-cultivated estates became transformed into modern, highly-commercial, specialized—and often quite efficient—farms. A second is that the efficiency of small farms is often the result of the weaknesses—rather than the strengths—of small-scale farming. Smallholder peasants on postage-stamp plots with low nonfarm income-earning possibilities may, because of their poverty, be forced to intensively cultilvate their plots for the lack of any other avenues for productively deploying family labor. Would, for example, the cultivator of a 0.5-ha sharecropped plot in the northeast of Brazil farm a larger owned plot of 5 ha just as intensively? The answer to this question remains unclear in the absence of more recent comparative data.

A third pressing issue relating to agrarian structure, rural poverty, and hunger is that of the impact of the increased prevalence of export-oriented production by medium- to large-scale farms in many regions of the Third World. A number of observers have documented this trend, especially in Third World zones proximate to First World markets (e.g., Central America, Africa; de Janvry, 1981; Sanderson, 1985). These trends have typically been explained in terms of the logic of labor-intensive horticultural production and land-extensive cattle production tending to move toward areas of more abundant labor and cheaper land. Also, in many Third World countries the saturation of domestic markets for grains has led large farmers to seek out more profitable investment opportunities, which typically revolve around producing for rich countries or rich classes in which the source of demand is more dynamic. Finally, given the Third World debt crisis and pervasive balance of payments problems, the need for export revenues has intensified over the past decade.

How might the growing trend to large-scale, export-oriented production in the Third World affect rural poverty and hunger? Two major concerns suggest themselves. One is that the shift of production away from basic food crops for domestic consumption toward luxury commodities for export will reduce food availability for the rural and urban poor. The other is that the superior profitability of export-oriented production might ultimately lead to larger farmers being able to appropriate the lands of their smaller neighbors.

Comprehensive assessments of these issues are scarce. Sisler and Blandford (1984), for example, have laid out a detailed set of considerations and have effectively identified the nature of the trade-offs involved, but without drawing clear conclusions. Bailey (1986) has explored the possible implications of the increased export orientation of Central and South American and Caribbean nations and has suggested that its impacts on peasants and food availability will be most severe in countries that are also net food importers (e.g., Haiti, Dominica, Trinidad/Tobago). He also has projected substantial adverse trade-offs in Mexico, Peru, and Costa Rica. Nonetheless, it is Bailey's estimate that the increased export orientation in the region will serve to sharpen the already high degrees of dualism in agriculture and in their economies more generally. For many countries, however, there will be strong pressures to encourage export-oriented production to deal with foreign debt and foreign exchange problems.

VI. FOOD POLICY, FOOD CONSUMPTION, AND NUTRITION

Virtually all Third World governments directly or indirectly intervene in their agricultural and food economies to accomplish particular social and political goals. Typical policy instruments have included overvalued exchange rates, subsidized credit to large farmers, minimum wages for urban workers, and food price controls, each of which has tended to keep food prices cheap. For the poor—particularly the urban poor—in many Third World countries these direct and indirect cheap food policies have raised food consumption standards above the levels that they otherwise would have been. But a wide range of analysts has suggested that Third World agricultural and food policies, by making food prices artificially cheap, have undermined agricultural development, exacerbated inequalities in agriculture, and prejudiced long-term food security (Bates, 1979; Timmer et al., 1983; de Janvry, 1981).

The tendency toward cheap food policies should not be considered merely an instance of policy myopia. Indeed, there are several understandable—in many respects reasonable—explanations for why these policies have been undertaken. In particular, it is essential to recognize the trade-offs involved between production-oriented and cheap food policies. Even Falcon (1984, p. 1186), an unabashed advocate of raising prices for farmers, is keenly aware of the nature of the trade-offs:

> In general, low-income countries tend to discriminate against the agricultural sector and to provide less than international prices to their farmers. . . . [R]aising prices to farmers in many countries is absolutely essential. However, it is more than sheer neglect or urban bias that keeps governments from making this change. Higher food prices also mean lower real incomes, especially for poorer people, who may spend up to 80 percent of their income on food. This basic pricing dilemma—short-term consumption losses versus long-run production gains—needs to be recognized for the very real problem it poses, even for the most responsible government.

Following Falcon, there has recently been an outpouring of major books and papers on agricultural price policy in which the underlying message is the importance of "getting prices right" (Timmer, 1986; see also Timmer et al., 1983) in order to provide adequate incentives to farmers to make investments and increase their production over the long term. Most such food policy analysts have blamed cheap food policies on the power of urban consumers and the tendency for Third World regimes to yield to this political pressure. For example, Timmer et al. (1983, pp. 271–272) argue that "in virtually all poor countries the strongest and most visible constraint on choosing among food policy alternatives is consumer pressure in urban areas to keep basic food prices cheap. The less developed the country, the more acute the problem."

Other research efforts in this area suggest additional factors involved in creating cheap food policies. Bates (1979), for example, agrees with Timmer *et al.* on the importance of political pressure from urban consumers, but he stresses as well that African states have strongly tended to exempt the crops cultivated by rich farmers from the food price ceilings and subsidized imports that plague peasant-produced crops.

De Janvry (1983) has more fundamentally challenged the model of urban-consumer-led pressure against "getting prices right." De Janvry agrees with Bates on the role played by farm lobbies in shaping the prices of luxury and agroindustrial commodities. But he suggests that this instrumental political role of rich farmers is apparently in decline as economic and fiscal crisis sharply limit the range of choices available to policy-makers. For de Janvry (1983), it is crucial to recognize that there is an inescapable logic of agricultural and development policy in underdeveloped countries. This logic is based on the fact that underdeveloped countries, because of their lack of capital and technology, are forced to compete in the international division of labor essentially on the basis of providing cheap labor. Cheap food, in turn, is essential in keeping labor cheap so that technologically inferior Third World industrialists and other producers will be able to compete with their First World counterparts. Thus, in de Janvry's view, the dilemma of "getting prices right" is more complex than many food policy analysts have portrayed. Increasing the prices of farm commodities will, as Timmer *et al.* suggest, increase hunger and unrest among nonfarm consumers. But increased farm product prices will also threaten to undermine the world-economic basis of developing-country accumulation—cheap labor.[9] More expensive food (along with terminating export taxes) will also tend to reduce the level of surplus that can be extracted from agriculture—surpluses that have heretofore been pivotal in investments in Third World industry.

De Janvry (1983) is thus substantially more pessimistic than Timmer *et al.* (1983) about whether use of pricing instruments to provide for long-term food production increases can be accomplished smoothly or equitably. De Janvry suggests that these policies can be successful only under

---

[9] For de Janvry, the significance of cheap labor as the basis of Third World accumulation is that this pattern of accumulation differs fundamentally from that which characterized the advanced industrial nations, i.e., ("articulated") accumulated driven by wages and peasant income as the principal source of demand. In Third World ("disarticulated") economies, the key dynamic of growth and accumulation lies in affluent local classes and especially in exports, rather than in wages and peasant incomes. Increased wages and peasant incomes, in fact, would tend to interrupt disarticulated accumulation by making it more difficult for Third World industry to compete in export markets on the basis of cheap labor costs.

two conditions. One is a situation where agricultural capital is relatively equitably distributed—a rare occurrence in the contemporary Third World. The other—a sharp political movement to a social-democratic political-economic structure in which peasants and workers share in productivity gains—is equally rare.

It is arguably the case at present that product pricing policies of the sort suggested by Timmer and colleagues have become far more prevalent in the past few years. These policy changes, however, rather than reflecting changing policy preferences of Third World state officials, appear to be largely due to pressure by international lending institutions and by the U.S. Agency for International Development. Judging from recent FAO data on the increased prevalence of hunger and malnutrition (*New York Times*, 1987), these policies are likely contributing to what both Timmer and de Janvry have predicted and feared: higher food prices (than would otherwise have occurred), declining real incomes for poor consumers, and increased hunger. Whether these policy reforms will be in place sufficiently long to evaluate their ablility to stimulate long-term productivity and income growth and reduction of hunger is difficult to say.

VII. DISCUSSION

In this chapter we have dealt with a broad range of issues bearing on the general matter of whether hunger is primarily caused by too many people, too little agricultural production, or ineffective mechanisms of food distribution. We have, without saying so explicitly up to this point, placed greater weight on the complex of "distribution" factors than on population growth or lack of production. This is not to diminish the ultimate importance of restraining population growth and ensuring that there is food to distribute. Our perspective is that population and food production factors are often given far more emphasis than is warranted, while socioeconomic phenomena are given too little emphasis.

We have also sought to convey the notion that alleviating world hunger is not a simple matter. Colonial legacies, unwise policies, the lack of appropriate agricultural technology, unequal (or dualistic) agrarian structures, population growth, authoritarian regimes, and many other factors render solutions to the hunger problem extremely difficult. Our broad conclusion about the importance of socioeconomic and distributional factors does not mean that solutions can be pursued along this single dimension. Population control, even if it may play little immediate role in reducing poverty and hunger, is nonetheless warranted because it will enhance social well-being decades into the future when, with hope, more direct measures to alleviate hunger have been implemented. The same can be said for increasing investments in agricultural research and for increasing agricultural productivity. However, if our interpretation of the evidence is correct, neither population control nor productivity increase will be as successful as it could be unless changes along the socioeconomic-distributive dimension are made.

APPENDIX: DATA SOURCES AND OPERATIONALIZATION OF VARIABLES FOR THE EMPIRICAL ANALYSIS OF FOOD ACCESS IN THIRD WORLD COUNTRIES

The operational definitions of the variables used in this analysis are given below along with the sources of data. The main source of data was IBRD (1986). Variables taken from this standard source will not be defined at length.

GNP per capita, 1984; gross domestic product growth, 1973–1984; agriculture growth, 1973–1984; population growth, 1973–1984; and infant mortality rate, 1984, were taken directly from IBRD (1986). Percent of income received by the lowest 20th percentile of households was also taken directly from IBRD (1986); these data pertain to various years between 1972 and 1982 and are available for only 26 countries. Change in calories per capita is the percentage change in daily calorie supply per capita during the period from 1974 (IBRD, 1978) to 1983 (IBRD, 1986).

Food production per capita, 1984–1985, is the total quantity of food produced in an economy in 1985 times the price weight (in constant dollars) at 1976–1978 average levels (USDA, 1986). This is divided by the mid-1984 population (IBRD, 1986). Change in food production per capita is an index measuring the change in average annual quantity of food produced per capita in 1982–84 relative to that in 1974–1976 (IBRD, 1986). Total agricultural exports per capita, 1984, is the total agricultural exports for 1984 in current dollars (FAO, 1985a) divided by the mid-year population (IBRD, 1986). Cereal exports per capita, 1984, measures the exports of wheat, rice, barley, maize, rye, and oats in 100 metric tons, divided by the mid-year population (IBRD, 1986); wheat flour is included in terms of wheat equivalents, and other flours are not included (FAO, 1985a).

Population per arable hectare, 1984–1985, divides the 1985 total arable area (FAO, 1985b)—including land under temporary crops, temporary meadows for mowing or pasture, gardens, and land temporarily fallow—by the mid-1984 population (IBRD, 1986). Population per arable hectare, 1975, is computed as above (FAO, 1976; IBRD, 1983). Change in population per arable hectare measures the percentage change in the two preceding variables from 1975 to 1984–1985.

REFERENCES

Bailey, J. J. (1986). "Politics and agricultural trade policies in the Western Hemisphere: a survey of trends and implications for the 1980s." *In* "Food, the State, and International Political Economy" (F. L. Tullis and W. L. Hollist, eds.). Univ. of Nebraska Press, Lincoln.

Barker, R., and Cordova, V. G. (1978). "Labor utilization in rice production." *In* "Economic Consequences of the New Rice Technology" (R. Baker and Y. Hayami, eds.). Int. Rice Res. Inst., Los Banos, Philippines.

Barker, R., and Hayami, Y., eds. (1978). "Economic Consequences of the New Rice Technology." Int. Rice Re. Inst., Los Banos, Philippines.

Barker, R., *et al.* (1985). "The Rice Economy of Asia." Resources for the Future, Washington, D.C.

Bates, R. H. (1979). "Markets and States in Tropical Africa." Univ. of California Press, Berkeley.

Baum, W. C. (1986). "Partners Against Hunger." World Bank, Washington, D.C.

Bayliss-Smith, T., and Wanmali, S. eds. (1984). "Understanding Green Revolutions." Cambridge Univ. Press, London and New York.

Berry, R. A., and Cline, W. R. (1978). "Agrarian Structure and Productivity in Developing Countries." Johns Hopkins Press, Baltimore, Maryland.

Blaikie, P. (1985). "The Political Economy of Soil Erosion in Developing Countries." Longman, New York.

Boserup, E. (1965). "The Conditions of Agricultural Growth." Allen & Unwin, London.

Bray, F. (1986). "The Rice Economies." Blackwell, Oxford.

Brown, L. (1987). Analyzing the demographic trap. *In* L. Brown *et al.*, "State of the World, 1987." Norton, New York.

Brown, L., *et al.* (1985). "State of the World, 1985." Norton, New York.

Buttle, F. H. (1986). On revisionism and pendulum swings: a review essay. *Rural Sociol.* **51**, 229–234.

Byres, T. J. (1972). The dialectic of India's green revolution. *South Asian Rev.* **5**, 99–116.

Chambers, R. (1983). "Rural Development." Longman, New York.

Cleaver, H. M., Jr. (1972). The contradictions of the green revolution. *Am. Econ. Rev.* **62**, 177–186.

Cleaver, H. M., Jr. (1982). Technology as political weaponry. *In* "Science, Politics, and the Agricultural Revolution in Asia" (R. Anderson, ed.). Westview Press, Boulder, Colorado.

Consultative Group on International Agricultural Research (CGIAR). (1985). "Summary of International Agricultural Research Centers: A Study of Achievements and Potential." CGIAR, Washington, D.C.

Dalrymple, D. G. (1985). The development and adoption of high-yielding varieties of wheat and rice in developing countries. *Am. J. Agric. Econ.* **67**, 1067–1073.

de Janvry, A. (1981). "The Agrarian Question and Reformism in Latin America." Johns Hopkins Press, Baltimore, Maryland.

de Janvry, A. (1983). Why do governments do what they do? The case of food price policy. *In* "The Role of Markets in the World Food Economy" (D. G. Johnson and G. E. Schuh, eds.). Westview Press, Boulder, Colorado.

de Janvry, A., and Dethier, J.-J. (1985). "Technological Innovation in Agriculture." World Bank, Washington, D.C.

DeWalt, B. R., and Barkin, D. (1987). Seeds of change: the effects of hybrid sorghum and agricultural modernization in Mexico. *In* "Technology and Social Change" (H. R. Bernard and P. J. Pelto, eds.). Waveland Press, Prospect Heights, Illinois.

Dorner, P. (1973). "Land Reform and Economic Development." Penguin, Baltimore, Maryland.
Eberstadt, N. (1981). Hunger and ideology. *Commentary* **72,** 40–49.
Eicher, C. K. (1986). Strategic issues in combating hunger and poverty in Africa. *In* "Strategies for African Development" (R. J. Berg and F. S. Whitaker, eds.). Univ. of California Press, Berkeley.
Evans, P. (1979). "Dependent Development." Princeton Univ. Press, Princeton, New Jersey.
Falcon, W. P. (1984). The role of the United States in alleviating world hunger. *In* "Agricultural Development in the Third World" (C. K. Eicher and J. M. Statz, eds.). John Hopkins Press, Baltimore, Maryland.
Farmer, B. H., ed. (1977). "Green Revolution?" Macmillan, New York.
Food and Agriculture Organization (FAO). (1970) "1970 World Census of Agriculture." FAO, Rome.
Food and Agriculture Organization (FAO). (1976). "FAO Production Yearbook," Vol. 30. FAO, Rome.
Food and Agriculture Organization (FAO). (1985a). "FAO Trade Yearbook," Vol. 39. FAO, Rome.
Food and Agriculture Organization (FAO). (1985b). "FAO Production Yearbook," Vol. 39. FAO, Rome.
Frankel, F. R. (1971). "India's Green Revolution." Princeton Univ. Press, Princeton, New Jersey.
Ghai, D., Lee, E., Maeda, J., and Radwan, S., eds. (1979). "Overcoming Rural Underdevelopment." Int. Labour Off., Geneva.
"Global 2000 Report to the President." (1980). U.S. Gov. Print. Off., Washington, D.C.
Goodman, D., and Redclift, M. (1982). "From Peasant to Proletarian." St. Martin's, New York.
Griffin, K. (1974). "The Political Economy of Agrarian Change." Harvard Univ. Press, Cambridge, Massachusetts.
Grindle, M. (1985). "State and Countryside." Johns Hopkins Press, Baltimore, Maryland.
Hayami, Y. (1981). Induced innovation, green revolution, and income distribution: comment. *Econo. Dev. Cult. Change* **30,** 169–176.
Hayami, Y., and Ruttan, V. W. (1985). "Agricultural Development," 2nd Ed. John Hopkins Press, Baltimore, Maryland.
Homem de Melo, F. (1986). Unbalanced technological change and income disparity in a semi-open economy: the case of Brazil. *In* "Food, the State, and International Political Economy" (F. L. Tullis and W. L. Hollist, eds.). Univ. of Nebraska Press, Lincoln.
Hopkins, R. (1986). Food security, policy options, and the evolution of state responsibility. *In* "Food, the State and International Political Economy" (F. L. Tullis and W. L. Hollist, eds.). Univ. of Nebraska Press, Lincoln.
Humphrey, C. R. and Buttel, F. H. (1982). "Environment, Energy, and Society." Wadsworth, Belmont, California.
Hyden, G. (1986). African social structure and economic development. *In* "Strategies for African Development" (R. J. Berg and J. S. Whitaker, eds.). Univ. of California Press, Berkeley.
International Bank for Reconstruction and Development (IBRD), The World Bank. (1978). "World Development Report 1978." Oxford Univ. Press, London and New York.
International Bank for Reconstruction and Development (IBRD), The World Bank. (1982). "World Development Report 1982," Oxford Univ. Press, London and New York.
International Bank for Reconstruction and Development (IBRD), The World Bank. (1983). "World Development Report 1983." Oxford Univ. Press, London and New York.
International Bank for Reconstruction and Development (IBRD), The World Bank. (1984). "World Development Report 1984." Oxford Univ. Press, London and New York.

International Bank for Reconstruction and Development (IBRD), The World Bank. (1985). "World Development Report 1985." Oxford Univ. Press, London and New York.
International Bank for Reconstruction and Development (IBRD), The World Bank. (1986). "World Development Report 1986." Oxford Univ. Press, London and New York.
Johnston, B. F., and Clark, W. C. (1982). "Redesigning Rural Development." John Hopkins Press, Baltimore, Maryland.
Johnston, B. F., and Cownie, J. (1969). The seed-fertilizer revolution and labor force absorption *Am. Econ. Rev.* **59**, 569–582.
Kahn, H. (1979). "World Economic Development." Morrow, New York.
Kenney, M., and Buttel, F. H. (1985). Biotechnology: prospects and dilemmas for Third World development. *Dev. Change* **16**, 61–91.
Koppel, B. (1985). Technology adoption among limited resource rice farmers in Asia. *Agric. Adm.* **20**, 201–223.
Koppel, B., and Oasa, E. (1987). Induced innovation theory and Asia's green revolution: a case study of an ideology of neutrality. *Dev. Change* **18**, 29–67.
Kumar, S. K. (1987). The nutrition situation and its food policy links. *In* "Accelerating Food Production in Sub-Saharan Africa" (J. W. Mellor *et al.*, eds.). Johns Hopkins Press, Baltimore, Maryland.
Lappe, F. (1986). "World Hunger." Inst. Food Dev. Policy, San Francisco, California.
Lappe, F., and Collins, J. (1979). "Food First." Ballantine, New York.
Latham, M. C. (1984). International nutrition problems and policies. *In* "World Food Issues" (M. Drosdoff, ed.). Program Int. Agric., Cornell Univ., Ithaca, New York.
Lipton, M., and Longhurst, R. (1985). "Modern Varieties, International Agricultural Research, and the Poor." World Bank, Washington, D.C.
Mamdani, M. (1972). "The Myth of Population Control." Monthly Review Press, New York.
Meadows, D. H., *et al.* (1972). "The Limits to Growth." Universe, New York.
Mellor, J. W., and Lele, U. J. (1973). Growth linkages with the new foodgrain technologies. *Ind. J. Agric. Econ.* **28**, 35–55.
Michaelson, K. L., ed. (1981). "And the Poor Get Children: Radical Perspectives on Population Dynamics." Monthly Review Press, New York.
Mooney, P. R. (1979). "Seeds of the Earth." Inter Peres, Ottawa.
Murdock, W. M. (1980). "The Poverty of Nations." Johns Hopkins Press, Baltimore, Maryland.
New York Times. (1987). World hunger found still growing. *New York Times* June 28, p. A2.
Oasa, E., and Jennings, B. H. (1982). Science and authority in international agricultural research. *Bull. Concerned Asian Scholars* **14**, 30–44.
Office of Technology Assessment (OTA). (1984). "Africa Tomorrow." OTA, Washington, D.C.
Ophuls, W. (1977). "Ecology and the Politics of Scarcity." Freeman, San Francisco, California.
Paulino, L. A. (1987). The evolving food situation. *In* "Accelerating Food Production in Sub-Saharan Africa" (J. W. Mellor *et al.*, eds.). Johns Hopkins Press, Baltimore, Maryland.
Pearse, A. (1980). "Seeds of Plenty, Seeds of Want." Oxford Univ. Press, London and New York.
Perelman, M. (1977). "Farming for Profit in a Hungry World." Allanheld, Osmun, Montclair, New Jersey.
Perlman, M. (1984). The role of population projections for the year 2000. *In* "The Resourceful Earth" (J. Simon and H. Kahn, eds.). Blackwell, Oxford.
Petersen, W. (1979). "Malthus." Harvard Univ. Press, Cambridge, Massachusetts.
Pinstrup-Andersen, P. (1982). "Agricultural Research and Technology in Economic Development." Longman, London.

Plucknett, D. K., and Smith, N. J. H. (1982). Agricultural research and third world food production. *Science* **217,** 215–220.

Poleman, T. T. (1981). Quantifying the nutrition situation in developing countries. *Food Res. Inst. Stud.* **18,** 1–58.

Population Reference Bureau. (1986). "World Population Data Sheet of the Population Reference Bureau, Inc." Popul. Ref. Bur. New York.

Presidential Commission on World Hunger. (1984). "Overcoming World Hunger: The Challenge Ahead." Presidential Commission on World Hunger, Washington, D.C.

Redclift, M. (1984). "Development and the Environmental Crisis." Metheun, New York.

Reutlinger, S., and Selowsky, M. (1976). "Malnutrition and Poverty." Johns Hopkins Press, Baltimore, Maryland.

Ruttan, V. W. (1977). The green revolution: seven generalizations. *Int. Dev. Rev.* **19,** 16–23.

Sanderson, S. E., ed. (1985). "The Americas in the New International Division of Labor." Holmes & Meier, New York.

Sen, A. K. (1981). "Poverty and Famines." Oxford Univ. Press, London and New York.

Sen, A. K. (1983). Economics and the family. *Asian Dev. Rev.* **1.**

Simon, J. (1981). "The Ultimate Resource." Princeton Univ. Press, Princeton, New Jersey.

Simon, J., and Kahn, H., eds. (1984). "The Resourceful Earth." Blackwell, Oxford.

Sisler, D. G., and Blandford, D. (1984). "Rice or rubber?"—the dilemma of many developing nations. *In* "World Food Issues" (M. Drosdoff, ed.). Program Int. Agric., Cornell Univ., Ithaca, New York.

Timmer, P. (1986). "Getting Prices Right." Cornell Univ. Press, Ithaca, New York.

Timmer, P., Falcon, W. P., and Pearson, S. A. (1983). "Food Policy Analysis." Johns Hopkins Press, Baltimore, Maryland.

U.S. Department of Agriculture (USDA). (1986). "World Indices of Agricultural and Food Production, 1976–85," Stat. Bull. No. 744. Econ. Res. Serv., USDA, Washington, D.C.

Wharton, C. R., Jr. (1969). The green revolution: cornucopia or Pandora's Box. *Foreign Aff.* **47,** 464–476.

14

Environment and Population: Crises and Policies

David Pimentel, Linnea M. Fredrickson,
David B. Johnson, John H. McShane, and
Hsiao-Wei Yuan

College of Agriculture and Life Sciences
Cornell University
Ithaca, New York

I. Introduction
II. Human Needs Worldwide
III. Standard of Living and Population Growth
IV. Population Growth
V. Per Capita Use of Resources in the United States and China
VI. State of the Environment in the United States and China
VII. Policy Decisions Concerning Environmental and Population Problems
 A. Pesticides
 B. Water Pollution
 C. Air Pollution
VIII. National Population Policies in the United States and China
IX. Conclusion
 References

I. INTRODUCTION

For the greater part of this century, leading scientists, public officials, and various organizations have been calling attention to the world's deteriorating environment and rapidly growing human population (Sanger, 1927; U. N., 1955, 1966, 1975, 1984a; Ehrlich and Holdren, 1971; Meadows *et al.*, 1972; NAS, 1975; Ridker, 1979; CEQ, 1980; Keyfitz, 1984; Demeny, 1986; Hardin, 1986). Statistics have been analyzed and world conferences assembled with resulting expressions of genuine concern for the future quality of human life. The conclusion is clear: If the environment and associated resources continue to deteriorate by mismanagement and the increasing demands of rapidly growing populations, any hope of achieving a quality life for all people will be impossible.

Yet most societies and governments appear unable to deal with these problems, and have poor records of effectively managing and protecting essential natural resources from over-exploitation by growing human numbers. Evidence suggests that this difficulty exists because nations have not developed cohesive policies stating their desired standard of living while clearly recognizing that this standard is interdependent with environmental quality and population density.

Most decisions concerning people, natural resources, and the environment currently being made in the United States and other nations appear to be *ad hoc* policies designed to protect or promote a particular aspect of human well-being and/or the environment. These policies are often adopted only after a crisis evolves from years of interspersed contention and indifference. Benjamin Franklin wrote in *Poor Richard's Almanac* that it is not until "the well runs dry, we know the worth of water." Little has changed since his day. Apparently, it is not until a human situation becomes intolerable that some corrective action is finally taken.

In this chapter we compare the rates of population growth and environmental degradation in the United States and China—one country without a population policy and one country forced by crises to develop one. We also examine the role that crises have played in forcing environmental decisions and policies in the United States. It is our aim, by examining the reluctance of people to assess the interdependency of populations, environment, and standard of living, to help societies and governments improve and integrate their decisions and policies in these interrelated areas before "the well runs dry." For human well-being, now and in the future, society should consider using its knowledge, wisdom, and technology to safeguard the environment and resources plus establish sound national population policies.

The reason we compare the United States and China when so many cultural, historical, and geographical differences separate the two is that both nations are populated by humans who have the same basic needs that can be provided only by the environment, and both nations have growing populations. For residents of the United States, it is useful to compare resource and population figures to speculate about the future. The wall China has erected and forced itself against can be, and presently is being, erected in the United States. Though the United States does not yet face such a wall, the trends of our growth, consumption, and continued mismanagement of natural resources would indicate that we will be against it in a half-dozen generations.

II. HUMAN NEEDS WORLDWIDE

At the very least, humans require food, water, shelter, fuel, and protection from disease in order to survive. All of these basic needs are obtained from the resources of our environment: land, water, fossil fuels, the atmospheric elements, and diverse life forms. Our standard of living depends on the abundance and quality of these resources, which in turn depends on technologies and how well humans are able to manage them.

Considering food, it is possible for an adult to live by consuming 10 kg of potatoes and 0.5 liter of milk per day (Connell, 1950). Nutritionally, this is reported to be a satisfactory diet for an adult. The Irish consumed a similar diet in the early 1800s (Connell, 1950). However, eating potatoes and milk for breakfast, lunch, and dinner would not be considered consonant with a high standard of living in the United States.

Concerning water, an adult must drink two to three liters of water per day (Pimentel et al., 1982). An additional minimum of two liters of water are necessary for cooking and washing. But, as of 1980, more than 25% of the developing world's urban population and more than 70% in rural areas lacked reasonable access to clean water (U. N., 1984b). Decent waste disposal facilities are not available in 47% of urban and 87% of rural areas in developing countries. Even though the percentage of people having access to sanitary water and waste disposal facilities increased in the 1960s and 1970s, the absolute number of people who do without has increased because of population growth (U. N., 1984b). Given a choice, most people would prefer clean water, bathroom facilities, and a modern sewage disposal system for health, cleanliness, and convenience.

A room the size of an ordinary living room (4 × 5 m) can suffice as shelter for 25 adults, as observed in some urban and rural areas of the world (Lappé, 1982). Whatever nations' desired standards of living may be, such housing would hardly seem adequate for comfort and health.

To provide food, water, and shelter, energy resources are essential. It is estimated that at least 9,000 kcal/day of food, fossil, and biomass fuel are required to meet a human's minimum needs for survival (Pimentel, 1984). For the average standard of living in the United States, each person uses 205,000 kcal/day, or 23 times the minimum.

In addition, energy is used to protect humans from disease. And disease control (reduced mortality) is the most important factor contributing to human population growth (NAS, 1971). For example, pesticides and medicines are used to control malaria and other diseases. Also, energy is used to purify and eliminate disease organisms from water and then pump it to consumers.

III. STANDARD OF LIVING AND POPULATION GROWTH

The gross national product (GNP) is sometimes used as a measure of the standard of living of a nation (although this figure is questionable). Based on this criterion, the United States per capita average is $14,080, whereas in China it is just $300 (PRB, 1986). The average for the world is $2,760. Another descriptive measure is the "physical quality of life index" (PQLI), which takes longevity, infant mortality, and literacy into account (WABF, 1982). Based on a scaled maximum of 100, PQLI rates the United States a 95 and China a 71; however, no measure is fully satisfactory for making a comparative assessment of the standard of living.

These figures merely illustrate what everyone knows: Among the world's nations, the United States has a high per capita standard of living. Because the U.S. standard of living depends upon its consumption habit, it appears to be quite fragile.

Several investigators have pointed out that it would be impossible to provide each person in the world (five billion humans) with the equivalent of $14,080/yr, 205,000 kcal/day of fossil energy, and 721 kg/yr of food like that consumed in the United States (Hardin, 1968; Ehrlich and Holdren, 1969; Pimentel and Pimentel, 1979; Hardin, 1986). Simply put, multiplying the American per capita use of resources by five billion people exceeds most basic world resources. This unbalanced situation worsens day by day because the world population is rapidly increasing at a rate of 1.7% per year—1,700 times faster than that of the first two million years of human existence. Such a growth rate adds more than 230,000 people a day to our world population, thus virtually assuring a continuing decline in the worldwide standard of living (and most likely assuring an eventual decline in the U.S. standard of living). Demographers project that the world population will reach 6.1 billion by the year 2000, approach 8.2 billion by 2025 (U. N., 1982), and reach about 12 billion by 2100. Even with the disturbing knowledge of these figures, there still seems to be no generally accepted way to limit this growth (NAS, 1975).

IV. POPULATION GROWTH

Although the average rate of world population growth is an alarming 1.7%, the rate of growth in some nations (e.g., Kenya) is as high as 4.2% (PRB, 1986). This is a population doubling time of just 17 years (the U.S. population doubling time is 99 years). With such a population growth rate, how can countries like Kenya provide the teachers, doctors, farmers, industrialists and others needed for a viable economy and society, let alone a reasonably comfortable standard of living?

In addition to the rate of increase, one must also consider that about 40% of the population in most developing countries is now within childbearing age (PRB, 1986). A young age structure will continue to contribute to rapid population growth for decades because the young still have reproductive lives ahead of them (Coale, 1984). For example, if 70% of the couples in China have just one child, the current population of slightly more than one billion (Coale, 1984) will reach 1.2 billion by the year 2000 (Wren, 1982; Zao, 1982) (Figure 14.1). China will have to deal with 200 million more people in slightly more than a decade—a population slightly less than the total U.S. population today.

The rate of population growth in China is now 1%, which is a pop-

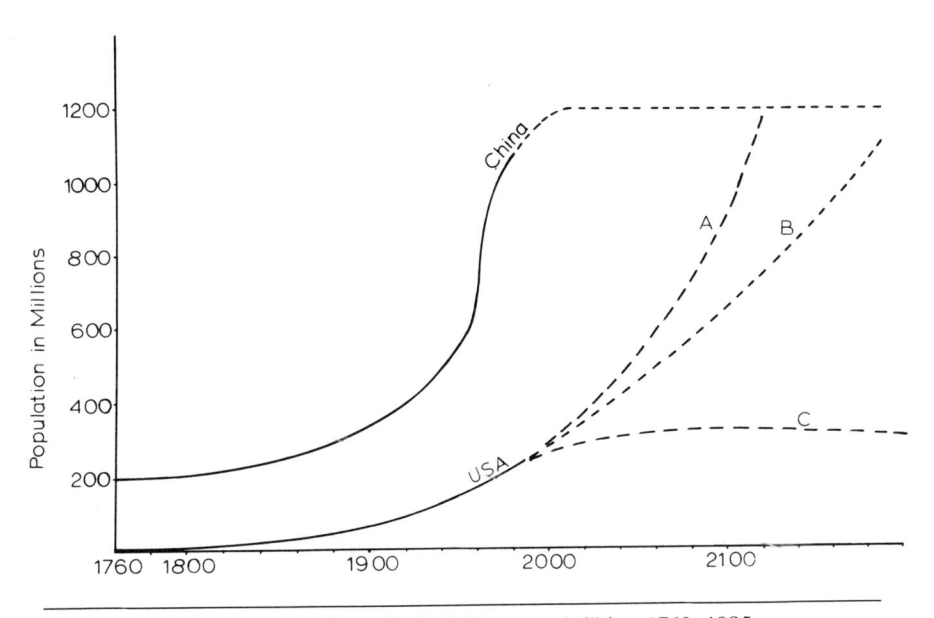

Figure 14.1 Populations of the United States and China 1760–1985 and projected for 1987–2180. The U.S. projection is based on current (B), projected high (A), and projected low (C) population growth rates. China's projection is based on its current national population policy.

ulation doubling time of 70 years (PRB, 1986) (Figure 14.1). Others report that China's population is growing at 1.4% (V. Smil, 1987 personal communication; E. B. Vermeer, 1987 personal communication). Although the United States population is often indicated as being relatively stable, it is actually growing at 0.7% [including legal immigration (PRB, 1986)] (see line B in Figure 14.1). The question is, do we want to leave our descendants the job of supporting twice today's national population in the next three to four generations?

V. PER CAPITA USE OF RESOURCES IN THE UNITED STATES AND CHINA

As previously mentioned, each living human requires certain resources, including food, water, shelter, fuel, and protection from diseases to survive. Approximately 1500 kg of agricultural products are used annually to feed each person in the United States (Table 14.1). China, however, makes do with only 594 kg/capita/yr. It is particularly notable that in the United States 801 kg/capita of the 870 kg of grain/capita we produce per year is fed to livestock, and people directly consume only 69 kg/yr. Thus,

Table 14.1 Foods and Feed Grains Consumed per Capita (Kilogram) per Year in the United States and China

Food/feed	U.S.[a]	China
Food grain	69	269[b]
Vegetables	112	204[c]
Fruit	63	11[d]
Meat and fish	103	25[d]
Dairy products	265	3[d]
Eggs	15	6[d]
Fats and oils	28	6[d]
Sugar	66	6[d]
Subtotal	721	530
Feed grains	801	64[b]
Total	**1,522**	**594**
Kilocalories/person/day	3,500	2,484[e]

[a]USDA (1985).

[b]Total grain production per capita in 1985 was 364 kg (CDAAHF, 1986). Based on unpublished data, it is estimated that 8.5% of the total grain production was used for seeds and industrial materials, 17.5% for feed, and 74% for food (D. Wen, 1987 personal communication).

[c]Estimated on the basis of total vegetable planting area (D. Wen, 1987 personal communication).

[d]CDAAHF (1986).

[e]Chinese Agricultural Academy (1986).

14. Environment and Population: Crises and Policies   369

92% of our grain (excluding exports) is used to produce meat, milk, and other animal products. In China, 17.5% of the grain is fed to livestock, and grains make up a large portion of the daily diet (Table 14.1). The average calorie consumption in the United States is 3,500 kcal/person/day, whereas in China the average is 2,484 kcal/person/day (Table 14.1).

To produce food for each person in the United States, a total of 1.9 ha of cropland and pastureland is used (USDA, 1985), whereas in China only 0.4 ha/person is used (Table 14.2). Thus, China uses one-quarter of the land and feeds its population adequately on a diet that uses one-third less food (Table 14.1). In the United States, about 192 million ha of actual cropland are planted to provide food for 240 million people (USDA, 1985). Considering that about 20% of U.S. food/feed crops are exported, the land tilled to feed an American is about 0.6 ha/year (Table 14.2). In China, however, with more than one billion humans, each person is fed using only 0.1 ha of cropland (Table 14.2). On 0.1 ha the diet must be essentially vegetarian (Table 14.1). It appears that China has nearly approached the carrying capacity of its land resources. Constrained by its population's huge demand on resources, it is clearly impossible for China to achieve a standard of living similar to that in the United States in the near future.

The amount of water used per capita in the United States is also significantly greater than in China—a total of 2.5 million liters/person/yr of water are pumped in the United States compared with only 0.46 million liters/person/yr in China (Table 14.2).

Table 14.2 Resources Utilized per Capita per Year in the United States and China to Supply Basic Needs

Resource	U.S.	China
Land		
Cropland (ha)	0.6[a]	0.1[b,c]
Pasture (ha)	1.3[a]	0.3[b]
Forests (ha)	1.3[a]	0.1[b,d]
	3.2	0.5
Water (liters $\times 10^6$/yr)	2.5[e]	0.46[c]
Fossil fuel		
Oil equivalents (liters)	8,000[f]	413[g]
Forest products (tons)	14[a]	0.03[c,d]

[a]USDA (1985).
[b]Wu (1981).
[c]Smil (1984).
[d]Vermeer (1984).
[e]USWRC (1979).
[f]DOE (1983).
[g]State Statistical Bureau PROC (1985).

Per capita use of fossil energy in the United States amounts to 8,000 liters of oil equivalent per year, which is about 20 times the level in China (Table 14.2). Industry, transportation, heating homes, and producing food account for most of the energy consumed in the United States (Pimentel and Hall, 1984). In China, most fossil energy is used for industry and food production (Kinzelbach, 1983; Smil, 1984).

Almost 500 times more forest products are used per capita in the United States than in China (Table 14.2). Paper and lumber products account for most of the forest resources used in both nations.

The quantities of forest products, energy, water, food, and land used in the United States, compared to the quantities used in China, serve to illustrate several things. First is the impact one billion people have on their living space and how a huge population can create a desperate situation. Second is the tremendous wealth we enjoy in the United States, and that no matter how far lifestyles remove people from an awareness of the land and natural resources, it yet remains that natural resources are the sole suppliers of our goods. Third, it is evident that without an awareness of population growth and natural resource dependency instilled in our population, it is possible that Americans could one day easily number one billion and suffer poverty.

It is evident that our present population size and level of resource consumption are seriously affecting our environment and happiness (Ehrlich and Ehrlich, 1987). The question is, do the people of this nation wish to increase its population size until the standard of living declines for our descendants? Or do we wish to control our population size and retain a high standard of living for future generations?

VI. STATE OF THE ENVIRONMENT IN THE UNITED STATES AND CHINA

Degradation of land, water, air, and biological resources is rampant in the United States and China and is continuing because of increasing populations, growing affluence, and lack of sustained concern for the environment. Land, a vital natural resource, is very often neglected. It is essential for food production and the supply of other basic human needs.

On U.S. cropland, soil erosion rates average 18 t/ha/yr (Lee, 1984). Severe soil erosion also occurs in China, averaging about 43 t/ha/yr on its cultivated land (Brown and Wolf, 1984). The degradation of soil by erosion is of particular concern because soil reformation is extremely slow. Under agricultural conditions, from 200 to 1,000 years are required to renew 2.5 cm or 340 t/ha of topsoil (Hudson, 1981; Lal, 1984a,b; Elwell, 1985; T. W. Scott, 1985 personal communication). Thus, the United States and China are losing soil 18 (0–200) to 43 t/ha (0–300) times faster than it is reformed, an alarming rate of environmental degradation.

In what used to be some of the productive regions of the United States, crop soil productivity has been reduced 50% by erosion (Follett and Stewart, 1985). The reason crop yields have increased in the United States and China during the past three decades despite erosion has been the growing use of fertilizers, irrigation, high-yielding crops, and other inputs that mask soil degradation (Wen and Pimentel, 1984). Thus, fossil energy, a nonrenewable resource, is being used to offset soil degradation; in the United States about 3 kcal of fossil energy are required to produce 1 kcal of food in the agricultural system (Pimentel and Pimentel, 1979). Were a primitive society to put more energy into growing and gathering food than they received consuming it, the people would starve. This policy supporting a 3:1 energy ratio has serious implications for the future. How long can such intensive agriculture be maintained on the croplands of the United States and China?

Erosion reduces soil productivity, not necessarily because the eroded soil leaves behind only bedrock, but because a poorer soil with little organic matter remains, and through a variety of mechanisms less water is available to plants (NSESPRPC, 1981; Pimentel et al., 1987). Water is the major limiting factor for all world crop production (FAO, 1979). Sufficient rain falls upon most arable agricultural land, but periodic droughts continue to limit yields in some areas of the world.

All crops require and transpire massive amounts of water. For example, a corn crop that produces 7,000 kg/ha of grain will take up and transpire about 4.2 million liters/ha of water during the growing season (Leyton, 1983). To supply this much water each year, about 10 million liters (1,000 mm) of rain must fall per hectare, and furthermore it must be evenly distributed during the year and growing season.

Of the total water currently utilized in the U.S. and China, agriculture consumes 81% and 87%, respectively (USWRC, 1979; Smil, 1984). Because more water is likely to be needed to support agricultural production in both nations, the extent and location of water supplies will become greater constraints than they are currently on increasing crop production.

Water consumption in the United States and China is projected to rise because of population growth and higher per capita use (USWRC, 1979; CEQ, 1983). The rapidly rising use of water in these nations is placing a growing demand on both surface and groundwater resources. In the United States, groundwater overdraft is estimated to be 25% higher than the replenishment rate (USWRC, 1979), though groundwater conditions vary widely across the nation. In arid Northern China, where most of its urban population resides, groundwater withdrawals exceed the sustainable supply by 25% (Brown, 1985). In addition to the concern for groundwater overdraft is the problem of groundwater pollution with various chemicals, including pesticides, in both the United States and China (Smil, 1984; Pimentel and Levitan, 1986).

Surface water pollution is also a problem in the United States and China, but it is particularly serious in China where 98% of the wastewater is dumped directly into rivers and lakes without treatment (Brown, 1985). In contrast, the United States treats 91% of all human wastewater to reduce the BOD (biological oxygen demand) before it is released into rivers and lakes (CEQ, 1985). The Clean Water Act of 1987 should further raise this percentage; however, toxins and nutrients (N, P, K, etc.) are still commonly released into rivers and lakes.

In addition, the water-holding capacity of reservoirs in both the United States and China is being reduced because siltation from soil erosion is filling the reservoirs. In the United States, the capacity of reservoirs has significantly decreased (Clark, 1985). In China, siltation of reservoirs is more severe than in the United States because of the more serious soil erosion problem. China was creating 260 million m^3 of water storage capacity per year in the early 1980s (considering only reservoirs of 1 million m^3 or more capacity), but 80 million m^3, or one-third, of the storage capacity was being lost annually through sedimentation (Kinzelbach, 1983).

Air pollution also persists. In the United States, an estimated 21 million metric tons (t) of SO_2 are released into the atmosphere, resulting in serious environmental problems in both the natural and agricultural environments (EPA, 1986). China is less heavily industrialized than the United States, but still releases 18 million t of SO_2 each year (Kinzelbach, 1983). In China's northern cities, where air pollution from the combination of low-quality coal in household stoves and industries is serious, lung cancer rates range from 17 to 31 cases per 100,000 humans per year (Kinzelbach, 1983). The national average in China is only 4–5 cases/100,000.

Chemical pollutants released to the air, water, and soil are also seriously affecting natural biota in the United States (Pimentel and Edwards, 1982) and China. In addition to toxic chemicals, the conversion of forests and other natural habitats to croplands, pastures, roads, and other modifications is resulting in greatly reduced biological diversity (McFarland *et al.*, 1985; Hanks, 1987).

VII. POLICY DECISIONS CONCERNING ENVIRONMENTAL AND POPULATION PROBLEMS

Human societies often wait until a crisis occurs or the "well runs dry" before decisions are made and policies established to protect environmental resources and limit population growth. This can be illustrated in part by examining four policies emerging from crises: Policies concerning pesticides, water, and air pollution in the United States, and the national population policies in the United States and China.

A. Pesticides

DDT and several other chlorinated insecticides were banned in the United States in 1972, about two years after the establishment of the U.S. Environmental Protection Agency. How long after pollution problems were first linked with DDT was a policy adopted to ban this chemical?

DDT was first used for insect control on crops in 1944 during World War II. Ecological problems with the insecticide were reported almost immediately. For example, when DDT was used to treat apple orchards for codling moths, mite populations increased and became a serious problem because of the destruction of their natural insect predators (Steiner *et al.*, 1944; Newcomer and Dean, 1946).

In Idaho and Wyoming, treatment of forests with DDT at 1.1, 2.8, 5, and 8.4 kg/ha either killed fish or radically altered the diets of several fish species populations by killing their usual prey (Adams *et al.*, 1949). As time passed it became evident that DDT could accumulate in fish 100,000-fold over the level that existed in the surrounding water. DDT was found to kill bird embryos, cause eggshell thinning in predaceous birds, and cause cancer in laboratory mice and rats (Pimentel, 1971).

After nearly 20 years of accumulated evidence demonstrating that DDT killed wildlife, that it was present in over half of the human food supplies, and that DDT was being concentrated in body tissues of humans, Rachel Carson's book *Silent Spring* (1962) appeared. It summarized most of the evidence and Carson called for action to protect public health and the environment. Her book prompted President Kennedy to appoint a special scientific commission to investigate the pesticide problem. The commission issued a report in 1963 stating that DDT and other persistent insecticides were a hazard to humans and the environment (PSAC, 1963). No action was taken by the President or by Congress. Subsequently, President Johnson appointed another scientific commission to investigate the growing pollution problem, including pesticide use, in the United States. The report of this commission appeared in 1965, and it too recommended that DDT and the other persistent insecticides be restricted in use (PSAC, 1965). Again, no policy action followed this report.

Four years later, in 1969, yet another report on persistent pesticides was prepared by the National Academy of Sciences for the U.S. Department of Agriculture, recommending that DDT and other persistent insecticides be restricted in their use (USDA, 1969). And still another investigation was prompted in mid-1969 when several millions of dollars worth of DDT-contaminated salmon canned in Michigan and mackerel canned in California had to be discarded. The canned fish contained DDT residues above the tolerance level of 5 ppm (USDHEW, 1969). Like its predecessors, this last study, sponsored by Secretary Finch of the U.S. Dept. of Health, Education and Welfare, recommended that DDT and other chlorinated insecticides be banned. But unlike prior reports, this report and the accompanying economic disaster finally prompted action. In 1972, DDT was banned—nearly 30 years after the first reports (1944) indicated that it was a serious environmental problem (Table 14.3).

Despite the damage caused by DDT, the failures to take action against it until the environmental situation had become quite serious (Pimentel, 1971) are also underlined by the improvements in public health and the environment after the ban. A ten-fold decrease in DDT was soon detected in the fat tissue of humans (Figure 14.2). DDT residues in lake trout caught in eastern Lake Superior were observed to decline from 1.05 ppm in 1971 to only 0.05 ppm by 1975 (Frank et al., 1978). DDT residues in brown pelican eggs collected in South Carolina declined from 0.45 ppm in 1968 to only 0.004 ppm in 1975 (Blus et al., 1979). Similar reductions in DDT residues in other wildlife were reported to be taking place in the environment (Pimentel, 1987).

The legislation on DDT and other chlorinated insecticides has improved the environment (Pimentel, 1987); however, public health and environmental problems from pesticides are still serious in the 1980s. For example, about 45,000 humans are poisoned annually in the United States from pesticides (Pimentel et al., 1980). In addition to the public health problem, the environmental hazards include fish, bird, and mammal kills; developing pesticide resistance in pests; destruction of natural enemies

Table 14.3 Legislative Actions Concerning Pesticides from 1910 to 1986

Year	Action
1910	Insecticide Act 1910 (protect consumers purchasing pesticides)
1947	Federal Insecticide, Fungicide, and Rodenticide Act (FIFRA) (focused on pesticide efficacy)
1964	FIFRA amended to place the burden of proof for effectiveness and safety on the registrant
1970	Reorganization Order creating Environmental Protection Agency and transferring all pesticide regulation responsibilities to EPA from USDA
1972	DDT and several other chlorinated insecticides banned
1972	Federal Environmental Pesticide Control Act
1975	FIFRA amended
1978	FIFRA amended (a general tightening of regulations)
1980	FIFRA amended
1986	FIFRA proposed amendment did not pass

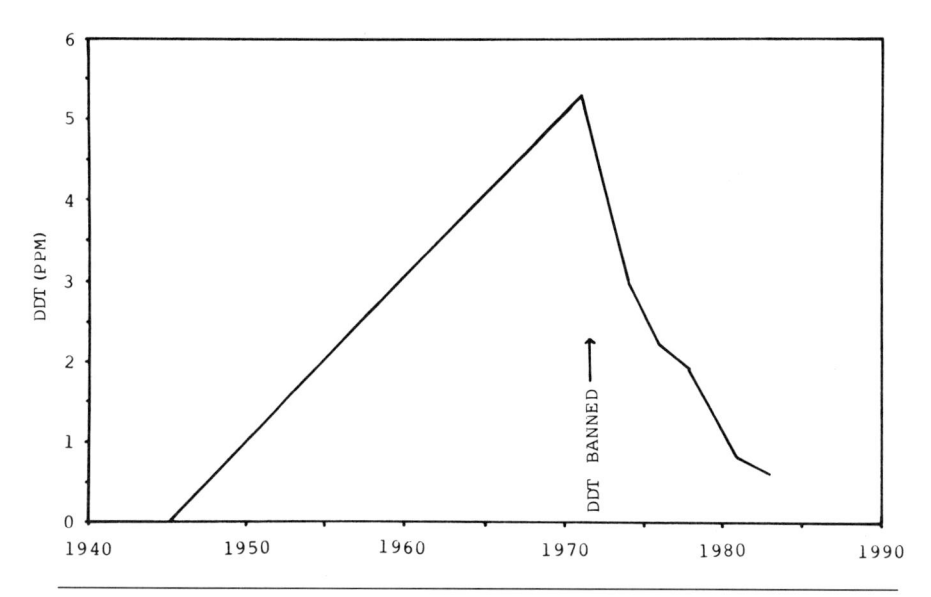

Figure 14.2 DDT (ppm) in fat tissue of whites and nonwhites age 0–14, 1945–1983 (Kutz et al., 1977; F. W. Kutz, 1986 personal communication).

and consequent outbreaks of pests; and pesticide contamination of foods (Pimentel et al., 1980). Almost ironically, recent evidence is suggesting that less than 0.1% of applied pesticides are actually reaching the target pests (Pimentel and Levitan, 1986). Thus, although progress has been made in controlling pesticide use, problems persist. Clearly, more refined legislation is needed. The question is, can it be created without the level of environmental degradation, losses, and outrage that seemed to have been required to create the current legislation?

B. Water Pollution

In the United States during the 19th and 20th centuries, growing numbers of people, increasing affluence, and little or no waste treatment caused water pollution to grow rapidly. By the mid-1800s many small surface waters produced a "stench so strong as to arouse the sleeping, terrify the weak, and nauseate and exasperate everybody" (Dworsky and Berger, 1979). Things became much worse before the 1972 Water Pollution Control Act was passed (Table 14.4).

Several bills were passed before 1972 dealing with water pollution. The first was in 1899, but it accomplished very little because if focused only on the effects of pollution on navigation (Arbuckle et al., 1983). Most

Table 14.4 Chronological List of Federal Legislation to Control Water Pollution and Improve Water Quality

Year	Legislation
1899	Rivers and Harbors Act
1912	Federal Water Pollution Control Act
1914	Interstate Quarantine Regulations of the National Quarantine Act of 1893. These regulations established the first municipal bacteriological data water quality standards.
1924	Oil Pollution Act
1948	Water Pollution Control Act
1956	Water Pollution Control Act Amendments
1958	Water Supply Act
1961	Water Pollution Control Act
1965	Water Quality Act
1966	Clean Water Restoration Act
1969	National Environmental Policy Act
1970	Water Quality Improvement Act
1972	Federal Water Pollution Control Act
1974	Safe Drinking Water Act
1976	Resource Conservation and Recovery Act
1977	Clean Water Act
1986	Safe Drinking Water Act Amendments

human and industrial wastes continued to be flushed into waterways. Human wastes in surface water were considered the major factor in the spread of typhoid (Cohn and Metzler, 1973), which caused 358 deaths per 100,000 people in 1900 (Figure 14.3).

Typhoid epidemics stimulated the public and Congress to press for investigations of polluted water supplies in the late 1800s and early 1900s. However, none of the introduced bills passed because this problem was considered to be in the domain of state and local governments (Dworsky, 1976). Ultimately, typhoid epidemics prompted efforts to cleanse water pumped from rivers and lakes, although little was done to reduce the dumping of wastes into the waterways (Cohn and Metzler, 1973). Still, the improvement in the quality of drinking water reduced the number of typhoid deaths from 35.8 to 2.5 per 100,000 by 1936 (Figure 14.3).

The first "true" Water Pollution Control Act was passed in 1948 (Table 14.4). This act covered administrative oversight, interstate cooperation, research, grants for pollution control programs, and the formation of the Water Pollution Control Administration and Advisory Board. It was largely ineffective, however, and along with most pollution control efforts did little to improve water quality (Arbuckle et al., 1983). Although progress was occasionally made with pollution control on the local level, the water pollution problem continued to grow nationally. For example, in 1959, the severely polluted Cuyahoga River in Ohio caught fire and burned for

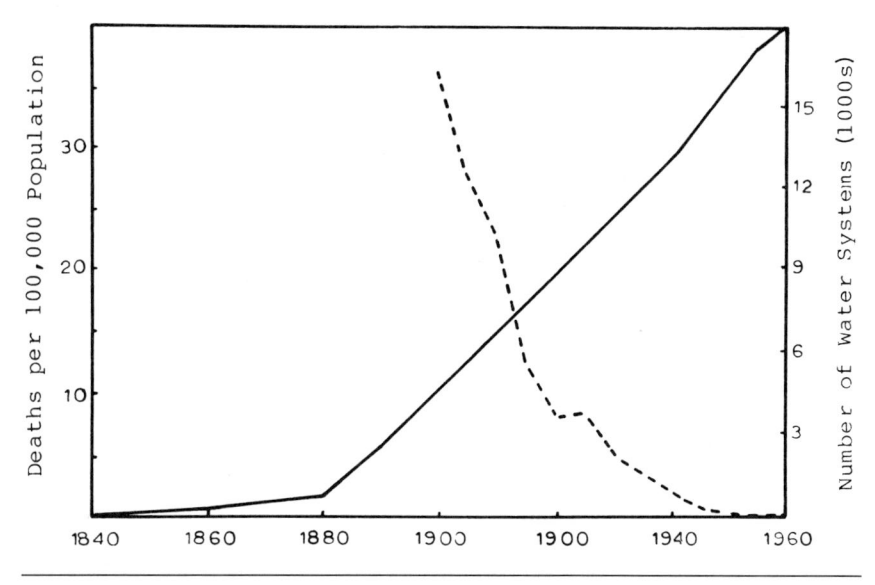

Figure 14.3 Decline in number of typhoid deaths associated with the rise in public drinking water treatment (Wolman and White, 1969).

eight days, a result of the 140 t of chemical, sewage, and garbage wastes that were being dumped into the river every day (Tufty, 1984). Municipalities, for the most part, refused to deal with pollution of waterways, despite provision of federal funds for pollution control by the 1948 Federal Water Pollution Control Act.

In 1963, an inadvertent pesticide release into the Mississippi River killed at least five million fish (Holmes, 1979). Similar fish kills were taking place in other waterways because pesticides and other toxic chemicals were being indiscriminately discharged into streams and lakes.

Growing public concern finally led Congress to pass the 1970 Clean Water Restoration Act, which dramatically increased funding for waste treatment facilities in the nation. Also in 1970, President Nixon created the Environmental Protection Agency, which was charged with enforcing the nation's environmental laws.

Concern for water pollution continues in the United States today, as demonstrated by the 1986 amendments to the 1974 Safe Drinking Water Act and the 1987 Clean Water Act (Table 14.4). The 1972 Water Pollution Control Act and subsequent legislation have substantially contributed to, and provided the impetus for, improving the quality of U.S. water. For example, clean stream water was nearly doubled from 272,000 miles of streams in 1972 to 488,000 miles 10 years later (CEQ, 1985).

Although real progress has been made since 1972 in cleaning U.S.

streams, water pollution still exists. Nearly 300,000 miles of streams still do not meet clean water standards because of pollution (CEQ, 1985). In addition, 95% of the rural households in the United States obtain their drinking water from groundwater, yet 66% of this groundwater does not meet federal drinking water standards (B. Brower, 1986 personal communication).

This history of water pollution crises and failed attempts to reduce pollution in the United States shows that a series of catastrophes was necessary before the public understood the magnitude of the problem. Indeed, water pollution had to make a great many citizens uncomfortable, ill, or worse before sufficient concern was generated for some action. Legislative action was slow and tentative and sometimes ineffective even with significant public concern. Additionally, laws and the public support that encourages them seldom keep up with growing new problems, let alone foresee future problems created by a growing population and its needs for resources. Everyone would agree that real progress has been made in cleaning up U.S. waters since 1972, but a great deal remains to be done to restore U.S. water quality (CEQ, 1985). It is hoped that the 1986 amendments to the Safe Drinking Water Act and the 1987 Clean Water Act will further improve water quality for public health and wildlife.

C. Air Pollution

Like water pollution, air pollution increased over time in the United States. For example, in 1948 an air pollution episode in Donora, Pennsylvania caused 20 deaths and 6,000 illnesses (Benarde, 1970). In addition, air pollution reduced U.S. crop and fish yields, caused forest dieback, and deteriorated paint, buildings, and other structures in the late 1940s (Arbuckle et al., 1983). Slowly, the growing problem and concern expressed by scientists and the public about air pollution convinced Congress to take action to reduce the adverse public health and other environmental effects of air pollutants. The resulting 1963 Federal Clean Air Act addressed the problem. Critical amendments to the Federal Clean Air Act came in 1970 and 1977.

The 1963 action was aimed primarily at controlling black smoke, particulate emissions, and open burning. Under the new law, particulate matter in the air slowly declined from 22.8 million t in 1940 to 7 million t in 1984 in the United States (EPA, 1986).

Although reducing particulate matter in the atmosphere was helpful, dangerous unseen air pollutants have continued to increase (Arbuckle et al., 1983). Sulfur dioxides and other invisible chemicals released into the atmosphere are causing problems that require long periods of time to become apparent. The evidence suggests that toxic air pollutants cause at least 1,700 cancer cases in the United States per year as well as noncancer health effects such as genetic mutations, birth defects, emphysema, and other respiratory and cardiopulmonary diseases (ALA, 1985).

A major obstacle in air pollution control is trying to identify the source of the air pollutants (Landau and Rheingold, 1971; Arbuckle et al., 1983). This problem exists today with acid rain. Generally, the northeastern United States blames the Midwest for its acid rain problem. Many people believe the increasing acidity is due to long-range atmospheric transport of sulfur emissions, principally from power plants (Arbuckle et al., 1983). The effects of air pollution were extended several hundreds of miles downwind when industries constructed tall smokestacks to dilute pollutant concentrations in order to improve ambient air quality.

Despite the insidious nature of the effects of most air pollutants on humans and their environment (Arbuckle et al., 1983), some progress has been made in reducing the quantities of a few air pollutants like particulate matter and sulfur dioxides (Figure 14.4). Others pollutants, however, have been increasing in the United States. Note in Figure 14.4 that nitrogen oxides have nearly doubled since 1940 despite efforts to control air pollution. How serious must these types of pollution become before suitable and effective air pollution controls are implemented?

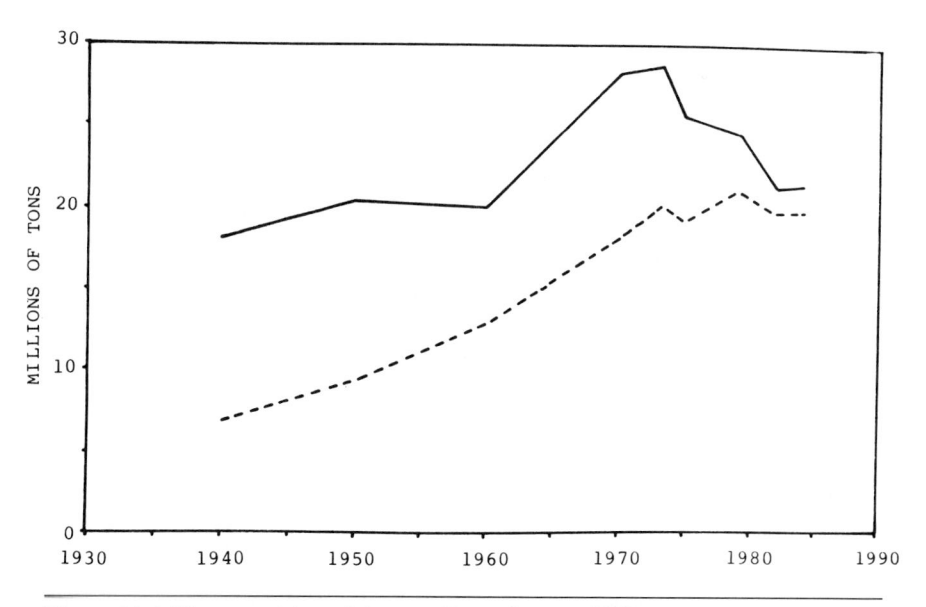

Figure 14.4 The quantities of three sulfur oxides (solid line) and nitrous oxides (dashed line), air pollutants released into the atmosphere in the United States annually, 1940–1984 (millions of tons per year) (EPA, 1986).

VIII. NATIONAL POPULATION POLICIES IN THE UNITED STATES AND CHINA

The United States does not have a national population policy; at best it has a tangle of federal and state laws, some encouraging and some discouraging population growth. Laws affecting sex education, availability of contraceptives, abortion, marriage, child care, and taxes form incentives of varying degree to either have children or not have them (Farrell, 1984). Even laws permitting apartment complexes to exclude residents with children may influence some family decisions. But whatever effect these laws may have on domestic population growth, they are designed for other purposes.

The only clear expression by Congress of concern for overpopulation refers to other nations. The Foreign Assistance Act of 1961 states as an objective that the United States will offer developing countries assistance to reduce the rate of population growth (22 USC §215lb (b)). The Agricultural Trade Development and Assistance Act of 1954 says that before entering agreements with developing countries for the sale of U.S. agricultural products, the United States should note whether these nations are carrying out voluntary programs to control population growth (7 USC §1709 (a)).

Congress is considerably more neutral when considering domestic population growth. For example, the National Environmental Policy Act of 1969 (NEPA) speaks of achieving a "balance between population and resources use which will permit high standards of living and a wide sharing of life's amenities . . ." (42 USC §433(b)(s)) but never suggests what that balance might be. The Wilderness Act of 1964 speaks of an increasing population and its threat to wilderness areas, but mentions only stringent preservation as a solution to the threat (16 USC §113(a)). It fails to suggest that slower population growth might help as well.

China, on the other hand, has struggled with several national population policies in the last 40 years or so. After reaching a density of 600 million in 1950, China decided that its population was too high for its available resources (Liu et al., 1981; Coale, 1984; Shei, 1985). China occupies about 960 million ha, which is approximately the same size as the United States, which covers 917 million ha (Rowe, 1984; USDA, 1985). The United States, however, has 190 million ha of cropland, whereas China has only half of this or 100 million ha (Vermeer, 1984; FAO, 1985; USDA, 1985). This amount of cropland and per capita availability of resources in China affects its standard of living.

The Chinese decided to adopt a national population policy soon after the 1953–1954 drought caused crop failure and thousands of deaths from famine (Liu et al., 1981; Coale, 1984; Shei, 1985). After agriculture recovered in 1956 and the population had grown beyond 600 million (Figure 14.1), the government proposed a new birth-control program (Liu et al., 1981; Coale, 1984; Shei, 1985). The government manufactured the contraceptives and made sterilization and abortion available to the people. The program failed, however, because only 22% of the married couples could obtain contraceptives and the new birth-control program was not enforced.

Another severe drought caused crop failures for three years (1958–1961), and again famines occurred in many parts of China (Coale, 1981, 1984; Liu et al., 1981; Ashton et al., 1984; Shei, 1985). Soon after the recovery from this crisis, another birth-control program was launched. This time a law was passed that forbade males from marrying before they were 20 years of age and females before the age of 18. The government encouraged people to have a maximum of two children, although no laws were passed to limit births. The goal was to reduce the rate of population growth from 2% to about 1%, or about the level in the United States today (0.7%) (PRB, 1986). Again, the program was generally a failure, especially in the rural areas (Table 14.5) where 80% of the people live.

When the disastrous Cultural Revolution took place in 1966–1968, the social structure of China was disrupted and the birth-control program was abandoned. Birth rates exploded, and the population growth rate reached a high of 2.8% (Liu et al., 1981; Coale, 1984; Shei, 1985; Zeng, 1985).

In 1969, after the Cultural Revolution, the birth-control campaign was

Table 14.5 Total Fertility Rates, Rural and Urban Populations of China, 1950–1981[a]

Year	Rural	Urban	Urban:rural	Year	Rural	Urban	Urban:rural
1950	6.0	5.0	.84	1966	7.8	3.1	.45
1951	5.9	4.7	.80	1967	5.8	2.9	.50
1952	6.7	5.5	.83	1968	7.0	3.9	.55
1953	6.2	5.4	.87	1969	6.3	3.3	.53
1954	6.4	5.7	.90	1970	6.4	3.3	.51
1955	6.4	5.2	.89	1971	6.0	2.9	.48
1956	6.0	5.3	.89	1972	5.5	2.6	.48
1957	6.5	5.9	.91	1973	5.0	2.4	.48
1958	5.8	5.3	.91	1974	4.6	2.0	.43
1959	4.3	4.2	.97	1975	4.0	1.7	.45
1960	4.0	4.1	1.02	1976	3.6	1.6	.45
1961	3.3	3.0	.89	1977	3.1	1.6	.51
1962	6.3	4.7	.76	1978	3.0	1.6	.52
1963	7.8	6.2	.80	1979	3.0	1.4	.45
1964	6.6	4.4	.67	1980	2.5	1.2	.46
1965	6.6	3.7	.57	1981	2.9	1.4	.48

[a] Source: Coale (1984).

resumed. Again, the government's objective was to reduce births, especially in the rural areas. The government added three rules to the national population policy: (1) marriage at about age 25 for women, approximately age 30 for men; (2) intervals between births of three to four years; and (3) no more than two children per couple in the cities and three in rural areas.

These policies led to a rapid reduction of the natural growth rates from 2.6% to 1.2% from 1970 to 1979 (Liu et al., 1981; Coale, 1984; Shei, 1985).

In 1980, the famous one child per couple birth-control rule was passed. The goal of the Chinese government was to limit the population to 1.2 billion by 2000 and eventually achieve a stable population at 1.2 billion (Landman, 1981; Coale, 1984). In 1983, the population policy was modified to encourage sterilization and use of IUDs for birth control (E. B. Vermeer, 1987 personal communication). The Central Committee of the Communist Party of China noted that "rapid population growth is confronting the whole nation with ever-increasing difficulties in food, clothing, housing, transportation, education, public health and employment, making it difficult to bring about a speedy change in the country's impoverished and backward state. Moreover, too rapid population growth not only creates difficulties in education and employment but will deplete energy, water, forest and other natural resources, aggravate environmental pollution and make the production conditions and living environment poor and very difficult to improve and modify" (CPIC, 1983). With resource shortages and a rapidly degrading environment due to overpopulation, the government and people of China had little or no alternative but to attempt to limit population growth.

Before the United States reaches a population crisis like that of China, a clear population policy from the U.S. Congress is needed. This is not an unreasonable expectation. The issue is important and fundamental to the nation's well-being. In the creation of a population policy the government should consider the damages that population pressure inflicts on the environment that feeds, shelters, and protects us. The optimum population selected should allow protection of our environment and maintenance of *the quality of life we desire now and in the future*.

Why has the United States failed to develop a policy on such an important issue? For several reasons. Population growth can be limited only by lowering the birth rate, reducing net immigration, and/or by increasing the death rate. In a moral society, intentionally increasing the death rate is considered untenable and is therefore dismissed as a possibility. Reducing net immigration has recently received attention, and the Immigration Act of 1986 has been enacted. However, more illegal immigrants enter the United States each year than legal immigrants. Both immigrant groups contribute to the high rate of population increase in the United States. Decreasing immigration would contribute to reducing U.S. population growth; however, immigration is not the main cause of U.S. population growth. It accounts for more than one-third of U.S. population growth, but most of U.S. growth is due to a high birth rate (USBC, 1985). A problem with reducing immigration is that in international politics it symbolizes a hypocritical "your end of the boat is sinking" attitude. By focusing all of our concern on immigration issues, we falsely assume that we can always prevent immigrants from invading our borders. Also as a result of not having a domestic population policy, some foreign countries whose population growth we would like to curb believe our commitment to world population control is an effort to maintain Western dominance.

But our lack of a population policy can best be explained by examining the last option open to Congress—to decrease the birth rate and encourage smaller families. It takes little imagination to envision the uproar that talk of a national population policy would create. Most U.S. citizens would take considerable umbrage if they were suddenly requested, let alone *required,* to have, for example, more than two children per couple. Besides several plausible Constitutional conflicts that could be raised (Barnett, 1982), one could safely predict fervent arguments from many religious denominations. Proposing compulsory reduction of the average family size would most likely be political suicide for legislators, while proposing incentives overtly aimed at producing a reduced growth rate might be received just as coolly.

Other explanations exist for why we do not yet have a national population policy. First, many people mistakenly believe that the United States' population growth has stopped or will soon do so (Barnett, 1982). Second, since some developing nations have considerably higher population growth rates than the United States, many people believe we should devote all our energy towards slowing their growth rates and ignore our own. A third is that, amazingly, some people still do not view the growing world population—let alone domestic population—as a source of concern (Simon, 1981), and others find the idea of a stable population dangerous (Cook, 1952; Carlson, 1985; Wattenberg, 1987).

But most importantly, few decisions are as personal as the decision to have children. For the government to step into such a private sphere seems, to many, a challenge to their liberty. Perhaps so. But it may be a mistaken liberty, a historical liberty enjoyed by those who lived when the world's population was much lower. For today the decision to contribute to the rapid population growth by having many children is a decision to lower the quality of life for these and future children. Thus, U.S. population growth may be on a collision course with much of what the Bill of Rights attempts to protect. Consider what John Stuart Mill wrote in *On Liberty* in 1859:

> To undertake this responsibility—to bestow a life which may be either
> a curse or a blessing—unless the being on whom it is to be bestowed

will have at least the ordinary chances of a desirable existence, is a crime against that being.

The United States is threatened with overpopulation, at least to the extent of retaining an adequate standard of living (Hardin, 1986). With the current U.S. population growth rate of 0.7% (PRB, 1986), the nation could well double its population in three to four generations (99 years), and in another three to four generations double again. Thus, in six to eight generations, at the current rate of growth, the United States could have a population similar to that of China, or more than one billion. The question is: Do we want to be responsible for the equivalent of China's population density and its related impacts on the environment and reduced standard of living in the United States? Because of the exponential nature of population growth, the decision to avert such a course *must* be made in this century. Those who live six to eight generations in the future will not be given the choice.

IX. CONCLUSION

Most environmental and population problems appear to develop gradually, with the public becoming more and more aware as they grow in intensity. Similarly, policy decisions made to correct the difficulties come about gradually. This has been true for the adoption of policies dealing with pesticide, air, and water pollution in the United States, as well as the national population policies of China. The gradual intensification of environmental and population problems and growing public awareness are usually accompanied by some catalytic event or crisis such as fish kills, numerous human illnesses and deaths, or droughts and famines before suitable policies or legislation are created.

Policy-making is indeed difficult because it deals with the future; perhaps a series of crises is necessary to highlight and simplify the decision-making process. After all, the complex nature of environmental and population problems as well as pressures on governments for their limited time and resources make environmental and population policy formulation extremely difficult. Equally important, difficulties in environmental and population policy-making exist because we believe we have a nearly unlimited right to manipulate environmental resources for personal gratification. The right to reproduce without governmental interference has historically been a liberty, and the responsible exercise of that right has never before been a factor in national policy.

While extensive freedoms are basic to American existence, individual decisions that affect pressures on natural resources must have a limit, as they influence the well-being of all other humans in society. Only with sufficient deterioration of resources in the United States did the people and the government decide that individuals and organizations no longer had the right to pollute the environment. Likewise, overpopulation in China due to unlimited procreation was finally recognized as incompatible with the availability of environmental resources for the well-being of everyone in society now and in their future.

While population pressures alone can decrease environmental quality and human well-being, it is clear that excessive resource use and poor management can intensify environmental and population problems. Science and technology can help make more effective use of resources and thus protect the environment, but only to a limited extent. To leave future generations relying solely on hoped-for advancements in science and technology to provide resources and protect the environment is an intentional neglect of our present-day responsibilities. Reliance on science and technology shows its limits even today—witness the suffering in Calcutta as just one small example. The World Bank estimated in 1980 that about 750 million people in this highly technological and sophisticated world lived in absolute poverty (U. N., 1984b).

Although complex and sometimes emotion-charged, cohesive policies are clearly needed by the people of the United States, China, and other nations to safeguard the environment and limit human numbers for human well-being. Ideally, people and their governments will make timely, effective policy decisions concerning environmental population problems: to know the worth of water before the well is dry. Enough starvation, pollution, erosion, flooding, deforestation, species extinctions, and other painful crises have been recorded for the U.S. Congress to begin the process of legislating a truly better standard of living for all people now and in the future.

ACKNOWLEDGMENTS

We thank the following people for reading an earlier draft of this article and for their many helpful suggestions: S. A. Briggs, A. Ehrlich, P. Ehrlich, C. A. S. Hall, G. Hardin, J. Holdren, W. J. Hudson, C. V. Kidd, P. Reining, R. Revelle, F. T. Sai, L. L. Severinghaus, V. Smil, E. B. Vermeer, K. E. F. Watt, D. Wen, and A. A. Wilson. And at Cornell University we thank D. Baer, R. W. Howarth, B. Knuth, and B. Wilkins.

REFERENCES

Adams, L., Hanavan, M. G., Hosley, N. W., and Johnston, D. W. (1949). The effects on fish, birds, and mammals of DDT used in the control of forest insects in Idaho and Wyoming. *J. Wildl. Manage.* **13**, 245–254.

American Lung Association (ALA). (1985). "Toxic Air Pollution: The Unseen Enemy." Am. Lung Assoc., New York.

Arbuckle, J. G., Frick, G. W., Hall, R. M., Miller, M. L., Sullivan, T. F. P., and Vanderver, T. A., Jr. (1983). "Environmental Law Handbook," 7th Ed. Gov. Inst., Rockville, Maryland.
Ashton, B., Hill, K., Piazza, A., and Zeitz, R. (1984). Famine in China 1958–61. *Pop. Dev. Rev.* **10**(4), 613–645.
Barnett, L. D. (1982). "Population Policy and the U.S. Constitution." Kluwer-Nijhoff, Boston, Massachusetts.
Benarde, M. A. (1970). "Our Precarious Habitat." Norton, New York.
Blus, L. J., Lamont, T. G., and Neely, B. S. (1979). Effects of organochlorine residues on eggshell thickness, reproduction, and population status of brown pelicans *(Pelecanus occidentalis)* in South Carolina and Florida. *Pestic. Monit. J.* **12**, 172–184.
Brown, L. R. (1985). "State of the World 1985." Norton, New York.
Brown, L. R., and Wolf, E. C. (1984). "Soil Erosion: Quiet Crisis in the World Economy," Worldwatch Pap. No. 60. Worldwatch Inst., Washington, D.C.
Carlson, A. C. (1985). The "population question" returns. *Persuasion At Work* **8**, 12–21.
Carson, R. (1962). "Silent Spring." Fawcett, Greenwich, Connecticut.
China Population Information Center (CPIC). (1983). "China: Population Policy and Family Planning Practice." China Popul. Inf. Cent., State Family Plann. Comm., Beijing.
China's Department of Agriculture, Animal Husbandry and Fishery (CDAAHF). (1986). "The Statistics of Agriculture, Animal Husbandry and Fishery in China in 1985." Agric. Press, Beijing. (In Chin.)
Chinese Agricultural Academy. (1986). Is it necessary to have 400 kg grains per capita in China? *Chin. Agric. Sci. (Zhonggue Nongye Kexue)* **5**, 1–7. (In Chin.)
Clark, E. H., II. (1985). The off-site costs of soil erosion. *J. Soil Water Conserv.* **40**, 19–22.
Coale, A. J. (1981). Population trends, population policy and population studies in China. *Popul. Dev. Rev.* **7**(1), 85–97.
Coale, A. J. (1984). "Rapid Population Change in China, 1952–1982." Natl. Acad. Press, Washington, D.C.
Cohn, M., and Metzler, D. (1973). "The Pollution Fighters." New York State Dep. Health, Albany.
Connell, K. H. (1950). "The Population of Ireland. 1750–1845." Clarendon Press, Oxford.
Cook, R. C. (1952). Why people refuse to face population problems. *Popul. Bull.* **8**, 4.
Council on Environmental Quality (CEQ). (1980). "The Global 2000 Report to the President," Technical Report, Vol. 2. U.S. Gov. Print. Off., Washington, D.C.
Council on Environmental Quality (CEQ). (1983). "Environmental Quality 1983, 14th Annual Report." U.S. Gov. Print. Off., Washington, D.C.
Council on Environmental Quality (CEQ). (1985). "Environmental Quality—Report of the Council on Environmental Quality." U.S. Gov. Print. Off., Washington, D.C.
Demeny, P. G. (1986). "Population and the Invisible Hand," Working Pap. No. 123. Cent. Policy Stud., Popul. Counc., New York.
Department of Energy (DOE). (1983). "Energy Projections to the Year 2010," DOE/DE-0029/2. Off. Policy, Plann. Anal., U.S. Dep. Energy, Washington, D.C.
Dworsky, L. B. (1976). "The Nation and its Water Resources," Mimeo. Dep. Environ. Eng., Cornell Univ., Ithaca, New York.
Dworsky, L. B., and Berger, B. B. (1979). Water resources planning and public health 1776–1976. *J. Water Resour. Plann. Manage. Div. WRI*, pp. 133–149.
Ehrlich, A. H., and Ehrlich, P. R. (1987). "Earth." Franklin Watts, New York.
Ehrlich, R. P., and Holdren, J. P. (1969). Population and panaceas: a technological perspective. *BioScience* **10**, 1065–1071.
Ehrlich, R. P., and Holdren, J. P. (1971). The impact of population growth. *Science* **171**, 1212–1217.

Elwell, H. A. (1985). An assessment of soil erosion in Zimbabwe. *Zimbabwe Sci. News* **19**(3/4), 27–31.
Environmental Protection Agency (EPA). (1986). "National Air Pollutant Emission Estimates, 1940–1984," EPA-450/4-85-014. Off. Air Qual. Plann. Stand., EPA, Research Triangle Park, North Carolina.
Farrell, L. (1984). Population policies and proposals: when big brother becomes big daddy. *Brooklyn J. Int. Law* **10**, 83–114.
Follett, R. F., and Stewart, B. A. (1985). "Soil Erosion and Crop Productivity." Am. Soc. Agron., Crop Sci. Soc. Am. and Soil Sci. Soc. Am., Madison, Wisconsin.
Food and Agriculture Organization (FAO). (1979). "Yield Response to Water," FAO Irrigation and Drainage Paper No. 3. FAO U. N., Rome.
Food and Agriculture Organization (FAO). (1985). "FAO Production Yearbook." FAO U. N., Rome.
Frank, R., Holdrinet, M., Braun, H. E., Dodge, D. P., and Sprangler, G. C. (1978). Residues of organochlorine insecticides and polychlorinated biphenyls in fish from lakes Huron and Superior, Canada—1968–76. *Pestic. Monit. J.* **12**, 60–68.
Hanks, J. (1987). "Human Populations and the World Conservation Strategy." Int. Union Conserv. Nat. Nat. Resour., Gland, Switzerland.
Hardin, G. (1968). The tragedy of the commons. *Science* **162**, 1243–1248.
Hardin, G. (1986). Cultural carrying capacity: a biological approach to human problems. *BioScience* **36**, 599–606.
Holmes, B. H. (1979). History of federal water resources programs and policies, 1961–1970. *USDA Misc. Publ.* No. 1379.
Hudson, N. W. (1981). "Soil Conservation," 2nd Ed. Cornell Univ. Press, Ithaca, New York.
Keyfitz, N. (1984). Impact of trends in resources, environment and development on demographic prospects. *In* "Population, Resources, Environment and Development," pp. 97–124. United Nations, New York.
Kinzelbach, W. K. (1983). China: energy and environment. *Environ. Manage.* **7**, 303–310.
Kutz, F. W., Yobs, A. R., Strassman, S. C., and Viar, J. F. (1977). Pesticides in people. *Pestic. Monit. J.* **11**, 61–63.
Lal, R. (1984a). Productivity assessment of tropical soils and the effects of erosion. *In* "Quantification of the Effect of Erosion on Soil Productivity in an International Context" (F. R. Rijsberman and M. G. Wolman, eds.), pp. 79–94. Delft Hydraul. Lab., Delft, Netherlands.
Lal, R. (1984b). Soil erosion from tropical arable lands and its control. *Adv. Agron.* **37**, 183–248.
Landau, N. J., and Rheingold, P. D. (1971). "The Environmental Law Handbook." Ballantine Books, New York.
Landman, B. C. (1981). China's one-child drive: Another long march. *Int. Fam. Plann. Perspect. Dig.* **7**, 102–107.
Lappé, F. M. (1982). "Diet for a Small Planet." Ballantine Books, New York.
Lee, L. K. (1984). Land use and soil loss: a 1982 update. *J. Soil Water Conserv.* **39**, 226–228.
Leyton, L. (1983). Crop water use: principles and some considerations for agroforestry. *In* "Plant Research and Agroforestry" (P. A. Huxley, ed.), pp. 379–400. Int. Counc. Res. Agrofor., Nairobi.
Liu, Z., Song, J., *et al.* (1981). "China's Population: Problems and Prospects." New World Press, Beijing.
McFarland, W. N., Pough, F. H., Cade, T. J., and Heiser, J. B. (1985). "Vertebrate Life," 2nd Ed. Macmillan, New York.

Meadows, D. H., Meadows, D. L., Randers, J., and Behrens, W. W., III. (1972). "The Limits to Growth." Universe Books, Washington, D.C.

Mill, J. S. (1975). "On Liberty" (David Spitz, ed.), 1st Ed. Norton, New York.

National Academy of Sciences (NAS). (1971). "Rapid Population Growth," Vols. 1 and 2. Published for NAS by Johns Hopkins Press, Baltimore, Maryland.

National Academy of Sciences (NAS). (1975). "World Food and Nutrition Study: Enhancement of Food Production for the U.S." Report of Board on Agriculture and Renewable Resources, Commission on Natural Resources, National Research Council. NAS, Washington, D.C.

Newcomer, E. J., and Dean, E. P. (1946). Effect of xanthone, DDT, and other insecticides on the Pacific mite. *J. Econ. Entomol.* **39**, 783–786.

NSESPRPC (National Soil Erosion–Soil Productivity Research Planning Committee). (1981). Soil erosion effects on soil productivity: a research perspective. *J. Soil Water Conserv.* **32**, 82–90.

Pimentel, D. (1971). "Ecological Effects of Pesticides on Non-Target Species." U.S. Gov. Print. Off., Washington, D.C.

Pimentel, D. (1984). Energy flows in agricultural and natural ecosystems. *In* "Options Mediterraneennes," pp. 125–136. Inst. Agron. Mediterraneo de Zaragoza, Zaragoza, Spain.

Pimentel, D. (1987). Is silent spring behind us? *In* "Silent Spring Revisited" (G. J. Marco, R. M. Hollingsworth, and W. Durham, eds.), pp. 175–187. Am. Chem. Soc., Washington, D.C.

Pimentel, D., and Edwards, C. A. (1982). Pesticides and ecosystems. *BioScience* **32**, 595–600.

Pimentel, D., and Hall, C. W., eds. (1984). "Food and Energy Resources." Academic Press, New York.

Pimentel, D., and Levitan, L. (1986). Pesticides: amounts applied and amounts reaching pests. *BioScience* **36**, 86–91.

Pimentel, D., and Pimentel, M. (1979). "Food, Energy, and Society." Arnold, London.

Pimentel, D., Andow, D., Dyson-Hudson, R., Gallahan, D., Jacobson, S., Irish, M., Kroop, S., Moss, A., Schreiner, I., Shepard, M., Thompson, T., and Vinzant, B. (1980). Environmental and social costs of pesticides: a preliminary assessment. *Oikos* **34**, 127–140.

Pimentel, D., Fast, S., Chao, W. L., Stuart, E., Dintzis, J., Einbender, G., Schlappi, W., Andow, D., and Broderick, K. (1982). Water resources in food and energy production. *BioScience* **32**, 861–867.

Pimentel, D., Allen, J., Beers, A., Guinand, L., Linder, R., McLaughlin, P., Meer, B., Musonda, D., Perdue, D., Poisson, S., Siebert, S., Stoner, K., Salazar, R., and Hawkins, A. (1987). World agriculture and soil erosion. *BioScience* **37**, 277–283.

Population Reference Bureau (PRB). (1986). "World Population Data Sheet." Popul. Ref. Bur., Washington, D.C.

President's Science Advisory Committee (PSAC). (1963). "Use of Pesticides." President's Science Advisory Committee, The White House, Washington, D.C.

President's Science Advisory Committee (PSAC). (1965). "Restoring the Quality of our Environment." Report of the Environmental Pollution Panel, President's Science Advisory Committee, The White House. U.S. Gov. Print. Off., Washington, D.C.

Ridker, R. G. (1979). Resources and environmental consequences of population and economic growth. *In* "World Population and Development: Challenges and Prospects" (P. M. Hauser, ed.), pp. 99–123. Syracuse Univ. Press, Syracuse, New York.

Rowe, J. S. (1984). Some observations on forestry in China, 1983. *For. Chron.* **60**(2), 96–100.

Sanger, M., ed. (1927). "Proceedings of the World Population Conference 1927." Albert Kundig, Geneva.

Shei, C. L. (1985). The stages of birth-control planning in China. *Demography* **3**, 49–54. (In Chin.)
Simon, J. L. (1981). "The Ultimate Resource." Princeton Univ. Press, Princeton, New Jersey.
Smil, V. (1984). "The Bad Earth, Environmental Degradation in China." M. E. Sharpe, Armonk, New York.
State Statistical Bureau PROC. (1985). "Statistical Yearbook of China 1985." Econ. Inf. Agency, Hong Kong.
Steiner, A., Arnold, C. H., and Summerland, S. A. (1944). Laboratory and field tests of DDT for the control of the codling moth. *J. Econ. Entomol.* **37**, 156–157.
Tufty, B. (1984). What about tomorrow? *Am. For.* **90**(8), 37–51.
United Nations. (1955). "Proceedings of the World Population Conference 1954," 5 vols. U. N., New York.
United Nations. (1966). "Proceedings of the World Population Conference," 4 vols. U. N., New York.
United Nations. (1975). "The Population Debate: Dimensions and Perspectives," Papers of the World Population Conference, Bucharest, 1974, 2 vols. U. N., New York.
United Nations. (1982). "World Population Trends and Policies. 1981 Monitoring Report, Vol. 1, Population Trends." U. N., New York.
United Nations. (1984a). "Report of the International Conference on Population." U. N., New York.
United Nations. (1984b). "Population, Resources, Environment and Development." U. N., New York.
U.S. Bureau of the Census (USBC). (1985). "Statistical Abstract of the United States: 1986," 106th Ed. U.S. Bur. the Census, Washington, D.C.
U.S. Department of Agriculture (USDA). (1969). "Committee on Persistent Pesticides." Div. Biol. Agric., Nat. Res. Counc., Washington, D.C.
U.S. Department of Agriculture (USDA). (1985). "Agricultural Statistics 1985." U.S. Gov. Print. Off., Washington, D.C.
U.S. Department of Health, Education, and Welfare (USDHEW). (1969). "Report of the Secretary's Commission on Pesticides and Their Relationship to Environmental Health." U.S. Gov. Print. Off., Washington, D.C.
U.S. Water Resources Council (USWRC). (1979). "The Nation's Water Resources 1975–2000," Summary and Vol. 1. U.S. Gov. Print. Off., Washington, D.C.
Vermeer, E. B. (1984). Agriculture in China—a deteriorating situation. *Ecologist* **14**(1), 6–14.
Wattenberg, B. J. (1987). "The Birth Dearth." Pharos, New York.
Wen, D., and Pimentel, D. (1984). Energy flow through an organic agroecosystem in China. *Agric. Ecosyst. Environ.* **11**, 145–160.
Wolman, A., and White, G., eds. (1969). "Water, Health and Society." Indiana Univ. Press, Bloomington.
"World Almanac and Book of Facts" (WABF) (1982). Newspaper Enterprise Assoc., New York.
Wren, C. S. (1982). China plans a new drive to limit birth rate. *New York Times* Nov. 7, p. 8.
Wu, C. (1981). The transformation of agricultural landscape in China. *In* "The Environment: Chinese and American Views" (L. J. C. Ma and A. G. Noble, eds), pp. 35–43. Methuen, New York.
Zao, Z. (1982). Report about the sixth five-year plan. *Renmin Ribao (People's Daily)* Dec. 14.
Zeng, Y. (1985). "Family Dynamics and Population Change in China." Netherlands Interuniv. Demogr. Inst., Voorburg.

15

Food Availability and Natural Resources

Carl W. Hall

Directorate for Engineering
National Science Foundation
Washington, D.C.

I. Introduction
II. Utilization of Resources
   A. Solar Energy
   B. Soil and Nutrients
   C. Water
   D. Fertilizer
III. Production of Food in the United States and the World
   A. Weather and Climate
   B. Plants and Animals
   C. Wastes
   D. Cultural and Religious Preferences
   E. Losses
   F. Systems for Producing Food
   G. Commercial Farms
   H. Infrastructure
   I. Other Protein Sources of Food
   J. Trends and a Look Ahead
IV. Summary
   References

I. INTRODUCTION

The global availability of food is increased by (1) increasing the production of plants and animals on the land, including reclaiming land; (2) increasing the production of fish and plants in the lakes, oceans, and constructed water-containing devices; and (3) utilizing wastes and by-products as food or as nutrients for plants, animals, and fish. This chapter will cover the impact of these actions on various natural resources and opportunities for increasing the availability of food.

Additional technological approaches for providing food are the use of single cell protein (SCP) from fungi, yeast, and bacteria; algae and non-microbial sources; flour from plants not now being so utilized; and oil from soybeans, peanuts, coconuts, cottonseed, and rapeseed. The overall strategy of increasing food availability should be toward a goal of maximizing use of renewable resources such as solar energy and minimizing use of non-renewable resources such as fossil fuels.

II. UTILIZATION OF RESOURCES

A. Solar Energy

Solar energy is the most important renewable natural resource for food production. The range of conversion efficiency for natural plants is from practically zero to 2%, with an average efficiency of about 0.1% for solar energy reaching the earth being utilized in plant production (Odum and Odum, 1981; Sorensen, 1979). Each year 150×10^9 tons of dry matter are fixed on the land and in the ocean by photosynthesis of which about one-half is in cellulose, not readily available for food (Janick et al., 1976; Saeman, 1977). Cultivated plants make up 4–5% of all organic matter.

The quantity of living matter on the earth is 20–40 trillion metric tons (t) of which land plants make up 3 trillion, land animals 1 trillion, and marine plants and animals 16 trillion t. (FAO, 1965). Of course, much of these resources are not edible but remain to be explored and possibly utilized for food.

The flow of carbon (Figure 15.1) is driven by energy from the sun and represents the steps by which solar energy from the sun is converted through the process of photosynthesis to plants for food. The hydrocarbon fuels such as coal, petroleum, and natural gas—even though primarily carbon and produced as the result of solar energy—are considered non-renewable resources. The formation of these energy resources which are easily converted to fuels occurred over at least 500,000 years. By contrast, carbohydrates are produced from solar energy during short periods, usually one year for annual plants or several years for trees, and are considered renewable energy sources. Both energy resources—the fossil fuels and

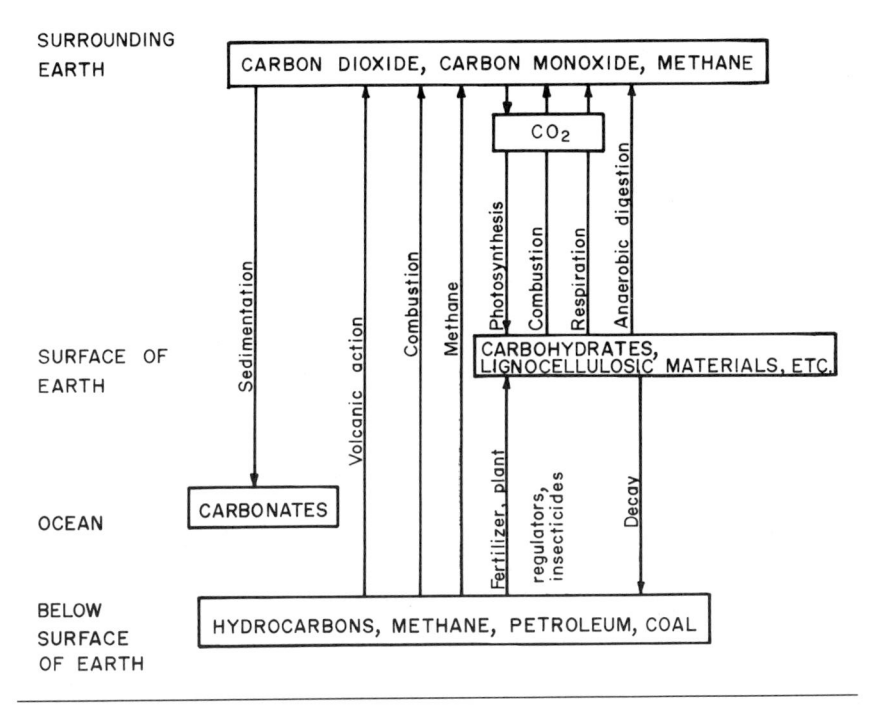

Figure 15.1 Flow of carbon (Hall, 1981).

hydrocarbons, and the biological materials, carbohydrates, and lignocellulose—produce carbon dioxide when burned. The cycles of events are parallel for both energy sources, the major difference being the time scale to complete a cycle.

Photosynthesis provides a process for returning the carbon of the carbon dioxide in the atmosphere back into the energy sources as described by the carbon cycle. Photosynthesis is dependent on (1) light intensity or solar energy, (2) temperature, and (3) concentration of carbon dioxide at the photosynthetic site, which is usually less than the concentration external to the plant. These relationships between the plants and environment can be exploited to increase food availability.

Plants vary greatly in their ability to convert solar energy to carbon. Those plants which are quite effective in the conversion are known as C_4 plants, as represented by sorghum, corn, and sugar cane. C_3 plants are the most common and include many of the plants involved in photosynthesis. CAM (crassulean acid metabolism) plants are slow growing, efficient users of water. They fix carbon dioxide in the dark and include plants such as cacti and pineapple, with some green plants capturing 15–22% of the solar energy (Janick *et al.*, 1976).

B. Soil and Nutrients

Soil is an important non-renewable natural resource used for providing food. Although soil can be rebuilt slowly (2.5 cm in 500 yr) by replacing plant food elements, soil is usually considered a non-renewable resource.

One of the principal losses of soil is through erosion by water, wind, and ice. One estimate places the annual loss at an average of 18 t/ha (7 t/acre) from cultivated land in the United States (Pimentel et al., 1987; OTA, 1982). In addition to the loss of soil itself, there is often leaching of nutrients into the deeper soil layers and water table and loss of runoff water. For some situations, although the soil is lost from one location it is moved from one site to another and used for food production, such as on periodically flooded land.

Soil loss is also caused by clear-cutting of trees, cultivation of steep slopes, lack of cover to proctect the soil, uncontrolled flooding, controlled flooding and improper irrigation.

Overgrazing of the land by animals is a serious problem in many parts of the world, with perhaps 60% of the land in United States overgrazed. If the environment cannot support the recovery and regrowth of plants, grasses, and trees, the land is left barren and susceptible to erosion and leaching. Grazing may be carried out on a continual, rotational, or seasonal basis depending upon the density of animals and the ability of the soil, water, and plant environment to replace the plants

Many of the misuses or overuses of the land occur when an attempt is made to increase the area of land available for food production. However, it does not necessarily follow that soil losses result from high yields. Soil loss is often the result of utilizing steep land for row crop or non-cover crop use. Approximately twice as much land is available worldwide for crop production as the 1.5 billion ha presently being used. New land is brought into production at costs that include expenditures for labor, water, energy, and fertilizer elements. Additional expenditures may be required to protect land with steep slopes or for land that may be flooded.

C. Water

On a global basis, water is a renewable resource necessary for life of plants, animals, and humans. For local situations, such as water in basalt cavities, water could be considered as a non-renewable resource. Water is basically a materials-handling substance that carries nutrients with it to and through the plants, soil, animals, and people. Considerable energy and cost are involved to convert contaminated or polluted water back to its original condition. Water is being removed from undergound natural storage at a rate exceeding replacement. Water removed from these sources often becomes polluted.

The total annual global precipitation is about 350×10^3 km^3 of which

275 × 10³ km³ falls over the oceans. Of the remainder, 64% of the water evaporates, leaving 28 × 10³ km³ for runoff or ground water (Rogers, 1986). As water is evaporated minerals are left behind, possibly deteriorating the soil and water.

About half of the water available world-wide is used to support human endeavors. The growth of population will increase the demand for water to drink, bathe, and cook in order to meet human needs and for recreational purposes. Likewise, the demand for food will increase the demand for water, either naturally or through irrigation systems. Irrigated soils, especially if they are mismanaged, can become excessively salty. Salt accumulates as a result of evaporation of excess water applied for irrigation, lack of proper drainage to remove water, and runoff of water deteriorated by accumulation of salt. In order to conserve the use of fresh water as used in irrigation supply systems, methods are needed to minimize large surface areas which are sites for evaporation. With evaporation of water the minerals concentrate in the soil and water, making them less useful for human, plant and animal needs.

D. Fertilizer

Commercial fertilizer is a non-renewable resource which is expensive and requires fossil fuel—gas or petroleum—for manufacture. Some of the elements of fertilizer are recycled in the ecosystem so that some fertilizer can be considered renewable. About 30% of the increased crop productivity in the United States in recent years is directly attributable to increased use of fertilizer. In developing countries, where less fertilizer is in use and often with less fertile soil, up to 50% more crop response to use of fertilizer is obtained.

The principal fertilizer elements are nitrogen, phosphorous and potassium, which require energy to produce (Table 15.1). The possibility of

Table 15.1 Energy to Produce Fertilizer and Chemicals for Agriculture[a]

To produce 1 kg of	Energy from fossil fuel (kcal)
Fertilizer	
Nitrate	14,700
Phosphorus	3,000
Potassium	1,600
Chemicals	
Herbicide	100,000
Insecticide	87,000
Fungicide	65,000

[a]Source: Pimentel and Hall (1984).

396   Carl W. Hall

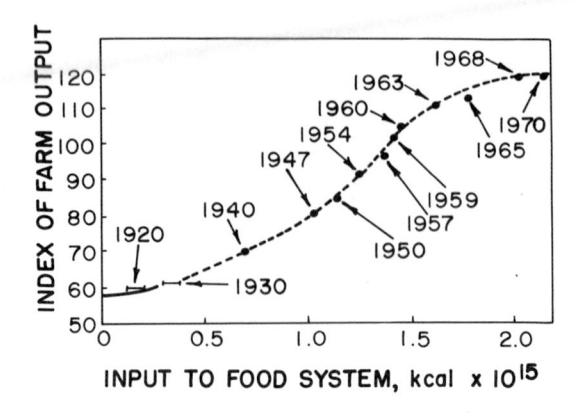

Figure 15.2 Farm output as a function of energy input to the U.S. food system (Steinhart and Steinhart, 1975). 1957–1959 index is 100.

getting non-leguminous plants to produce nitrogen as is done by legumes offers considerable promise. The results of the developments in biotechnology offer the possibility of using a biological approach to reduce energy requirements by reducing the need for nitrogen fertilizer, and perhaps other elements. As shown in Figure 15.2, the use of fertilizer in the United States has reached a point of diminishing return; the use of additional fertilizer does not provide a proportionate response in plant yield of food crops.

III. PRODUCTION OF FOOD IN THE UNITED STATES AND THE WORLD

The percentage of the dry weight of a plant that provides useful food, such as grain or seed is known as the harvest index. The harvest index can vary from as little as 23%, such as for soybeans, to as much as 67% for dry beans. In general, the harvest index has been increasing slightly as varieties are developed to shift energy conversion from stems, roots, and male parts to the edible portion (Zelitch, 1975). The leading edible crops produced in the world are wheat, rice, corn, potato, and barley (Figure 15.3). In the United States, the leading edible crops are corn, wheat, and soybeans (Figure 15.4).

Considerable potential exists for increasing the yields of these edible crops. As an example, the average yield of corn over the recent years is 6,500 kg/ha with a record yield of 19,525 kg/ha; the average yield of wheat is 2,200 kg/ha with a record yield of 7,715 kg/ha; and an average yield of soybeans of 1,600 kg/ha with a record yield of 7,000 kg/ha (Wittwer, 1975).

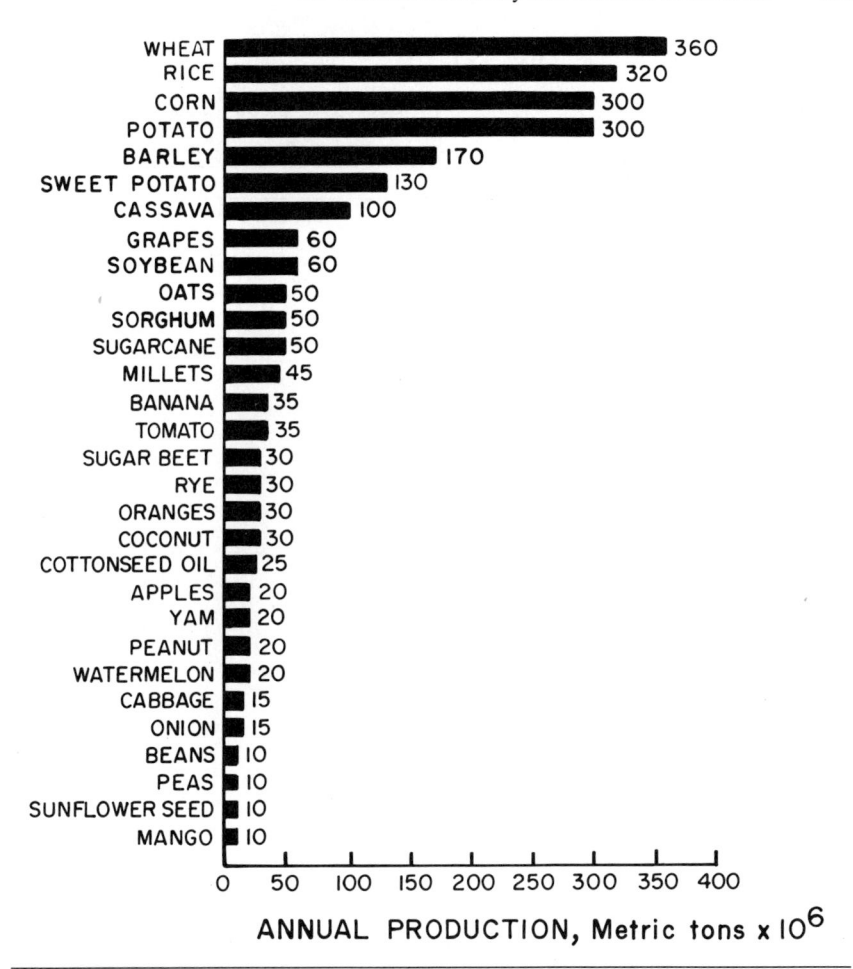

Crop	Annual Production (Metric tons × 10⁶)
WHEAT	360
RICE	320
CORN	300
POTATO	300
BARLEY	170
SWEET POTATO	130
CASSAVA	100
GRAPES	60
SOYBEAN	60
OATS	50
SORGHUM	50
SUGARCANE	50
MILLETS	45
BANANA	35
TOMATO	35
SUGAR BEET	30
RYE	30
ORANGES	30
COCONUT	30
COTTONSEED OIL	25
APPLES	20
YAM	20
PEANUT	20
WATERMELON	20
CABBAGE	15
ONION	15
BEANS	10
PEAS	10
SUNFLOWER SEED	10
MANGO	10

Figure 15.3 World annual production of leading edible crops (Vietmeyer, 1986).

Thus, there is a demonstrated production capability of at least four times the present average yield of these crops under near ideal conditions when not limited by availability of water, and where appropriate nutrients are provided (Wittwer, 1979).

A. Weather and Climate

Weather refers to those periodic changes in atmospheric conditions which occur daily, weekly, and monthly throughout the world and reported as the condition of temperature, humidity, rainfall, and wind speed and direction at a particular time. Climate refers to the general weather conditions

PRODUCTION, Tons x 10^6

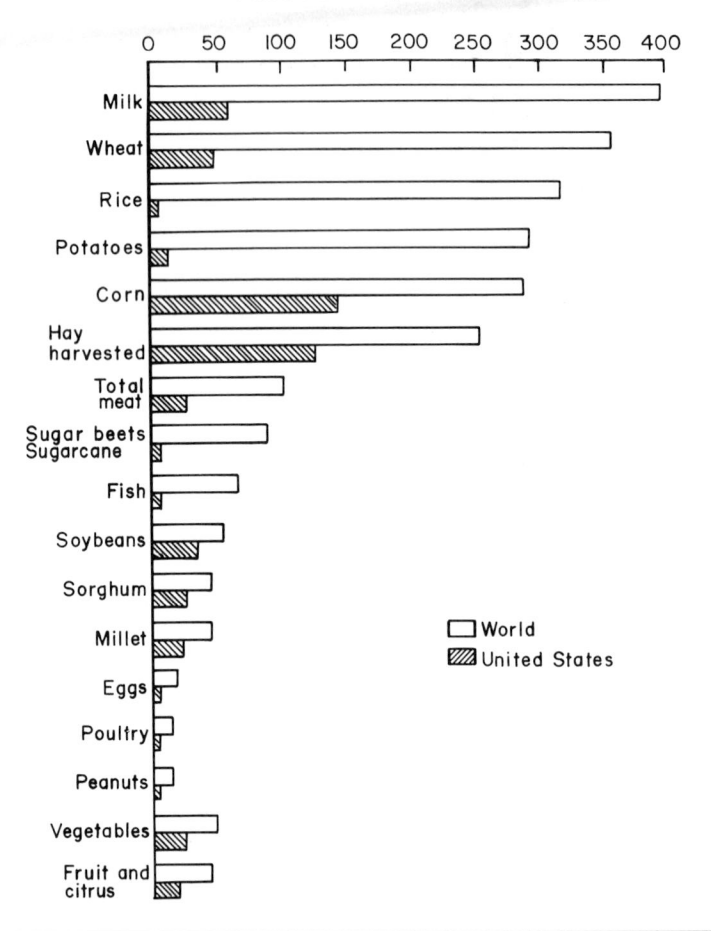

Figure 15.4 United States and world annual production of food and feed (Hall, 1977).

and represents the conditions under which plants, animals, and humans must live on a long-term basis. The term microclimate is used to represent the climate in the vicinity of plants, animals, and humans.

Conflicting predictions are made regarding major changes in global climate. With the addition of carbon dioxide to the atmosphere on a global scale resulting from increased industrial and combustion operations, one could predict that the temperature surrounding the earth would increase as the result of the greenhouse effect (Chou *et al.*, 1977). Also, climatic trends have taken place over the years as the result of external factors not fully explained at present. The difficulty of determining the temperature

15. Food Availability and Natural Resources 399

surrounding the earth is obvious when one considers the magnitude of the surface of the earth and the difficulty of placing sensors in appropriate positions to obtain temperatures representing average global values. Most of the instruments from which the weather and climate are monitored have been near cities where increased temperatures occur. These measurements could represent local temperatures and not provide the basis for a meaningful average temperature over the surface of the earth.

Climate changes occur which cause a shift in locations of rainfall and drought. Climatic changes are usually slow and subtle, and not easily predictable.

The jet stream is acknowledged as one of the phenomenon that causes weather conditions to move north or south in the northern hemisphere. The information on the jet stream is used for weather predictions. On a global basis, the total heat received is about the same on an annual basis. The El Niño Current, which occurs along the western coast of South America, has had a tremendous influence on weather conditions affecting the climate around the world (Kerr, 1986; *Science News*, 1987). Whether the El Niño Current has changed the average global temperature or moisture is unknown, but the weather patterns have moved and have changed the climate and food producing capabilities of the areas affected.

In the future, plants may be developed to withstand greater extremes in temperature. Some plants have "antifreeze-type" fluids which might be incorporated into other food producing plants so that production will occur during colder weather and thus provide longer seasons, making land

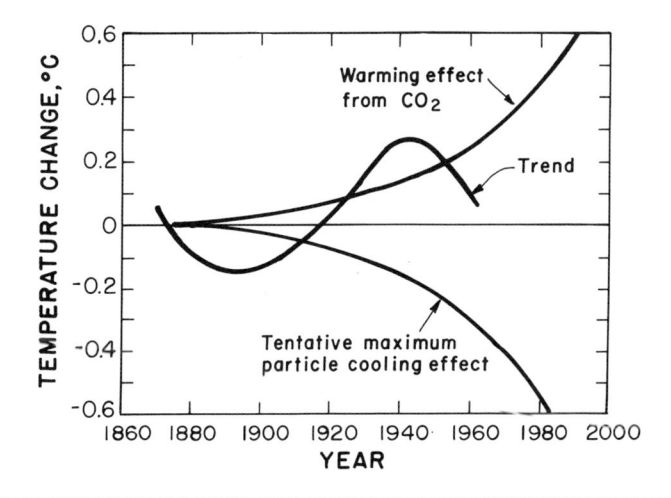

Figure 15.5 Effect of CO_2 and particulates on global heating and cooling (Chou *et al.*, 1977).

and ocean resources available for food production that are now considered too cold. A temperature increase, such as from an increase in atmospheric carbon dioxide, would cause reduced rainfall and thus affect crops. The carbon dioxide from the atmosphere may serve as a plant fertilizer and may tend to reduce the fears of a catastrophic increase in world temperature as a result of increased carbon dioxide production. At the same time the particulate level would need to be reduced to prevent excessive cooling, if in fact the particulate production is a primary influence on the temperatures over the surface of the earth (Figure 15.5)

Based on a study of long-term trends, if the mean temperature in the northern hemisphere declined 0.1°C per decade, the growing season could be shortened and yields decreased. A recent comprehensive article by Schneider (1986) reports that the earth is 0.5°C warmer than 100 years ago (Pimentel and Pimentel, 1979).

B. Plants and Animals

Plants will be developed and animals managed to enhance those characteristics which will decrease the negative impacts on the environment. Major changes have occurred in plants and only minor changes have occurred in animals to adapt to different environments. Generally climate has predominated and it is doubtful whether that relationship will change in the near future. Thus, plants and animals must be grown in the environment available. In many years of the world the production of plants and animals is increased by controlling the environment, principally temperature and humidity. In addition, for plants, the carbon dioxide and for food products the nitrogen level can be controlled to increase productivity.

Efforts will continue to reduce chemical use, but the global trend is to increased chemical use as growth regulators for control of pests and diseases. Crops can be designed to be resistant to many of the pests and diseases now being controlled primarily by chemicals. Inter-cropping with certain plants can be used as an approach to decrease the losses from pests and diseases, although some crops used in inter-cropping could harbor undesirable pests. Further, crops can be bred to increase hardiness to temperature variations, cold and heat, to reduce water needs by being more drought resistant, and to minimize loss of the food part of the plant, such as seeds and tubers during growth and after maturity. Plants are bred to ripen uniformly at the best time for harvest and dessicants are used to reduce foliage at time of harvest to reduce food losses.

Animals will continue to have a major role in diets by providing appropriate proteins and fats. In the United States, we consume excessive fat, with about 43% of our calories provided in fat. The trend to reduce fat consumption will continue in industrialized areas and thus animals will be bred that produce meat with low fat content. Vital and important pro-

Table 15.2 Conversion of Carbohydrates by Animals to Produce Edible Protein, Percent[a]

	Protein	Energy
Poultry (eggs)	25.5	17.5
Cattle (milk)	24.6	16.5
Poultry (broilers)	22.5	11.0
Swine (pork)	13.5	13.5
Cattle (beef)	4.0	13.5
Sheep (lamb)	3.9	2.0

[a]Source: Janick et al. (1976).

teins can be provided by animal meat and milk. Swine and poultry are bred for a high conversion rate of carbohydrates to proteins (Table 15.2). The ruminant animals are important in providing food as converters of grasses and forages that are not edible or digestible by humans. The fish is also an excellent converter of low-grade waste and feed to provide important proteins and minerals for the human body.

C. Wastes

World-wide, the waste produced annually per person is approximately 100 kg of dry weight, while in the United States nearly 1 metric ton is produced per person. The organic portion of these wastes is potentially available for producing food. On a practical basis, the waste would be used indirectly for producing food through plants, microorganisms, and animals. These wastes can be refined to produce food, but the processing is considered too expensive to accomplish from the standpoint of materials handling and energy.

Using a rule of thumb, about 1 kg of biomass plant materials is produced for each kilogram of edible grain, or approximately 1.2 billion metric tons on a world-wide basis. Only a portion of these wastes or crop residues can be economically salvaged and used for feed or food. The weight of the food portion of a plant has been gradually increasing for many plants. Less of the plant biomass produced is now available as a residue for other uses. Further, some of the plant residues need to be returned to the soil in order to maintain fertility and, more important, to protect the soil against erosion and maintain soil organic matter. The extent of the recycling of plant residues needed to maintain productivity varies greatly, depending upon soil and climate and crop.

Ruminant animals consume a large portion of the plants, and these animals can return fertilizer nutrients to the soil. From 60–90% of the fertilizer nutrients consumed (nitrogen, phosphorous, and potassium) are excreted by animals, but the wastes must be handled and managed properly to be returned to the food producing system.

Another category of waste is loss of food resources by pests such as birds, insects, and rodents. A loss is material produced for food but not available for use in the food chain, and thus differes from the natural wastes produced by plants, animals, and people. The availability of food to the consumer is decreased as a result of the losses that occur before, during, and after harvest. Most of these losses are considered as wasted effort and energy. In addition, during some activities such as processing, the quality of food may be reduced by lost food components such as vitamins and minerals. The amount of energy to recover a loss is usually greater than the energy embodied in the product lost. A major factor in determining whether or not wastes are recovered or prevented is the cost-benefit relationship.

Not to be overlooked in the recovery of wastes from people and animals is the potential for disease transmission to humans, and the impact on the environment. Garbage, for example, must be heat-treated or pasteurized if fed to swine to produce pork.

D. Cultural and Religious Preferences

Cultural and religious beliefs affect food availability and use. Some foods are used almost universally, and other potential foods are almost universally avoided. In the long range, these cultural and religious views will not change quickly, even when people are faced with starvation. Not only are some sources of food from plants and animals avoided as a whole, but some populations refuse to eat certain nutritious parts of those plants or animals. In addition, some groups of people avoid certain animal products, thus decreasing food availability to that group. Swine, beef, and horsemeat, common foods in many parts of the world, are avoided entirely by others. No section of the world is exempt from avoiding eating some plant or animal product or component which might be used as food. Pigeons, rabbits, insects, certain sea creatures, and snakes are but a few of the examples of potential edible food avoided by many. Insects eaten in some parts of the world include grasshoppers, crickets, caterpillars, and cutworms (Tufts, 1987).

E. Losses

In some parts of the world as much as 50% of the food production is lost before is reaches the consumer (Hall, 1977). Of particular concern is the loss of cereal grain and root or tuber products destined for food consumption. The world-wide losses on the average are probably about 25% of production.

Pests such as insects, rodents, and birds cause significant losses before and after harvest in grains, fruits, nuts, and animal products. Some losses can be prevented by eliminating or reducing these pests. In addition, methods of handling, packaging, and storing with pest-resistant materials and containers with appropriate temperature control can reduce these losses. Packaging or storing in non-toxic gases provides an additional approach for reducing losses. In areas of the world where the product is consumed at or near the location of production and there is not a large commercial market, the losses are greater than where the product is prepared for a market- oriented commercial enterprise. These losses must be considered before and at the time of harvest, transportation, storage, processing, and preservation. Often different techniques for preventing losses are involved in each of these functional operations. The value of the product saved must usually equal or exceed the cost of preventing the loss, except as the secondary costs and savings are considered. It is usually not economical to expend the extra money, effort, and energy to prevent all the losses. In commercial systems it is probably economical to prevent about one-half of the loss and increase the availability of food to consumers.

The losses of cereal grain, losses which are probably as well documented as for any food commodity, are estimated at about 10% of the production in which the world production is 1.2 million metric tons (Hall, 1977). Grains make up approximately one-half of the world food supply. Tropical roots provide stable food for about 500 million people or one-tenth of the world population, of which cassava or manioc is the most important root crop followed by yams, sweet potatoes, plantain, and taros or dasheen. The losses of these crops in the tropics is 25–50% of the production (Hall, 1977). The losses of fruits and nuts run as high as 30–40%, and the loss of vegetables often exceeds 50%.

F. Systems for Producing Food

The potential for increasing food availability is greatly dependent on the systems used for providing the food. The potential is limited for increasing food availability in certain closed cultures particularly where barter or exchange is a basis of obtaining food rather than monetary resources. For purposes of discussion, three major systems of food production and marketing are considered.

1. Traditional, Subsistence Agriculture Local Systems

These are systems where approximately 65% or more of the people make their livelihood from farming activities, and thus are more properly called farming in contrast to agriculture. Productivity and incomes are usually low. The external energy supplied is small, but there is a high overall energy efficiency. Changes in production processes take place very slowly, and most of the food crops which are sold are transported short distances. Basic grains, root crops, and pulses make up the bulk of the diets of the people. Marketing systems and facilities expand to meet only limited increases of marketable food supplies, based primarily on local consumption rather than extensive markets.

2. Transitional, Production-Oriented National Systems

In these systems, urban population is a larger portion, at least 50%, of the national population than the rural population. The production of food crops is a priority sector in the development plans and economy of the country. Commercial agriculture in food crops is increasing and capital is moving into farming and agribusiness. With increases of income and shifts in population more food crops move greater distances to markets.

3. Transcontinental, Market-Oriented Global Systems

With these systems the urban population is usually 75% or more, and commercial farming becomes the dominant producer of food crops for internal use as well as for export commodities. The food supply is characterized by a shift from major activities in production to most of the effort and cost in distribution. More attention is given to nutrition as a measure of the value of the commodity. Improvements in incomes and new technologies are incorporated into agriculture as well as other sectors of the economy. More food is processed and packaged and a strong marketing component exists as compared to subsistence and production oriented systems.

G. Commercial Farms

Highly industrialized or commercial farms usually contribute the greatest load on the natural resource system. Wastes in water and air, packaging materials, and chemicals need to be handled by the external environment. Chemical and biological treatments are used to increase production. Resources are used intensively and environmental impacts are recognized. As the producers begin to share in these environmental costs and come under more control from others in society, the cost of impacting the environment will be paid by the producer, and indirectly by the consumer.

Most of the emphasis on increasing food availability has been on increasing the production and productivity of land based production systems. More and more emphasis will be given to increasing the systems of food production on or in water bodies of lakes, oceans, and artificial water containment bodies for aquaculture and mariculture. Plants and fish raised in these water bodies will be used to increase food availability. Currently, less than 3% of world food comes from aquatic systems.

The traditional subsistence farming operations are usually small in acreage, but high in labor, as contrasted to the commercial, large, market-oriented production units. Entirely different management approaches are required for these contrasting systems.

H. Infrastructure

In order to support various systems for food availability for consumers the existence of physical, economic, political, and educational infrastructures are important. The physical infrastructure of transportation, communication, and energy must be available. Economic policies must provide access to financial resources, permit formation of cooperatives and businesses, and encourage a stable financial structure. A stable country, politically, with potential markets is needed. An educational system is required where people at all levels of production can gain knowledge for appropriate decisions, use of financial resources, and technical information.

The political and government structure should assure that safe, healthful, and nutritious products are produced. A system of grades and standards is needed to provide the basis for exchange, particularly for exported materials. Quality standards must be adopted to assure confidence among trading partners.

I. Other Protein Sources of Food

Potentially large quantities of protein are not now included in the world food supply because of the presence of certain natural toxicants in some plants (Hall, 1977). Techniques have been partially developed for removing or reducing these toxicants, including gossypol in cottonseed, tannins in sorghum grains, and glucosides in rapeseed.

Many food materials are nutritious but not acceptable to consumers because of flavor, color, texture, or stability. These products might be used as components for fabricated or manufactured foods. An example of a fabricated food is the treatment and preparation of vegetable protein meat analogues to replace meat cuts.

Synthetic foods may be used to increase availability of food. Synthetic food refers to assembling parts of odd components to provide an acceptable food. Components, some of which might be considered unedible by themselves, could be a part of a synthetic food, thereby increasing food availability. Various approaches are represented in Figure 15.6.

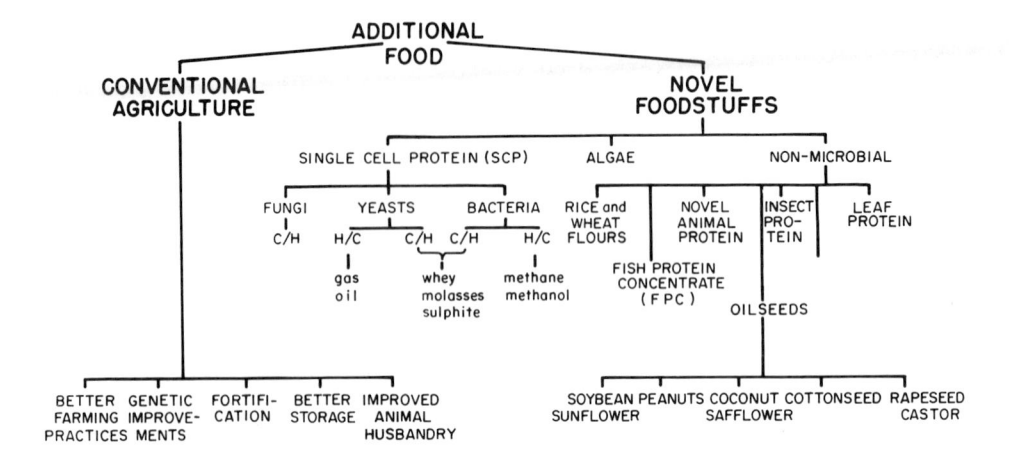

```
                           ADDITIONAL
                              FOOD
    ┌──────────────────────────┴──────────────────────────┐
CONVENTIONAL                                            NOVEL
 AGRICULTURE                                         FOODSTUFFS
                  ┌──────────────────────┬──────────────────────┐
            SINGLE CELL PROTEIN (SCP)   ALGAE            NON-MICROBIAL
         ┌────────┬──────────┐                      ┌────┬──────┬──────┬──────┐
       FUNGI   YEASTS     BACTERIA    RICE and    NOVEL  INSECT       LEAF
         │    ┌──┴──┐    ┌──┴──┐       WHEAT     ANIMAL  PRO-         PROTEIN
        C/H  H/C   C/H  C/H   H/C      FLOURS    PROTEIN TEIN
              │     └─┬──┘     │         │
             gas    whey    methane   FISH PROTEIN
             oil   molasses  methanol  CONCENTRATE
                   sulphite              (FPC)
                                                 OILSEEDS
```

Figure 15.6 Techniques for increasing protein production (based on Birch, 1976).

J. Trends and a Look Ahead

Many of the developments required to meet needs for food over the next 10 years are already in place. These include (1) major concern for preserving the environment of the soil, water, and atmosphere with a move to biological control and recycling of materials; (2) movement of developing countries to a market-oriented economy and development of the needed associated infrastructure; (3) increases in food availability due to increasing production by using existing resources and incorporating new technologies; (4) use of marine resources for production of plants and fish for food; (5) emphasis on providing a variety of protein products to meet the needs of many different cultures; and (6) increased processing to produce and preserve quality foods to permit long-term storage and long-distance transportation.

IV. SUMMARY

The natural resources and knowledge base are driving forces to meet the food and nutrition needs of a larger population. Pockets of starvation and the need for improved nutrition will continue in various degrees. The challenge is to reduce malnutrition in the world, which will involve cooperative efforts among many segments of society. The move to larger systems of production associated with sophisticated marketing has put increased demands on the natural resources. It is not necessary to destroy the environment to increase productivity but current practices are damaging or destroying the environment. Management approaches and incorporation of technology will help minimize impacts on the natural resources, if implemented, and perhaps enhance the environment as food availability is increased. There will be increased use of plants that are disease and pest resistant or tolerant to drought, freezing, and salt concentration. Other methods for increasing food availability include reducing losses, utilizing wastes, removing toxic materials from would be protein foods, and eventually using synthetic and fabricated foods to provide additional food over a larger area of the earth.

The utilization of increased production is highly dependent not only on material resources, but also on the natural, physical, economic, and educational infrastructures.

REFERENCES

Birch, G. G., Parker, K. J., and Morgan, J. T. (1976). "Food from Wastes." App. Sci. Publ., London.
Chou, M., Harmon, D. P., Jr., Kahn, H., and Wittwer, S. H. (1977). "World Food Prospects and Agricultural Potential." Praeger, New York.
FAO. (1965). "Report of the World Food Congress," Vol. 2, Washington, D.C., 1963. FAO, Rome.
Hall, C. W. (1977). Food availability to consumers. In "NRC Report of study on World Food and Nutrition," Study Team 5, Vol. 3, pp. 1–99. Natl. Res. Counc., Washington, D.C.
Hall, C. W. (1981). "Biomass as an Alternative Fuel." Government Institutes, Inc., Rockville, Maryland.
Hardy, R. W. F., and Havelka, U. D. (1975). Nitrogen fixation research: A key to world food. In "Food: Politics, Economics, Nutrition and Research" (P. H. Abelson, ed.), pp. 178–188. Am. Assoc. Adv. Sci., Washington, D.C.
Janick, J., Noller, C. H., and Rhykerd, C. L. (1976). The cycles of plant and animal nutrition. Sci. Am. **235**(3), 75–86.
Kerr, R. A. (1986). Another try at forecasting El Nino. Science **232**, 185.
Odum, H. T., and Odum, E. C. (1981). "Energy Basis for Man and Nature." McGraw-Hill, New York.
Office of Technology Assessment (OTA). (1982). "Impacts of Technology on U.S. Cropland and Rangeland Productivity." U.S. Gov. Print. Off., Washington, D.C.
Pimentel, D., and Hall, C. W. (1984). "Food and Energy Resources." Academic Press, Orlando, Florida.
Pimentel, D., and Pimentel, M. (1979). "Food, Energy and Society." Arnold, London.
Pimentel, D., et al. (1987). World agriculture and soil erosion. BioScience **37**, 277–283.
Rogers, P. (1986). Water. Technol. Rev. **89**(8), 31–43.
Ruttan, V. (1986). Increasing productivity and efficiency in agriculture. Science **231**, 781.
Saeman, J. F. (1977). Energy and materials from the forest biomass. Symp. Clean Fuels, IGT, Chicago, Ill.
Schneider, S. H. (1986). Climate modeling. Sci. Am. **256**(5), 72–80.
Science News. (1987). Warming up on El Nino Sci. News **131**, 55.

Sorensen, B. (1979). "Renewable Energy." Academic Press, New York.
Steinhart, J. S., and Steinhart, C. E. (1975). Energy use in the United States food system. *In* "Food: Politics, Economics, Nutrition and Research" (P. H. Abelson, ed.), pp. 85–94. Am. Assoc. Adv. Sci., Washington, D.C.
Tufts University. (1987). *Diet Nutr. News.* **4**(11), 8.
Vietmeyer, N. D. (1986). Lesser-known plants of potential use in agriculture and forestry. *Science* **232**, 1379–1384.
Wittwer, S. H. (1975). Food production and the resource base. *In* "Food: Politics, Economics, Nutrition and Research" (P. H. Abelson, ed.), pp. 85–94. Am. Assoc. Adv. Sci., Washington, D.C.
Wittwer, S. H. (1979). Future technological advances in agriculture and their impact on the regulating environments. *BioScience* **29**, 603–610.
Zelitch, I. (1975). Improving the efficiency of photosynthesis. *In* "Food: Politics, Economics, Nutrition and Research" (P. H. Abelson, ed.), pp. 171–177. Am. Assoc. Adv. Sci., Washington, D.C.

16

Food as a Resource

Marcia Pimentel

Division of Nutritional Sciences
Colleges of Human Ecology and Agriculture and Life Sciences
Cornell University
Ithaca, New York

I. Introduction
II. Patterns of Human Population Growth
III. Food and Dietary Patterns
   A. Nutritional Standards
   B. Nutrient Needs over the Life Cycle
   C. Differing Energy and Protein Intakes
   D. Protein Quality
   E. Vegetarian Diets
   F. Changes in U.S. Food Consumption
   G. Sociocultural Aspects of Food Selection
IV. Major Nutritional Problems
   A. Assessment of Nutritional Status
   B. Famine, Hunger, and Malnutrition
   C. Overnutrition
V. Trends in Food Production
   A. Grain Production
   B. Beef Production
   C. Fish Production
   D. Forests
VI. Resources Used in Food Production
   A. Land Resources
   B. Water Resources
   C. Energy Resources
   D. Food Needs for Future Generations
VII. Planning Future Policy
   A. Safeguarding Environmental Resources
   B. Science and Technology
   C. Accessibility of Food Supply
   D. Population
References

I. INTRODUCTION

Food is a basic necessity of human life. Lack of food—hunger—is now widespread throughout the world, even in some regions where food is abundant. Hunger is a precondition for malnutrition, which causes poor health and even death, and reduces productivity, income, and buying power. All of this lowers the quality of human life and sets in motion the cycle of hunger once again for another generation.

Lack of sufficient food is widespread and growing. Worldwide, the number of hungry people increased in the last decade, 1970 to 1980, from 15 million to 475 million (Lewis, 1987). Nowhere are there signs that the number of hungry people will diminish in the future, particularly as the world population spirals upward. Nor are there many signs that humans have faced the problem and started to solve it.

Obviously the rapid increase in the human population will compound and intensify the task of adequately feeding people. Many diverse factors are involved in growing, distributing, and making food accessible to all who need it. Many of the earth's most valuable resources will be used to produce more food and to keep up with the increased food needs. Furthermore, the nutritional content of the foods eaten and the wide array of sociocultural factors that influence individual dietary patterns will affect the outcome. These aspects of the food–population equation must be understood before action can be taken to ensure that future generations will be adequately fed.

II. PATTERNS OF HUMAN POPULATION GROWTH

For 99.9% of about 2 million years that humans have inhabited the earth, the maximum world population was less than 5 million—fewer than the current population of New York City. Population growth during this time was only about 0.001% per year. During most of this period, humans were hunter–gatherers and depended on their immediate environment for food and other basic needs. Humans managed to stabilize their numbers at a level that conformed with the availability of their basic resource needs (Douglas, 1966; Harris, 1977).

When people began to cultivate food crops about 10,000 years ago, some of the limitations and uncertainties imposed by nature were reduced. The larger, more stable food supply resulting from agricultural methods (Brown et al., 1985) contributed to and supported the slow, steady growth of the world population.

In 1650, the world population was only about 545 million (Deevy, 1960). But shortly after 1700, coinciding with the industrial revolution and the availability of cheap fossil fuels, the explosive increase in world population numbers started. A similar pattern occurred in the United States, where the population was about 4 million in 1790 and expanded to 240 million by 1985 (PRB, 1986).

The ability of humans to secure a relatively abundant, stable, and nutritious food supply, thanks to improved agricultural and technological ideas, contributed to the rapid increase in human numbers. Equally important has been the effective control of disease, which lowered the usual high death rates that previously had balanced high birth rates.

Because fossil fuels are the basis of fertilizers and pesticides (1 gallon of oil produces 1 lb of DDT), their availability was instrumental in decreasing disease and death. The dramatic reduction in death rates that occurred in Mauritius following the eradication by DDT of malaria-carrying mosquitoes illustrates this. Death rates fell from 27 to 15 per 1,000 over the period from 1945 to 1950. Because fertility (birth) rates did not decrease, an explosive increase in the human population followed (Corsa and Oakley, 1971). Recent history documents similar changes in population growth in Guatemala and Mexico where medical technology and availability of medical supplies have significantly reduced death rates.

The U.S. population, as mentioned, stands at 240 million and is growing at a rate of 0.7% (PRB, 1986). Based on this rate of increase, it is expected to double in 99 years or in three to four generations. Within six to seven generations, at the current rate of increase, the U.S. population density could be as great as that of present-day China. Is it not time for the United States to examine policies and for individuals to decide what standard of living and quality of environment they desire for themselves and their children? A sound population policy would help insure that future generations will live as we do today. With four times as many people, the prospect of a high standard of living for future generations is not encouraging.

Although the population density of the U.S. is the highest it has ever been, it is low compared with other nations.

The current world population stands at a high of more than five billion (PRB, 1986). What is more alarming than the numbers is the average 1.7% annual growth rate—a rate, 1,700 times greater than occurred in the first 2 million years of human existence. Such a growth rate adds more than 270,000 people a day to the world population. Demographers project that numbers will reach 6.1 billion by the year 2000, approach 8.2 billion by 2025 (U. N., 1982), and reach 12 billion by 2100.

Not considered in these predictions is the recently observed change in the world birthrate, which rose slightly in 1986. This increase reverses the slow decline that started in 1960 and upon which predictions were made (Dunn, 1987). The increase occurred mainly because China relaxed its stringent policy of one child per couple, causing the birthrate to rise from 1.8 to 2.1%. If this increase continues in China (and elsewhere), the world population could easily reach 7 billion, not 6 billion, by the year 2000.

Or consider the population in Bangladesh where about 100 million people are crowded into an area the size of the state of Wisconsin (Mydans, 1987). One-fifth of the area is delta land where the Ganges and Brahmaputra rivers join. This land area frequently shifts as the rivers flood and encroach on the land mass, making agricultural production and life there uncertain. If current trends continue, the population of Bangladesh is expected to rise to 160 million by the year 2000, further stressing land use.

Brown and Jacobson (1986) report that increases in population rates are associated most often with nations that have declining incomes. For example, they calculate that the 1986 increase in birth rates in Nigeria and Kenya, where per capita incomes are declining, are 3.0 and 4.2, respectively. This is significantly higher than in the United States and China, which have an increase rate of 0.7 and 1.0, respectively. As mentioned, this latter number may be increasing again as China seems to be relaxing its draconian rule of one couple, one child.

Again the question arises, how will so many more people be fed adequate diets?

III. FOOD AND DIETARY PATTERNS

The number of people who must be fed dictates the quantity of food that must be produced and distributed if optimum standards of health and well-being are to be maintained. Beyond production needs, consideration should also be given to the diverse factors that interact and result in different eating patterns for populations as well as for individuals.

Basically, specific nutrients are essential for human life. The plant and animal foods that humans produce and eat convey these nutrients to individuals for the development of their genetic potential and the maintenance of their optimum health.

Major nutrients are classified as carbohydrates, fats, amino acids (which are the building blocks of proteins), vitamins, and minerals. These nutrients are needed in varying amounts for human metabolism, growth, reproduction, and other vital activities of life. Therefore, it is critical that the human food supply contain adequate amounts of essential nutrients and be of high nutritional quality.

A. Nutritional Standards

Various guides have been compiled by nutritionists to serve as standards for evaluating how well food intakes or food supplies meet nutritional needs of individuals or population groups. The guide established by the

Food and Agriculture Organization (FAO) recommends a daily energy intake of 3,000 kcal for a 65-kg moderately active male and 2,200 kcal for a 55-kg moderately active female. The recommended protein intake, consisting of animal and plant materials, is 41 g/day per person (FAO, 1973).

In the United States, the National Research Council established a guide, called the Recommended Dietary Allowances (RDA), now in its ninth revised edition. The RDAs are "the levels of intake of essential nutrients considered on the basis of available scientific knowledge to be adequate to meet the known nutritional needs of practically all healthy persons" (NAS, 1980). In the RDA, the recommended energy intake is specific for age, weight, height, and for light physical activity. Thus, for males 23–50 years of age, who weigh 70 kg and are 70 in. tall, the recommended intake is 2,700 kcal and 56 g protein; for females in the same age bracket, but who weigh 55 kg and are 64 in. tall, the intake is 2,000 kcal and 44 g protein/day (NAS, 1980). The RDA protein recommendations are corrected for the efficiency of protein utilization, depending on whether the protein is of animal or plant origin.

If an individual does not meet the suggested RDA, it does not follow that the person is malnourished. This is because the RDAs are set high enough to cover a wide variety of individual needs. The allowances for nutrients are set much above the requirements that would prevent the development of a specific nutritional deficiency, e.g., scurvy, pellagra.

B. Nutrient Needs over the Life Cycle

Nutritional guides and standards also reflect changes in the amounts and kinds of nutrients that individuals need during their lifespan (NAS, 1980).

For example, a young child who is rapidly growing requires more calories per kilogram of body weight than an adult. In addition, the young often need additional calcium for bone growth, as well as other nutrients, during this stage of the life cycle. The pregnant female needs to increase energy, protein, and certain vitamin and mineral intakes to insure her continued health and that of her offspring. Taking these special needs of the pregnant female into consideration, the RDA for protein increases from 46 to 76 g/day, while vitamin C increases from 60 to 80 mg, and calcium increases from 800 to 1,200 mg/day.

Activity, or lack of it, also influences nutrient needs. Thus, an adult male engaged in heavy work, such as sawing wood or tilling soil by hand, may burn from 400–600 kcal/hr above the requirements for basic metabolism, and therefore may need 4,000–5,000 kcal of food energy and additional vitamins per day to maintain such strenuous labor. In contrast, sedentary people do not need to consume large quantities of energy and frequently must curtail their kilocalorie intake while endeavoring to meet their needs for the other vital nutrients.

Over the entire lifespan, illnesses of various kinds and severity change nutritional needs (Anderson, 1979). For example, the nutrient needs of an individual who is infected with common parasites like ascarids or parasitic worms will increase. In such cases the increase in nutrients helps offset the food needs of the parasite population and the losses that occur if the parasites interfere with the normal digestion and absorption process. Sometimes more nutritious food is needed to counteract the blood and other body fluid loss associated with parasitic infections. All too often these types of problems occur in population groups that are already being stressed by malnutrition or starvation.

C. Differing Energy and Protein Intakes

In comparing human diets, distinct differences are immediately apparent between those typical of industrialized nations and those of developing nations. For example, in the United States about 3,600 kcal and 102 g of protein are consumed per person per day (USDA, 1987). In contrast, the majority of the world population consumes about 2,300 kcal and 61 g protein per person per day (CEQ, 1980).

Further differences emerge when the type of protein consumed is analyzed. In the United States, over two-thirds of the protein eaten is of animal origin (USDA, 1987). Although the per capita grain use is high in the United States, only about 10% is consumed directly as food. The remainder is cycled through the livestock system to provide the beef, pork, and poultry foods that Americans prefer (Pimentel *et al.*, 1980). However, in developing nations, about 80% of the protein consumed is of plant origin, and the remainder is of animal origin (Hudson, Chapter 11). In China, the plant protein content of the diet is slightly lower than this, averaging 70%, with animal products and fish making up the remainder (Pimentel *et al.*, Chapter 14).

D. Protein Quality

Protein itself is not an essential nutrient, but it contains necessary amino acids and nitrogen. The value of a food as a protein source depends not only on the amount of protein it contains, but on its specific amino acid content. Of the 22 amino acids that are used in body processes, eight are called essential because the human body has no way to make them. They include lysine, tryptophan, threonine, valine, leucine, isoleucine, phenylalanine, and the sulfur-containing amino acids, methionine and cystine. The other amino acids can be synthesized by body cells from various other ingested materials.

The essential amino acids must be in the food, ready for use in metabolism. They must also be available in appropriate amounts. For most efficient use in body processes, all eight essential amino acids should be ingested at the same time.

The quality of a protein food is judged by its essential amino acid content. Animal protein foods like meat, milk, and eggs are considered high quality or *complete* proteins because they contain all the essential amino acids in appropriate amounts. Although gelatin is animal protein, it is an incomplete protein because it lacks one essential amino acid.

Plant protein foods like legumes, cereals, and nuts contain substantial amounts of protein, but all the essential amino acids are not present in appropriate amounts. For this reason, plant proteins are classified as *partially complete*. Quantities of the amino acids containing sulfur as well as lysine, tryptophan, and threonine are those that are most often limited in plant foods.

To effectively use plant proteins, their amino acid composition needs to be taken into account (Table 16.1). Threonine content is not listed in the table because if the needs of the other three are met, its supply is generally adequate. Of the legumes, soybeans have the best amino acid content, and by eating large servings, the need for essential amino acids can be met. But peanuts, another legume, have an amino acid content similar to that of nuts and seeds, which have a less satisfactory amino acid content than legumes as a group.

One simple way to overcome the amino acid deficiencies of plant proteins is to combine them with small amounts of animal proteins. Such

Table 16.1 Amino Acid Content of Major Protein Foods[a]

Food groups	Amino acids[b] Good source of	Poor source of
Legumes	Lysine	Tryptophan; S-C
Soybeans	Lysine;[c] tryptophan[c]	S-C
Dry beans	Lysine[c]	Tryptophan; S-C
Nuts and seeds	Tryptophan; S-C	Lysine
Peanuts	Tryptophan	Lysine; S-C
Sesame seed	Tryptophan;[c] S-C[c]	Lysine
Cereals and grains	Tryptophan; S-C	Lysine
Corn meal	S-C	Lysine; tryptophan
Whole wheat flour	Tryptophan; S-C	Lysine
Wheat germ	Lysine[c]	Tryptophan; S-C
Rice	Tryptophan; S-C	Lysine
Eggs	Lysine;[c] tryptophan;[c] S-C[c]	—
Meat, fish, poultry	Lysine;[c] tryptophan; S-C	—
Dairy	Lysine;[c] tryptophan;[c] S-C[c]	—

[a] Reprinted with permission from Devine and Pimentel (1985).

[b] S-C: sulfur-containing amino acids.

[c] Superior source.

combinations of foods as cereal and milk, macaroni and cheese, grits and sausage, and arroz con pollo (rice and chicken) illustrate this commonly used strategy.

The other successful alternative is to combine plant proteins in such a way that the amino acid deficiencies in each are supplemented (Figure 16.1). Thus, legumes, which are excellent sources of the essential amino acid lysine, supplement or complement the lack of lysine in cereals. At the same time, the cereals make up the tryptophan deficiency of legumes. In this way, a satisfactory protein quality is achieved. If each food were the sole source of protein, quality would be poor. Note that the ability to supplement between grains and nuts and seeds is not as strong as that between legumes and either grains or nuts and seeds. In contrast, legumes as a group provide an excellent supplement to the other two major plant groups (Lappé, 1982).

Examples of protein complementarity are found in eating patterns around the world, and many were established long before scientists discovered the importance of "essential" amino acids in human nutrition. Combinations of food are as varied as rice and beans, peanut butter sandwiches, baked beans and brown bread, and pulse and rice. Yet all these are basically supplementary combinations of legumes and cereals.

Mention needs to be made of other nutritional concerns associated with complete vegetarian diets. First is the lack of the essential vitamin B_{12}, which is not found in plant foods. The exceptions are specially culturedyeasts and fermented soybean products (tempeh and miso) of the Far East. These contain B_{12} in varying amounts. When milk products and all meats are eliminated, then the minerals calcium, iron, and zinc

Legumes
Peas, Lentils
Soybeans
Kidney beans
Navy beans

High Lysine

Grains ⟷ **Seeds & nuts**
Rice
Corn
Wheat
Rye
Sesame, Sunflower
Pecans, Peanuts
Almonds, Walnuts

High Tryptophan & sulfur-containing

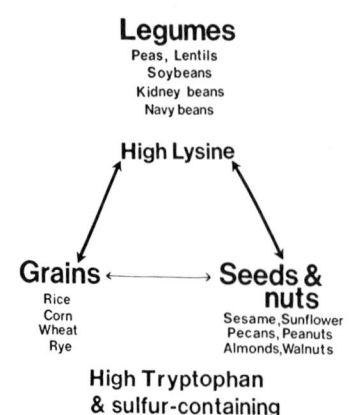

Figure 16.1 Supplementarity of plant proteins.

and the vitamins riboflavin and D are especially difficult, although not impossible (except vitamin D), to obtain from all plant foods. Children and women experience more problems than men because usually they do not eat the large amounts of vegetable proteins needed to supply these important nutrients.

E. Vegetarian Diets

In industrialized nations, vegetarian diets are often selected or favored because of individual religious beliefs, concern about environmental pollutants, or land use. Plant protein food sources are also more economical than animal sources and therefore are used alone or in combination with animal proteins to stretch the food budget. They also have nutritional advantages because they contain no cholesterol and are lower in saturated fat than animal proteins.

In contrast, large sectors of the world population, especially in developing countries, must rely on plant foods to meet their energy, protein, and other nutritional needs. Worldwide, as a group cereals are the most important food for people. Thus, depending on the country or region, wheat, rice, barley, millet, and corn constitute the mainstay of the diet. Legumes and tubers also contribute, as do very small amounts of meat, fish, or eggs.

For example, in rural areas of Central America, where corn is the staple food, laborers commonly consume about 500 g of dry corn and 100 g of dry black beans per day, which provide about 2,120 kcal and 68 g of protein daily (E. Villagran, 1974 personal communication). In this diet, corn and beans supplement each other in essential amino acid patterns and together improve the quality of the two incomplete proteins consumed. Tandon *et al.* (1972) report a survey of 12 rural villages in India, in which the average daily consumption per family member was between 210 and 330 g of dry rice and wheat, 140 ml of milk, and 40 g of dry peas, lentils, and beans. This food intake provided about 1,400 kcal and 48 g of protein per day.

When these contrasting consumption patterns are compared with existing nutritional guidelines, the U.S. average protein intake is excessively high, while the two predominantly vegetarian diets discussed meet the FAO protein allowance. However, the calorie intakes for the vegetarian diets are low, especially for physically active male adults.

F. Changes in U.S. Food Consumption

In the United States, shifts have occurred over time in the per capita consumption of major foods (Table 16.2). Note that these data represent the disappearance of food from commercial market channels and, as such,

Table 16.2 Civilian per Capita Consumption of Major Food Commodities[a,b]

	1960	1979[c]	1985[d]
Meats (retail)	134.1	147.1	144.4
Fish (edible weight)	10.3	13.2	14.5
Eggs (pounds)	42	35.8	32.4
Chicken/turkey	34.0	61.6	69.3
Cheese	8.3	17.6	22.4
Fluid milk (pounds)	321.0	283.2	245.1
Ice cream	18.3	17.5	27.2
Total fats and oils	45.3	57.7	67.2
Fruits (fresh)	90.0	81.3	88.2
Fruits (processed)	22.6	19.4	14.8
Fruits (frozen and juice)	9.2	12.3	19.9
Vegetables (fresh)	96.0	94.2	81.4
Vegetables (canned)	45.7	55.7	32.6
Vegetables (frozen, except potatoes)	6.9	11.5	12.0
Potatoes	87.9	75.0	125.3
Wheat flour	118.0	120.0	122.5
Rice	6.1	9.2	9.3
Edible beans (dry)	7.3	6.4	—[e]

[a] Quantity in pounds, retail weight. Data on calendar-year basis except for dried fruits, fresh citrus fruits, and rice, which are on a crop-year basis.
[b] Data from U.S. Department of Agriculture (1981).
[c] Preliminary, USDA (1981).
[d] USDA (1987).
[e] Not available.

overstate actual consumption. However, trends are clear and illustrate the kinds of changes that can occur in the food supply of a country.

In the 25-year time span from 1960 to 1985, per capita consumption of meat, chicken, and cheese substantially increased while egg and milk consumption declined. Fat consumption, based on disappearance figures, increased over 24% per capita as use of oils increased with the consumption of more salad dressings and deep-fat-fried foods.

Increases also occurred in consumption of fruits and vegetables. Their year-round availability increased because of improved transportation, storage, and preservation. Per capita consumption of legumes remained relatively unchanged, again emphasizing the low use of vegetable protein sources in this country.

Because of these shifts in the food supply, consumption levels of many nutrients also changed (Marston and Raper, 1987). For example, greater use of citrus fruits, especially the availability of frozen orange juice, increased the availability of ascorbic acid. More vegetables, particularly dark green and improved deep yellow types, accounted for the gain in vitamin A. The decrease in egg consumption meant a decline in per capita intake of cholesterol.

Looking to the future, it is impossible to project the kinds and magnitudes of change that will undoubtedly occur in per capita consumption of major foods in the United States and in other countries.

G. Sociocultural Aspects of Food Selection

Although much is known about the biology, chemistry, absorption, and metabolism of specific essential nutrients, information about why individuals do or do not eat certain foods is not so precise. Knowledge about nutritional needs or dietary advice from a nutritionist does not always influence an individual's choice of food.

Individuals make dietary choices from a variety of foods and are influenced by many different factors in making their choices (Hertzler et al., 1982). At times, the availability of the food in the marketplace and its cost dictates selection. For most population groups and individuals, factors like prestige, convenience, customs, religion, and social practices also play an important role. Even within the same cultural group, community or family, individual likes and dislikes often determine which foods are eaten. Palatability aspects such as flavor, color, texture, and mouthfeel influence personal choice. Sometimes individuals associate special food choices with social celebrations and emotional feelings. Krondl and Law (1982) have organized many of these factors in what they term a "cultural anthropology framework for food selection study." Specifically, they classify as "endogenous determinants" factors that relate to each individual, such as age, sex, and inherited characteristics, while "exogenous determinants" include those existing in the environment that influence life, such as culture and economics. The third part of their framework, "perception," acknowledges the importance of beliefs about foods and palatability characteristics in influencing individual food patterns. However, in situations where food is scarce, people must eat whatever is available just to survive and these factors do not operate.

Messer (1984) suggests that a "combination of ecological and economic constraints limit the range of choices and create other conditions affecting health and nutrition, while within such constraints, cultural tastes and values, as well as ideas of adequate nutrition and health beliefs and practices dictate actual food and behavioral choices." Because food selection ultimately determines the nutritional status of an individual, understanding the sociocultural dimensions of food selection is vital in long-range planning programs that endeavor to improve or change food selection to improve nutritional status.

IV. MAJOR NUTRITIONAL PROBLEMS

Worldwide the availability of food has always played a major role in the health, quality of human life, and the very survival of humans. Conditions range from the severe food scarcity of famines to chronically inadequate food supplies to an overabundance of food. All conditions are associated with the major nutritional problems that now exist in various areas of the world. All aspects of the world food situation must be understood in order to make plans to adequately feed the growing world population.

A. Assessment of Nutritional Status

The full dimensions of the present world food situation are difficult to assess on an individual level because accurate data are scarce and because assessing nutritional status is not always complete or standardized in the scientific studies available. Ideally, an assessment should include data on clinical or visual physical signs of disease such as biochemical analyses of tissue, blood, or urine (e.g., hemoglobin levels), anthropometric measurements (e.g., the relationship of height and weight or skinfold thickness), and an evaluation of the dietary intake, preferably over a long period of time.

Table 16.3 summarizes the type of data needed for the assessment of nutritional status as set forth by the World Health Organization (1963). Even to the uninitiated, it is obvious that such complete assessments take competent scientists much time in the field and in the laboratory and are expensive to carry out.

B. Famine, Hunger, and Malnutrition

Famines, which are characterized by a severe shortage of food, have occurred throughout history and are associated with massive numbers of human deaths. Reports indicate that "there is famine in various developing nations, and death rates from famines are reported rising in at least 12 and perhaps 20 nations, largely in Central Africa and Southern Asia" (NAS, 1975). Specific examples of devastation are numerous. For example, Sen (1980) reports that approximately three million humans perished in the 1943 Bengal famine, while an estimated 1.5 million lives were lost in a more recent famine in Bangladesh (Mellor, 1986).

More recently, world attention has been focused on the famines occurring in sub-Saharan Africa. Print and television have graphically presented the horror of these famines. Yet, compared to former famines, the mortality in the Sahel famine in Ethiopia of about 200,000 is considered by some to be relatively small (FAO, 1984).

Table 16.3 Data Needed for the Assessment of Nutritional Status[a]

Sources of information	Nature of information obtained	Nutritional implications
Agricultural data Food balance sheets	Gross estimates of agricultural production Agricultural methods Soil fertility Predominance of cash crops Overproduction Food imports and exports	Approximate availability of food supplies to a population
Socioeconomic data Information on marketing, distribution, and storage	Purchasing power Distribution and storage	Unequal distribution of available foods between the socioeconomic groups in community within family
Food consumption patterns Cultural anthropological data Dietary surveys	Lack of knowledge, erroneous beliefs and prejudices, indifferences Food consumption	Low, excessive, or unbalanced nutrient intake
Special studies on foods	Biological value of diets Presence of interfering factors (e.g., goitrogens) Effects of food processing	Special problems related to nutrient utilization
Vital and health statistics	Morbidity and mortality	Extent of risk to community Identification of high-risk groups
Anthropometric studies	Physical development	Effect of nutrition on physical development
Clinical nutritional surveys	Physical signs	Deviation from health due to malnutrition
Biochemical studies	Levels of nutrients, metabolites, and other components of body tissues and fluids	Nutrient supplies in body Impairment of biochemical function
Additional medical information	Prevalent disease patterns including infections and infestations	Interrelationships of state of nutrition and disease

[a] From World Health Organization (1963).

Mellor (1986) suggests that future famines will occur in areas of subsistence crop production, areas with poor transportation and inadequate communication, and where political structures are unsteady. In reference to the latter, Mellor suggests that democracies, because they have "broadly based political constituencies and relative freedom of speech and press" are best able to prevent famine. Unfortunately, Mellor's descriptions do not seem to characterize the vast majority of developing nations.

The general connotation of hunger is not consuming enough food to support the proper functioning of the human body. The reasons that hunger occurs are many and complex in nature. In attempting to understand them, Kates (1987) has classified the following causes: outright food shortages occurring in a given area; hunger caused by poverty and the inability to afford food, even though the supply is plentiful; and hunger of individuals not receiving food that is generally available to the household or family group. Indeed, patterns of food distribution within the family differ, and in particular infants and women in some cultures receive both less food and/or less nutritious food than others in the group.

Hunger exists everywhere, even in industrialized nations like the United States where the overall food supply is plentiful. In fact, a recent report by the Physicians' Task Force on Hunger in America (1985), sponsored by the Harvard School of Public Health, documented the reappearance and escalation of hunger in this country. The study estimates that about 20 million Americans now suffer from hunger. Growing poverty plus inadequate governmental and societal support systems (e.g., food stamps, school lunch programs, and neighborhood or church soup kitchens) are identified as major causes of this bleak picture.

When hunger persists over long periods of time, chronic malnutrition occurs. Mild malnutrition, although debilitating to an individual, does not present obvious clinical symptoms. Yet whenever mild malnutrition occurs and people's food intake is marginal they are more vulnerable to natural disasters, delayed or inadequate harvests, and loss of income. Eventually, the continued lack of food and essential nutrients leads to medically identifiable changes in physical and mental health.

Although metabolic adjustments occur in humans who have low caloric intakes, reduction in their physical activity is responsible for the major adaptation (Scrimshaw, 1986). First, family and community activities are curtailed and, lastly, work. This leads to decreased productivity and, especially in agricultural areas, can ultimately decrease food production. Obviously, decreased activity will adversely affect the economic status of the individual or family and their ability to secure sufficient food.

How extensive and how serious a problem is malnutrition? Based on the FAO 1973 standard, Reutlinger and Alderman (1980) estimate that some 800 million humans are deficient in their caloric intake. Latham (1984) estimates that about one billion people are more malnourished and warns that the problem is growing in severity. Caliendo (1979), reviewing the many worldwide assessments of malnutrition that have been made, concludes that although there is a lack of detailed information about the precise assessment of the extent of malnutrition, "it is clear that problems of nutritional deprivation touch the lives of a large proportion of the world's population."

The extent of protein-energy malnutrition (PEM) in developing nations, characterized by marasmus and kwashiorkor among infants (Reddy, 1981), is a major on-going problem. Protein-energy malnutrition occurs when energy intake is inadequate. In such cases, protein is metabolized to make up the energy shortfall because the energy needs of the human body have a higher priority than protein needs. Each year, severe cases of PEM are responsible for approximately 10 million deaths in the infant to four-year age group (Latham, 1984). The survivors face a lasting effect on their ability to learn to be productive in adulthood (Cravioto and DeLicardie, 1976). Diverse environmental factors impinge on infant health and ultimately on adult productivity. Once in motion, events seem to cycle so that there is no chance for improvement, either in economic status or individual health (Figure 16.2).

Besides dietary deficiencies in energy and protein, insufficient intakes of vitamin A, iodine, and iron all cause serious health problems for the chronically malnourished (Wilson *et al.*, 1979; Feachem, 1987; Scrimshaw, 1986). Vitamin A deficiency leads to blindness and, for many people, death (Latham, 1984). Iodine deficiencies cause goiter and cretinism, while low iron intakes are responsible for anemia, especially in pre-menopausal females.

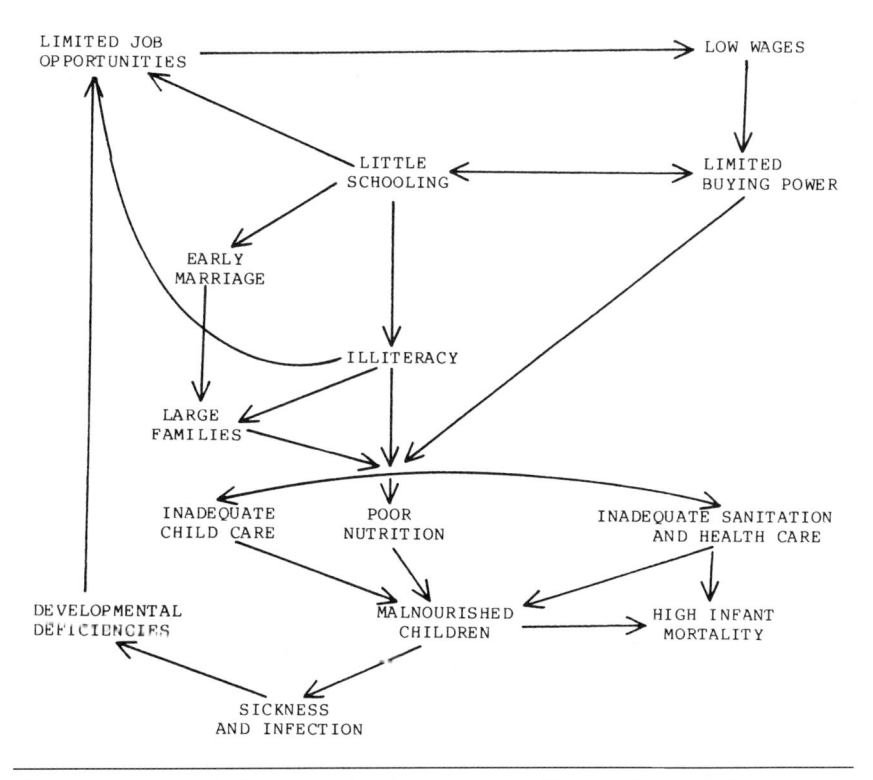

Figure 16.2 Interrelationships of poverty, malnutrition, and economic status. [After Cravioto and DeLicardie (1976).]

With regard to these specific nutrients, fortification of staple foods with well-absorbed forms of the deficient nutrients can do much to alleviate the problems. For example, the basic staple foods like salt, sugar, coffee, tea, and monosodium glutamate (MSG) now are being used for iron fortification (Zoller *et al.*, 1980). Such programs must be government sponsored, with advice from nutritionists who can not only identify nutritional deficiencies but assist in selecting the staple that is to be fortified.

Although average life spans have lengthened over those of previous decades, it is generally accepted that malnutrition decreases the body's ability to resist disease (Scrimshaw *et al.*, 1968; Wilson *et al.*, 1979; Chandra, 1981). Fever, diarrhea, and infections weaken malnourished people, thereby further stressing their already poor nutritional status.

C. Overnutrition

Malnutrition has another side—that of overnutrition, as typified by many people in industrialized nations who can afford all the food they want and eat more than they need. Overweight and obesity that result are caused, in large measure, by not balancing energy (kilocalorie) intake with the decreasing energy requirements of an increasingly sedentary population (U.S. Senate, 1977).

Obesity, as well as arteriosclerosis, hypertension, and certain cancers are prevalent in the U.S. and have been associated with diets high in calories, high in fats (especially saturated fats) and high in cholesterol. Nutritionists seem to agree that diet is a major factor in the incidence of these chronic diseases, but many other factors, such as genetic predisposition, activity, stress, and smoking also influence their incidence in the U.S. population.

Increasing the food supply will not necessarily reduce the malnutrition that now exists. Solving this problem is vastly complex and involves increasing the purchasing power of the poor so they can purchase enough food to meet their needs (Briggs and Calloway, 1984). Certainly using interventions like food fortification to ease manifest malnutrition should become more widespread. Scrimshaw (1986) focuses on the problem when he warns, "prevention of famine and hunger is not just a technological issue, but a moral, political and social one." Starting now, solutions that address the basic reasons behind malnutrition must be found before the problems become further intensified because of increased human numbers.

V. TRENDS IN FOOD PRODUCTION

Before 1800 almost all countries produced sufficient food for their own people. Subsequently, with the development of efficient modes of transportation, some areas became major food producers for their urban areas and also for other nations. By the mid-1850s great shifts in food production and consumption patterns occurred (Gilland, 1986). Britain, Belgium, Germany, and more lately Japan have, as Gilland notes, increased "their prosperity by utilizing the natural resources of other countries." At present, many of the countries that are food suppliers are the developing nations. These raise cash crops for export to industrialized nations that need food.

Even with the disparities in production, ample food now is being produced in the world so that each person could have a nutritious diet. This assumes that the food could be distributed in an equitable way among nations and to individual people within each nation. It also assumes that the economic status of individuals enables them to purchase the necessary food.

A. Grain Production

Although per capita grain production in the world declined during the late 1970s, grain production began to increase again in the 1980s (Table 16.4). This is due to increases in land under cultivation, increased fertilizer and pesticide inputs, more land under irrigation and especially favorable weather conditions in most parts of the world.

The inputs of fertilizers, pesticides, and irrigation water helped offset grain production declines that would have followed soil erosion problems that have escalated worldwide (Pimentel *et al.*, 1987).

B. Beef Production

Beef production peaked in the mid-1970s (Table 16.4). In the United States increased grain prices have been a factor in the subsequent slow decline since then because animal production became less profitable. In the future less grain may be available for animal production but will have to be consumed directly as human food to support the increased human population. Also, animal production may be reduced as valuable pasture and rangelands deteriorate because of overgrazing by large animal populations.

C. Fish Production

Fish production has declined since the 1970 peak because of overfishing or outstripping the fish resources of the ocean (Table 16.4). Catches have not surpassed the 1970 levels despite the greater energy inputs associated with larger fishing vessels and new fish-finding devices (Rochereau and Pimentel, 1978).

Table 16.4 World per Capita Production of Fish, Beef, and Grain[a]

Year	Fish (kg)	Beef (kg)	Grain (kg)
1950	8.4	—	251
1955	10.5	—	264
1960	13.2	9.3	285
1961	14.0	9.6	273
1962	14.9	9.8	288
1963	14.7	10.7	282
1964	16.1	10.1	292
1965	16.0	9.9	284
1966	16.8	10.2	304
1967	17.4	10.4	303
1968	18.0	10.7	313
1969	17.4	10.7	311
1970	<u>18.5</u>	10.6	309
1971	18.3	10.4	330
1972	16.8	10.6	314
1973	16.8	10.5	332
1974	17.7	11.0	317
1975	17.2	11.3	316
1976	17.7	<u>11.6</u>	337
1977	17.3	11.5	330
1978	17.3	11.4	351
1979	16.9	10.9	331
1980	16.1	10.5	324
1981	16.6	10.1	369
1982	16.6	9.5	369
1983	16.5	9.6	350
1984	17.4	9.6	379
1985	17.5	9.5	<u>380</u>

[a] Source: Food and Agriculture Organization (1986a,b); Population and Vital Statistics (U. N., 1986); and after Brown and Shaw (1982) from 1950–1980. Peak years are underlined.

D. Forests

Forests are an essential resource for future economic development, and their status indirectly influences food production. They provide lumber for housing, pulp for paper and, most important, biomass for fuel. Biomass energy is the major source, about 80%, of the fuel used by the poor people of the world for heating and cooking (Pimentel et al., 1986). Approximately half of the cooking/heating biomass is woody material and the remainder is 33% crop residues and 17% dung.

Forests help control erosion and water runoff on steep slopes and in this way help conserve vital soil and water resources needed by agriculture. Because rates of erosion and sedimentation are high, the reservoirs and dams of many nations are being rapidly filled with sediments. Where this happens the generating capacity of hydroelectric plants and irrigation water capacity of the dam is reduced and the sediments have to be removed at high cost to users.

Worldwide about 11.6 million ha of forestland are being cleared each year (FAO, 1982). Most of this deforestation is the result of expanding agricultural needs. Almost all or about 10 million ha of cleared land are needed for agriculture each year to keep production levels high. Of this, about 6 million ha of the forestland replaces agricultural land that has severely degraded soils, while the other 4 billion ha is used for increased food production (Pimentel *et al.*, 1986).

As more food is needed, more agricultural land will be required and more forest will be cleared to meet this need. As mentioned, deforestation increases soil erosion rates, especially on sloping land, and diminishes soil fertility. Concurrently, with less wood available, more crop residue and dung will have to be burned for fuel. This will further decrease soil fertility by exposing soils to erosion and by removing valuable nutrients that could be recycled. Continuous loss of soil productivity will necessitate removing more forests, and the disastrous cycle accelerates.

VI. RESOURCES USED IN FOOD PRODUCTION

Increased production of food to meet future needs will require increased use of land, water, energy, and other biological resources. To gain some insight into the potential "carrying capacity" of the earth to supply sufficient nutritious food for a growing population, the following estimates are made about the amounts of food that can be produced, based on current technologies and the natural resources that are now available.

A. Land Resources

At present there is sufficient land to feed a world population a satisfactory diet, if current technologies and inputs of fertilizer, irrigation, pesticides, and hybrid seeds are used. This also assumes that food resources could be distributed adequately so that the more than one billion humans who presently are malnourished would receive at least their minimum nutritional needs.

But would there be sufficient arable land to feed the current world population of about five billion a U.S.-type diet? Presently in the United States about 160 million ha are cultivated in crops (USDA, 1985). Based on more than 240 million Americans, this averages to 0.7 ha of land per capita that is used to produce the typical U.S. diet.

World arable land resources are approximately 1.5 billion ha (Buringh, Chapter 4), and, based on a world population of five billion, the per capita land available is only 0.3 ha. Therefore, present arable land supplies are insufficient to feed the current world population the typical U.S. diet, even assuming that sufficient fossil energy supplies for fertilizer and pesticide supplies were available everywhere in the world. Of course, not everyone in the world desires to eat a typical U.S. diet, but the above examples clearly illustrate that land will be an important constraint in future food production.

Another factor influencing the amount of available agricultural land is the continuous expansion of urban areas and associated roads into these areas. Not only has food-producing land been covered with structures, but food production has been pushed farther distant from major markets. This move increases both time and cost associated with distribution. Furthermore, when good land is lost, new land, often less fertile than that which was lost, must be cleared for production. Then to sustain crop yield on poorer land, more fossil-based inputs are used, and in this way the supply of non-renewable resources is further diminished.

As important as the amount of land available for agriculture is its quality. Soil quality slowly deteriorates with use unless carefully managed. Misuse and outright loss of soil by wind and water erosion hastens soil degradation.

At present, soil erosion rates in the United States average 18 t/ha/yr, but erosion rates are much higher in other nations. For example, in India and China where nearly 40% of the world's population live, erosion rates are about 30 and 40 t/ha/yr, respectively (Pimentel *et al.*, 1987). A "sustainable" erosion rate for agricultural soils is considered to be about 1 t/ha/yr. Soil reforms very slowly, and high erosion rates quickly diminish soil fertility.

As a result of these high erosion rates, plus the waterlogging and salinization that result from irrigation and other soil degradation factors, an estimated 6 million ha of arable land now are abandoned each year (UNEP, 1980). At the same time, to keep the land that is in use productive, more fertilizers, pesticides, irrigation, and other energy-intensive inputs are being employed. So far, intervening with these fossil-based inputs has been affordable in the industrialized nations, and to a lesser extent in the less developed areas. Both cost and availability of the non-renewable fossil energy resources are major factors that agriculturists will have to contend with in future decades.

B. Water Resources

Water is the single most limiting factor in crop production, now and in the future. Even with sunlight-energy for photosynthesis and fertile land, plants will not grow, or animals survive without water. Plants require enormous amounts of water as they grow. For example, a corn crop yielding 6,500 kg/ha of grain will transpire about 4.2 million liters of water just during the growing season of the crop (Leyton, 1983). To supply this much water to the corn crop requires the application of about 10 million liters of water per hectare. In addition, this water must be applied at relatively even rates during the growing season if the crop is to have a maximum yield.

Most, or about 80%, of the water consumed by human societies is used directly for agriculture. Needs of industry and society account for the remainder. This high demand by agriculture for water will continue and no doubt grow in the future as more food is produced.

When it was suggested earlier that land resources for crops might be doubled, the assumption was that a significant amount of the "new" land could be irrigated. This is an unrealistic assumption because irrigation has two major constraints. First, there must be an adequate source of water available, and second, large amounts of energy are expended to move enormous amounts of water to croplands. In the United States, about one-fifth of all the energy expended for direct on-farm use, is for pumping irrigation water (USDA, 1974).

Sufficient rainfall is needed to sustain agricultural production. That not used directly by growing plants is stored in rivers, lakes, man-made reservoirs, and in the vast aquifers located beneath the earth's surface. When rainfall is scant or use excessive, reserves diminish, water tables drop, wells dry, and water sources needed for irrigation are depleted. For example, in arid areas of California, the present overdraft of groundwater is 75% greater than the rate of recharge (USWRC, 1979). If this continues, the water supply will be depleted and crops now under irrigation will not be able to be grown there.

Consider that as many as 80 countries, in which almost 40% of the world population reside, are now experiencing serious water shortages because of drought (Kovda *et al.,* 1978). Prolonged drought continues to be one of the major causes of the Ethiopian famines. Drought is presently disrupting Indian agricultural production, especially in the grain belt.

Industrialized nations, like the United States, can afford to irrigate arid areas and be affluent in their use of water for industry and individual use. At present, the amount of water pumped per capita on a global basis is less than one-third the amount withdrawn in the United States per capita (CEQ, 1980). With a doubling of the world population by 2000, estimates are that world agricultural production will need to use about 64% of all water withdrawn from aquatic systems (Biswas and Biswas, 1985). This will necessitate major changes in water usage, within communities and countries as well as between countries where river and lake water storage is shared. Ultimately, water availability may be the prime factor that limits food production. Arid lands will be least able to maintain, let alone increase, food production in the coming decades.

C. Energy Resources

In the previous analysis of land, fossil energy resources were assumed to be unlimited. That is, only arable land was limited but fertilizers, fuels, and pesticides could be used to enhance the yields. Future projections, however, must be based on a limited and more expensive fossil energy supply.

The following example illustrates this energy constraint. Seventeen percent of the total energy used each year in the United States is expended in the food system (Pimentel, 1980). Specifically, each year about 1,500 liters of gasoline equivalents are used for food production, processing, distribution, and preparation per capita. In the production sector, energy is used to make and apply fertilizers and pesticides, for irrigation, and to power the machinery needed for planting and harvesting.

When this example is expanded to include the world population of five billion, the equivalent of 7,500 billion liters of gasoline equivalents would be expended to feed them the high protein-calorie diet of the United States for one year.

Based on this rate of use, how long would it take to deplete the known world petroleum reserves of 113,700 billion liters (Linden, 1980)? Assuming that 76% of the raw petroleum can be converted into gasoline (Jiler, 1972), this would provide a usable reserve equal to 86,412 billion liters of gasoline equivalency. Therefore, if petroleum were the only source of energy for food production, and if all petroleum reserves were used only to feed the present world population, the reverse would last less than 12 years.

These estimates indicate that the present world population already has exceeded the capacity of arable land and energy resources to provide all with a U.S. diet, produced with U.S. technology. Note that these estimates were based on known arable land and petroleum resources. When potential arable land and possible petroleum reserves are included, the projection improves. Also, the current world population figures were used in this analysis. Estimates based on various combinations of population size, desired dietary standards, and production technology are possible and will result in slightly different projections. This example, however, focuses on three major factors—land, water and fossil fuel—that will limit food production in the future.

D. Food Needs for Future Generations

The degradation of land, water, and forests is already having a major impact on the productivity of these basic resources. Fortunately, at present, their productivity is being maintained through the increased input of fossil energy for fertilizers, pesticides, and irrigation. But it will be a challenge to meet the future needs of the rapidly expanding human population. Food production in all countries, but especially in the developing nations where populations are escalating the fastest, must increase at a greater rate than ever before.

One estimate (D. E. Bauman, 1982 personal communication) predicts that "an amount of food equal to all the food produced so far in the history of mankind will have to be produced in the next 40 years" to fulfill human food needs. This opinion further confirms the staggering impact the rapidly growing world population is having on food production and natural resources.

A NAS report (1977) recommended that developing countries increase food production by 3–4% per year until the year 2000. Is this a realistic expectation, considering that according to the USDA (1986), the actual annual increase in food production has been only 2.1% during the last 10 years?

In this same report, NAS (1977) also targeted the following eight basic food sources for increase: rice, wheat, corn, sugar, cattle, sorghum, millet, and cassava. Currently, these foods provide about 70–90% of all the calories and 60–90% of the protein consumed in the developing countries of the world. For this reason, changes in dietary patterns will probably be less drastic in developing nations than in the industrialized nations, which rely heavily on the costly high animal-protein diets. Such a diet modification would be nutritionally beneficial because intakes of saturated fat and cholesterol would also be decreased. But such a change is not easy as dietary patterns are deeply ingrained in the habits of all humans. Changes, except in times of outright food shortages, are met with considerable resistance.

Even if individual dietary patterns can be modified to include less animal products and more plant foods like grains, overall food production must be greatly increased above present levels. The message is clear—more food, much more food, will have to be grown to sustain the human population of the future, and it must be distributed in an equitable way.

VII. PLANNING FUTURE POLICY

The many changes that have occurred since the early 1900s when most nations were self-sufficient in food have been discussed. Today the great majority of the world's 183 nations are major food importers, underscoring a growing disparity in food resources (Swaminathan, 1983). The food supply problem has persisted, "and in some cases worsened despite an increased pace of development" (Latham, 1984). Given these ominous trends, a sustainable agriculture system is essential to all programs designed to improve world food security and development.

The approaches to providing adequate food supplies for future generations must include protecting the environment, developing new technologies, and improving accessibility of food supply, while limiting the rate of the human population growth.

A. Safeguarding Environmental Resources

Hardin (1986) reminds us, "there is no hope of ever making carrying capacity figure as precise as, say, the figure for chemical valence." Nonetheless, we must consider and put some faith in the best estimates of availability of land, water, and fuels and projected population growth. Even if, in the course of time, our present estimates prove overly conservative, they represent the present knowledge base upon which we can rely as we plan for our children and generations beyond.

The environmental resources for food production, including land, water, energy, forests, and other biological resources must be conserved and protected if food production is to continue to grow. Especially over the past four decades, humans have allowed vital environmental resources to be rapidly degraded because they were either careless or ignorant about their importance. Even in less developed nations this degradation has been offset with fertilizers and irrigation. These aids are based on fossil energy, which is a nonrenewable resource. Clearly, this is a dangerous and disastrous policy to pursue in the future.

B. Science and Technology

Recent decades have witnessed many exciting and productive technological advances that have helped increase food supplies. For example, the advances in plant genetics focused on some major crops have been successful in raising their "harvest index." The formulation and use of agricultural chemicals, pesticides, and fertilizers, have helped increase yields of food and fiber crops per hectare. Improved processing of foods has enabled the food supply to be safely extended beyond harvest time. The growing transporation network in most countries has facilitated the movement of foods from production sites to far distant markets. In the industrialized nations, this has meant a more abundant, more nutritious, and a safer food supply. People living in developing nations, however, have not been as fortunate, even though successful plant breeding products like high-yielding rice have benefited millions in the Far East.

The new technology of genetic engineering or biotechnology offers great promise in raising crop and livestock production, while making more efficient use of some natural resources. This will be especially true if, for example, rice, wheat, corn, and other cereal grain crops can be "engineered" to fix nitrogen by symbiosis like legumes do. Nitrogen fertilizer is one of the nutrients essential for plant growth and the one that requires the largest fossil energy input to produce. Thus, cereal grains that could fix nitrogen would be a major scientific breakthrough, but estimates are that this breakthrough will not occur before 2100 (Pimentel, 1987). In contrast, some expectations of genetic engineering, like growing plants with little or no water, are without sound scientific foundation.

Even if many of the promises of biotechnology are forthcoming, it is essential that quality soil, water, and biological resources are available. As some biotechnologists have said, without good soil and water resources biotechnology will be a failure (NAS, 1987). Undoubtedly, biotechnology and other new technologies will help conserve resources and facilitate increased food production. Sufficient, reliable energy resources will be developed to replace most of the fossil fuels now being rapidly depleted. These new sources will probably be more costly in terms of dollars and their environmental impact. Energy obtained from the sun, from fission, perhaps from fusion, and from the wind will become more economically viable in the future than they are today. But relying solely on new technological advances is depending and hoping that the "lottery" of science will pay off. The real concern is that these developments may not materialize as rapidly as needed to meet future food and other needs. One has only to observe the millions of homeless, malnourished masses of people in Calcutta and Mexico City to recognize science and technology have an enormous challenge now and for the future.

C. Accessibility of Food Supply

Another problem associated with the human food supply that must be solved now, before the situation becomes more acute, is insuring the accessibility of food to all humans. It is shameful that, although present food production is plentiful, millions of people do not have enough to eat. Many inequities exists within countries and between countries. The poor of all nations, including those in the United States, all too often cannot obtain sufficient food for themselves and their children. Many poor countries must export agricultural products to rich nations because this is their only way to pay for the imports of certain essential items like oil and gas. Lappé and Collins (1986) point out that there is more to the hunger problem than human density. Or, put another way, high population numbers do not always correlate with hunger or the inability of a country to feed its people. They cite distribution of such resources as "land, jobs, food, education and health care" as major contributors to present and future hunger problems. All these factors interact and influence the production, distribution and accessibility of the food supply.

D. Population

Operating together, food production, distribution, and availability, as well as individual dietary patterns, determine whether or not humans are able to meet their basic nutritional requirements. Inevitably, large population numbers with great food needs will exert intense pressure on all sectors of the system to deliver the necessary food resources. The future outlook for achieving a workable balance between the escalation of human population and world food resources is not encouraging. If and when human numbers surpass the capacity of world resources to sustain them, then a rapid deterioration of human health and social structure can be expected. We humans are no different from other forms of life—ultimately nature will control our numbers.

REFERENCES

Anderson, R. M. (1979). The influence of parasitic infections on the host population growth. *In* "Population Dynamics" (R. M. Anderson, B. D. Turner, and L. R. Taylor, eds.), pp. 245–281. Blackwell, Oxford.

Biswas, M. R., and Biswas, A. K. (1985). The global environment. Past, present, and future. *Resour. Policy* **3**, 25–42.

Briggs, G. M., and Calloway, D. H. (1984). "Nutrition and Physical Fitness," 11th Ed. Holt, New York.

Brown, L. R., and Shaw, P. (1982). "Six Steps to a Sustainable Society," Worldwatch Pap. No. 48. Worldwatch Inst., Washington, D.C.

Brown, L. R., and Jacobson, J. L. (1986). "Our Demographically Divided World," Worldwatch Pap. No. 74. Worldwatch Inst., Washington, D.C.

Brown, L. R., Chandler, W. U., Flavin, C., Pollock, C., Postel, S., Starke, L., and Wolf, E. C. (1985). "State of the World 1985." Norton, New York.

Caliendo, M. A. (1979). "Nutrition and the World Food Crisis." Macmillan, New York.

Chandra, R. K. (1981). Marginal malnutrition and immunocompetence. *In* "Nutrition in Health and Disease and International Development" (A. E. Harper and G. K. Davis, eds.), pp. 261–265. Alan R. Liss, New York.

Corsa, L., and Oakley, D. (1971). Consequences of population growth for health services in less developed countries—an initital appraisal. *In* "Rapid Population Growth," Research Papers, Vol. II, pp. 368–402. Natl. Acad. Sci., Johns Hopkins Press, Baltimore, Maryland.

Council on Environmental Quality, (CEQ). (1980). "The Global 2000 Report to the President," CEQ and Dep. State, Vol. 2. U.S. Gov. Print. Off., Washington, D.C.

Cravioto, J., and DeLicardie, E. R. (1976). Malnutrition in early childhood and some of its later effects at individual and community levels. *Food Nutr.* **22**(4), 2–11.

Deevey, E. S., Jr. (1960). The human population. *Sci. Am.* **203**, 195–204.

Devine, M. M., and Pimentel, M. H. (1985). "Dimensions of Food," 2nd Ed. AVI, Westport, Connecticut.

Douglas, M. (1966). Population control in primitive groups. *Br. J. Sociol.* **17**, 263–273.

Dunn, W. (1987). China's baby boom heard worldwide. *USA Today* Apr. 14.

Feachem, R. G. (1987). Vitamin A deficiency and diarrhoea. *Trop. Dis. Bull.* **84**(3), R2–R16.

Food and Agriculture Organization (FAO). (1973). "Energy and Protein Requirements; Report of a Joint FAO/WHO Ad Hoc Expert Committee," FAO Nutr. Meet. Rep. Ser., No. 52. FAO U. N., Rome.
Food and Agriculture Organization (FAO). (1982). "1981 Production Yearbook." FAO U. N., Rome.
Food and Agriculture Organization (FAO). (1984). "World Food Report," Vol. 31. FAO U. N., Rome.
Food and Agriculture Organization (FAO). (1986a). "Yearbook of Fishery. Statistics. Catches and Landings," Vol. 58. FAO U. N., Rome.
Food and Agriculture Organization (FAO). (1986b). "FAO Production Yearbook," Vol. 39. FAO U. N., Rome.
Gilland, B. (1986). On resources and economic development. *Popul. Dev. Rev.* **12**(2), 295–305.
Hardin, G. (1986). Cultural carrying capacity: a biological approach to human problems. *BioScience* **36**, 599–606.
Harris, M. (1977). Murders in Eden. *In* "Cannibals and Kings: The Origins of Cultures" (M. Harris, ed.), pp. 89–93. Random House, New York.
Hertzler, A. A., Wenkam, N., and Standal, B. (1982). Classifying cultural food habits and meanings. *J. Am. Diet. Assoc.* **80**, 421–25.
Jiler, H. (1972). "Commodity Yearbook." Commodity Res. Bur., New York.
Kates, R. W. (1987). The world hunger. R.I. Med. J. **70**(2), 65–68.
Kovda, V. A., Rozanov, B. G., and Onishenko, S. K. (1978). On probability of droughts and secondary salinisation of world soils. *In* "Arid Land Irrigation in Developing Countries" (E. B. Worthington, ed.), pp. 237–238. Pergamon, Oxford.
Krondl, M., and Law, D. (1982). Social determinants in human food selection. *In* "The Psychobiology of Human Food Selection" (L. M. Barker, ed.), pp. 139–69. AVI, Westport, Connecticut.
Lappé, F. M. (1982). "Diet for a Small Planet," 10th Ed. Ballantine, New York.
Lappé, F. M., and Collins, J. (1986). "World Hunger." Grove, New York.
Latham, M. C. (1984). International nutrition problems and policies. *World Food Issues* **2**, 55–64.
Lewis, P. (1987). World hunger found still growing. *New York Times* June 28.
Leyton, L. (1983). Crop water use: principles and some considerations for agroforestry. *In* "Plant Research and Agroforestry" (P. A. Huxley, ed.), pp. 379–400. Int. Counc. Res. Agrofor, Nairobi.
Linden, H. R. (1980). Importance of natural gas in the world energy picture. *Int. Inst. Appl. Syst. Anal. (Conf. Proc.), Laxenburg, Austria.*
Marston, R., and Raper, N. (1987). Nutrient content of the U.S. food supply. *Nat. Food. Rev. (Econ. Res. Serv., USDA)* **36**, 18–23.
Mellor, J. W. (1986). Prediction and prevention of famine. *Fed. Proc.* **45**, 2427–31.
Messer, E. (1984). Anthropological perspectives on diet. *Annu. Rev. Anthropol.* **13**, 205–249.
Mydans, S. (1987). Life in Bangladesh: a race bred by disaster. *New York Times* June 21.
National Academy of Sciences (NAS). (1975). "Population and Food: Crucial Issues." NAS, Washington, D.C.
National Academy of Sciences (NAS). (1977). "World Food and Nutrition Study." NAS, Washington, D.C.
National Academy of Sciences (NAS). (1980). "Recommended Dietary Allowances," 9th Ed. NRC–NAS, Washington, D.C.
National Academy of Sciences (NAS). (1987). "Agricultural Biotechnology." Nat. Acad. Press, Washington, D.C.
Physicians' Task Force on Hunger in America (1985). "Hunger in America." Wesleyan Univ. Press, Middletown, Connecticut.

Pimentel, D., ed. (1980). "Handbook of Energy Utilization in Agriculture." CRC Press, Boca Raton, Florida.

Pimentel, D. (1987). Down on the farm: genetic engineering meets technology. *Technol. Rev.* **90,** 24–30.

Pimentel, D., Oltenacu, P. A., Nesheim, M. C., Krummel, J., Allen, M. S., and Chick, S. (1980). Grass-fed livestock potential: energy and land constraints. *Science* **207,** 843–848.

Pimentel, D., Wen, D., Eigenbrode, S., Lang, H., Emerson, D., and Karasik, M. (1986). Deforestation: interdependency of fuelwood and agriculture. *Oikos* **46,** 404–412.

Pimentel, D., Allen, J., Beers, A., Guinand, L., Linder, R., McLaughlin, P., Meer, B., Musonda, D., Perdue, D., Poisson, S., Siebart, S., Stoner, K., Salazar, R., and Hawkins, A. (1987). World agriculture and soil erosion. *BioScience* **37,** 277–283.

Population Reference Bureau (PRB). (1986). "World Population Data Sheet." Popul. Ref. Bur., Washington, D.C.

Reddy, V. (1981). Protein energy malnutrition: an overview. *In* "Nutrition in Health and Disease and International Development" (A. E. Harper and G. K. Davis, eds.), pp. 227–235. Alan R. Liss, New York.

Reutlinger, S., and Alderman, H. (1980). The prevalence of calorie deficient diets in developing countries. *World Dev.* **8,** 399–411.

Rochereau, S., and Pimentel, D. (1978). Energy tradeoffs between Northeast fish production and coastal power reactors. *J. Energy* **3,** 575–589.

Scrimshaw, N. S. (1986). Consequences of hunger for individuals and societies. *Fed. Proc.* **45,** 2421–26.

Scrimshaw, N. S., Taylor, C. E., and Gordon, J. E. (1968). Interactions of nutrition and infection. *W. H. O. Monogr. Ser.* No. 57.

Sen, A. (1980). Famines. *World Dev.* **8,** 614–621.

Swaminathan, M. S. (1983). Our greatest challenge—feeding a hungry world. *In* "Perspectives and Recommendations. Chemistry and World Food Supplies: The New Frontiers" (G. Bixler and L. W. Shemilt, eds.), Chemrawn II, pp. 25–46. Int. Rice Res. Inst., Los Banos, Philippines.

Tandon, B. N., Ramachandran, K., Sharma, M. P., and Vinayak, V. K. (1972). Nutritional survey in rural population of Kumaon Hill area, North India. *Am. J. Clin. Nutr.* **25** 432–436.

United Nations. (1982). "World Population Trends and Policies. 1981 Monitoring Report, Vol. 1, Population Trends." U. N., New York.

United Nations. (1986). "Population and Vital Statistics Reports," Vol. 38. Stat. Off. U. N., New York.

United Nations Environment Programme (UNEP). (1980). "Annual Review." U. N. Environ. Programme, Nairobi.

U.S. Department of Agriculture (USDA). (1974). "Energy and U.S. Agriculture: 1974 Data Base," Vols. 1 and 2. Fed. Energy Adm. Off. Energy Conserv. Environ., State Energy Conserv. Programs, Washington, D.C.

U.S. Department of Agriculture (USDA). (1981). "National Food Research," Econ. Res. Serv., Summer, NFR-15. USDA, Washington, D.C.

U.S. Department of Agriculture (USDA). (1985). "Agricultural Statistics 1985." U.S. Gov. Print. Off., Washington, D.C.

U.S. Department of Agriculture (USDA). (1986). "World Indices of Agricultural and Food Production, 1976–85," Econ. Res. Serv., Stat. Bull. No. 744. USDA, Washington, D.C.

U.S. Department of Agriculture (USDA). (1987). "National Food Review." Econ. Res. Serv., Winter–Spring, NFR-36. USDA, Washington, D.C.

U.S. Senate. (1977). "Dietary Goals for the United States," Select Committee on Nutrition and Human Needs. U.S. Gov. Print. Off., Washington, D.C.

U.S. Water Resources Council (USWRC). (1979). "The Nation's Water Resources. 1975–2000," Vols. 1–4, Second National Water Assessment. U.S. Gov. Print. Off., Washington, D.C.

Wilson, E. D., Fisher, K. H., and Garcia, P. A. (1979). "Principles of Nutrition," 4th Ed. Wiley, New York.

World Health Organization. (1963). Expert Committee on Medical Assessment of Nutritional Status. *W. H. O. Tech. Rep. Ser.* No. 258.

Zoller, J. M., Wolinsky, I., Paden, C. A., Hoskin, J. C., Lewis, K. C., Lineback, D. R., and McCarthy, R. D. (1980). Fortification of non-staple food items with iron. *Food Technol.* **34,** 38–47.

17

Population Growth and the Poverty Cycle in Africa: Colliding Ecological and Economic Processes?

A. R. E. Sinclair and Michael P. Wells

Department of Zoology
University of British Columbia
Vancouver, British Columbia, Canada

I. The African Paradox
II. Population
    A. Population Increase
    B. Mortality
    C. Reproduction
    D. The Demographic Trap
III. The Ecological Crisis
    A. Pastoralism, Overgrazing, and Desertification
    B. The Cycle of Agriculture and Declining Crop Yields
    C. Forests and the Fuelwood Crisis
IV. The Economic Decline
    A. African Economies—The Agricultural Sector
    B. Economic Explanations for Africa's Agricultural Crisis
V. Foreign Aid
    A. The Impact of Donor and Lender Policies
    B. The Policy Environment
    C. How Will Agricultural Production Be Increased?
VI. The Poverty Cycle and the Way Ahead
    References

I. THE AFRICAN PARADOX

Average world food production per capita has been increasing over the past 30 years (Mellor and Gavian, 1987). In the developed countries, grain production per capita rose 1% per year (1950–1980), while in developing countries it rose 0.4% per year (Barr, 1981). A recent conference, "Science and Technology in the World Food Crisis" (Guelph, Ontario, October 1986), largely attributed these gains to the Green Revolution and related technological advances in agricultural production. A popular viewpoint at this conference was that technology would continue to lead to productivity improvements in agriculture and that there was little evidence of a world food crisis.

In stark contrast to this mood of optimism is the situation in sub-Saharan Africa, where overall per capita food production has been declining since 1961 (Figure 17.1): by 0.4% in the 1960s and by 1.5% in the 1970s. Between 1960 and 1976 total food output increased by 50% in a few African countries, but at the same time the population increased by 75%. Thus population has outpaced the production of food (Grigg, 1985). Increased food imports and food aid donations have not compensated for this increase in population, so that per capita food consumption dropped by 0.4% in the 1970s. There is consensus among observers of African development that this situation is unlikely to improve in the near future and is likely to become considerably worse.

If the conclusions of the Guelph conference are correct, then why is Africa so different? There are three problems with the Guelph conclusions: (1) Agricultural "experts" confined themselves to their own specific areas of technical expertise and failed to question the assumption that technology

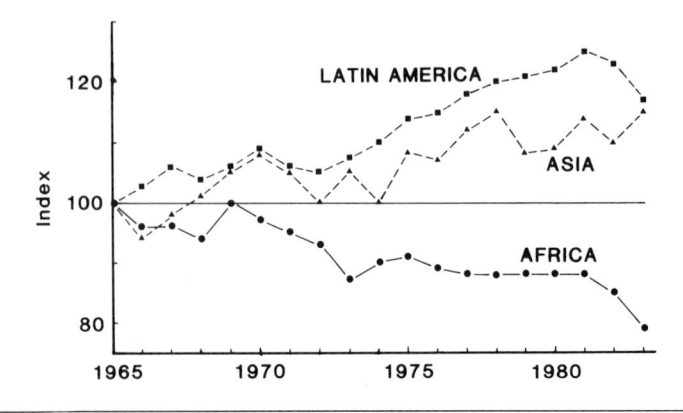

Figure 17.1 Index of food production per person, relative to a starting point set at 100 as the average of years 1961–1965 (World Bank, 1984).

transfers automatically lead to greater food production. The consequences of technological advances on social and ecological systems were given minimal consideration. Can advances in crop production developed in Asia, for example, be transferred to other areas, Africa in particular? (2) The issue of rapid human population growth and its implications for food production was not addressed. It was implicitly assumed that advances in food production will remain ahead of population increases. Is this a reasonable assumption in the light of both historical events to the contrary and current processes in Africa? (3) It is questionable whether the production of more food will automatically lead to the alleviation of hunger [see Griffin (1987), and, for a contrasting view, see Mellor et al., (1986)]. This important argument is not within the scope of our current thesis but must be addressed in the formulation of agricultural development policies.

In this chapter we examine the various interrelated causes of the food crisis in Africa. We contend that the roots of the food crisis can be found in the interactions between (1) human population increase, (2) declining economic performance, (3) the degradation of ecological resources, and (4) the activities of international development agencies. It is the relationship between these processes and food production which form the basis of our chapter. We suggest that the lessons to be learned from rapid human population growth in Africa should act as an early warning for what could happen later in other parts of the world.

II. POPULATION

A. Population Increase

Africa stands out from other developing areas of the world in that it has both high and increasing average annual population growth rates (currently 3.1%, Figure 17.2). In contrast, both Asia and South America have lower growth rates (1.8% and 2.3%, respectively), and the United Nations (1985) has predicted a continued decline in these rates. Since 1950 the population of Africa has increased two and a half times, to 583 million in 1986 (Brown, 1987). This dramatic population increase was the principal contributor to the drop in food production per capita shown in Africa (Figure 17.1). Correspondingly, in the rest of the world, falling rates of population growth have contributed to increasing food production per capita.

It is important, therefore, to analyse and seek insights into rapid population growth in Africa. We have looked for key explanatory variables while recognizing that (1) there are probably no simple cause and effect relationships, and (2) social and economic data compiled at a national level for African countries are frequently unreliable. We have analysed two sets of demographic data for individual countries. Table 17.1, from the World Bank (WB84), includes data up to 1982, whereas Table 17.2,

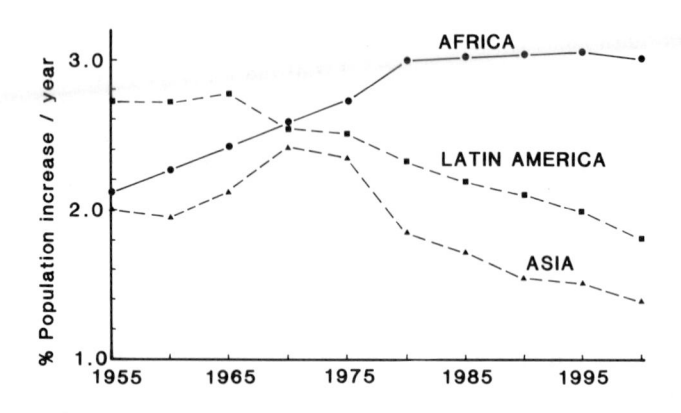

Figure 17.2 Average annual population increase for 5-year periods ending in the year shown, extrapolated to the year 2000. [From United Nations (1985).]

from a 1986 report of the World Resources Institute and International Institute for Environment and Development (WR86), includes data up to 1985. Both tables are ranked in order of the rates of population growth. There are a few obvious anomalies: Somalia appears to have increased its population growth considerably in Table 17.2, while both Cameroon and Congo People's Republic have lower rates. Table 17.2 also includes one more country, Equatorial Guinea.

We examined the countries of continental Africa south of the Sahara plus Madagascar, but excluded four countries, Mozambique, Angola, Uganda, and Ethiopia, which have experienced chronic civil war over the past decade; we have done this to avoid the confounding effects of war on birth and death rates. We used both analyses of variance and regression techniques in an attempt to identify those demographic features, as independent variables, which are related to, and might account for, differences in the rates of population growth between the African countries. Table 17.3 shows whether the relationship is positive or negative, the percentage of the variance that is accounted for by each variable (r^2), and the significance level (P).

In both sets of data crude death rates (deaths per thousand per year) account for a large amount of the variability in population growth. The WR86 crude birth rates were also strongly associated with population growth, but the WB84 birth rates accounted for a very small amount of the variance. However, when both crude death and birth rates were included, over 80% of the variance was accounted for. This is to be expected, since population growth was partially calculated from these data. The important point is that variations in death rates appear to have had a stronger effect on population growth than variations in birth rates.

Table 17.1 Demographic, Ecological, Economic, and Health Indicators (1980–1982) for 34 Sub-Saharan Countries Ranked in Order of Their Population Growth[a]

Country	1	2	3	4	5	6	7	8	9	10	11	12	13
Guinea Bissau	2.3	6.5	31	0.2	−19.8	28	50	88	0.22	8840	980	170	32.0
Gambia	2.3	6.5	46	5.6	−18.0	27	49	74	1.27	12310	1770	360	34.0
Burkina Faso	2.4	6.5	36	−1.5	−20.1	21	48	95	0.95	48510	4950	210	12.7
Somalia	2.4	6.5	47	0.2	−12.3	25	48	60	0.46	14290	2330	290	38.7
Guinea	2.4	6.5	50	1.8	−22.6	27	49	89	0.23	17110	2570	310	7.1
Sierra Leone	2.4	6.5	50	−0.2	−20.6	27	49	81	0.15	16220	1890	390	7.8
Chad	2.5	5.5	37	−6.6	−27.7	21	42	95	0.24	47530	5780	80	5.3
Mauritania	2.6	6.0	27	−14.3	−28.3	19	43	73	0.11	14350	2080	470	3.0
Gabon	2.6	4.5	22	7.7	−37.0	17	35	93	0.03	3030	—	—	0
Mali	2.8	6.5	27	−3.2	−23.0	21	48	83	0.38	22130	2380	180	9.4
Central African Republic	2.8	5.5	23	−3.9	−35.4	17	41	104	0.08	26430	1720	310	0.9
Lesotho	2.8	5.8	17	0	−35.8	15	42	84	0.47	18640	4330	510	24.4
Sudan	2.9	6.6	23	−3.4	−29.9	18	45	87	0.32	8930	1430	440	9.3
Burundi	3.0	6.5	24	2.9	−23.7	19	47	96	1.54	45020	6180	280	2.1
Senegal	3.1	6.5	34	0	−22.5	21	48	93	1.22	13800	1400	490	12.8
Madagascar	3.2	6.5	23	−0.1	−33.0	18	47	94	0.16	10170	3660	320	8.5
Niger	3.3	7.0	27	0.7	−24.5	20	52	88	0.31	38790	4650	310	12.1
Zaire	3.3	6.3	20	−4.1	−34.2	16	46	87	0.13	14780	1920	190	3.0
Benin	3.3	6.5	23	−2.5	−32.2	18	49	100	0.33	16980	1660	310	2.2
Togo	3.3	6.5	25	−2.7	−17.6	19	49	89	0.49	18100	1430	340	1.7
Malawi	3.4	7.8	29	0.2	−15.7	23	56	99	0.55	40950	3830	210	0.3
Tanzania	3.5	6.5	18	0.8	−33.4	15	47	88	0.42	17560	2980	280	13.1
Liberia	3.5	6.5	16	−0.3	−30.6	14	50	88	0.06	9610	1420	490	21.0
Nigeria	3.5	6.9	20	−4.7	−35.6	16	50	92	2.00	12550	3010	860	0
Cameroon	3.5	6.5	16	21.2	−30.7	15	46	102	0.13	13990	1950	890	1.2
Rwanda	3.6	8.3	25	0.9	−27.4	20	54	105	1.40	31510	9840	260	2.3
Zambia	3.6	6.8	20	−2.2	−36.5	16	50	87	0.16	7670	1730	640	16.6
Botswana	3.6	6.5	13	−14.3	−44.6	11	44	73	0.10	9480	1250	900	6.9
Ivory Coast	3.7	7.0	23	−2.7	−28.2	17	48	107	0.18	21040	1590	950	0.1
Congo, Peoples Republic	3.8	6.0	10	6.8	−46.0	10	43	81	0.05	5510	790	—	0.2
Ghana	3.9	7.0	15	−1.8	−35.7	13	49	72	0.51	7630	780	360	3.8
Swaziland	3.9	7.0	27	−1.9	−39.4	13	51	107	0.41	7670	1010	940	0
Kenya	4.4	8.0	13	0.2	−47.9	12	55	88	0.62	7890	550	390	6.3
Zimbabwe	4.4	8.0	14	−1.8	−25.0	12	54	87	0.38	6580	1190	850	0

[a]Source: World Bank (1984). Key to indicators:

1. Population increase %/year, 1980
2. Mean female fertility
3. Child death rate per thousand
4. % change crude birth rate, 1960–1982
5. % change crude death rate, 1960–1982
6. Crude death rate per thousand, 1982
7. Crude birth rate per thousand, 1982
8. Food production/capita 1980–1982 Index
9. Density (# km^2/cm rain)
10. Population/doctor
11. Population/nurse
12. GNP/capita ($US), 1982
13. Food aid, 1982 (kg/capita)

Table 17.2 Demographic, Ecological, Economic, and Educational Indicators (1980–1985) for 35 Sub-Saharan Countries Ranked in Order of Their Population Growth[a]

Country	1	2	3	4	5	6	7	8	9	10	11	12	13	14
Gabon	1.64	34.6	18.1	4.67	112	119	—	—	4.3	0	1520	6634	103	72
Sierra Leone	1.77	47.4	29.7	6.13	200	84	26	20	4.6	0.8	1371	3330	95	96
Guinea Bissau	1.91	40.7	21.7	5.38	143	97	58	36	4.5	1.0	752	6154	86	88
Gambia	1.94	48.4	29.0	6.39	193	93	32	22	4.3	1.4	1003	3000	76	112
Equatorial Guinea	2.15	42.5	21.0	5.66	137	—	—	—	4.3	−0.6	—	2451	—	—
Chad	2.28	44.2	21.4	5.89	143	75	19	10	6.1	1.3	539	4456	96	100
Central African Republic	2.29	44.7	21.8	5.89	143	95	41	28	4.5	0.6	549	3283	94	99
Guinea	2.33	46.8	23.5	6.19	159	81	25	16	5.4	1.5	878	7293	91	93
Burkina Faso	2.34	47.8	22.2	6.50	149	81	11	08	4.7	2.1	521	4444	96	95
Lesotho	2.53	41.7	16.4	5.79	110	103	72	85	7.4	2.3	844	15000	73	73
Cameroon	2.54	43.2	17.8	5.79	117	93	64	56	6.6	0	917	2407	83	98
Congo Peoples Republic	2.59	44.5	18.6	5.99	124	111	—	—	3.7	1.9	566	6524	96	89
Senegal	2.66	47.7	21.2	6.5	141	99	29	22	4.9	1.2	616	3485	67	95
Burundi	2.67	47.6	20.9	6.44	137	102	15	12	4.4	2.6	1045	7115	92	98
Mali	2.78	50.2	22.4	6.70	149	74	19	13	5.3	2.2	742	9463	107	96
Madagascar	2.80	44.4	16.5	6.09	67	111	54	50	5.7	2.1	1685	5899	90	98
Niger	2.82	51.0	22.9	7.10	140	105	14	10	6.9	2.1	372	6949	115	104
Sudan	2.86	45.9	17.4	6.58	118	99	36	29	6.3	1.6	470	3500	85	100
Benin	2.86	51.0	22.5	7.00	149	94	40	25	7.3	0.5	678	7347	97	100
Togo	2.86	45.4	16.9	6.09	113	94	75	53	5.7	2.2	924	11231	92	95
Mauritania	2.93	50.1	20.9	6.90	137	95	23	15	8.0	0.7	274	1871	95	54
Zaire	2.94	45.2	15.8	6.09	107	96	59	47	5.2	1.3	821	6876	96	96
Swaziland	3.03	47.5	17.2	6.50	129	109	81	81	8.7	1.3	1294	2446	114	151
Liberia	3.16	48.7	17.2	6.90	112	98	45	33	5.6	1.7	1158	3908	90	87
Malawi	3.23	52.1	19.9	7.00	165	96	44	36	7.6	2.7	1164	3997	100	104
Ghana	3.25	47.0	14.6	6.50	98	72	52	44	5.2	2.0	636	6696	68	98
Zambia	3.31	48.1	15.1	6.76	101	90	67	61	6.2	0.8	1637	3635	73	89
Nigeria	3.34	50.4	17.1	7.10	114	104	—	—	5.8	2.7	675	10005	96	93
Ivory Coast	3.44	46.0	18.0	6.70	122	115	49	36	5.9	1.8	904	3972	109	106
Rwanda	3.46	51.1	16.6	7.3	110	91	45	43	6.8	3.3	1279	9137	111	98
Botswana	3.46	50.0	12.7	6.50	79	106	73	79	8.0	2.5	229	5385	64	52
Zimbabwe	3.50	47.2	12.3	6.60	70	91	56	52	5.8	2.8	936	4514	69	114
Tanzania	3.52	50.4	15.3	7.10	98	105	62	58	8.1	2.8	964	10716	96	98
Somalia	3.71	46.5	21.3	6.09	143	90	23	17	6.2	2.6	611	10841	68	80
Kenya	4.12	55.1	14.0	8.12	82	88	73	68	7.4	3.5	1472	7900	85	97

[a] Source: WRI/IIED (1986). Key to indicators:

1. Population increase %/year, 1980–1985
2. Crude birth rate per thousand
3. Crude death rate per thousand
4. Mean female fertility
5. Infant mortality per thousand
6. Calories/day as % of requirement
7. Total in school as % of school age population, 1980
8. Females in school as % of females of school age, 1980
9. Urban population increase %/year
10. Rural population increase %/year
11. Grain production (kg/ha), 1982–1984
12. Tuber production (kg/ha), 1982–1984
13. Food production/capita Index, 1982–1984
14. Calories produced as % of total supply

Table 17.3 Regression Analysis of Percentage Population Growth per Year in Each Country as the Dependent Variable with Various Independent Variables

Independent variable	Sign	$r^2(\%)$	P
World Bank data			
Crude deaths/1000	−	61.7	.001
Crude births/1000	+	16.4	.01
Child mortality/1000	−	54.8	.001
Total fertility	−	33.7	.001
% Drop in death rate	−	29.5	.001
GNP	+	26.0	.01
Crude deaths + births/1000		92.9	.001
Child mortality + fertility		81.4	.001
World Resources Institute data			
Crude deaths/1000	−	42.6	.001
Crude births/1000	+	40.3	.001
Infant mortality/1000	−	31.8	.001
Total fertility	+	53.8	.001
% City population increase	+	34.1	.001
% Rural population increase	+	43.7	.001
% Female education	+	15.6	.05
% Total population education	+	12.3	.05
Crude deaths + births/1000		82.8	.001
Fertility + infant mortality		76.1	.001

Of more interest is mean female fertility (the average number of children born during a woman's lifetime). This has a strong positive association with population growth: countries with high population growth rates have high fertility rates (Figure 17.3). Of even greater importance is child mortality (deaths per thousand children per year, ages 1–4) which alone accounts for 55% of the variance in population growth: low child mortality leads to high growth rates (Figure 17.4). A similar effect is seen with infant mortality rates (as above, ages 0–1).

We examined the influence of ecological factors on population growth (indices of food per capita and population density in Table 17.1; and grain and tuber production, food per capita, calories as a percentage of requirement, and calories of food grown as a percentage of the total supply in Table 17.2. None of these showed any relationship with population growth. This is interesting for two reasons: (1) it suggests that there are as yet no detectable density-dependent feedbacks from the ecological variables onto the population (that is, despite declining food production, the populations are not yet affected), and (2) the common perception that mortality is increasing is not reflected in the WB84 data. Between 1960 and 1982 all crude death rates declined, and this is significantly associated with population growth (Table 17.3).

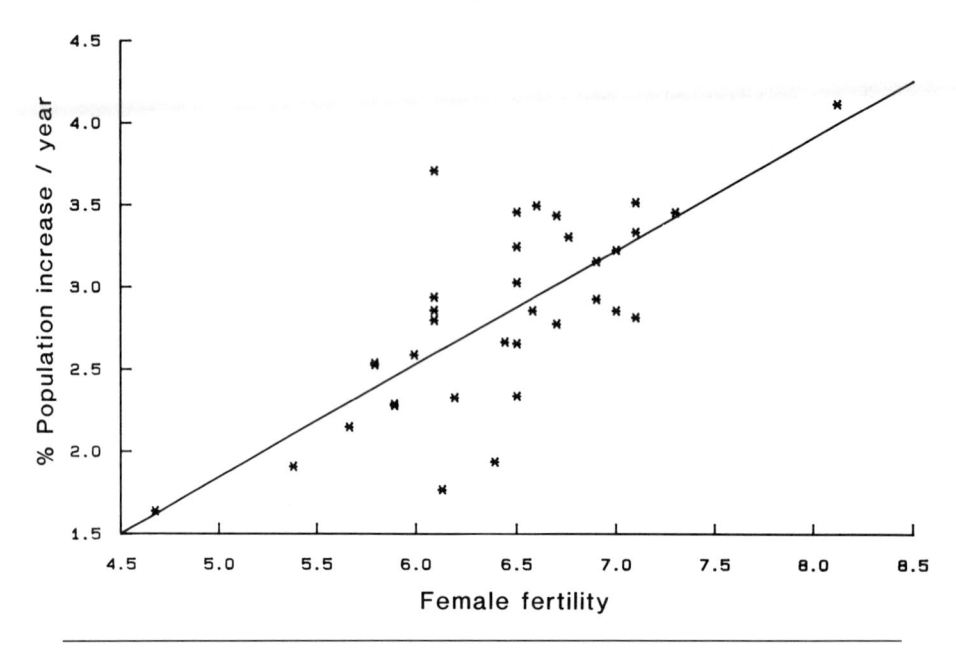

Figure 17.3 Annual population increase for individual countries (y) as a function of life-time female fertility (x). ($y = -1.60 + 0.69x$)

We recognize that these ecological variables may be too crude to reflect what is happening to the populations and that improved census techniques may be causing a bias by inflating population growth rates. However the same bias applies to crude death rates, predicting a positive correlation between population growth and mortality if the results are due to census improvement. In fact we find the opposite relationship, suggesting that these trends are real. Nevertheless, better data and information on trends within countries are needed.

The socio-economic indices we examined were Gross National Product (GNP) per capita, food aid per capita, population per doctor and per nurse (Table 17.1), the proportion of the school age population in school for both sexes (total education) and for females alone (female education), and the urban and rural population growth rates (Table 17.2). Of these, rural population growth rates were most closely associated with overall population growth, while urban population growth had a lesser but still significant correlation. The GNP and both education indices were all positively associated with population growth but their effects were small. Neither of the health indicators showed any relationship with growth.

To summarize, variations in population growth are clearly linked to crude death rates, and to child and infant death rates. Fertility and crude

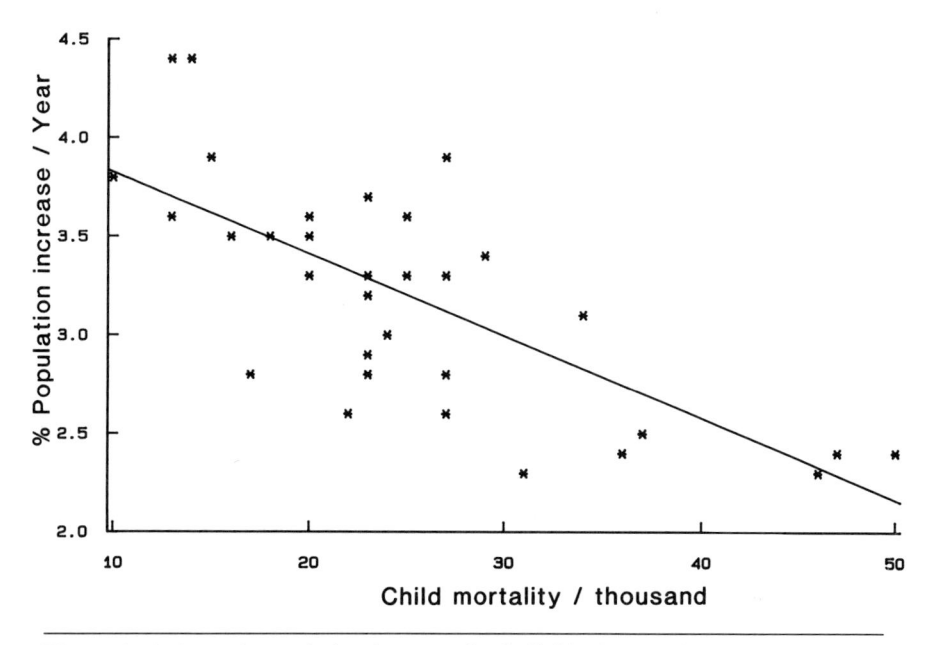

Figure 17.4 Annual population increase for individual countries (y) as a function of child (1–4 years) mortality rate per thousand (x). (y = 4.24 − 0.042x)

birth rates have smaller effects but are still important. Dividing the population into urban and rural components showed that rural population changes were more closely associated with overall growth than were changes in urban populations. These data sets did not reveal any association between population growth rates and ecological factors relevant to food production.

B. Mortality

Since mortality strongly affects population growth, we examined which demographic variables contributed to mortality and which external factors (education, income levels, and health care—which we assumed could be represented by the relative abundance of doctors and nurses) were correlated with it.

Variations in crude death rates (Table 17.4) could be largely accounted for by variations in either infant mortality (88% of the variance, WR86, Figure 17.5) or child mortality (79%, WB84); i.e., variations in the mortality of infants and children explain most of the variation in crude death rates. Of the external variables that could affect crude death rates, by far the

Table 17.4 Regression Analysis of Mortality Variables with Independent Variables

Dependent variable	Independent variable[a]	Sign	r^2(%)	P
Crude mortality per thousand	Child mortality (WB)	+	78.8	.001
	Infant mortality (WRI)	+	88.3	.001
	% Female education (WRI)	−	51.9	.001
	% Total education (WRI)	−	47.6	.001
	GNP (WB)	−	28.4	.001
	Food production/capita (WRI)	+	9.9	.05
	Population/doctor (WB)	+	8.7	.05
	Child mortality + GNP		83.2	.001
% Drop in crude mortality	Child mortality (WB)	+	51.2	.001
	Population/doctor (WB)	+	16.3	.01
	GNP (WB)	−	15.1	.05
	Child mortality + population/doctor		50.3	.001
Infant mortality	% Female education (WRI)	−	41.6	.001
	% Total education (WRI)	−	37.1	.001
	Food production/capita (WRI)	+	10.7	.05
	% Rural population increase (WRI)	−	9.1	.05
	Fertility (WRI)		0.0	
Child mortality	Food aid/capita (WB)	+	14.3	.01
	GNP (WB)	−	12.9	.05

[a]WB, data from World Bank. WRI, data from World Resources Institute and International Institute for Environment and Development.

most important were female education (52% of the variance) and total education (48%); countries with higher values for education had lower crude death rates. Those countries with higher GNP per capita (28% of the variance, Figure 17.6) and more doctors (9%) also had lower crude death rates, but the influence of these two variables when included in a multiple regression with female education was negligible.

The decline in crude death rates since 1960 (WB84 data) is explained best by changes in child mortality (51% of the variance). The number of doctors (Figure 17.7) and GNP per capita both explained a small but significant amount of the variance, but again their effect in the multiple regression with child mortality was trivial.

Infant and child mortality appear to be the main factors influencing population death rates and, indirectly, rates of population increase. What factors affect these mortalities? Female education is the most important variable explaining infant mortality (Figure 17.8) (there were no appropriate data to examine child mortality). Countries with higher rates of rural population increase and higher GNP had lower infant and child mortality, but the effects were small. More surprising was that countries with higher food production per capita (WR86) and higher food aid (WB84) also had

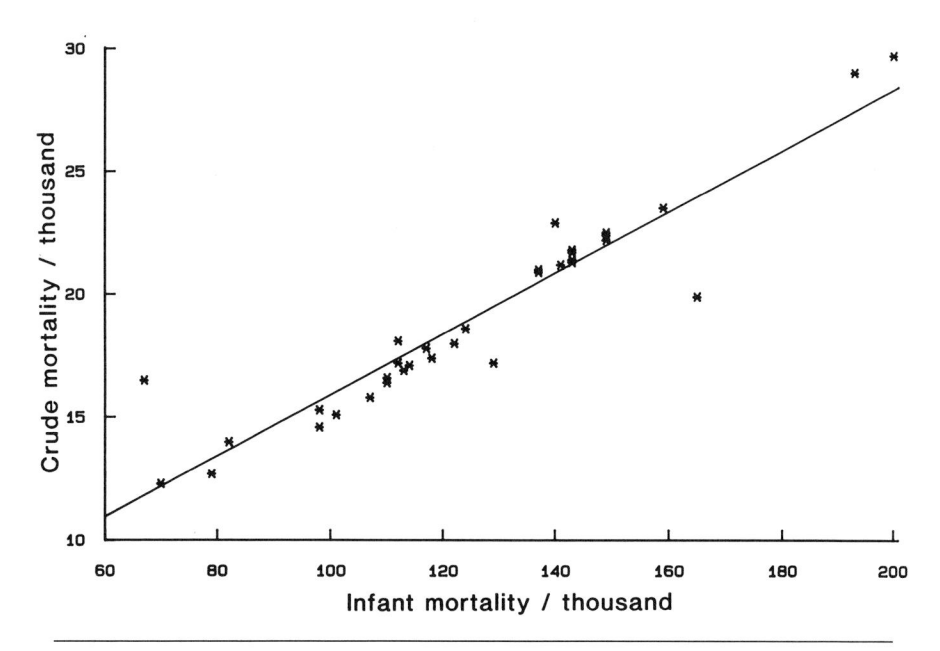

Figure 17.5 Annual crude mortality rate (deaths per thousand population) for individual countries (y) as a function of infant (0–1 year) mortality rate per thousand (x). (y = 3.53 + 0.125x)

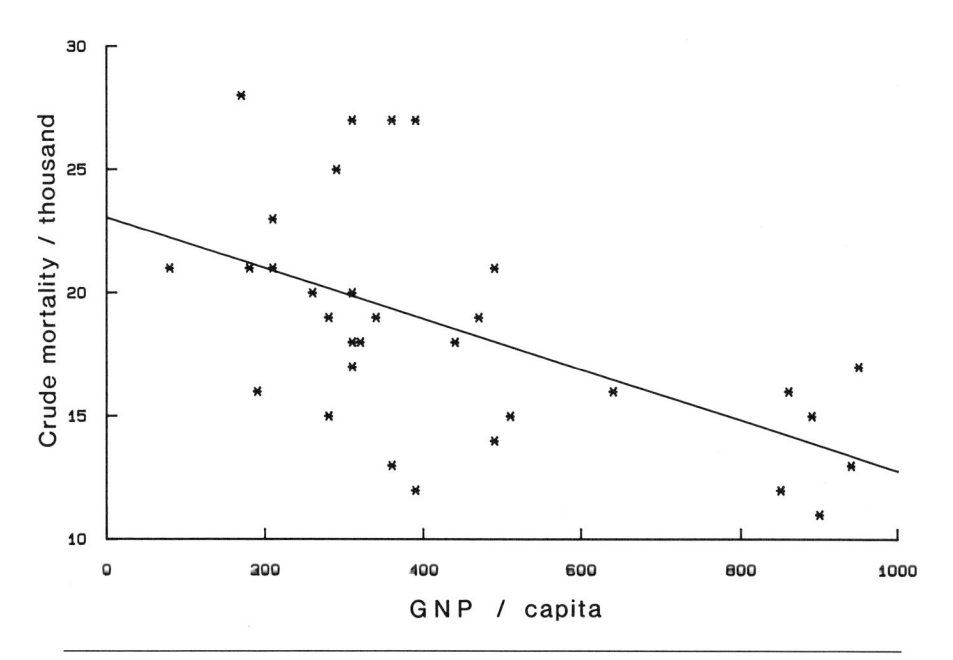

Figure 17.6 Annual crude mortality rate for individual countries (y) as a function of their gross national product per capita (x). (y = 23.1 − 0.0103x)

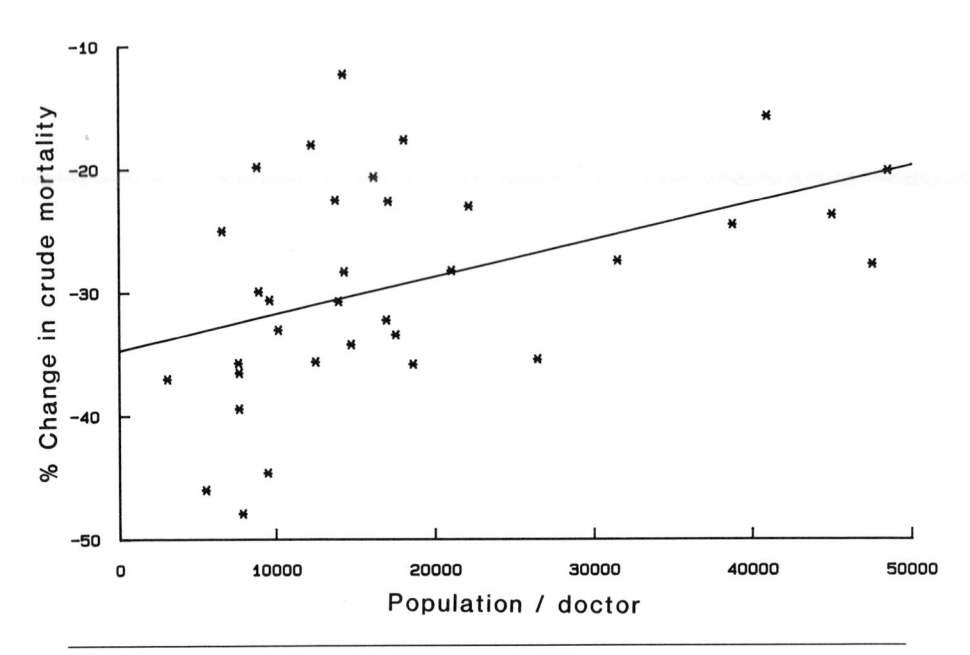

Figure 17.7 The percentage drop in crude mortality rate for individual countries over the period 1960–1982 (*y*) as a function of the number of people per doctor (*x*). ($y = -34.7 + 0.000301x$)

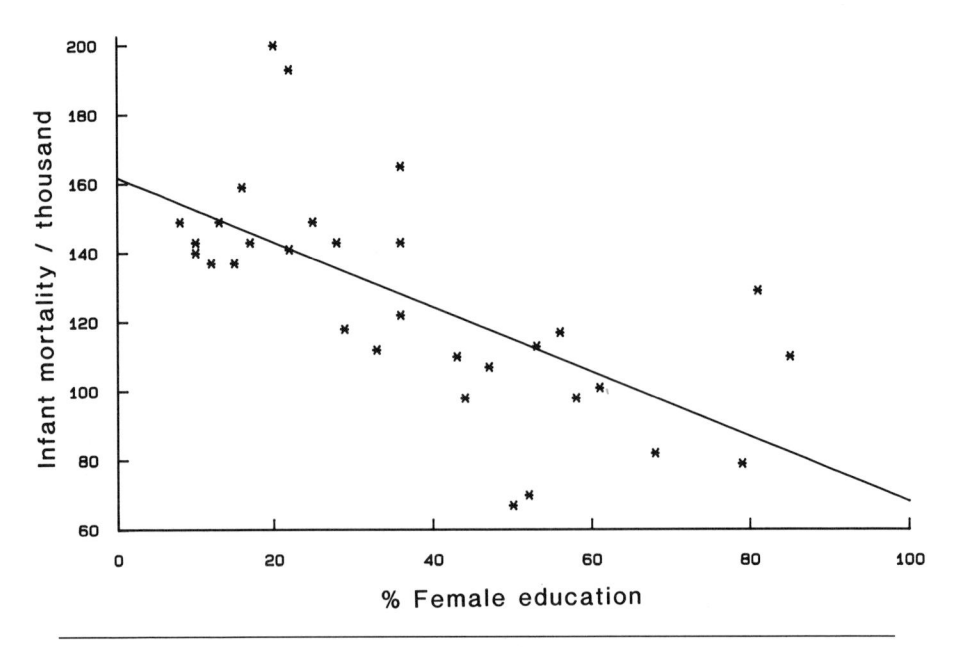

Figure 17.8 The infant mortality rate for individual countries (*y*) as a function of the proportion of females of school age in school (*x*). ($y = 162 - 0.937x$)

higher infant and child mortality. One interpretation is that food aid goes to countries experiencing high mortality. We found no association between early age mortality and female fertility, nor with ecological variables such as population density, and grain and tuber production.

In summary, crude mortality rates are largely accounted for by mortality in early ages (infant and child mortality). In turn these mortalities appear to be highly positively associated with lack of female education, slightly negatively associated with GNP per capita, and positively associated with food production per capita and food aid volume.

C. Reproduction

The crude annual birth rate per thousand can almost entirely be explained by female fertility (89% of the variance, WR86), as we should expect (Table 17.5). Fertility, as noted above, accounted for part of the population growth in the WR86 data, but in the WB84 data fertility was more constant across countries and accounted for little of the population growth.

High fertility was associated with high rates of urban population increase (26% of the variance WR86), and rural population increase (34%). Fertility and population density were also positively associated to a small extent (10%, WB84). No relationships were detected between fertility and other ecological, health, education or economic variables. In particular, fertility remained high irrespective of the degree of female education. Fertility, therefore, appears to be a relatively constant demographic feature insensitive to external factors.

Table 17.5 Regression Analysis of Reproduction, Urban, and Rural Population Variables with Independent Variables

Dependent variable	Independent variable[a]	Sign	r^2(%)	P
Crude births	Fertility (WRI)	+	89.4	.001
per thousand	Fertility (WB)	+	85.0	.001
	% Urban population increase (WRI)	+	21.9	.001
Fertility	% Rural population increase (WRI)	+	34.3	.001
	% Urban population increase (WRI)	+	25.6	.001
	Density (WB)	+	9.8	.05
% Rural population	Fertility	+	34.3	.001
increase/year	Infant mortality	−	9.1	.05
% Urban population	Fertility	+	25.6	.01
increase/year	% Female education	+	25.8	.01
	Fertility + female education		42.6	.01

[a] WRI, data from World Resources Institute and International Institute for Environment and Development. WB, data from World Bank.

D. The Demographic Trap

Brown (1987) has recently reemphasized an idea of the demographer F. Notestein that demographic change is related to the effect of economic and social progress on population growth. We would add to this the ecological effects of natural resource degradation and over-exploitation. Notestein identified three stages of demographic transition. In the first stage birth and death rates are high, and the population either grows slowly or not at all. The second stage occurs when living conditions improve. As health and education improve, death rates fall but birth rates remain high and the population increases rapidly. Finally, in the third stage, economic improvements cause people to reduce their desire for large numbers of children and the birth rate falls to approximately that of the death rate. This theory would place most of the developed countries at this last stage, while most of the African countries are in the second stage.

Brown's (1987) thesis is that the poorer African countries are unable to escape from the second stage because declining ecological and economic conditions are continually undermining their attempts to raise living standards. Consequently, populations continue to expand, but the more they do so the more conditions become worse as a result of the population impacts. This is what Brown calls the "demographic trap." Obviously this situation cannot continue indefinitely, and Brown speculates that we might see, for the first time in history, countries sinking back into the first stage. The appearance of AIDS in a significant proportion of the prime working age groups (groups not normally prone to high mortality) in several African countries might well contribute to a reversal in the demographic progression.

We next examine the ecological consequences of the high population growth rates.

III. THE ECOLOGICAL CRISIS

In general, rapidly increasing human populations have been associated with large-scale ecological degradation in Africa, through processes which have now been well described (see, e.g., Timberlake, 1985; Brown *et al.*, 1986), if not particularly well understood. We have chosen to seek insights into these processes from the perspective of ecosystem dynamics. An ecosystem is made up of interacting components—climate, soil, vegetation, and animal (including human) populations. These components can fluctuate in quantity and quality as a result of disturbances from outside (e.g., climatic changes and man-made disturbances). Provided disturbances are not too large, the system can recover, and this ability to recover is termed "resilience" (Holling, 1973). Some ecosystems, usually those with high productivity, fluctuate naturally and have high resilience, while others, such as those in semi-arid regions, are fragile and have low resilience. If disturbances are too great, it is possible that an ecosystem will not return to its original configuration, but instead will change to another state, where it will remain even if the cause of the original disturbance has ceased (Sinclair, 1981). Therefore, we envision an ecosystem with more than one stable state and which requires some external disturbance to move between these multiple equilibria.

Ecologically, Africa differs from other tropical areas of the world in being largely arid or semi-arid. It is, therefore, relatively fragile (or "unresilient") and cannot withstand major disturbances to the same degree as other tropical areas. It can absorb only small increases in grazing, crop production, and tree harvesting due to population increases before experiencing major shifts.

A. Pastoralism, Overgrazing, and Desertification

Africa has more rangeland (778 million ha) than any other continent (Asia 645 million ha, South America 550 million ha), and grasslands form a greater proportion of the total vegetation types (47%) in Africa than elsewhere (Asia 39%, South America 32%) (FAO data, cited in Wolf, 1986).

The African dry rangelands lie in the Sahel strip and its extension into Kenya and Tanzania, the southwest arid zone of Botswana, Namibia, and parts of Zimbabwe. They support between 15 and 25 million people, mainly pastoralists. The most common pastoralism practices involve some form of seasonal migration with livestock herds to follow rainfall and take advantage of fresh grasses (Breman and de Wit, 1983; Sinclair and Fryxell, 1985). Migration has several advantages in arid, marginal environments: (1) it allows larger populations of livestock and humans to live in a given area than if they were sedentary; (2) grasslands are free from grazing for a portion of the year, allowing plants to grow, reproduce, and set seed; and (3) it represents an adaptation to conditions of extreme climatic variability, not only within years but between years, by allowing people to escape the worst of periodic but persistent droughts.

Migration strategies range from true nomadism with continuously moving herds in parts of the Sahel, to more traditional regular migration routes shown by the Masai and other tribes in East Africa. Sometimes the routes cross international boundaries. For example, herds in Mauritania cross into Mali and Senegal during the dry season, and sometimes remain there during drought years. This system seems well adapted to African conditions and there is historical evidence that it has persisted for several thousand years (Dumont, 1978). We are in general agreement with Horowitz (1986) that traditional pastoralist migration systems represent rational adaptations to their environments. However, we argue below that when these practices are disturbed by economic and political influences, traditional animal husbandry systems can lead to a process of ecological degradation and economic decline.

Two processes have undermined Sahelian pastoral migration systems—sedentarization, and human and livestock population increases. Sedentarization resulted from several factors operating together. Starting in the early 1960s, the newly independent governments encouraged pastoralists to settle. This avoided the complications of populations moving across international boundaries, allowed taxation and regulation, and facilitated provision of education, health care, and veterinary and other social services.

At the same time foreign development organizations were beginning their work. These organizations tend to have a predisposition towards well-defined short-term technical projects rather than long-term social programs. They identified a need for the provision of year-round water supplies (to them this was obvious, otherwise people would not be wandering around following rain) and thousands of boreholes and wells have been constructed in the Sahel since the late 1950s. Formerly nomadic tribes settled around the new wells and their livestock began to graze the same area year round. As we know from other studies (Sinclair and Fryxell, 1985; Fryxell et al., 1988), changing a migratory pastoral system to a sedentary one leads to four problems: (1) livestock no longer have access to the high quality but ephemeral food supplies in the drier areas of their range: (2) the year-round grazing results in the death of the grasses and the collapse of the pasture—in other words overgrazing; (3) as described above, human populations have increased as a result of improved health care and education, while access to land has decreased through settlement and through the incursion of permanent agriculture (see below); and (4) livestock populations, already above carrying capacity for a sedentary system, further increased as a result of improved veterinary services.

In Kordofan, Sudan, it is estimated that the livestock population increased six-fold between 1957 and 1977 (Tinker, 1977). In 1950 the livestock population of Africa was 295 million; by 1983 it had increased 75% to 518 million. Between 1950–70, cattle increased most (2.15% per year), but sheep and goats increased most (2.36%, 2.19% per year, respectively) during the period 1970–83, possibly reflecting a deterioration in the quality of rangelands as the smaller animals can live on sparser food and tolerate drier conditions than can cattle (Brown and Wolf, 1986). Periodic "droughts," during which huge numbers of animals died, have not contained this increase.

The increase in livestock numbers was due partly to the improved veterinary aid that was available in settlements and partly to cultural and economic pressures (Tinker, 1977). Grainger (1982) describes how in past centuries pastoralists used salt, arms, and gold with which to trade during times of drought. Salt is no longer a valuable commodity and the rest can no longer be traded easily. Thus tribes now tend to use *numbers* of livestock as the way to save for the future: to quote Grainger "overgrazing of land is often of less concern than having enough animals to maintain viable herds in the aftermath of drought." This cultural attitude, of numbers as an insurance policy, is also reflected in the attitude of women towards the number of children they produce—often they have more in case some of them die (Chandler, 1986; Wolf, 1986).

All of these factors—sedentarization, well-digging, health care, education, increasing populations of humans and livestock, overgrazing, agricultural expansion, and cultural practices—act in the same direction, reinforcing each other, to create a progressively out-of-balance system leading to desertification (Sinclair and Fryxell, 1985).

Desertification is a global problem. Grainger (1982) has estimated that 30 million km^2 out of 47 million km^2 of the world's arid or semi-arid land are threatened with desertification, leading to a greater than 50% drop in vegetative productivity. This is approximately 88% of the world's available rangeland.

The process of degradation and denudation (desertification does not mean the development of sand dunes, although this does occur in some areas) can begin around wells with the constant trampling and overgrazing, and then spreads out along livestock transit routes. In the Sudan for instance, Tinker (1977) reported that each village at a waterhole was surrounded by an area of sand extending for 20 km or more, and even for 40 km along the migration routes several hundred meters wide. As more wells were put in, the circles of denudation fused to form larger areas with little ground vegetation and eventually the whole area was affected. Timberlake (1985, p. 94,) says, "Wells and other watering points have a bleak record in the history of African rangeland management. . . . Most are dug without consulting the pastoralists, and without fitting the new water sources into pastoralist routes and strategies. . . . Aid agencies have competed in digging wells and boreholes, and have not consulted among themselves." Wolf (1986, p. 73) notes, "Development assistance to pastoralists has sometimes fostered the deterioration and vulnerability it was intended to reverse. . . . The concentration of herds around wells in the dry season creates a circle of trampled, barren land called a 'sacrifice area' that may be as much as several kilometers in diameter." Some estimates put the rate of advance of the denuded areas in the Sahel as 5–15 km per year (Tinker, 1977; Sinclair and Fryxell, 1985; Mellor and Gavian, 1987). Desertification makes an already fragile system even more so. Cattle herds and people may be able to exist on the now sparse overgrazed vegetation but even a small decrease in rainfall can create a crisis. In past years (before the 1950s) such fluctuations were easily tolerated. Although the drought in 1984 was not the most severe in the past 50 years, the famine it precipitated was. These abnormally low rainfall years, therefore, may trigger crises but they are not the fundamental cause of them. These crises are essentially man-made (Sinclair and Fryxell, 1985).

In the past 20 years there has been a steady decline in the average annual rainfall in the Sahel region, while for the previous 100 years the average had remained steady. Some authors consider the recent decline is due to intrinsic global meteorological changes as opposed to local weather events strongly influenced by man-made actions. Nicholson (1986) has reviewed the historical evidence for dry periods in Africa. She points out that over the past ten thousand years there have been dry periods lasting hundreds or even thousands of years. There was a period during the last century (around 1830) when low rainfall lasted several decades. On the basis of these precedents she argues that the present rainfall decline, which has spanned two decades, must also be due to meteorological changes. She says, "Neither the length nor severity of recent droughts nor the rainfall pattern preceding them is unique for African droughts. Therefore, the characteristics of recent African droughts do not advance the argument that human pressures are responsible for the current drought episode." She notes that drought years (i.e., those with abnormally low rainfall) and wetter years in West Africa are correlated with similar events in East and South Africa, suggesting global meteorological causes rather than local human actions. So she concludes that the fundamental cause of African droughts is through large scale meteorological events.

We recognize that there are two fundamentally distinct rainfall events shown by the Sahel data over the past century: (1) the sporadic and unpredictable very low rainfall years (precipitation about 75% of the long-term mean) which we define as a "drought"; and (2) the declining mean annual rainfall since the late 1960s which we call the "drying trend." Unfortunately these two different processes have been confused in the literature and lumped under the one heading of drought.

It is the drying trend and not the "droughts" which is the new feature in the Sahel rainfall record over the past 100 years. Whether this drying trend has been caused by large scale (global) meteorological events, as suggested by Nicholson (1986), remains an open question. There are, however, several problems with Nicholson's argument. (1) Her weather hypothesis is untestable using the precedence argument; if one looks far enough back one can always find a similar period to the present to blame on global climatic events. Her conclusions, based on these methods, are self-fulfilling and so not very informative. (2) There is the problem of equifinality; similar effects, such as periods of lower mean rainfall, can be produced by different means, i.e., just because there was a dry period in the 1830s caused by global weather events, does not mean the present drying trend must necessarily be caused by the same climatic events. (3)

It is necessary to distinguish between extreme years (droughts) and drying trends. No one would disagree that extreme dry years are caused by weather changes, and these are correlated across Africa. It is the declining rainfall trend over the past two decades in the Sahel that is at issue—and this trend is not seen in Nicholson's (1986) data in either East or South Africa. Thus, by her own argument, it is unlikely to be caused by large-scale meteorological events. We do not deny the existence of global meteorological causes for the drying trend, but we have concluded that the rainfall record itself cannot yet be interpreted in this way. It would be most unfortunate if the weather hypothesis were to promote a fatalistic "do nothing" policy, simply because one cannot change the climate. At best this would lead to a policy of accepting and adapting to the dry conditions. The problem is, What if this interpretation is wrong and the real cause has human origins? The above policy of adaptation would exacerbate the problem by ignoring the effects of human actions.

In fact there is accumulating evidence that the recent rainfall decline may be man-made. In the 1970s Otterman (1974) and Otterman et al. (1975) predicted that extensive denuded land can result in lower rainfall which in turn results in further desertification. Thus another positive feedback cycle is produced on a scale of decades. At the time that Otterman made his predictions it was not known that a downward trend in rainfall was occurring, but his hypothesis predicted it. Subsequent events have supported his prediction. Clearly more weather data are needed, but if the human impact hypothesis is true it has profound implications for the whole ecosystem. It would mean that a new system state (in the sense of Holling, 1973) has been reached, one of extreme aridity. It would also mean that aid in the form of agricultural and animal husbandry technology based on past experience would be out of context and ineffective. Instead, long-term vegetative regeneration would be required to cause the ecosystem to revert back to its state prior to 1950.

One of the first significant steps in recognizing the ecological crisis in sub-Saharan Africa occurred with the UN Conference on Desertification (United Nations, 1977). The conference declaration assigned The United Nations Environment Program (UNEP) the responsibility for assessing the problem of desertification and designing and executing plans and projects to mitigate the deterioration of agricultural lands. In 1984, UNEP commissioned an evaluation of their programs in the Sudano–Sahelian region. For all 19 countries in the region, representing about 280 million people, the evaluation concluded that the desertification trend had continued unabated and field projects had not led to any notable achievements.

Desertification, famines, and mortality in Africa's arid regions have thus accompanied declining food production per capita (Figure 17.1). How have these processes been affecting the human populations in these regions? Eight Sahelian countries (Senegal, Burkina Faso, Mauritania, Mali,

Niger, Chad, Sudan, Somalia) have experienced these effects more than other countries. In these countries the human population annual growth rate has been significantly lower (2.75% per year versus 3.3% for all sub-Saharan African countries, WB84 data, $P < .05$) and the average annual child mortality rate higher (3.2% versus 2.3% average for Africa, $P < .05$). The most significant socioeconomic factor related to these differences appears to be female education. This is much lower in the Sahel (15.5% of the eligible female population) than in Africa as a whole (37.6%, $P < .01$). The Sahel countries also have lower levels of GNP and higher food aid per capita but neither are statistically significant influences. Fertility, food production per capita, population density, and other ecological variables are the same in these countries as in other countries. Higher mortality in the Sahel therefore appears to be related to deficiencies in education and health care rather than negative ecological feedbacks from desertification.

B. The Cycle of Agriculture and Declining Crop Yields

1. The Sahel System

African croplands form only 11% of the total land area (183 million ha) but produce most of the food. The soils are old, sandy or lateritic, high in iron oxides and generally poor in nutrients. Nevertheless, traditional agricultural practices have produced sustainable crop yields and exhibited considerable resilience, at least until recently. In the savanna areas (500 mm rain or more per year) south of the Sahel crop farmers developed a symbiotic relationship with pastoralists from the north. Food crops of sorghum and millet were grown during the rains and then harvested. In the dry season the pastoralists' cattle, migrating south, were allowed to feed on the crop stubble and use the water supplies. In return, feces and urine provided nutrients for the fields.

This system has been broken down by increased production of crops for export (commonly referred to as "cash crops"). Cotton, especially in Sudan, and peanuts in the western Sahel were two of the most important cash crops (Franke and Chasin, 1980; Grigg, 1985). Traditional systems of food crop production had allowed some land to lie fallow for a few years, to be used only by cattle. With cash crops replacing food crops, fallow areas began to be used for crops, causing fallow periods to decline and the displacement of cattle. Since the 1960s, when cash crops expanded rapidly, export earnings have declined as international markets for African export crops have steadily deteriorated. The response was increased pressure on marginal croplands. Timberlake (1985) states, "African governments feel the need to grow more cash crops in much the same way African peasants feel the need to have more children. If children are dying, more—not less—children are needed. And if crop prices are falling, more—not less—cash crops are needed."

The pressure to grow more cash crops has caused a greater area of land to be used rather than more intensive production. The increased area turned over to cash crops for export has frequently been at the expense of food production for local markets: either food is not grown and farmers are forced to use their declining cash returns to buy food, or they are forced to plant in marginal land which is more fragile; i.e., more susceptible to erosion, as in Ethiopia where 50–100 t/ha/yr are lost (Newcombe, cited in Brown and Wolf, 1986), or in drier country more prone to drought. In Kondoa Province, Tanzania, for example, gully erosion has caused the abandonment of agriculture on 150,000 ha of land that was once forest (Timberlake, 1985). Population increases have increased the demand for arable land and reinforced this trend.

International aid agency projects introduced new varieties of both cash and food crops that have higher yields and shorter growing seasons, the latter allowing planting further into the Sahel. But these new varieties require high amounts of fertilizer which have to be purchased with cash. Faced with falling earnings and little credit, most peasant farmers gave up applying fertilizer during the 1970s.

Thus one has what Timberlake calls the "cash crop squeeze." More land for cash crops means less for food production. The demands of increasing populations force farmers to reduce fallow times, and eventually yields fall. Cattle are excluded so their nutrient inputs are lost. High yield plant varieties demand still more fertilizer and, when this is not available, they generally produce *less* than the original varieties. Populations forced onto marginal land not only find low productivity but are more susceptible to the effects of climatic fluctuations. The index of food production per capita in the eight Sahelian countries fell by 23.5% between 1964 and 1982 (calculated from WR86 data), compared to 8.2% in the other African countries. Thus, while the Sahel used to be a significantly better food producing area of Africa ($P < .05$), productivity has now dropped to a greater extent than other areas as a consequence of population and economic pressures on the ecosystem: the resilience of the system has been reduced to a point where it is prone to collapse from minor climatic disturbances.

2. The Woodland System

In those areas of Africa dominated by woodlands or bushlands, the traditional agricultural system is that of shifting cultivation. Shifting agriculture starts in an uncultivated area by clearing the woodland and burning, to release nutrients into the soil. Grain crops such as maize, and root crops such as cassava and yams are then grown for a few years. In wetter areas there is a sequence of low crops in the first year followed by taller ones such as bananas which can outgrow weeds. These latter crops require less manual labour and tree seedlings can become established. After a few years soil nutrients become depleted and the area is then abandoned. It is left to regrow for 5–20 years before the cycle repeats itself.

Shifting cultivation occurs in the forests of west and central Africa. Bush fallowing, in which fallows are shorter and neither forest nor woodland become reestablished, occupies most of west Africa and reaches east to Ethiopia. Perhaps three-quarters of Africa's cultivated land is under shifting cultivation or bush fallowing. Permanent agriculture, where fallows are nonexistent or very short, occupies little area but supports most of the population. It occurs in the Kenyan highlands, eastern Madagascar, northern Nigeria, Burundi and Rwanda (Grigg, 1985). In central Africa (e.g., Zambia and Zimbabwe) the predominant vegetation type is nearly closed canopy *Brachystegia* or *Mopane* woodland. The soils are sandy, and most of the nutrients are locked up in the trees. In west Africa trees and soils are different but the system is essentially the same.

With increasing populations fallow periods have progressively shortened, finally leading to permanent cultivation. Sometimes there is intercropping with beans or other indigenous legumes. Total production drops under these circumstances, because soils are depleted of nutrients and fertilizer is too expensive to apply. Further population expansion forces people into unsuitable habitats: for example, the steep montane forest slopes on the east African mountains (e.g., Mt. Meru, Mt. Kilimanjaro) are being taken over for cultivation as are the few remaining forest patches on the edge of the Boma plateau in Sudan. Soils in these areas are rich but the high rainfall on the steep slopes soon leaches and erodes the soil so that the forest cannot regenerate once the area is abandoned. Grigg (1985) concludes that the traditional methods of farming, with long fallows of natural vegetation to restore fertility, have been undermined by the growth of population, and no adequate solution has been found. The introduction of large scale farming using the plough and other heavy machinery has led to some well-documented disasters. Use of the plough and tractor to increase crop area requires clear-cutting of vegetation and makes natural fallows difficult to restore. Use of chemical fertilizers is inefficient because they are leached out in the high temperatures and rainfall of forest areas, or they are not absorbed in the semi-arid areas (Grigg, 1985).

One can see an evolutionary progression in the development of agricultural systems around the world, starting with the relatively primitive shifting cultivation of central Africa and ending with the sophisticated permanent cultivation, irrigation, multiple crops per year and application of fertilizer which one sees in Indonesia and southeast Asia. In Africa, farmers are being forced towards the Asian style of permanent cultivation but without the resources of high soil fertility, water and fertilizer. The introduction of Green Revolution high-yield crop varieties only makes the situation worse because these varieties cannot survive under African conditions. This is the main reason why the Green Revolution has worked in some places in Asia but is not readily transferable to Africa. The trend towards dependency on single crops, instead of the multiple crops found in indigenous systems, is a classic example of loss of resilience: failures in these crops due to diseases, pests or drought allow no fall-back to other crops, i.e., there is no buffer to unexpected events such as drought.

With the limited resources available to farmers in Africa (low soil fertility, water availability, and labor), shifting cultivation was probably the optimum food producing technique. Population pressure has forced these methods to change, with a consequent decline in food production per capita. In Kordofan, Sudan, originally the richest agricultural province, the productivity of the land has fallen between 50–85% (1961–73). In the same period, yields of maize fell 54%, sorghum 55%, sesame 77%, peanuts 78%, and millet—the staple food of the region—87% (Tinker, 1977). Productivity has since fallen still further.

C. Forests and the Fuelwood Crisis

Forests in Africa, as elsewhere in the tropics, are disappearing at an alarming rate. Although data are tentative, one estimate puts the loss of African tropical moist forests at 1–3 million ha/yr (J. Lanley, FAO, cited in Timberlake, 1985), and open forest at 2–3 million ha/yr from a total area of 688 million ha (Wolf, 1986). Most of the tropical moist forest occurs in Zaire (60%), the rest occurring largely in west Africa.

Loss of tropical moist forests occurs through logging to some extent; Liberia is cutting about 8% per year, an unsustainable rate. The Ivory Coast lost 67% of its forest from 1956–77. In Nigeria most of the forests have already gone, timber exports have been banned, and the country will soon be importing timber. In contrast, other countries have lost less of their forests: Cameroon loses 1% per year, while Gabon and Zaire are nearly untouched by logging (Timberlake, 1985).

Although logging is responsible for some of the decline in forests, agricultural expansion has caused far greater losses. In the Ivory Coast agriculture destroyed 4.5 times as much as logging, and Timberlake (1985) estimated a similar ratio in other parts of Africa. Agricultural production in Africa has generally increased only by expanding the area under crops at the expense of woodlands, and not by increasing the productivity of the land already planted. Thus Guinea Bissau, Burkina Faso, and Senegal have been losing 30,000–80,000 ha/yr of forest to peanuts and other cash crops. In Tanzania, forests have been lost to tobacco, in Ghana to rice, in Kenya to tea and wheat.

One of the main uses of trees in the dry woodlands, savannas and croplands is as fuelwood for cooking. African staple foods have to be prepared and cooked at every meal—they cannot usually be prepared ahead of time and eaten cold. Thus maize flour is cooked into a form of paste which cannot be kept for later meals. Few Africans have access to bread which can be kept. Some foods, such as beans, have to be cooked for several hours before they become edible. These features of the food supply mean that cooking fires have to be used frequently, and in many cases fires are kept going all day. Thus the demand for fuelwood is relatively high, and has been driven even higher by oil price increases. Kerosene stoves were formerly common in urban areas, but kerosene is now too expensive for many countries to import and is generally unavailable. Thus the populace has turned increasingly towards fuelwood.

Shifting agriculture and low population densities formerly allowed enough tree growth to meet fuelwood demands. The trend towards intensive agriculture, higher populations and loss of forests has caused the demand for fuelwood to exceed supply. Tinker (1977) describes what happened in Sudan. In the Kordofan region there is a zone of *Acacia senegal* from which gum arabic is made, and this industry forms part of the cycle of shifting agriculture. The small *Acacia* trees were cleared by burning, and sesame, maize, sorghum, and millet were planted in the sandy soil. Crops were harvested for 4–10 years, by which time the soils were exhausted. Fields were abandoned and young *A. senegal* reoccupied the area. After about 8 years the trees were ready for tapping off the gum. This lasted for another 10 years by which time the trees began to die and the cycle repeated itself. The fallow period was thus about 20 years. Each farmer had fields in different stages of this rotation. The system broke down in the 1960s under pressure of population. Cultivation was extended for several years, further depleting the soil, and overgrazing in fallow periods prevented the establishment of seedlings. Finally *A. senegal* no longer returned in the fallow period, and it was replaced by other *Acacia* species which were not useful. Without the gum as a cash crop, farmers were forced to cultivate continuously until the soils became useless sand.

A World Bank study in 1980 estimated that in 11 of 13 west African countries fuelwood demand exceeded supply. The degree to which demand exceeded supply varied. In Mauritania and Rwanda demand was ten times the sustainable yield from the remaining forests; in Kenya the demand was five times the supply; in Tanzania, Nigeria, and Ethiopia it was 2.5 times; and in Sudan it was two times the supply (Brown and Wolf, 1986). In many of the drier areas of Africa there are no trees left and cowdung, originally used as fertilizer, is now dried and used as fuel for cooking. Crop residues can play an important role in controlling soil erosion as well as being an important source of nutrients in the absence of alternative fertilizers (D. Pimentel, personal communication).

Thus we see another feedback reinforcing ecological degradation: population pressure leads to a decline in trees due primarily to demands for cropland and fuelwood. Lack of fuelwood leads to further soil depletion, when dung is used as fuel, and eventual desertification. The population is forced to move into a new area and the cycle of land degradation repeats itself. Essentially the three processes of overgrazing, overcultivation, and overharvesting of trees are all linked inextricably to rapid human population growth, and all lead to declines in food production and ecological degradation. We next ask, How have economic and social factors contributed to this ecological collapse, and how in turn have ecological processes had socioeconomic implications?

IV. THE ECONOMIC DECLINE

During the last two decades, the economic situation and prospects of the African nations as a whole have deteriorated sharply. Income levels and food security have become increasingly threatened by high levels of human population growth, accelerating resource degradation, stagnant agricultural productivity, and the inability to generate foreign exchange to finance imports. Economic factors appear to have contributed significantly to Africa's declining situation, although the extent of this contribution, exactly how it is manifested, and the appropriate remedial actions are all far from clear.

A. African Economies—The Agricultural Sector

African economies are closely linked to the rest of the world through trade. Throughout the twentieth century, exports of primary commodities and (particularly since independence) imports of fossil fuels, food, and manufactured goods have comprised the bulk of this trade. The importance of trade is illustrated by the 21% average share of exports in the total output of goods and services, or gross domestic product (GDP), during 1983. This trade became increasingly imbalanced between 1973 and 1983, as exports declined in value by an average of 5.0% annually while imports grew at 3.3% (World Bank, 1986b). This deterioration can partially be explained by unfavorable movements in the terms of trade (the relative level of export to import prices weighted for all traded goods) during this period. Estimates by the World Bank suggest that losses suffered as a result of declines in terms of trade for all countries in sub-Saharan African outside Nigeria amounted to $15 billion over the period 1980-85, an amount approximately equivalent to the increase in their external debt (cited in FAO, 1986).

The decline in export revenues has been accompanied by a rapid escalation of debt as the African nations have been the recipients of significant financial flows from commercial and concessionary lenders. Between 1970 and 1980, African external debt rose by over 21% each year and several countries increased their debt levels ten times or more (World Bank, 1986b). Export earnings are the principal source of financing for external payments, whether for imports or loan repayments. The ratio of annual debt service (interest plus capital repayments) to export revenues is therefore critical. By 1984, debt service repayments had increased to 22.2% of exports (scheduled repayments had originally been higher). As a greater share of already declining export revenues were allocated to debt service, the amount of foreign exchange available to purchase imports became severely limited. Given likely export prospects, the current level of external debt for many African states appears to be unsustainable.

An analysis of the composition of external debt for the 25 poorest African countries reveals that official lenders (primarily multilateral agencies) plus the International Monetary Fund (IMF) account for over 80% of the amount outstanding at the end of 1984, while commercial lenders account for about 13% (World Bank, 1986b). Since neither payments to the IMF nor debt service due to other multilaterals can be rescheduled, the scope for debt relief is limited. The IMF and the World Bank, whose loans comprised about 14% of the amount owed by the 25 poorest nations, play an important role in shaping the views of the industrialized countries towards economic development in Africa, and they can wield considerable influence over the domestic economic policies of the individual African countries who they advance funds to. This influence is unlikely to diminish if, as seems inevitable in the near future, African economies continue to decline.

Agriculture is Africa's most important industry, employing approximately 70–80% of the labor force. Agriculture accounted for 29% of GDP in 1983, down from 41% in 1965 (World Bank, 1986b). In the 25 poorest nations, the agricultural sector accounted for 40% of GDP and 68% of export earnings in 1983 (World Bank, 1986a). The performance of the agricultural sector is therefore a critical component of African economies and a focus of international efforts designed to combat rural impoverishment.

Although large farms can be found in Zimbabwe, Zambia, Kenya, Ghana, and the Sudan, the great majority of farms in Africa are small. In Ethiopia the average farm size is less than 5 ha; in Tanzania 83% of farms are less than 3 ha, in Malawi 80%. Farms in southern Nigeria are also extremely small (Grigg, 1985). These small farms provide most of Africa's food output and a substantial proportion of its export crops.

Arable and permanent cropland covered about 143 million ha during 1981–1983 (WRI/IIED, 1986), an increase of almost 15% since 1964–66. Increases in area have come not only from the expansion of settlement into unoccupied, often marginal, areas, but also from the reduction of fallow periods and an increase in the period for which crops are grown

17. Population Growth and Poverty Cycle in Africa   **465**

between fallows (Grigg, 1985). The area of cropland per person, however, has declined rapidly throughout the continent (Figure 17.9). The average area of productive cropland per capita has been predicted to fall by a further 50% by the year 2000 (WRI/IIED, 1986). The accuracy of this estimate is not important; it is the trend of decreasing cropland per person which clearly has severe implications for agricultural production per capita and for future food security.

Roots and tubers comprised about 62% of major crop production in 1982, and cereals about 25% (Figure 17.10). Within and between individual African countries, crop yields have varied considerably during the last two decades (Figure 17.11). A number of countries have lower crop yields today than they did 30 years ago, however. Over 27% of the African population live in countries where the cereal yield per hectare declined between 1964–66 and 1982–84; over 20% of the population live in countries where the cereal yield declined one-tenth or more. Similarly, about 24% of the population inhabit countries where root and tuber yields have declined over the same period; about 15% of these in countries where the yield has fallen by one-tenth or more (calculated from WRI/IIED, 1986). Grigg (1985) has estimated that 90% of the increased cereal output between

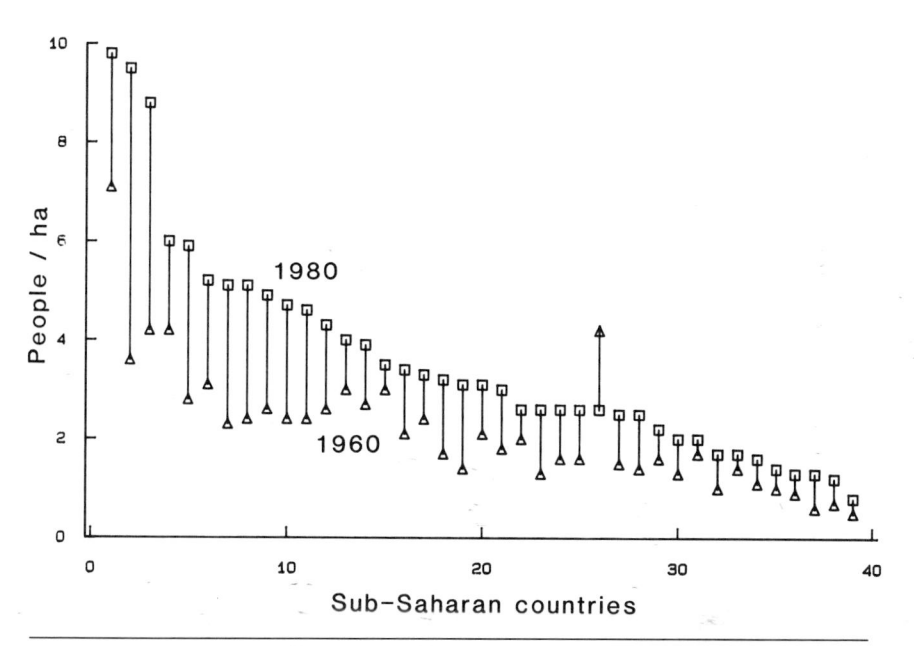

Figure 17.9 Ratio of human population numbers to area of cultivated land in 1980 and in 1960 for 39 countries. [Calculated from WRI/IIED (1986).]

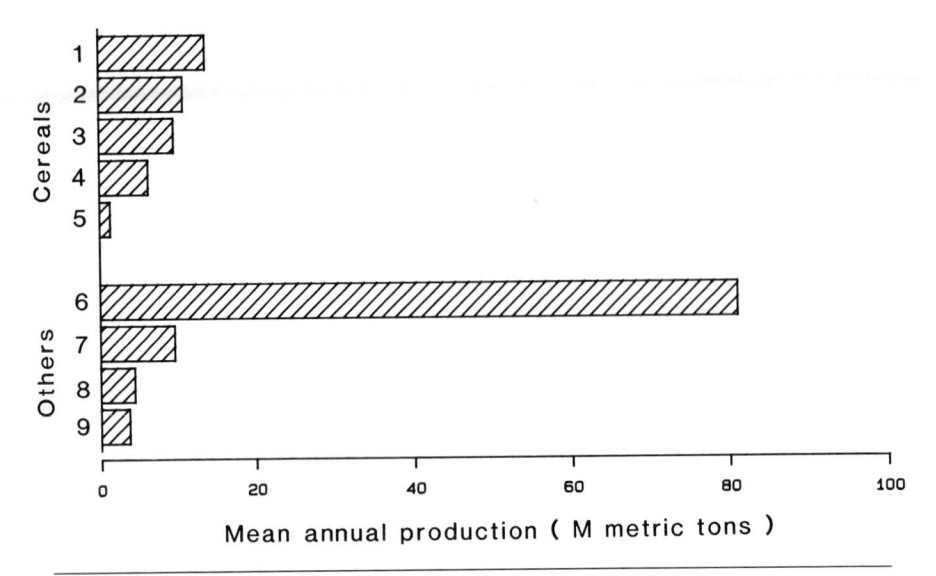

Figure 17.10 Mean annual production of major crops, 1980–1982 (World Bank, 1984): (1) maize, (2) sorghum, (3) millet, (4) rice, (5) wheat, (6) roots and tubers, (7) other crops, (8) groundnuts, (9) other oilseeds.

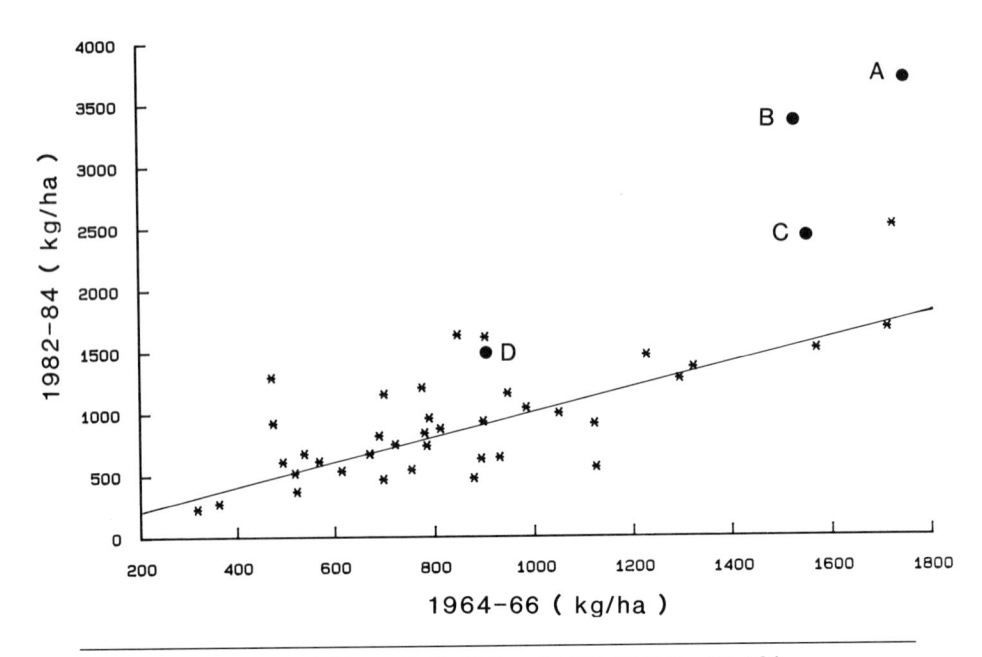

Figure 17.11 Cereal yields in 1964–1966 and 1982–1984 for 39 African countries and four other cereal producers (WRI/IIED, 1986): (A) China, (B) Indonesia, (C) Argentina, (D) India.

1960 and 1975 came from area expansion, and only 10% from yield increases.

The total agricultural labor force (persons actually employed in agriculture) increased by almost 50% to 148 million between 1965 and 1983. This is equivalent to an annual average growth rate of 2.1%. Labor productivity varied considerably between countries over this period. Agricultural production per agricultural worker increased by at least 10% in countries containing 64% of the African population, but decreased in countries containing 19% of the population (calculated from WRI/IIED, 1986).

The substantial majority of African agricultural crops are consumed locally. Virtually no roots or tubers, and only about 2% of cereals, are exported. Major crop exports in 1982 are shown in Figure 17.12. Export crops are almost exclusively grown for overseas markets, following their establishment during the colonial period. With few exceptions (tea, tobacco, cereals and sugar), there has been a precipitous drop in volume of cash crop exports over the last decade or so (Figure 17.13). Although falling agricultural commodity prices have clearly been a factor (Figure 17.14), and have contributed significantly to Africa's declining terms of trade, the African share of world markets for many agricultural products

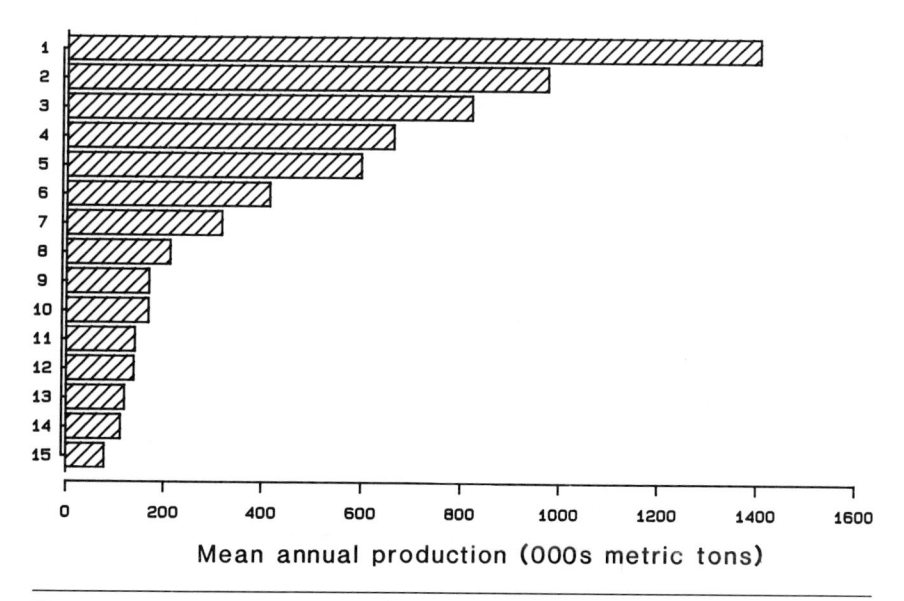

Figure 17.12 Mean annual production of the major agricultural exports, 1980–1982 (World Bank, 1984): (1) sugar, (2) coffee, (3) cocoa, (4) cereals, (5) oilseeds, (6) cotton, (7) palm nut and oil, (8) bananas, (9) tea, (10) tobacco, (11) rubber, (12) groundnuts, (13) sisal, (14) groundnut oil, (15) sesame seed.

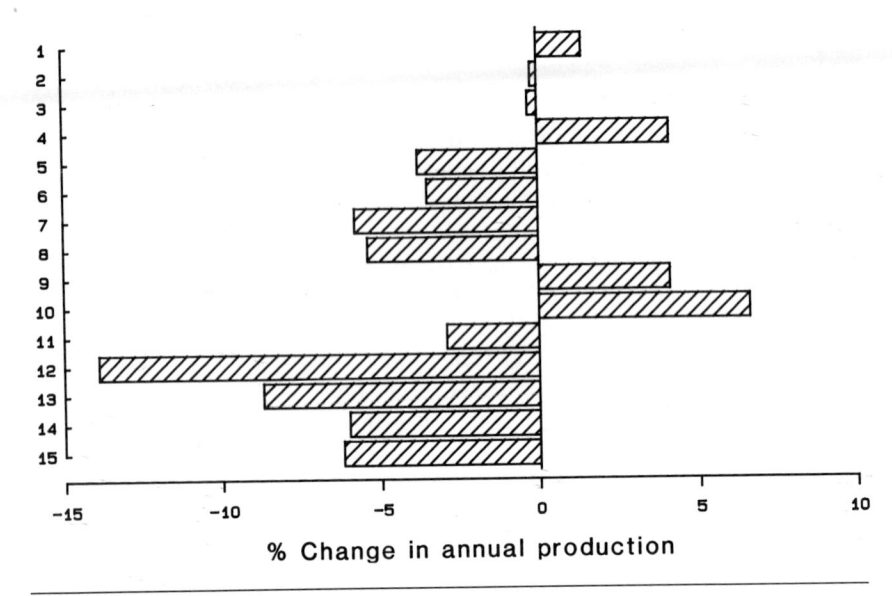

Figure 17.13 Changes in the annual production of agricultural exports between 1969–1971 and 1980–1982 (World Bank, 1984). Numbers refer to crops as in Fig. 17.12.

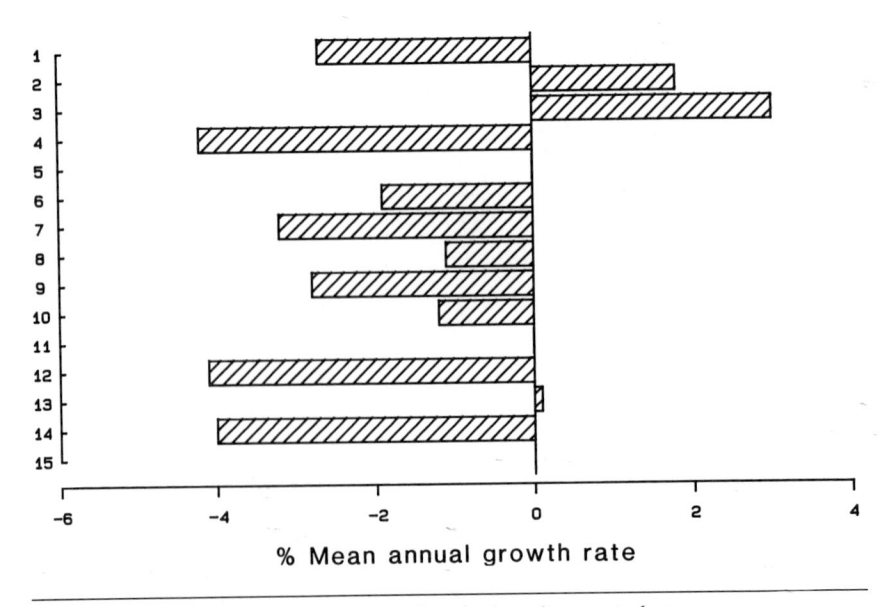

Figure 17.14 Changes in the prices of agricultural exports between 1970 and 1982 (World Bank, 1984). Numbers refer to crops as in Fig. 17.12.

has been severely eroded (Table 17.6), suggesting that other factors are also relevant. Groundnuts experienced the most spectacular collapse, with an average annual decrease in export volume of almost 14%, reducing Africa's world market share from 69% to 15% between 1970 and 1983. Of the 16 major export crops (Table 17.6), Africa's world market share increased only in tea, cereals, and tobacco during this period.

By the end of the 1970s the economies of many African nations were heavily dependent on cash crop exports, and nine countries were dependent on just one crop for over 90% of their income: coffee for Burundi, Rwanda, and Ethiopia; peanuts for Gambia, Guinea-Bissau, and Senegal. Other nations over 50% reliant on one single crop included: Mauritius (sugar); Uganda and Kenya (coffee); Chad, Sudan, and Mali (cotton); and Ghana (cocoa). Between 1974–1976 and 1982, the area devoted to major crop exports crops grew by 11.4% (Timberlake, 1985).

Agricultural (food) imports increased in value from $749 million in 1961–1963 to $6,833 million in 1980–82 (World Bank, 1984). Cereals comprised about 34% of agricultural import values in 1980–1982 compared to 15% 20 years earlier (World Bank, 1986b). The increase in cereal imports has corresponded with declining levels of domestic cereal production per capita.

Food aid imports in 1981 were about 2.3 billion metric tons in grain equivalent (World Bank, 1986b), compared to average purchased cereal imports of about 6.1 billion during 1980–1982 (World Bank, 1984). According to the FAO (1986), commercial imports and food aid accounted for an average of 17.2% of the total cereal supply between 1979–81. This

Table 17.6 Major Sub-Saharan Export Crops: World Market Share[a]

Crop	1969–1971 (%)	1982–1984 (%)
Sugar	5.6	4.9
Coffee	29.3	23.8
Cocoa	75.9	63.1
Cereals	<2.0	<2.0
Oilseed cake and meal	8.3	1.7
Cotton	15.5	10.8
Palm kernel oil	54.8	14.0
Palm kernels	82.2	70.0
Bananas	6.5	2.9
Tea	14.4	19.7
Tobacco	8.2	11.0
Rubber	6.8	4.5
Groundnuts	69.1	15.0
Sisal	59.7	53.9
Groundnut oil	57.6	38.3
Sesame seed	75.3	25.6

[a] Source: World Bank (1986b).

figure varies considerably between regions, from 14.2% in the Sudan–Sahel zone to 23.3% in west Africa and 28.6% in central Africa. The role played by imported cereals (including food aid) becomes more striking if they are compared with the volume of cereals marketed. In 1980, imported cereals accounted for 50% of marketed cereals (62% in the Sudan–Sahel, 48% in west Africa, and 66% in central Africa).

Distinct variations between rural and urban food consumption patterns are apparent. The ratio of roots and tubers to cereals volume in rural food consumption was approximately 1.8 between 1979–1981, compared to approximately 1.1 in urban areas. Although more than three times as many people live in rural areas, rural populations consume over five times the volume of roots and tubers as do city dwellers (FAO, 1986).

As foreign exchange earnings have dwindled, correspondingly larger proportions have been allocated to debt service, and to oil and food imports. Purchases of agricultural inputs have become particularly low. Imports of agricultural machinery and pesticides fell in real value terms by 14% and 17%, respectively, between 1978 and 1982 (FAO, 1986). Few of these inputs are produced in Africa. While about ten countries produce fertilizers, production costs have tended to be prohibitively high.

B. Economic Explanations for Africa's Agricultural Crisis

The declining economic performance and prospects of Africa's agricultural sector have led to intense debates as to its causes and potential remedies. We have elaborated on these arguments at some length, largely because the policy responses of lenders and donors have been significantly conditioned by explanations for the decline which are dominated by economic considerations. Institutional lenders and donors have exerted considerable influence over African agricultural development during the last two decades and seem likely to continue to do so.

The various economic explanations and their proponents have tended to be classified into two schools, internalist and externalist. Internalists generally argue that the basic cause of agrarian breakdown can be found in the economic policies pursued by African governments since independence. Basic economic policy reforms are their prescription. Externalists, on the other hand, emphasize adverse features of the international economic environment faced by countries which are heavily dependent upon agricultural exports. They believe that internal policy reforms will be inadequate to trigger economic recovery unless this external environment can be made more hospitable. There now appears to be a growing consensus that both internal and external economic factors have made a substantial contribution to the current weaknesses of the African agricultural sector, many of which are rooted in the policies pursued by the colonial governments prior to independence.

The recent development literature includes many summaries of this debate, and only the most critical arguments will be reviewed here. The importance of this debate must not be underestimated, however. Many African countries will undoubtedly continue to depend on large-scale financial inflows for the foreseeable future. The sources of these funds will be aid donors, the multilateral development banks (primarily the World Bank and the African Development Bank), the International Monetary Fund (IMF) and the UN agencies, among others. The IMF and the World Bank, whose policies have been extremely influential are showing an increasing tendency to impose conditions of domestic policy reform upon their loan packages. Future agricultural development in Africa and a reversal of the present trajectories towards greater human misery will depend largely on the appropriateness of these policies and the effectiveness with which they are implemented. Although the viewpoints and policies of the large institutional lenders and donors have increasingly tended to resemble one another recently, the nature of the underlying problems and the appropriate policy steps remain highly controversial.

The discussion of the internalist perspective which follows is based on the work of Lofchie (1986, 1987). The basic internalist proposition is that African governments have intervened in rural markets in ways that pose fundamental disincentives to agricultural production. The most influential proponents of this viewpoint have been Bates (1981) and Berg (World Bank, 1981). Both argue that, since independence, African governments have adopted a set of economic policies that have effectively reduced the economic incentives of agricultural producers by shifting the internal terms of trade against the countryside. These policies include agricultural price controls, overvalued exchange rates, inefficient and extortionary official marketing boards, and a strategy of import substitution.

Bates argues that agricultural pricing controls have been used to suppress the farmgate prices of agricultural commodities far below levels that would have prevailed if a free market had been allowed to operate. The payment of low prices to producers of export crops was initiated by colonial governments that viewed the agricultural sector as a source of tax revenue (arising from the difference between amounts paid to producers and sales proceeds at world market prices) and foreign exchange, an approach enthusiastically continued by the subsequent independent governments. The suppression of food prices is widely agreed to be a response to political pressure to keep food prices low for urban consumers (Bates, 1981; de Wilde, 1984).

A second major policy with adverse effects on agricultural productivity is the widespread tendency towards currency overvaluation. Currency overvaluation tends to lower the living costs of urban consumers by reducing the prices of imported goods, including foodstuffs, automobiles, household applicances, and luxury items. The foreign exchange which finances these imports has typically been generated by agricultural exports. The amount of local currency that the producers of agricultural exports are able to earn is reduced in direct proportion to the magnitude of currency overvaluation.

It has become increasingly common for western grains such as wheat, corn, and rice to be less expensive in African markets than the equivalent items, or equivalent staples, produced by local farmers. Sometimes the price structure of imported grains is lowered still further, particularly when they enter the country on concessional terms as food aid, leading to potential competition between food aid and locally-produced foodstuffs. For example, in Nigeria imported maize cost $US 315 per ton in 1983, compared to about $US 1,200 for locally grown maize delivered to Lagos (FAO, 1986). It is argued that producers become discouraged, reducing local production for urban markets and launching a cycle of ever-increasing dependency on food imports.

Official agricultural marketing boards are frequently granted control over the purchase, processing and sale of specific agricultural commodities. The marketing board system was originated during the colonial era to conduct the international sale of export crops, although they are now prevalent in domestic food production. In many cases, the agricultural marketing boards are also entrusted with functions such as credit provision, the distribution of inputs (such as fertilizer, capital equipment, and pesticides), and the conduct of research on improved methods of husbandry. There is a widespread consensus that agricultural marketing boards in Africa are generally characterized by waste, inefficiency, mismanagement and corruption. The proportion of revenue from crop sales absorbed in internal operations frequently leaves little income from which to pay producers. In many cases such payments have been made only partially or delayed, sometimes for years.

Industrialization through import substitution has also had a negative effect on the agricultural sector. Following independence, agricultural marketing board reserves were used as a source of venture capital for the new industrial sector. Import-substituting industries in Africa have represented a continuing burden on agricultural export earnings and have limited capital reinvestment in the agricultural sector, gradually eroding its capital base and diminishing its capacity to respond to changing conditions of demand in the world market.

These policies in combination, the internalists argue, have provided a fundamental disincentive to increase agricultural production and have caused innumerable farmers to abandon official state-controlled markets for the informal marketplace. Once local producers have withdrawn from markets dominated by artifically cheapened imports and have become accustomed to trading their goods in informal or illegal cross-border markets, it becomes difficult to recapture their involvement in the official economy.

Driven by powerful coalitions of urban-based interest groups, African political leaders have adopted a set of economic measures that cater to the economic needs of city dwellers while steadily impoverishing rural social classes, especially smallholder farmers. These policies have weakened the ability of society and of the household to escape extended breakdowns in production, such as occurred during recent prolonged droughts. Observers with an internalist viewpoint generally agree that the starting point for agricultural recovery must be the implementation of a set of measures that will shift the internal terms of trade back towards the rural sector and improve real economic incentives for agricultural producers.

Externalists tend to assign primary responsibility for the crisis to features of the international economic system, particularly the declining terms of trade for primary agricultural exporters. A host of other external shocks to the African countries include the increase in the price of oil since 1973. This drastically increased the cost of agricultural production since many agricultural inputs are petroleum derivatives. A rapidly growing tendency toward protectionism by industrial nations has discouraged initial efforts towards industrialization based upon export-processing. The scarcity of foreign exchange associated with declining commodity prices has been further compounded by the low demand elasticities of Africa's key agricultural commodities. This means that African countries as a whole cannot compensate for low price levels by increasing the volume of agricultural goods they market. A World Bank study of anticipated demand for Africa's agricultural exports over the next decade suggests that the world consumption of the majority of these products will increase annually by only about 3% or less for the foreseeable future. Only palm oil is expected to benefit from a growth in demand approaching 5% per year. It therefore seems likely that the majority of countries will continue to experience declining export revenues.

V. FOREIGN AID

A. The Impact of Donor and Lender Policies

Policies of the international lenders and donors have had an extremely significant effect on African agricultural development. Following independence, international donors interested in rural development saw the critical problem as how to improve the productivity of the food-producing sector so that food imports would not continue to be a drain on financial reserves needed for industrialization (Lofchie, 1986). Agriculture was regarded as a backward sector and these agencies identified the traditional food production methods of peasant food-producers as the principal constraint on this improvement.

The roots of this viewpoint lie in the colonial era. Africa's colonial governments had generally pursued a set of agricultural policies that were designed to promote the production of exportable crops (cash crops), often at the expense of food production. The growing of cash crops was frequently assigned the best arable land, often on highly favorable terms, and was generously subsidized with physical infrastructure, extension services, and marketing assistance. Food production, in contrast, generally took place on small-scale farms where use of the hand-held hoe was prevalent and even animal-drawn cultivation was rare.

The goal of rural development, as perceived by innumerable professional experts and academic observers, was to introduce modern agricultural practices throughout the countryside. The policy chosen to achieve this goal was the creation of large-scale farms as part of a "project approach," characterized by intensive research into areas such as new methods of husbandry, high-yielding seeds, and the most economic methods of implementing these techniques. The idea underlying this strategy was that once peasant farmers had been made aware of the benefits to be derived from agricultural innovation, their traditional resistance to change would be overcome and sweeping improvements in the process of food production would take place (Lofchie, 1986).

As well as failing to attain the ambitious goals of its proponents, the project approach contributed immeasurably to a worsening of economic and ecological conditions in rural areas. The development literature now documents countless examples of projects which not only fell short of their objectives but frequently resulted in seriously harmful side-effects (see, e.g., Adams, 1985; Cassen, 1986; *Ecologist,* 1985; Timberlake, 1985). The reasons why the project generally failed include the following (Harrison, 1987).

1. Projects frequently ignored the constraints imposed by local environmental conditions; constraints which had shaped the development of traditional agricultural methods. For example, fragile soils that should have been disturbed as little as possible were subjected to tractor plowing or machine clearing methods.

2. Failed projects often ignored the constraints faced by African farmers themselves. The most decisive of these, in addition to the traditional culture, society, economy and technology, are shortages of cash and labor, and reluctance to take on any avoidable risk. Many development projects have required high labor inputs, significant cash outlays or risked lower returns in bad years than the farmers' own practices.

3. The economic and managerial realities of African governments have tended to be ignored, leading to the design of projects that involved high

recurrent costs, high imports, or high levels of expert or government input. As soon as the donor financing ceased and the national governmemt had to take over, such projects became vulnerable to the import and public spending cuts that inevitably accompany Africa's endemic foreign exchange and budget crises, the breakdown of vehicles and equipment, or the shortage of skilled manpower.

Despite these well-documented deficiencies, the project approach has remained popular with donors. African governments welcome the vast inpouring of resources that projects makes available and the resulting opportunities for patronage. The preference for large projects may also reflect the bureaucratic character of some of the donor agencies. Like other large-scale bureaucracies, they are often committed to the expenditure of large sums of money within very short periods of time, and they therefore frequently find it impractical to fund numerous small-scale projects.

It now seems clear that development following the project model has done little, if anything, to prepare Sahelian agricultural systems for the effects of climatic variability. When drought occurred, only massive programs of relief food aid were able to limit the extent of human starvation. Recent famines provided dramatic and incontrovertible evidence of the failure of the project approach, and have substantially altered the present policy environment as it affects the African rural sector (Lochie, 1986).

B. The Policy Environment

A broad range of influential actors take an interest in Africa and its problems. These include the African national governments themselves (both individually and under the umbrella of the Organization for African Unity); multilateral lending organizations (such as the World Bank and the African Development Bank); bilateral organizations (such as the United States Agency for International Development and the Canadian International Development Agency); various agencies of the United Nations Organization (in particular the Food and Agriculture Organization, U.N.D.P., U.N.I.C.E.F., and the World Health Organization); governments of many nations outside Africa; private voluntary organizations (e.g., Oxfam, CARE, and Save The Children); nongovernmental organizations and independent research groups (such as the World Resources Institute, Earthscan, the International Institute for Environment and Development, the Worldwatch Institute, etc.); and last, but certainly not least, the Western media.

The responses of these actors to the African "crises" have been conditioned by their individual perceptions of the nature of the problems, and by their particular interests and agendas. There is general consensus that drastic action is required, although agreement on appropriate policies and how they might be implemented is limited. Policies which are carried out on a large scale will undoubtedly continue to be based largely on institutional perceptions of the nature of the crisis. An understanding of these policies, and their underlying assumptions concerning the working of African social, economic, and ecological systems, is therefore essential to any discussion of development in Africa, whether it be human, natural resource or economic development—all of which, we argue, are tightly interlinked.

In terms of financial resources, the multilateral and bilateral organizations are extremely influential, and they have tended to play a leading role in the dialogue on African development. The extent to which their diagnoses have been correct, and to which their treatments have relieved or exacerbated the living conditions and future prospects for rural Africans, is fiercely debated. The magnitude of their influence would appear to be undeniable, however, and their policies and actions therefore merit careful attention.

Various international groups have concluded studies of resource conditions in Africa and identified the causes of environmental degradation (see, e.g., FAO, 1986). But the results of these studies have been of a general nature. Recommendations for specific actions to reverse recent trends on a country or subregional scale are still lacking. For instance, general concepts such as environmental sustainability have not yet been translated into feasible field projects and other initiatives. Therefore, a program of action for the future has not yet emerged (M. T. El-Ashry, personal communication).

The Berg Report (World Bank, 1981) represents a fundamental shift away from the project approach that dominated donor approaches to rural development throughout the 1960s and 1970s. Whereas the project orientation was based on the premise that African governments would have to play a major role in the dissemination of innovative agricultural practices, as well as in providing the proper climate for economic growth, the Berg Report argues that the free market provides the most efficient stimulus to economic progress. Its analysis of the root causes of Africa's agricultural decline is internalist and its arguments resemble those of Bates. The influence of the Berg Report now extends far beyond the World Bank to encompass a very large proportion of the national and international donor organizations that operate in Africa. A large number of donor governments and agencies now insist that African governments shift their own policies to allow greater scope for market forces as a condition of further assistance. The IMF, in particular, now tends to insist that policy reforms of the sort advocated by Berg be implemented as a condition of financial assistance.

Another highly influential policy document was published at the beginning of the current decade. "The Lagos Plan of Action" (LPA) was signed by the African heads of state, and thus represents an African perspective on what steps need to be taken to put the continent on a proper development path (Organization for African Unity, 1981). The LPA, however, like the Berg Report, is concerned almost entirely with questions of economic development for Africa (Browne and Cummings, 1984). Food self-sufficiency is identified in the LPA as a key priority and policy prescriptions for increasing agricultural production are proposed. These include incentives to farmers and other methods consistent with the internalist viewpoint. The plan called for increased investment in agriculture but did not explicitly identify the interrelationship between natural resource management and sustainable development.

Neither the LPA, the Berg Report, nor subsequent World Bank policy documents include a comprehensive policy approach which combines economic and environmental strategies (M. T. El-Ashry, personal communication). In particular, the problem of expanding agricultural production while slowing or reversing degradation of the natural resource base has yet to be addressed within integrated development policies for the African nations. Although recent reports from the World Bank (1984, 1986b) have explicitly recognized problems of expanding human populations and natural resource degradation, appropriate policies have yet to emerge.

Our interest in these approaches to policy is primarily in pointing out what they omit. It should be noted, however, that many social scientists bitterly contest the conclusions and policy prescriptions of both the Berg report and the LPA (see, e.g., Berry, 1984; *African Studies Review,* 1984; Lawrence, 1986).

C. How Will Agricultural Production Be Increased?

Economic policy reforms are being introduced within the African continent, some voluntarily and some as a result of conditions imposed by lenders and donors. The general intent of many of these policies is to stimulate farmers to increase production by setting food and export crop prices at appropriate levels, devaluing exchange rates, reorganizing or dispensing with marketing boards, and investing considerably more money and attention in the agricultural sector. The new policies are based on the assumption that these measures will encourage farmers to increase both production and productivity through the implementation of modern farming methods and new technologies. Indeed, supporters of the free market approach appear to have almost unbounded faith in this assumption, suggesting that "given much needed incentives and realistic policy directives, physical and biological constraints are not factors that farmers are unable to transcend" [Asante (1986), citing Bates (1981) and Tarrant (1980)].

The Office of Technology Assessment of the U.S. Congress published a report emphasizing the technical and institutional factors that constrain development in the west African Sahel (OTA, 1986). One of the report's major conclusions was that "unless there is appreciable technological change [in methods of food production], environmental degradation and high population growth rates will make it increasingly difficult to reduce the region's poverty." This document reports that while programs to develop and disseminate have improved, appropriate technologies for rainfed agriculture have so far failed to improve on traditional approaches. Programs to develop irrigation have proven costly, slow, and of questionable economic vitality. Livestock and range management programs have offered little improvement over existing systems, and investment in the environment, particularly fuelwood reserves, has been low, costs have been high, and success limited.

A crucial characteristic of Sahelian ecology that was underestimated in most agricultural development efforts is the diversity of soils and microclimates. Rainfall variability is extreme, especially in arid areas. Fertility and other soil characteristics can also vary over short distances. Pest problems, too, are highly location-specific. Improved technical packages were often developed on the basis of assumed "average" rainfall and soil characteristics of a given area—a concept of little practical use when high variability means that the "average" rarely corresponds to the actual conditions faced by farmers and herders. More appropriate research and development programs require knowledge of the degree of diversity and the frequency or probability of any particular combination of environmental conditions occurring (OTA, 1986).

Poverty is the fundamental reality facing the vast majority of Sahelian farmers. By conducting a combination of agricultural activities (including migration of individual family members), farm households spread the risk rather than concentrate it. Farmers and herders in the Sahel lack cash and tend to minimize purchased inputs. These high-risk avoidance and low-input preferences in farmer decision-making have not been well understood. Many new technologies had low adoption rates because they increased risks; those seeds, practices, and animals breeds which were adopted, while providing higher yields, were more susceptible to drought, pests, and diseases than traditional approaches.

The lack of appreciation for the great diversity in production systems in the Sahel, each a response to a specific set of ecological and socioeconomic factors was a key design weakness of many livestock and crop development activities. Substantial differences exist between households, ethnic groups, and castes in access to land, labor and capital and services, gender and age group roles, cultural designations of "acceptable" occupations, traditions of communal economic activity, the importance of agriculture as opposed to other income sources, and the rights to surpluses. These differences, rather than their "average," often determine responses to proposed or available technologies.

The belief that environmental constraints will not limit future agricultural production, and that appropriate technological innovations can be implemented broadly enough to provide sustainable growth in production appear, at best, optimistic. We are unable to find any basis for supposing that the degradation of the natural resource base which supports African agriculture will not continue with increasing intensity, even if the liberal economic policies succeed in their immediate goal of increasing food production in the immediate future.

The disappointing results of technology transfer in the Sahel and elsewhere in Africa are clearly illustrated by multiple failed attempts to increase cereal crop production. Improved varieties of wheat, rice, and corn were at the heart of the Green Revolution of the 1960s in Asia and Latin America, and it was assumed that better varieties of millet and sorghum, which represent over 70% of Sahelian cereal production, would do the same for the Sahel. But success has been limited in finding varieties that perform significantly better than those already in use by farmers in any but the high rainfall areas. In particular, the failure to appreciate adequately the low fertility and other characteristics of Sahelian soils was a reason for the poor performance of many of the millet and sorghum varieties introduced from India and elsewhere (see, e.g., Ohm and Nagy, 1985).

Research directed towards improving food-crop production in Africa's humid tropics has been carried out at the International Institute of Tropical Agriculture (IITA), Ibadan, Nigeria, since 1967. The center for research into dry regions, the International Crop Research Institute for the Semi-Arid Tropics (ICRISAT), began research in the Sahel in 1981. This comparatively late start was accompanied by an initial approach which did not take account of Africa's harsh environmental constraints. The best research programs at IITA, ICRISAT, and other African research institutions are now testing crop varieties and farming techniques without— as well as with—pesticides and fertilizers, and under stresses of drought, sandstorms, poor soils, pests, and diseases. The crops are tested on actual farms as well as on controlled research sites (Harrison, 1987). A number of varieties and techniques with the potential to boost agricultural productivity under African environmental and social conditions have recently been identified, particularly with respect to soil management (Lal, 1987). While this research is encouraging, the dissemination of any advances will present a huge challenge. It would be rash to assume that such advances could rapidly be translated into agricultural production gains capable of feeding Africa's growing population, or even of significantly arresting the decline in food production per capita.

VI. THE POVERTY CYCLE AND THE WAY AHEAD

Africa's problems are complex and exhibit considerable spatial variation. Therefore it would be foolish to claim we have covered all of them by pointing to a limited number of cause and effect relationships. However, effective policies designed to improve the well-being of rural Africans must

480   A. R. E. Sinclair and Michael P. Wells

be implemented in the face of complexity and variability. We have approached this problem by drawing attention to some major trends within the African continent which have received insufficient attention from policymakers. These trends have become apparent to observers who have not limited themselves to the perspective of a single viewpoint (economist, ecologist, agriculturalist, demographer, etc.).

We have attempted to show how African agricultural systems are affected by, and in turn affect, population changes, natural resource systems, economies, and the activities of foreign development agencies. This is summarized in Figure 17.15, where solid lines represent inter-sectoral relationships which have been given priority by the academic literature and by policymakers. The broken lines in Figure 17.15 represent those relationships which, in our opinion, continue to be given minimal attention but which are fundamental to an understanding of the current crisis. We have emphasized the following features of the African crises:

1. There has been a continent-wide explosion of human populations, primarily attributable to declines in infant and child mortality since the

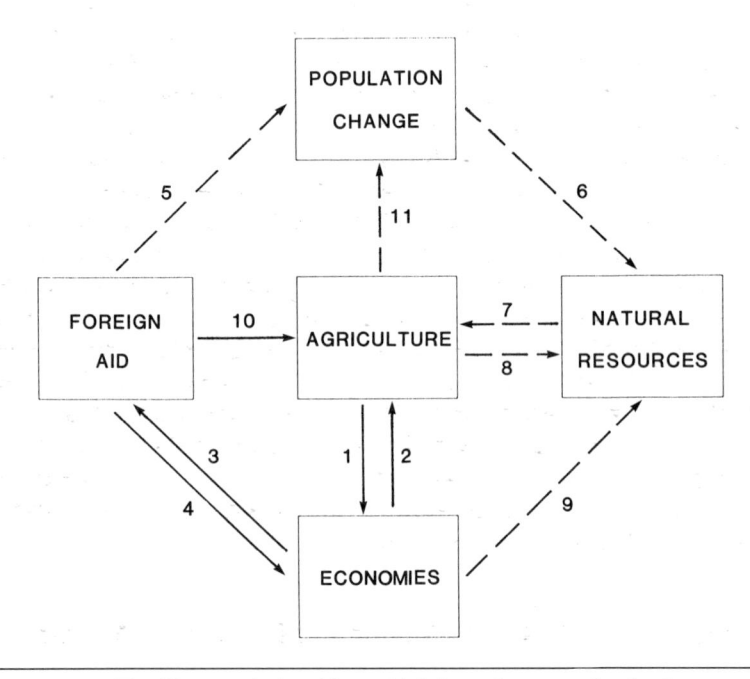

Figure 17.15 Significant relationships which have been emphasized by policymakers (solid lines) and those which have received insufficient attention (dashed lines). Numbered relationships are discussed in the text.

early 1960s. Fertility rates have remained constant and high. These features appear to be independent of nations' wealth, environments, and political ideologies.

2. The drop in childhood mortality correlates positively with female education, suggesting that knowledge of hygiene, nutrition, and sanitation may be positively influenced by education. As a health indicator, the number of people per doctor showed a weak correlation with the reduction in childhood mortality, suggesting an effect from direct health care.

3. These population increases have had serious negative consequences on African ecosystems (arrow 6, Fig. 17.15). Overpopulation has lead to overgrazing of rangelands, deforestation, soil erosion, reduction of soil fertility through overcropping and loss of nutrients, and overharvesting of fuelwood. These problems feed back directly onto the rural poor, forcing them further to degrade the ecosystems upon which they depend.

4. The economic situation facing African nations has declined since the early 1970s, due to both external and internal influences, triggering a debt crisis. The desperate need to increase foreign exchange earnings has stimulated the expansion of export crop production into marginal lands, former forests and former grazing lands (arrows 1, 2, 7, and 8).

5. The natural resource base has been directly impacted through the overexploitation of forests for timber and fuelwood, for which expensive imported fuel cannot be substituted (arrow 9).

6. Declining agricultural production (both food and cash crops) affected the economics of African countries, and this in turn has influenced the nature of aid provided by foreign lenders and donors (arrow 3). In various forms the aid policies were designed to stimulate the economy of African countries directly (arrow 4) (e.g., the "trickle down" hypothesis) or to help agricultural production through innovations (e.g., the Green Revolution) or mechanization (arrow 10).

7. In many cases foreign aid to agriculture has been ill-advised and led to unforeseen problems. It has caused breakdowns of agricultural systems (e.g., when farm machinery no longer runs), loss of soil fertility (e.g., use of plant varieties that require high soil nutrient content when no fertilizer is available) and, eventually, declines in food production. Thus the population now grows less food for itself and more for export than it once did (arrow 11). In short the population has become progressively vulnerable.

8. This completes the cycle of poverty. It involves aspects of ecology, agricultural production, the economy, and foreign aid, producing a positive feedback loop to the population.

9. The effects of falling food production per capita are not apparent in the demographic data sets (one would expect higher infant mortality, lower fertility), at least partly because they are buffered by the effect of foreign aid (arrow 5). Food aid directly mitigates starvation in the short term, and this aid is applied continuously in some countries. Other aid in the form of health care, sanitation, and education appears to have the consequence of lowering childhood mortality. This reduction of mortality may have triggered the cycle described above by causing the population explosion.

10. Many of Africa's problems—ecological, agricultural, and economic—are clearly linked to the recent population increase. To address these problems without simultaneously tackling the population problem would appear naive at best. In particular, it would seem futile to continue to direct huge development resources to increase food production for an increasing population, for it is axiomatic that the population will always catch up with the supply. Increasing food production is a strategy which buys time; it is not a long-term solution.

11. Historically, development policies have been based on a perception that saw Africa's problems as those indicated by the solid arrows in Figure 17.15, namely an interaction between economic factors and agriculture (food and cash crops), and that the problems could be solved by policies directed at these two areas. The ecological problems have been seen as a separate issue and the population changes were either considered to be incidental or too difficult to handle politically. In short, international development institutions and the African governments did not give appropriate emphasis to the interactions indicated by the broken arrows in Figure 17.15; and, without an understanding of these connections, the positive feedback loop stimulating population increase while exacerbating impoverishment—the poverty cycle—goes unnoticed.

There are many serious consequences of this increasing poverty and human destitution. One is that the unique natural habitats, wildlife, and other indigenous species of Africa will be steadily lost; extinctions will occur on a massive scale. Finally, we note that we were unable to detect any negative consequences of the ecological decline on rates of population increase. We predict that such negative consequences will become increasingly visible in the near future unless effective measures reduce the population increase.

ACKNOWLEDGMENTS

We thank M. T. El-Ashry, C. S. Holling, D. Pimentel, R. Powell, and B. H. Walker for helpful critiques of our manuscript. Our work was funded by the Canadian Natural Sciences and Engineering Research Council and the New York Zoological Society (to A. R. E. S.), and by the University of British Columbia and the International Development Research Centre, Ottawa (to M. P. W.).

REFERENCES

Adams, P. (1985). "In the Name of Progress: The Underside of Foreign Aid." Doubleday/Energy Probe, New York.
African Studies Review. (1984). Vol 27, No. 4.
Asante, S. K. B. (1986). Food as a focus of national and regional policies in contemporary Africa. In "Food in Sub-Saharan Africa" (A. Hansen and D. E. McMillan, eds.), pp. 11–24. Lynne Rienner, Boulder, Colorado.
Barr, T. N. (1981). The world food situation and global grain prospects. *Science* **214**, 1087–1095.
Bates, R. H. (1981). "Markets and States in Tropical Africa: The Political Basis of Agricultural Policies." Univ. of California Press, Berkeley.
Berry, S. (1984). The food crisis and agrarian change in Africa: a review essay. *Afr. Stud. Rev.* **27**, 59–112.
Breman, C. T., and DeWit, C. T. (1983). Rangeland productivity and exploitation in the Sahel. *Science* **221**, 1341–1347.
Brown, L. R. (1987). Analyzing the demographic trap. In L. R. Brown et al., "State of the World 1987," pp. 3–19. Norton, New York.
Brown, L. R., and Wolf, E. C. (1986). Reversing Africa's decline. In L. R. Brown et al., "State of The World 1986," pp. 177–194. Norton, New York.
Brown, L. R., Chandler, W. U., Flavin, C., Pollock, C., Postel, S., Starke, L., and Wolf, E. C. (1986). "State of The World 1986." Norton, New York.
Brown, L. R., Chandler, W. U., Flavin, C., Jacobson, J., Pollock, C., Postel, F., Starke, L., and Wolf, E. C. (1987). "State of The World 1987." Norton, New York.
Browne, R. S., and Cummings, R. J. (1984). "The Lagos Plan vs. the Berg Report." Afr. Stud. Res. Program, Howard Univ., Washington, D.C.
Cassen, R, ed. (1986). "Does Aid Work?" Oxford Univ. Press, London and New York.
Chandler, W. U. (1986). Investing in children. In L. R. Brown et al., "State of The World 1986," pp. 159–176. Norton, New York.
de Wilde, J. C. (1984). "Agriculture, Marketing and Pricing in sub-Saharan Africa." Afr. Stud. Cent. and Crossroads Press, UCLA, Los Angeles, California.
Dumont, H. J. (1978). Neolithic hyperarid period preceded the present climate of the central Sahel. *Nature (London)* **274**, 356–358.
Ecologist. (1985). Vol. 15, Nos. 1 and 2.
FAO. (1986). "African Agriculture: The Next 25 Years." FAO, Rome.
Franke, R. W., and Chasin, B. H. (1980). "Seeds of Famine." Allenheld, Osmun, Montclair, New Jersey.
Fryxell, J. M., Greever, J., and Sinclair, A. R. E. (1988). Why are migratory ungulates so abundant? *Am. Nat.* (in press).
Grainger, A. (1982). "Desertification: How People Make Deserts, How People Can Stop and Why They Don't." Earthscan, London.
Griffin, K. (1987). "World Hunger and The World Economy." Macmillan, New York.
Grigg, D. (1985). "The World Food Problem 1950–1980." Blackwell, Oxford.
Harrison, P. (1987). "The Greening of Africa." Penguin Books, New York.
Holling, C. S. (1973). Resilience and stability of ecological systems. *Ann. Rev. Ecol. Syst.* **4**, 1–23.

Horowitz, M. M. (1986). Ideology, policy, and praxis in pastoral livestock development. *In* "Anthropology and Rural Development in West Africa" (M. M. Horowitz and T. M. Painter, eds.), pp. 251–272. Westview Press, Boulder, Colorado.
Lal, R. (1987). Managing the soils of sub-Saharan Africa. *Science* **236,** 1069–1076.
Lawrence, P., ed. (1986). "World Recession and the Food Crisis in Africa." Curry, London.
Lofchie, M. F. (1986). Africa's agricultural crisis: an overview. *In* "Africa's Agrarian Crisis: The Roots of Famine" (S. K. Commins, M. F. Lofchie, and R. Payne, eds). Lynne Riennar, Boulder, Colorado.
Lofchie, M. F. (1987). The decline of African agriculture: an internalist perspective. *In* "Drought and Famine in Africa: Denying Famine a Future" (M. H. Glantz, ed.). Cambridge Univ. Press, London and New York.
Mellor, J. W., and Gavian, S. (1987). Famine: causes, prevention, and relief. *Science* **235,** 539–545.
Mellor, J. W., Delgado, C., and Blackie, M. J., eds. (1986). "Accelerating Food Production Growth in Sub-Saharan Africa." Johns Hopkins Press, Baltimore, Maryland.
Nicholson, S. E. (1986). Climate, drought, and famine in Africa. *In* "Food in Sub-Saharan Africa" (A. Hansen and D. E. McMillan, eds), pp. 107–128. Lynne Rienner, Boulder, Colorado.
Ohm, H. W., and Nagy, J. G., eds. (1985). "Appropriate Technologies for Farmers in Semi-Arid West Africa." Purdue Univ. Press, Lafayette, Indiana.
Organization for African Unity. (1981). "The Lagos Plan of Action for the Economic Development of Africa 1980–2000." Int. Inst. Labor Stud., Geneva.
OTA. (1986). "Continuing the Commitment: Agricultural Development in the Sahel." U.S. Congr., Off. Technol. Assess., Washington, D.C.
Otterman, J. (1974). Baring high-albedo soils by overgrazing: a hypothesized desertification mechanism. *Science* **186,** 531–533.
Otterman, J., Waisel Y., and Rosenberg, E. (1975). Western Negev and Sinai ecosystems: comparative study of vegetation, albedo, and temperatures. *Agro-Ecosystems* **2,** 47–59.
Sinclair, A. R. E. (1981). Environmental carrying capacity and the evidence for overabundance. *In* "Problems in Management of Locally Abundant Wild Mammals" (P. A. Jewell and S. Holt, eds), pp. 247–258. Academic Press, New York.
Sinclair, A. R. E., and Fryxell, J. M. (1985). The Sahel of Africa: ecology of a disaster. *Can. J. Zool.* **63,** 987–994.
Tarrant, J. R. (1980). "Food Policies." Wiley, New York.
Timberlake, L. (1985). "Africa in Crisis: The Causes, the Cures of Environmental Bankruptcy." Earthscan, London.
Tinker, J. (1977). Sudan challenges the sand-dragon. *New Sci.* **73,** 448–450.
United Nations. (1977). "Desertification: Its Causes and Consequences" (Secretariat, U. N., ed.). Pergamon, Oxford.
United Nations. (1985). "World Population Prospects," pp. 44–45. U. N., New York.
Wolf, E. C. (1986). Managing rangelands. *In* L.R. Brown *et al.*, "State of the World 1986," pp. 62–77. Norton, New York.
World Bank. (1981). "Accelerated Development in Sub-Saharan Africa." World Bank, Washington, D.C.
World Bank. (1984). "Toward Sustained Development in Sub-Saharan Africa." World Bank, Washington, D.C.
World Bank. (1986a). "World Development Report." World Bank, Washington, D.C.
World Bank. (1986b). "Financing Adjustment with Growth in Sub-Saharan Africa 1986–90." World Bank, Washington, D.C.
World Resources Institute and International Institute for Environment and Development (WRI/IIED). (1986). "World Resources 1986: An Assessment of the Resource Base that Supports the Global Economy." Basic Books, New York.

18

Food and Fuel Resources in a Poor Rural Area in China

Wen Dazhong

Institute of Applied Ecology
Chinese Academy of Sciences
Shenyang, China

I. Introduction
II. An Overview of Kazhou County
   A. Location and Natural Conditions
   B. Brief History of Agricultural Development
III. The Agroecosystem: Food and Fuel Production and Consumption System
IV. Energy Flows in the Kazhou Agroecosystem
V. Assessment of the Kazhou Agroecosystem
   A. Energy Inputs
   B. Harvested Biomass
   C. Food
   D. Household Fuel
VI. Strategies for Improving Food and Household Fuel Supplies in Kazhou
   A. Improving Food Supply
   B. Improving Household Fuel Supply
VII. Conclusions
References

I. INTRODUCTION

China is the most populous country in the world with a total population of 1,046 million. China has only 100 million ha of arable land for crop production to feed this enormous number of people (NSBC, 1986). On average each person has only 0.1 ha of arable land for food production, which is one-third of the world average, one-sixth of United States' average, and one-twentieth of Canada's average (He, 1986). Producing enough food to feed its population has been the prime goal of Chinese agriculture. Since 1949, grain production has steadily increased in China. In 1985, 379 million t (metric tons) of grains were produced in China, 3.4 times the amount produced in 1949 (Figure 18.1). However, the Chinese population doubled from 541 million in 1949 to 1,046 million in 1985 (NSBC, 1986).

Although grain production increased more than population growth, the annual yield of consumable grains per person increased only slightly. During the period 1952–1980, the annual yield of consumable grains per person only increased a total of 1.6 kg each year (Figure 18.1). The 362 kg of annual yield per person in 1985 in China is still less than the 1984 world average of 430 kg, and is only one-fourth of that in the United States

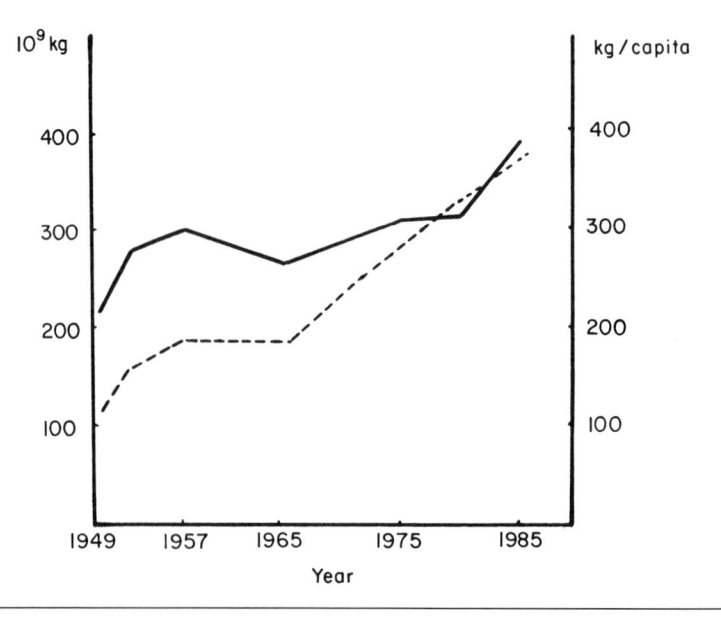

Figure 18.1 Total annual cereal grain production (dotted line) and annual average quantities of grains per capita (solid line) in China from 1949 to 1985. [Based on He (1981, 1983, 1986).]

18. Food and Fuel Resources in Rural China   **487**

(He, 1986). Meat consumption per person in China in 1983 was only 14 kg, one-eighth of that in the United States (He, 1986).

Based on the second survey of Chinese nutrition in 1982 (SGG, 1985), the daily food energy intake averages 2,484 kcal per person, and the daily protein intake is 66.7 g per person. Even though these nutrition intakes could meet the nutrient standards established by the Food and Agriculture Organization (FAO, 1973), food consumption levels in China are still lower than those in most developed countries.

Currently, 840 million Chinese people or 80% of the total population live in rural areas. Among this group, 317 million people are engaged in agriculture (He, 1986). These people produce food for their families and for the people in cities and towns. The current food consumption per person in the rural areas is 267 kg of cereal grains, 10.5 kg of meat, 1.7 kg of fish and 1.8 kg of eggs, which is lower than the level in cities and towns (He, 1986).

Because of uneven distributions of arable land, water, and other natural resources and some socioeconomic resources, about 80 million people in the rural regions, including more than 200 counties, are still living in poverty. Their annual incomes are less than 200 yuans per person (about $50), and the consumable cereal grains are less than 200 kg per person or about 1,900 kcal of daily food energy per person and 58 g of protein (Fei, 1986; Liu, 1986; Peng, 1986).

Household fuel is a basic requirement for life in China. China currently consumes 619 million t of coal equivalents of commercial or fossil energy annually, which accounts for only 7.4% of world commercial energy consumption. Commercial energy consumption per person in China is 0.61 t of coal equivalents, which is less than one-third of the world average (NSBC, 1986). However, the commercial energy consumption per person in rural China is only 0.13 t of coal equivalents, or less than one-tenth the world average (Zhang, 1985).

Most commercial energy in rural China is used for agricultural production, and little is used for household fuel. Rural people have to rely on locally produced biomass as their primary household fuel. Total biomass energy production in rural China is estimated to be about 814 million t of air-dried biomass. This provides about 3×10^{15} kcal of energy annually, about half of which is used as peasant household fuel (Table 18.1). Each peasant family consumes an average of 2.37 t or 8.5×10^6 kcal of air-dried biomass fuel annually. Biomass fuel is burned in traditional stoves with about 10–15% heat efficiency. Thus, the daily effective household energy input provided by biomass is about 3,000 kcal per peasant family. The total daily effective household energy input, including biomass and fossil energy, is about 3,500 kcal per peasant family per day. Based on a recent analysis of the needs in rural Chinese areas, an average 4,500 kcal of effective energy per peasant family per day is needed (Zhang, 1985).

Table 18.1 Annual Biomass Production and Consumption of Household Fuel in Rural China[a]

	Biomass production			Biomass fuel consumption		
Source	10^6 t (air dried)	10^{12} kcal	%	10^6 t (air dried)	10^{12} kcal	%
Crop residues	362	1160	40	240	770	49
Dung	252	959	33	13	49	3
Wood	200	798	27	180	740	48
	814	2917	100	433	1559	100

[a] Adapted from Wang (1985).

Clearly, the household fuel supply of rural China is less than 80% of current need. Because of the uneven distribution of biomass resources in rural areas, household fuel for some peasants meets only 40–50% of their needs.

It is essential to find some ways to provide peasant families with more food and fuel. This chapter is a case study of improving food and household fuel supplies in Kazhou county, one of the 200 poverty-level counties in China. The situation in the Kazhou agroecosystem was assessed by employing energy analysis techniques. Several strategies for improving food and household fuel supplies in this county are discussed.

II. AN OVERVIEW OF KAZHOU COUNTY

A. Location and Natural Conditions

Kazhou County lies between 40°46′ and 41°33′ north latitude, and between 119°27′ and 120°3′ longitude. The county is located in the western area of Liaoning province in northeastern China (Figure 18.2). The total area of the county is 2,238 km². The county is principally lower mountains and hills, and level terrain accounts for one-third of the total area. Annual precipitation is 450–500 mm, and the annual average temperature is 8.3°C. The frost-free period is 140 days. The soil of this county is of poor quality and is highly susceptible to erosion.

B. Brief History of Agricultural Development

About 200 years ago, the forest area in Kazhou County was preserved by the Ching dynasty as a restricted area, and some Mongolian herdsman utilized only the grassland. Even though soil and climatic conditions were

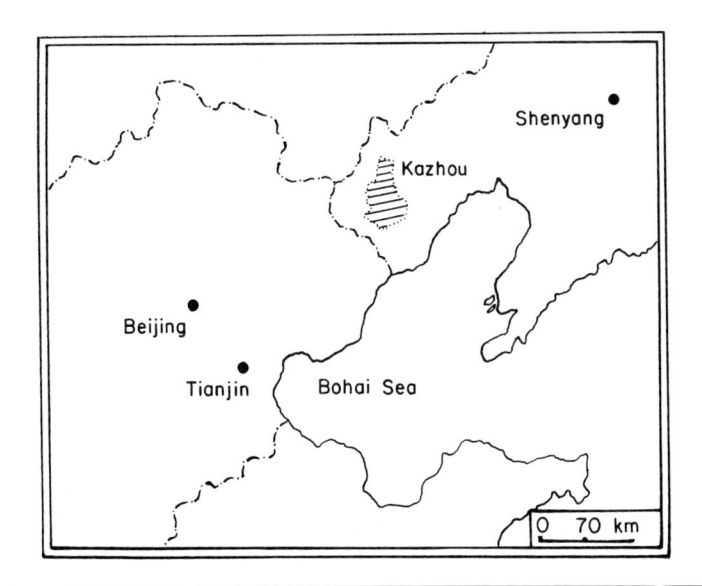

Figure 18.2 Location of Kazhou County in Northeastern China.

not ideal for vegetation growth, Kazhou was covered with forests and some grasses.

During the Chianlong Empire of the Ching dynasty (1736–1795), Kazhou became an area to station troops and their horses. This use damaged some forests (Nan and Xu, 1983). After 1829, a large number of Han people from the Central Chinese Plain immigrated to the Kazhou area. Some of the forests were cut for building and fuelwood, and some forestland and grassland areas were cleared for crops. After 1851, the restricted area was opened, more forests were cleared, and most of the grasslands were used to grow crops (Nan and Xu, 1983).

Currently, the county has 340,000 inhabitants with 70,200 families living in 236 villages. The population density is 162 persons/km^2. This dense population has had a serious impact on this fragile ecosystem. About 53% of the total area is suffering from serious soil erosion. At least 5.4 million t of soil are being washed off the area annually, equivalent to a soil loss of 24 t/ha/yr. About 1.23 million t of soil are eroded from crop fields annually, which is equivalent to a loss of 980 t of nitrogen or 15.5 kg N/ha (Zhao and Yu, 1984).

Although the residents of Kazhou have made some efforts to reduce this severe environmental degradation, serious soil erosion and frequent droughts are major threats to agricultural development. The Kazhou inhabitants still do not have enough food and fuel, and some cereal grains have to be imported into the county to support them.

III. THE AGROECOSYSTEM: FOOD AND FUEL PRODUCTION AND CONSUMPTION SYSTEM

Table 18.2 shows current land use in Kazhou county. Most of the county is being used for crop production, forests, orchards, and forage grasses. Only 6.6% of the total area is for buildings and industries, and 5.9% of the total area is unavailable barren land. The agriculture, forest, and other subsystems are examined next.

1. Crops. Crops are the major part of the Kazhou agroecosystem. The 63,000 ha of cropland accounts for 28.3% of the total area of the county (Table 18.2). About 70% of the cropland is on slopes, which contributes to heavy soil erosion and poor soil fertility. Only one-fifth of the crop area can be irrigated. During 1981–83, cereal grains accounted for 82% of total crop planting. Of the cereals, corn accounts for 30%, millet 27%, sorghum 26%, and soybeans 3%. Cash crops account for 15% of the total crop area. Of this area, cotton accounts for 34% and peanuts for 57%. Green manures are grown on 3% of the total crop area.

2. Forests. Almost no original forests exist. Most of the 63,500 ha of forest area is plantations. Pine *(Pinus tabulaeformis)* plantations account for 53%, and locust *(Robinia pseudo-acacia)* plantations account for 21% of the forest area; these are located on the slopes of most mountains and hills. Poplar *(Populus* spp.) plantations account for 9% of the forest area and are planted on river banks and flood plains. Mountain apricot *(Armeniaca sibirica)* plantations account for 11% of the forest area. Most (69%) of these plantations are young, while middle-aged plantations account for 30%, and mature plantations make up less than 1% of the total.

Table 18.2 Land Use in Kazhou County of Liaoning Province, China, 1981–1983[a]

Land type	10^3 ha	% of total area
Cropping fields	63.3	28.3
Tree plantations	63.5	28.4
Orchards	6.2	2.8
Grassland	10.0	4.4
Natural shrub-grassland	52.7	23.6
Unavailable land	13.2	5.9
Other used land	14.8	6.6
	223.7	100.0

[a] Based on Zhao and Yu (1984).

3. Orchards. A total of 6,350 ha of orchards account for 2.8% of the total land area of the county (Table 18.2). There are about 1.2 million apple trees, which are 61% of the fruit trees grown. The remaining fruit includes pear trees and grapevines.

4. Grasslands. Currently, the Kazhou people sow 10,000 ha of grass to feed their livestock and protect soil from erosion. Sweet clover and *Astraylus adsurgens* are used to establish grassland. Some poor cropland and uncultivated slopes are planted to grass.

5. Natural shrubs-grasses. About 52,800 ha of shrub-grassland make up the major natural vegetation (Table 18.2). However, overharvesting of biomass and overgrazing have seriously reduced the productivity of these natural areas.

6. Livestock. Livestock in Kazhou consists of 4,700 horses, 4,900 mules, 9,300 donkeys, 23,100 oxen, 150,700 pigs, 55,100 sheep, 53,600 rabbits, and 2.45 million poultry. Most of the horses, mules, donkeys, and oxen are used for draft animal power in agriculture.

7. Residents. About 340,000 people live in this agroecosystem, of which 86,000 are agricultural laborers. These people produce the food and are also the primary consumers.

IV. ENERGY FLOWS IN THE KAZHOU AGROECOSYSTEM

An assessment of energy flow through the Kazhou agroecosystem will give a better understanding of its structure and function. This analysis should help us design appropriate strategies for improving the system. Energy flows in the Kazhou agroecosystem include inputs into the system, outputs from the system, and flows between subsystems.

1. Inputs. Energy inputs into this system include solar energy and fossil fuel. The efficiency of converting solar energy into biomass energy in an agroecosystem depends on the management of the system. Fossil energy inputs in the form of machines, fertilizers, pesticides, etc., represent some of the management alternatives in modern agriculture. In addition to these inputs, some cereal grain and coal is imported into Kazhou for its residents.

2. Outputs. Energy outputs from this agroecosystem are in two forms. Some agricultural products are exported outside of Kazhou. Balanced against the heat energy that dissipates from Kazhou is the solar and coal energy imported into Kazhou.

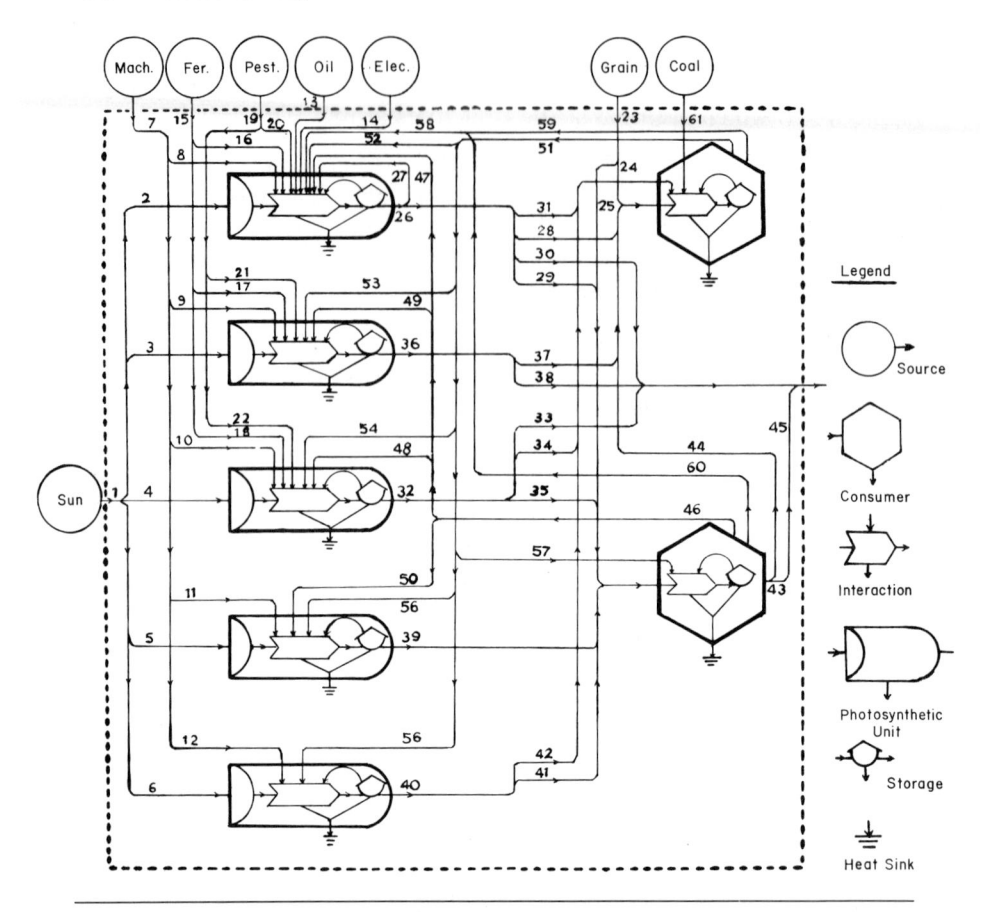

Figure 18.3 Interrelationships of the various components of the Kazhou agroecosystem and energy flows within the system.

3. Flows between subsystems. All subsystems in an agroecosystem are interconnected. For example, some grain and forage flows from these systems to livestock, and in turn draft animal power flows into cereal and forage systems. Clearly, the various systems are highly interconnected and dynamic.

For this analysis, the energy flows were calculated using data provided by the Statistics Bureau of Kazhou County in 1981–1983 and some estimates by the author. Energy flows for the Kazhou agroecosystem are shown in Figure 18.3 and Table 18.3.

Table 18.3 Average Annual Energy Flows through Kazhou Agroecosystem, 1981–1983

Item	Quantity/yr	10^9 kcal/yr	Code number in Figure 18.3
Total photosynthetically active solar radiation[a]		849,699	1
Crop fields		274,843	2
Tree plantations		275,711	4
Orchards		26,906	3
Grassland		43,396	5
Shrub-grassland		228,843	6
Total fossil energy inputs[b]		358.67	
Machinery and tools	796,112 kg	14.34	7
Crop fields	781,162 kg	14.06	8
Tree plantations	4,326 kg	0.08	10
Orchards	3,988 kg	0.16	9
Grassland	1,402 kg	0.03	11
Shrub-grassland	234 kg	0.01	12
Diesel fuel for crops	$3,097 \times 10^3$ liters	35.35	13
Electricity for irrigation	$4,730 \times 10^3$ KWh	13.54	14
Synthetic fertilizers	4,257.88 t	49.29	15
Nitrogen			
For crops	3,737.59 t	44.85	16
For plantations	41.69 t	0.50	18
For orchards	290.00 t	3.48	17
Phosphorus for crops	113.60 t	0.34	16
Potassium for crops	75.00 t	0.12	16
Insecticides and herbicides	190.00 t	16.51	19
For crops	182.25 t	15.84	20
For tree plantations	3.75 t	0.32	22
For orchards	4.00 t	0.35	21
Coal for household fuel	46,000 t	230.00	61
Cereal grain input to the system[c]	28,110 t	109.63	23
Consumed by humans	22,000 t	85.80	24
Fed to animals	6,110 t	23.83	25

Table 18.3 *(continued)*

Item	Quantity/yr	10⁹ kcal/yr	Code number in Figure 18.3
Total harvested biomass		903.46	
Harvested crop biomass		548.00	26
Cereal grains[c]	54,630 t	213.06	27
For seeds	2,815 t	10.98	28
For food	31,740 t	123.79	29
For animal feed	20,075 t	78.29	
Products of cash crops and other crops[c]		22.58	30
Exported		15.22	28
For human use		7.36	
Harvested crop residues[d]	47,763 t (air dried)	312.36	31
For household fuel		167.17	29
For animal feed		132.73	27
Used as organic fertilizers		12.46	32
Harvested biomass, tree plantations[e]		221.50	33
For household fuel	32,596 t (air dried)	130.38	33
Thinning timber for local use	13,000 m³	29.42	34
Apricot seed output	51.63 t	0.26	35
Tree leaves for animal feed		61.44	36
Harvested biomass from orchards[c]	6,654 t (wet)	3.79	37
Fruits for local residents	1,901 t (wet)	1.08	38
Fruit output from the system	4,753 t (wet)	2.71	
Harvested biomass from grassland	10,526 t (air dried)	40.00	39
For animal feed[f]	23,730 t (air dried)	90.17	40
Harvested biomass from shrub-grassland	19,351 t (air dried)	73.53	41
For animal feed	4,379 t (air dried)	16.64	42
For household fuel			
Meat and eggs produced[c]	5,818.80 t	24.68	43
For food	1,924.40 t	8.57	44
Exported	3,894.40 t	16.11	45

Using animal power[g]		46
To crop fields	48.40 × 10⁶ hr	89.60 47
To tree plantations	45.54 × 10⁶ hr	84.29 48
To orchards	0.77 × 10⁶ hr	1.43 49
To grassland	1.79 × 10⁶ hr	3.32 50
Using human labor	0.30 × 10⁶ hr	0.56 51
To crop fields	206.84 × 10⁶ hr	37.05 52
To orchards	126.00 × 10⁶ hr	22.57 53
To tree plantations	12.56 × 10⁶ hr	2.25 54
To grassland	10.60 × 10⁶ hr	1.90 55
To shrub-grassland	1.47 × 10⁶ hr	0.26 56
For raising animals	2.25 × 10⁶ hr	0.40 57
Human and animal manure to crop fields	53.96 × 10⁶ hr	9.67 58
Human manure[h]	43,598 t (dry)	169.91 59
Animal manure[i]	17,212 t (dry)	68.86 60
	26,386 t (dry)	101.05

---

[a]The physiological solar radiation during the growing season in Kazhou county is 43.4 kcal/cm² (Zhao and Yu, 1984).

[b]The quantities are based on statistical data provided by the Statistics Bureau of Kazhou county. The following energy values were used for calculating energy flows: machinery and tools, 18,000 kcal/kg; diesel fuel, 11,414 kcal/l; electricity, 2,863 kcal/KWh; nitrogen fertilizer, 12,000 kcal/kg; phosphorus fertilizer, 3,000 kcal/kg; potassium fertilizer, 1,600 kcal/kg; pesticides and herbicides, 86,910 kcal/kg (Pimentel, 1980).

[c]Based on statistical data provided by the Statistics Bureau of Kazhou county.

[d]Estimated from grain and cash crop products.

[e]Estimated from the survey report of Kazhou forestry (Lei, 1983).

[f]Based on Zhao and Yu (1984).

[g]It is estimated that one draft animal worked 200 days each year and 8 hr each day.

[h]It is estimated that about 20% of total human manure was lost.

[i]Estimated from the difference between the total feed energy and the total digestible feed energy.

495

V. ASSESSMENT OF THE KAZHOU AGROECOSYSTEM

A. Energy Inputs

The annual solar incident radiant energy in Kazhou County is 140.6 kcal/cm^2. The annual physiological solar incident radiant energy with the wavelength of 0.38–0.71 μ is 68.8 kcal/cm^2, and the physiological solar incident radiant energy just during the growing season (April to September) is 43.4 kcal/cm^2 (Zhao and Yu, 1984). It is estimated that the maximum crop photosynthetic efficiency for solar energy could be more than 10% (Deng and Feng, 1981). If 2% efficiency was reached in Kazhou County, the corn yield would be 8,000 kg/ha, or eightfold the current cereal grain yield in the county. Obviously, solar energy in Kazhou County is beneficial to agriculture.

About 359×10^9 kcal of fossil fuel energy were imported into Kazhou annually. The household fossil fuel accounted for 64% of the total fossil fuel use, while the remainder (36%) was for agricultural production. Of this 36%, farm machinery and tools accounted for 12%, oil 28%, electricity 12%, fertilizer 38% and insecticides 13%. Most (86%) of the fossil energy used for agriculture was used in crops. The fossil energy input into the crop systems was 1.97 Mkcal/ha, which is 77% of the national average in China (Wen, 1986, 1987).

B. Harvested Biomass

The total annual harvested biomass from the Kazhou agroecosystem was 903×10^9 kcal (Table 18.3). About 61% of the total harvested biomass came from crops, 25% from the forest system, and the remainder from the rest of the system. Thus, although the crop area accounted for one-third of the total land area, the harvested biomass from crops accounted for nearly two-thirds of the total harvested biomass. Obviously, the crop system is an important component of the agroecosystem.

1. Crop Yields

Total annual harvested biomass from the crop system was 548×10^9 kcal (Table 18.3); 43% of this total was cereal grains and other economic products, while the remainder was crop residues. Cereal grain energy accounted for 90% of the total crop products. However, cereal grain yield was only 1,052 kg/ha, one-third of the national average grain yield per hectare. Clearly, Kazhou county has one of the lowest grain yields in China.

2. Forest Production

Under the environmental and management conditions at Kazhou, forests grow slowly. The aboveground net productivity in pine plantations was only 5.7 Mkcal/ha/yr or 1,424 kg (air-dried biomass)/ha/yr with 1.5 m^3/ha/yr of timber volume increment. The plantation of locust, which is considered a fast-growing tree, produced only 10.8 Mkcal/ha/yr or 2,699 kg air-dried biomass/ha/yr with 1.9 m^3/ha/yr of timber volume increment. Fuelwood plantations accounted for only 4% of the total forest area. Because of fuelwood shortages, the residents have to overprune and collect almost all forest litter to meet their fuel needs. However, overpruning and litter collection destroy the ecological functions of plantations as well as reducing soil and water quality and forest productivity. About 28% of the total area of the county has been covered with plantations, but the poor quality environment reduced their productivity.

3. Orchard Yields

A total of 3.8 × 10^9 kcal or 6,554 t of fruit were harvested from the orchards annually (Table 18.3). Apple yield was only 1,057 kg/ha/yr, which is only one-third of the national average yield. Poor management was the major reason for low fruit yield.

4. Biomass Production of the Natural Shrub-Grass Systems

A total of 90 × 10^9 kcal or 23,730 t of air-dried biomass were harvested from 53,000 ha of natural shrub-grass areas each year. The natural shrub-grassland acccounted for 23% of the total area of the agroecosystem, but produced 10% of the total harvested biomass in the agroecosystem. Thus, the yield was only 450 kg/ha of air-dried biomass. Most of this biomass was used as animal feed and household fuel.

Cultured grass produced only 4% of the total biomass in Kazhou. More livestock feed could be produced if more cultured grassland were established.

C. Food

1. Animal Feed

The agroecosystem supplied a total of 346 × 10^9 kcal of animal feed (Table 18.3). Some imported cereal grains were also fed to livestock. Thus, the total animal feed provided to livestock was 370 × 10^9 kcal. It is estimated that about 300 × 10^9 kcal of digestible energy was needed for the livestock, but current production provided only 233 × 10^9 kcal. Thus, the livestock were fed 13% digestible energy suitable for humans.

A total of 24.7×10^9 kcal of animal products were produced (Table 18.3). The output:input ratio is 0.06:1. A large amount of the feed was consumed by draft animals, which produced no animal products.

2. Food for Residents

The people of Kazhou consumed 227×10^9 kcal of food energy each year (Table 18.3), or 1,800 kcal/day per person. About 92% of total food energy came from cereal grains, 4% from animal products, and 4% from vegetables and fruits. About 62% of total food energy was produced in the agroecosystem, and the remainder (38%) was imported into the agroecosystem. According to the Chinese standards of food consumption set by the Health Institute of the Chinese Academy of Medical Science (1981), the average daily consumption of food energy per person in China is about 2,400 kcal. The food energy consumption of Kazhou residents was only 1,800 kcal/day per person or three-fourths of the Chinese average. Thus, the Kazhou agroecosystem supplied only 47% of the needed 2,400 kcal/day per person.

D. Household Fuel

The agroecosystem produced about 314×10^9 kcal of biomass energy to be used as household fuel, which accounted for one-third of the total harvested primary biomass. About 52% of the biomass household fuel came from crop residues, 41% from branches and leaves of plantations, and 7% from other sources (Table 18.3). Each peasant family (average of five persons) burned only 4.19 Mkcal of biomass fuel each year. According to the recent living standards in rural China, the annual household fuel consumption averages 11.0 Mkcal/yr per peasant family (Zhang, 1985). In general, the peasant families in northeastern China need about one-third more fuel than the national average because of more heating energy required during the longer winter season. Thus, the Kazhou agroecosystem supplied only 30% of the needed household fuel.

About 46,000 t of coal produced in some small coal mines in or near the county were consumed for household fuel each year. On average, 600 kg or 3.0 Mkcal/yr per peasant family of coal were consumed. Thus, the total household fuel consumption per peasant family was 7.19 Mkcal/yr, which is only about half of their needed household fuel.

Because of the serious fuel shortage in Kazhou, the residents must reduce cooking to about once a day in the summer and live in rooms with low temperatures in the winter. The unrefrigerated food and cool conditions could harm people's health. In order to obtain biomass fuel, Kazhou residents have to collect as much biomass from the agroecosystem as is available, causing serious soil erosion and water degradation problems.

VI. STRATEGIES FOR IMPROVING FOOD AND HOUSEHOLD FUEL SUPPLIES IN KAZHOU

A. Improving Food Supply

Improving the food supply is the major need of Kazhou. One approach would be to import food from other regions of China. However, transporting large amounts of food from other agricultural regions would be very costly, and the people do not have funds to pay for the imports. Producing more food in the agroecosystem itself would be the best way to improve the supply. Based on the above analysis, the following approaches might be considered:

1. *Reduce domestic animal numbers.* A large amount of nutritious biomass is being lost when primary biomass is fed to livestock and converted into secondary biomass. Thus, one means of improving food supply (grains) for the Kazhou people would be to feed fewer cereal grains to livestock and more to humans. For example, the total amount of cereal grain fed to hogs was 11.2 million kg (Table 18.3). If all of this grain were used by humans instead of fed to hogs, the total food calorie consumption could be raised from 1,800 kcal/capita/day to 2,150 kcal/capita/day.

Note, about two-thirds of total pork production, or 3.0 million kg of pork, were exported outside of the Kazhou system. It is estimated that producing 1 kg of pork requires at least 2.5 kg of edible cereal grains plus some forage. Feeding a few pigs roughage and no grains would improve human food supplies.

2. *Stabilize industrial energy inputs and develop improved dry-land agricultural techniques.* Obviously, increasing crop yields in the crop system will be the basic way to improve food supply to the residents in Kazhou. How can this be done? One approach would be effective use of industrial energy. For example, the average use of nitrogen fertilizer in this system was 59 kg/ha, which was about 80% of the Chinese national average. However, producing 1 kg of cereal grains in the crop system consumed 0.056 kg of nitrogen fertilizer or twice as much nitrogen as required on average in China (Wen, 1986, 1987). Clearly, cereal grain production in the Kazhou was much lower than it should be.

Figure 18.4 shows the fluctuations in annual crop yields and fertilizer use during the last 25 years in Kazhou. No close correlation between crop yields and fertilizer use is apparent. For example, fertilizer use in 1978 was the same as that in 1983, but the cereal grain production in 1978 was double that in 1983. Also, the fertilizer consumption in 1976 was almost the same as that in 1981, but the cereal grain production in 1976 was threefold that in 1981.

In other parts of China there is a close correlation between fertilizer

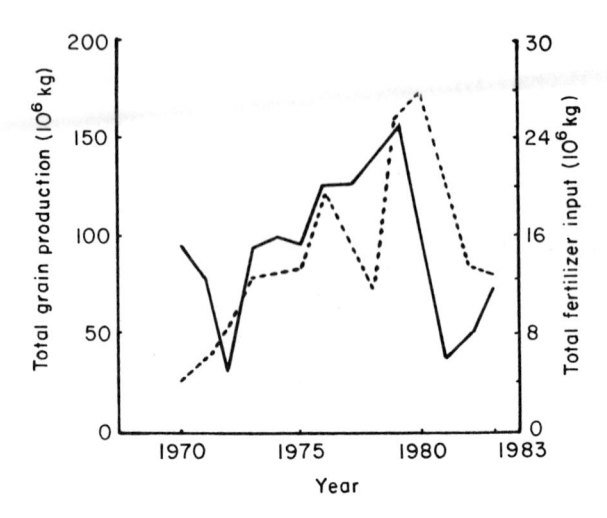

Figure 18.4 Total annual cereal grain production (solid line) and total annual synthetic fertilizer inputs in commodity quantities (dotted line) in Kazhou agroecosystem from 1970 to 1983. Based on data provided by the Statistics Bureau of Kazhou County.

use and increased cereal grain yield (Wen and Pimentel, 1984). In Kazhou this is not true. Increasing fossil energy inputs including fertilizers in Kazhou crop production in recent years did not appear to increase grain production, so they obviously were not the limiting factor. It appears that climatic fluctuation, especially reduced rainfall during certain years, was the cause of reduced grain yields, and high yields were due to heavy rainfall.

These data suggest that instead of increasing the inputs of commercial fertilizers, pesticides, and fossil energy inputs it would be more profitable to improve water and soil management practices to conserve moisture for crop use. Earlier it was mentioned that soil erosion was a serious problem in Kazhou; therefore, sound soil and water conservation techniques should be implemented. Water loss and shortages are the primary cause of reduced crop yields when severe soil erosion takes place (Pimentel et al., 1987).

Thus, grain yields in dry-land agriculture in Kazhou could probably be increased if sound soil and water conservation practices were implemented. Increasing the level of organic matter in the soil would also increase the water-holding capacity of the soil. Conserving soil and water and increasing soil organic matter will require that crop residues remain on the land and that they not be burned as fuel. Thus, the people will have to decide whether they want biomass fuel (crop residues) or food.

B. Improving Household Fuel Supply

As mentioned earlier, there is a shortage of household fuel in Kazhou. A few small coalpits in the county have limited production but not enough to meet local fuel requirements. Also, it is impossible to import a large amount of fossil fuel from other areas into Kazhou because of a lack of transportation and especially because of a lack of funds in this poor county. Thus, the only way to increase household fuel will be to develop a productive biomass fuel system and effectively use these biomass resources.

1. Improving Cooking and Heating Stoves

Traditional cooking and heating stoves (that have efficiencies of only 10–15%) are still used in Kazhou County. These inefficient stoves waste enormous amounts of heat energy and aggravate the household fuel shortage. Using more efficient stoves instead of these traditional stoves would improve the household fuel situation. For example, with the use of new stoves with 25% of heat efficiency instead of 10–15%, about 7.5 Mkcal/yr/family of fuel would meet the residents' needs instead of 14 Mkcal/yr/family. Thus, the current household fuel supply in the system would almost meet these needs. Efforts are currently being made to improve cooking and heating stoves in Kazhou.

2. Developing Fuelwood Plantations

Using crop residues for fuel reduces crop yields, as indicated previously. Effective use of these residues will reduce the need for fertilizers and improve soil water holding capacity. It might be more profitable to maintain soil quality and gain stable land productivity by employing sound soil and water conservation practices rather than purchasing fertilizers and pesticides. Then only about 4.5 Mkcal/yr/family in biomass fuel would be needed to be produced from trees. It is estimated that the total aboveground biomass yield of the Kazhou plantation system is 519×10^9 kcal/yr (Table 18.4), which corresponds to 7 Mkcal/yr of biomass per peasant family. The aboveground biomass yield in the plantation system is more than the 4.5 Mkcal/yr/family needed, assuming that about 2.5 Mkcal/yr/family in crop residues is replaced.

Recent figures indicate that only 43% of the total aboveground biomass yield in forests was harvested and 25% of this biomass was used as fuel. The other aboveground biomass in the forests was stored in the stands themselves. The pine plantations accounted for 60% of the total forest plantation area. However, the average timber volume increment of the pine plantations was 1.45 m^3/ha/yr. The locust plantation area accounted for 23% of the total plantation area, and the average timber volume in-

Table 18.4 Annual Aboveground Biomass Yield in Kazhou Plantations[a]

Plantation	Area (ha)	Boles	Branches	Leaves	Total (10^9 kcal)
Pine	33,647	101.38	33.80	56.32	191.50
Locust	13,140	75.09	25.03	41.72	141.84
Poplar	5,573	75.24	25.08	41.32	141.64
Other	4,087	23.36	7.79	12.98	44.13
	56,447	275.07	91.70	152.34	519.11

Annual aboveground biomass yield (10^9 kcal)

[a] Estimated based on annual timber growth data of plantation (Lei, 1983) and the proportions of boles, branches, leaves, and roots in a tree calculated as the net productivity in temperate zone (Dazheng, 1986).

crement of the locust plantations was 1.9 m³/ha/yr. In fact, these plantations could not be expected to have higher yields on seriously eroded hillsides. If the pine and locust plantations were both converted into fuelwood plantations, they could produce 5 Mkcal/yr/family of fuelwood. This plus the coal and other biomass supply could increase total household fuel to 8.5 Mkcal/yr/family, which is more than the 7.5 Mkcal/yr/family of fuel needed with improved stoves.

It would be easy to convert these pine and locust plantations into fuelwood plantations. First, replacing pine trees on slopes with *Hippophae rhamnoides, Lespedeza bicolor,* and other legume or nonlegume nitrogen-fixing shrubs would eventually establish shrub fuelwood plantations. Harvesting locust once every three years will convert the locust plantation into a fuelwood plantation.

Through these plantation reforms, the people in Kazhou will have about 30,000 ha of land area or 15% of the total area of the system will be new fuelwood plantations.

From the viewpoint of improving the environment, establishing fuelwood plantations on the slopes would be more effective in protecting the land from erosion than the current timber plantations. Once fully established, good timber plantation cover will control soil erosion, but fuelwood trees are planted densely and thus are more effective than timber plantations in controlling erosion.

Because of the current fuelwood shortages, most of the leaves, branches, and litter in timber plantations are harvested as fuel. This practice greatly reduces the growth rate and soil-protecting function of the timber plantations. If more fuelwood plantations were established, more fuelwood would be produced. The densely planted fuelwood plantations effectively protect the soil from erosion on the slopes. Several studies report that the soil erosion rates on slopes covered with shrub fuelwood plantation reduce erosion 75–86% compared with erosion on bare slopes (Pan and Zhao, 1983).

The proposed strategies for improving food and household fuel supplies in Kazhou would help the people fully and effectively use their natural resources. Efforts are needed to implement these proposed strategies. Suitable technical and management practices and methods will help guarantee that the proposed strategies will be effective.

VII. CONCLUSIONS

People in some poor areas of China face food and fuel shortages because of dense populations and poor environmental resource management. About 200 years ago, Kazhou County was inhabited only by roving Mongolian herdsmen. This small nomadic population generally had little impact on the ecosystem. After the population density increased and the forests were cut and a major portion of the land was converted to agriculture, serious soil erosion and land degradation problems developed. Currently, severe food and fuel shortages exist, and several policies are needed to help improve food and fuel supplies.

Effective population control will relax the pressure on the ecosystem. Currently, the population of China is 1,046 million (NSBC, 1986) and is expected to reach more than 1,200 million by the year 2000 despite the full adoption of the "one child per couple" program. This will result in further population impacts. Some regional development strategies for effective use and protection of natural resources are needed to increase the supply of food and fuel.

The first step for regional development in areas like Kazhou with environmental degradation and shortages of food and fuels is to find ways to improve soil, water, energy, and biological resource management. Improving environmental resource management should help increase food and fuel supplies. The people in Kazhou have made great efforts to establish timber plantations on eroded slope land during the past three decades in order to conserve water and soil from erosion and produce timber. However, because of the household fuel shortage, people have harvested leaves and branches from the tree plantations. As a result, the plantations have grown slowly and been less effective for water and soil conservation. Thus, any regional development strategy in Kazhou and similar regions should combine economic and social considerations with ecological management considerations.

In recent decades, the destruction of forests and pastures to extend cropland to produce more cereal grains in some regions of China has led to some natural resource degradation and environmental problems. It is necessary to adapt crop, forest, and grass production to each environmental site. This rational use of natural resources and protection of the environment will help improve food and fuel supplies.

Human and other biological and physical/chemical components in a given county constitute a complex ecological and economic system. Any proposed strategy for improving food and fuel supplies and for economic development would influence the whole system. Thus, it is essential to assess the total system comprehensively before recommendations are made. For example, an energy analysis of the Kazhou agroecosystem helped make such a total system analysis to make better use of resources to improve food and fuel supplies.

Because natural and social conditions vary from one region to another in China, no common model for improving food and fuel supplies can be used for the entire country. However, some basic principles mentioned above should be followed to be sure that the specific strategies for improving food and fuel supplies in a given region will be effective. Rational strategies for improving food and fuel supplies for poor counties like Kazhou can be made through careful analyses and studies.

REFERENCES

Dazheng, Z., ed. (1986). "Forestry." Chin. For. Press, Beijing. (In Chin.)

Deng, G., and Feng, X. (1981). Solar energy resource and photosynthetic productivity. In "The Comprehensive Survey Reports of Taoyuan County" (The Comprehensive Survey Team of Taoyuan County, ed.), pp. 186–191. Changsha. (In Chin.)

Fei, X. (1986). About concept, reasons, and future of poverty regions. *Agric. Modern. Res. (Nongye Xiandaihua Yanjou)* **6,** 1–4. (In Chin.)

Food and Agriculture Organization (FAO). (1973). "Energy and Protein Requirements," FAO Nutr. Meet. Rep. Ser., No. 52. FAO U. N., Rome.

He, K., ed. (1981). "Agricultural Almanac of China in 1980." Agric. Press, Beijing. (In Chin.)

He, K., ed. (1983). "Agricultural Almanac of China in 1982." Agric. Press, Beijing. (In Chin.)

He, K., ed. (1986). "Agricultural Almanac of China in 1985." Agric. Press, Beijing. (In Chin.)

Health Institute of the Chinese Academy of Medical Science. (1981). "Food Nutrition." Peoples Sanit. Press, Beijing. (In Chin.)

Lei, C. (1983). "Report of Forestry Survey in Kazhou County." Compr. Surv. Team Kazhou County, Liaoning. (In Chin.)

Liu, X. (1986). A summary of some viewpoints in Conference of Economic and Cultural Development of Poverty Regions in China. *Prob. Agric. Economy (Nongye Jingji Wenti)* **10,** 58–61. (In Chin.)

Nan, Y., and Xu, W. (1983). "Report of Vegetation Survey of Kazhou County." Compr. Surv. Team Kazhou County, Liaoning. (In Chin.)

National Statistics Bureau of China (NSBC). (1986). "Statistics Almanac of China in 1985." Chin. Stat. Press, Beijing. (In Chin.)

Pan, M., and Zhao, J. (1983). The effects and cultivation techniques of *Caragana korshinskii*. *Chin. Water Soil Conserv. (Zhonggue Shuitu Baochi)* **4,** 59–61. (In Chin.)

Peng, F. (1986). Developing agriculture in poverty mountain regions in China. *Agro-Tech. Economy (Nongye Jishu Jingji)* **4,** 7–9. (In Chin.)

Pimentel, D., ed. (1980). "Handbook of Energy Utilization in Agriculture." CRC Press, Boca Raton, Florida.

Pimentel, D., Allen, J., Beers, A., Guinand, L., Linder, R., McLaughlin, P., Meer, B., Musonda, D., Perdue, D., Poisson, S., Siebert, S., Stoner, K., Salazar, R., and Hawkins, A. (1987). World agriculture and soil erosion. *BioScience* **37,** 277–283.

Study Group of Grain-Economy, Chinese Agricultural Academy (SGG). (1985). Some considerations of improving Chinese food composition. *Agro-Tech. Economy (Nongye Jishu Jingji)* **8,** 34–36. (In Chin.)

Wang, C. (1985). The outline of energy resources in China. *Energy Inf.* **11,** 1–4. (In Chin.)

Wen, D. (1986). Energy intensifying in agriculture of China. *Agric. Modern. Res. (Nongye Xiandaihua Yanjiou)* **6,** 28–30. (In Chin.)

Wen, D. (1987). Energy intensification of Chinese agroecosystem and its improvement: I. The analysis of energy intensification of Chinese agroecosystem. *J. Ecol. (Shengtaixue Zhazi)* **3,** 1–5. (In Chin.)

Wen, D., and Pimentel, D. (1984). Energy use in crop systems in northeastern China. *In* "Food and Energy Resources" (D. Pimentel and C. W. Hall, eds.), pp. 91–120. Academic Press, New York.

Zhang, C. (1985). Saving and developing rural energy. *In* "Agricultural Almanac of China in 1984" (K. He, ed.), pp. 374–375. Agric. Press, Beijing. (In Chin.)

Zhao, L., and Yu, Y. (1984). "Report of Agricultural Division of Kazhou County." Kazhou Off. Agric. Div., Liaoning. (In Chin.).

Index

A

Accessibility of food, 433
Accidents, chemical, 211
Acidification, 113
Adaptive agriculture, 51, 58
African crisis, 480, 481, 482
African paradox, 440
Agrarian structure, 350
Agriculture
  chemicals, 193, 194
  gene-based, 51
  history, 76, 410
  history, China, 488, 489
  stages of development, 76, 77
  system, 303
  traditional, 403
  transcontinental, 404
  transitional, 404
Agroecosystems, 195, 196, 205, 206, 302, 303, 316, 490, 491, 497
Agroforestry, 180
Agroindustrial complex, 242
Air pollution, 378, 379
Alkalization, 109
Amelioration, 107
Amino acid contents, 415, 416
Anabaena, 7
Animal production, 425
  reproduction, 45, 452

Aquaculture, 5, 263
Arable cropland, 72, 73, 77
Arable land, 37, 38
  area, 86, 112, 130
Atmospheric dust, 100
Atmospheric nitrogen, 198
Azotobacter, 7

B

Balance, 2
Bankruptcy, 302
Beef production, 425, 426
Bioconcentration, 219
Biogeochemical cycles, 7, 8
Biological carrying capacity, 34
Biological diversity, 42
Biological nitrogen fixation, 59
Biological resources, 309, 310
Biomass, 6, 9, 10, 11, 12, 16, 20, 36, 37, 42, 91, 94, 107, 145, 177, 278, 309, 401, 487, 488, 496, 497, 501, 502
Biome, 64, 74
Biota, 42, 62
Biotechnology, 22, 23, 433
Birds, 220
Birth rates, 44
Buffalo gourd, 55

Index

C

Calorie, 339
CAM (crassulean acid metabolism), 393
Capacity of earth population, 277
Capital recovery factors, 257
Carbohydrate conversion, 401
Carbon dioxide, 93, 398, 399
Carbon flow, 393
Carrying capacity, 177, 180
Cash crops, 458, 459
Cattle
  population, 92
  raising, 57
Cereal
  crops, 74
  yields, 466, 500
Cheap food, 354, 355, 356
Chemical use, 400
Child and infant death rates, 446, 447
Child mortality, 447
China, 364–385, 486, 488, 489
Citrus juices, plant, 244
Climate, 288, 289, 291, 397
Climatic bias, 165
Climatic change, 119
Cogeneration systems, 248, 251, 259
  steam-injected gas turbine, 239
Commercial farms, 404
Compaction, 104, 106
Comparative advantage, 293, 297
Compost, 199
Conservation, 175, 179
Constraints, 173
Consumers, 2
Consumption of food, 368
Converting salt to fresh water, 155
Cornstarch, 53
Costs of electricity, steam, 249
Crop budgets, 147
Crop residues, 13, 14
Crop–water–soil relationship, 145
Crop yields, 15, 16, 34, 466
Cropping systems, 118
Crops
  food, 396
  mean annual production, 466
  modern varieties, 342–349
  potential yields, 51
  production, 468, 469
  production systems, 403
Crude birth rates, 451

Crude death rates, 447
Cultivation, 74
Cultural preferences, 402
Cultural revolution, 381
Cycle, agriculture and yield, 458

D

Death rate, 337
DDT, 373, 374
DDT equivalent, 221
Decomposers, 2
Deforestation, 20, 64, 88, 89, 90, 92
Degrading land loss, 177
Demographics, 336, 443–445, 452
Desertification, 100, 101, 166, 167, 453, 455
Developing countries, chemicals, 212
Diet, 36, 38
  deficiencies, 422, 423
  patterns, 411
  requirements, 86
Diversity, 15, 17
Drainage water, 156
Drought, 41, 167, 381
Drying trend, 456

E

Earth, living matter, 392
Ecology, 64, 165, 304
Economics of agglomeration, 334
Economy
  decline, 463, 470, 471
  development, 33
  efficiency, 147
  performance, 255
  and population, 279
  reform, 477
Ecosystems, 2, 3, 7–9, 16, 88, 96
Edible portions of plants, 396
Education indicators, 444
Eelgrass, 54
Efficiency
  of irrigation, 159
  of water use, 145, 146
El Niño Current, 399
Electricity, 238
Electrification, 263
Endemic disease, 167

Index   509

Energy, 19, 20, 34–36, 131, 238, 247, 249–251, 257, 302
  alternatives, 251
  flows, China, 491–495
  inputs, 262, 396
  intake, 414
  intensive, 288
  resources, 430
Environment, 60, 364, 406, 407
  concerns, 284
  degradation, 115, 118, 169, 302
  policy, 372, 431, 475
  resources, 432
  risks, 223
Erosion, 39, 40, 102, 107, 290, 291, 370, 394, 489, 502
Evapotranspiration, 109, 146
Evolution, 6, 7
Evolutionary feedback, 4, 5
Export crops, 469
Extinction, 3, 50, 62, 63, 65

F

Failed projects, 474, 475
Famine, 330, 420, 421
Farming, 263
  mulch, 109
Farm land, 142
Feed mixing, 246
Fertility, 127, 445, 451
Fertilizer, 15, 59, 61, 78, 80, 114, 127, 192, 194–197, 302, 381, 382, 395
  human, 381, 382
Fire, 90, 91
Fish, 57, 58
  production, 425, 426
Flood, 35
Food, 11, 16, 18, 21–26, 35, 392
  access, 327
  availability, 269
  conservation, 368
  consumption, 354, 417, 418
  contamination, 220
  crops, 75, 86
  future needs, 430, 431
  losses, 402, 403
  policy, 475
  processing, 149
  processing plant, 242, 243

  production, 424, 440
    U.S., 398
    world, 396, 397, 424
  religious preferences, 402
  sociocultural aspects, 419
  supply, 418
Foreign aid, 473
Forests, 41, 63, 64, 461
  products, 370
  resources, 20, 21
  yields, 15
Fruits, 55
Fuel
  biomass, 488
  fossil, 262
  household, China, 487, 498, 501
  pipeline, 254
  wood, 461, 501, 502

G

Genetic variability, 52, 60
Germ plasm, 51, 53
GNP, 240, 281, 286, 336–338, 357, 446, 448
Grain
  equivalents, 78
  production, 277, 425, 426, 486
  trade, 280
  yield, foreign, 289
Grassland, 80
Grazing, 91, 93
Greenhouse gases, 91
Green revolution, 52, 75, 130, 326, 341–349
Groundwater, 142
Growth rate, 32

H

Habitat, 217
Harvesting methods, 265
Health risk, 201, 213, 214
Herbicides, 216
Herbivore/parasite, 4, 5
Host, 4–6
Hunger, 75, 296, 331–333, 420–422
Hydrology, 165, 189
  cycles, 142–145, 152
Hydroponics, 278

510   Index

I

Infant mortality, 330, 339, 451
Infiltration, 105
Infrastructure, 237, 242, 243, 255, 405
Investments, 255
Irrigation, 40, 41, 111, 112, 125, 142, 143, 148
  systems, 160

L

Labor, 35, 268, 270, 467
  resources, 270
  use, 268, 490, 491
  work force, 467
Land, 17, 37–39, 70–76
  degradation, 87, 128–131, 166, 172
  loss, 74
  and population, 87
  productivity, 176
  resources, 71, 427, 428
    global, 70
  use, 71, 86
Laterization, 103, 104, 124
Leaching, 113–115, 204
Legislation, water, 376
Legumes, 59, 198, 416
Limits to growth, 332
Livestock, 79, 454

M

Malnutrition, 24, 25, 327–329, 420
Malthus, 330–335, 401
Management, 94, 108, 112, 121, 123, 144, 147, 205, 303
Manpower, 264, 265
Manufacturing accidents, chemical, 211
Manure, 13, 199, 203, 310, 311
Marine resources, 406
Maximum production, 79, 80
Meat processing, 243
Mechanization, 262–265, 269
  characteristics, 271
  effects, 105
Microclimates, 119
Migration, 453
Monteagudo (Bolivia), 240, 241
Mortality, 447, 448

Mortality variables, 448
Municipal sludge, 199

N

Natural gas, 236, 237, 258, 259
Natural resources, 32, 392, 393
  nonrenewable, 394
New foods, 54
New technologies, 32
Nitrogen, 7, 8, 13, 14, 23, 59, 192–206, 225, 302, 309–311, 395
  cycle, 226
No-till, 314
Nutrition, 354, 487
  problems, 420
  standards, 412
Nutrients, 13, 116, 117
  life cycle, 413
  needs, 413
  organic matter, 108, 109
  recycling, 310

O

Oncogenetic risk, 222
Organic matter, 43, 108, 109, 305, 310
Overgrazing, 394, 453
Overnutrition, 424
Overpopulation, 174

P

Parasitic insects, 4
Pastoralism, 453
Peasants, 360, 361
Pesticides, 193, 194, 207, 208, 212–225, 292, 302, 373–375
Pests, 14, 307, 308
Phosphorus, 8, 9
Photosynthesis, 2, 61, 145, 176, 393
Phytomass, 176
Plant production, 392
  species, 50, 62, 63
Plantation, fuelwood, 501–503
Poisoning, pesticides, 213
Pollination, 216
Pollution, 65, 100, 153

Population, 23–24, 32–35, 72–74, 86, 87, 142, 216, 276–279, 336–338, 364–368, 380–385, 434, 465
 capacity, 178, 181
 growth, 410, 411, 441–451
 policy, 372, 480
Porosity, 112
Poverty, 478, 479
 cycle, 479
 and nutrition, 423
Power, 264, 266
 hydroelectric, 253
Physical quality of life index, 366
Precipitation, 144, 145
Predators, 5
Protein, 54, 327–329, 401, 414, 417
 intake, 414, 417
 production and manufacture, 406
 quality, 414, 415
 sources, 405, 406

R

Rainfall, 40, 167, 169, 429, 456, 457
 decline, 459
RDA (Recommended Dietary Allowances), 413
Recharge, 169
Reciprocating engines, 252
Reducers, 2
Refrigerated transport, 256
Reserves, gas and oil, 236
Resources, per capita, 368, 369
Respiration, 5
Ridge planting, 306, 307, 311, 314
Roots and tubers, 465
Ruminant animals, 401
Runoff, 144, 305
Rural development, 174, 262, 263

S

Salinity, 109, 110, 124, 145, 155, 156
Salt-tolerant plants, 61, 156, 157
Savannah, 90
Scale of mechanization, 269
Science and technology, 432
Sedimentation, 454
Self-sufficiency, 293

Shelter, 365
Shifting agriculture, 459, 460
Slaughterhouse, 243
Sociocultural aspects, 419
Soil
 degradation, 87, 94, 95, 315
 erosion, 14, 17, 18, 81, 89, 93, 95–98, 121, 123, 130, 304–306
 loss, 39, 394
 maps, 70
 nutrients, 305
 organisms, 217
 toxicity, 209, 210
Solar energy, 37, 392, 393
 living matter, 392
Somalia, 54
Standards of living, 366
Strategies, food and fuel, 499, 504
Subsistence, 403
Surface mulch, 117
Synthetic food, 405
Synthetic nitrogen, 192–197, 202–204, 225
Systems of production, 403

T

TDS (total dissolved solids), 155
Technology transfer, 44
Thornthwaite aridity index, 97
Tilapia, 58
Tillage system, 121, 272
Topsoil loss, 103, 302
Tractorization, 77, 263
Transpiration, 19
Transportation, 255
Trophobiosis, 215
Tropics, 62–64
Typhoid, 376, 377

V

Varieties, *see* Modern varieties
Vegetable dehydration, 88
Vegetable oil plants, 245
Vegetables, 5–6
Vegetarian diets, 417
Vision for year 2020, 293
Vitamins, 413

W

Wastes, 401
Water, 13, 18, 40, 41, 142, 164, 394, 395
  availability, 164, 167, 177, 180, 187
  barrier, 183, 186
  competition, 183, 186
  contamination, nitrogen, 201, 203, 218
  costs, 147
  for crops, 371
  cycle, 175
  deficiency, 172, 173
  degradation, 154
  human, 365
  losses, 146, 160
  pollution, 153, 375–378
  quality, 152, 156
  requirements
    animals, 148, 149
    plants, 145–147
    processing, 153
  resources, 428
  runoff, 117
  storage, 146
  supply, 165, 253
  treatment costs, 254
  usable, 142
Waterlogging, 109
Wax gourd, 55
Weather, 60, 397
Weathering, soil, 104
Weeds, 225, 306
  losses, 225
Wild corn, 52, 53
Wind erosion, 97–99, 122
Woodland, 459
Worker productivity, 265, 267
World
  food production, 282, 283, 287, 396, 397
  hunger, 327, 331
  population, 278

Y

Yields, 32, 36, 72–76, 102, 277, 341, 342